Bifurcation and Chaos in Complex Systems

EDITED SERIES ON ADVANCES IN NONLINEAR SCIENCE AND COMPLEXITY

Bifurcation and Chaos in Complex Systems

Edited by

JIAN-QIAO SUN
University of Delaware, Newark, USA

ALBERT C.J. LUO
Southern Illinois University, Edwardsville, USA

ELSEVIER

AMSTERDAM · BOSTON · HEIDELBERG · LONDON · NEW YORK · OXFORD
PARIS · SAN DIEGO · SAN FRANCISCO · SINGAPORE · SYDNEY · TOKYO

Elsevier
Radarweg 29, PO Box 211, 1000 AE Amsterdam, The Netherlands
The Boulevard, Langford Lane, Kidlington, Oxford OX5 1GB, UK

First edition 2006

Library of Congress Cataloging-in-Publication Data
A catalog record for this book is available from the Library of Congress

British Library Cataloguing in Publication Data
A catalogue record for this book is available from the British Library

ISBN-13: 978-0-444-52229-0
ISBN-10: 0-444-52229-8

Transferred to Digital Printing 2010

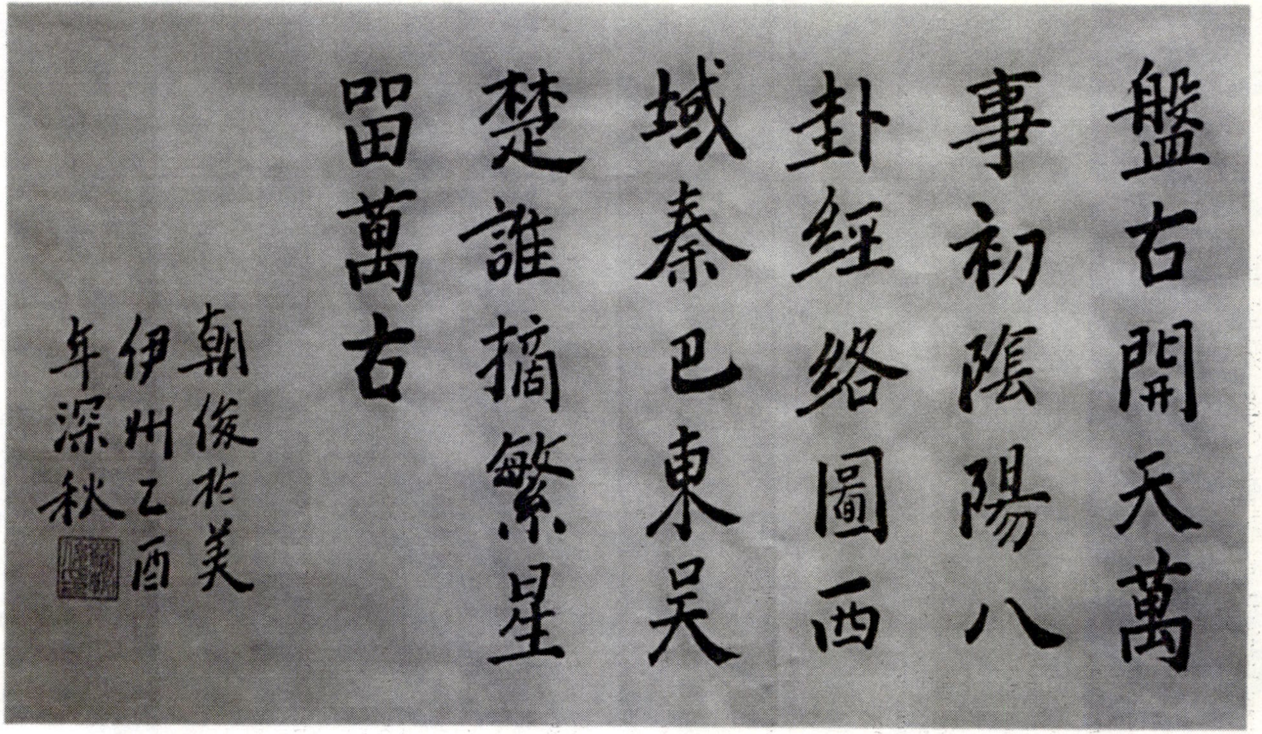

盤古開天萬
事初陰陽八
卦經絡圖西
域秦巴東吳
楚誰摘繁星
留萬古
朝俊於美
伊州乙酉
年深秋

[illegible] nonlinear [illegible] and [illegible] potential [illegible] for [illegible] and [illegible] global nonlinear models [illegible] dynamic [illegible] traditional linear thinking. Nonlinear science [illegible] dynamics [illegible] have [illegible] topics for [illegible] and continue to [illegible] scientists [illegible] all over the world. The [illegible] has [illegible] systems including civil, mechanical, electrical, [illegible] systems, we are [illegible] of complex nonlinear dynamical systems.

[illegible] chapters [illegible] the state of the art [illegible] of [illegible] systems. [illegible] authors of the [illegible] the first two chapters [illegible] and [illegible] chapter presents a [illegible] of [illegible] chaos in [illegible] nonlinear dynamical systems. The fifth chapter presents [illegible] in [illegible] systems [illegible]. [illegible] description of each chapter is presented in the [illegible].

[illegible] bifurcations of nonlinear dynamical systems [illegible] differential equations [illegible] investigations of limit cycles and [illegible] control of [illegible] and chaos [illegible] theory [illegible] method [illegible] is developed. Efficient methodologies and software based on [illegible] computations are presented for normal form computations. An efficient method based on the concept of modes and [illegible] is presented to identify the parameter values leading to chaos. The method provides necessary conditions for the existence of chaos [illegible] to systematically generate chaotic systems. Many interesting examples are included in the chapter, such as bifurcations [illegible] Hopf bifurcations, Hopf bifurcations [illegible] and chaos [illegible]. The [illegible] determined the limit cycles in the [illegible] nonlinear dynamical systems. The [illegible] also used the normal form to represent [illegible] Hamiltonian systems, [illegible]

Preface

Modeling complex natural and social dynamic phenomena has been an intellectual pursuit of mankind for centuries. Aided with modern computers, we can now develop global nonlinear models of dynamical systems and obtain solutions that defy traditional, linear thinking. Nonlinear sciences, chaotic dynamics and bifurcations have been active research topics for decades, and continue to stimulate interests from researchers all over the world. The research has found applications in a wide range of dynamical systems including civil, mechanical, electrical, control, biological, ecological, economic and financial systems. We are proud to present to readers this volume containing recent research results on bifurcation studies of complex nonlinear dynamical systems.

The book contains five chapters describing the state of the art of bifurcation studies of nonlinear systems. The contributing authors of the book are all active researchers in this interesting subject area. The first two chapters deal with theoretical issues of bifurcation analysis in smooth and nonsmooth dynamical systems. The third chapter presents a numerical method for global bifurcations. The fourth chapter studies bifurcations and chaos in time-varying, parametrically excited nonlinear dynamical systems. The fifth chapter presents bifurcation analyses of modal interactions in distributed, nonlinear, dynamical systems of circular thin von Karman plates. A brief description of each chapter is presented in the following.

Chapter 1 of the book considers bifurcations of nonlinear dynamical systems governed by ordinary differential equations, difference equations and time delayed differential equations. In-depth investigations of limit cycles and chaos, Hopf bifurcation control, chaos control and chaos synchronization are presented. In particular, a method unifying the center manifold theory and the method of normal forms is developed. Efficient methodologies and software based on symbolic programming are presented for normal form computations. An efficient method based on the concept of modes and "mode competition" is presented to identify the parameter values leading to chaos. The method provides necessary conditions for the existence of chaos, and can be used to systematically generate chaotic systems. Many interesting examples are included in the chapter, such as bifurcations of limit cycles, Hopf and double Hopf bifurcations, Hopf bifurcation control, chaos control and chaos synchronization. The author and his co-workers determined the limit cycles in the Hilbert 16th problem, which is an unsolved mathematical difficult problem in nonlinear dynamical systems. The author also used the normal form to represent polynomial Hamiltonian systems possessing

80 limit cycles. In such Hamiltonian systems, the appearance, forming and disappearance of limit cycles are presented.

Chapter 2 presents a rigorous mathematical framework for dealing with bifurcation problems of nonsmooth dynamical systems which do not satisfy the Lipschitz condition. The grazing flow in the vicinity of the discontinuous boundary is investigated thoroughly. The concept of accessible and inaccessible subdomains is introduced. The local singularity and tangency of a flow on the separation boundary is investigated. The necessary and sufficient conditions for the local singularity and tangency are provided. Examples of the grazing flows in piecewise linear systems and friction-induced oscillators are studied in detail.

Chapter 3 studies global bifurcations of complex nonlinear dynamical systems by using the cell mapping methods. In addition to deterministic nonlinear systems, stochastic and fuzzy nonlinear dynamical systems are also investigated. The response of stochastic and fuzzy dynamical systems is far more difficult to obtain. This is because stochastic and fuzzy responses must be handled by a method which can describe the global properties of nonlinear dynamical systems. The cell mapping methods are created just for this purpose. The global bifurcation analysis amounts to studying changes of the topology and probability or possibility distribution of stable attractors, boundaries of the domains of attraction and unstable solutions, as the control parameter varies. This chapter contains recent research results on a variety of complex global bifurcations including chaotic boundary and interior crises, indeterminate crisis, double crises and codimension two bifurcations of deterministic, stochastic and fuzzy nonlinear dynamical systems. Furthermore, a study of the effect of bifurcation on the semiactive optimal controls is also presented.

Chapter 4 presents a general methodology for the bifurcation analysis of nonlinear ordinary differential equations with periodic coefficients. The analysis is based on construction of dynamically equivalent normal forms involving normal form reduction and the center manifold theory. With the help of the Lyapunov–Floquét transformation and time periodic center manifold theory, the reducibility conditions for the original time-varying, nonlinear systems are developed. In addition, the resonance conditions are obtained for time-varying parametrically excited systems. A periodically parametrically excited pendulum is studied to illustrate the methodology.

Chapter 5 presents a bifurcation analysis of modal interactions in distributed, nonlinear dynamical systems. The nonlinear oscillations with modal interaction of circular von Karman plates subject to harmonic excitations are presented for illustrations. The method of multiple scales is applied to this analysis, and correct solvability conditions are derived. Several special cases of the plate are investigated, including a study of modal interactions of circular plates with internal resonance on an elastic foundation; and an examination of the effect of the number of nodal diameters on nonlinear interactions in asymmetric vibrations of circular, thin plates.

It is our hope that the book presents a reasonably broad view of the state of the art of bifurcation analysis of complex nonlinear dynamical systems, and will further stimulate interest in this important subject among scientists, engineers and students.

November 2005

Albert C.J. Luo
Edwardsville, Illinois

Jian-Qiao Sun
Newark, Delaware

Contents

Dedication v
Preface vii

Bifurcation, Limit Cycle and Chaos of Nonlinear Dynamical Systems 1
Pei Yu

1. Introduction 2
2. Bifurcation of limit cycles 5
3. Bifurcation control and chaos synchronization 73
4. Competitive modes 92
5. Conclusions 120
   Acknowledgement 121
   References 121

Grazing Flows in Discontinuous Dynamic Systems 127
Albert C.J. Luo

1. Introduction 127
2. Domain accessibility 129
3. Discontinuous dynamic systems 131
4. Oriented boundary and singular sets 132
5. Local singularity and grazing flows 141
6. Piecewise linear systems 150
7. Friction-induced oscillators 158
8. Conclusions 187
   Appendix 187
   References 189

Global Bifurcations of Complex Nonlinear Dynamical Systems with Cell Mapping Methods 191
Ling Hong and Jian-Qiao Sun

1. Introduction 192
2. Cell mapping methods 193
3. Crises in deterministic systems 198
4. Bifurcations of nonlinear systems with small random disturbances 212
5. Fuzzy bifurcations 230

6. Effect of bifurcation on semiactive optimal controls 258
References 274

Bifurcation Analysis of Nonlinear Dynamic Systems with Time-Periodic Coefficients 279
Alexandra Dávid and S.C. Sinha

1. Introduction 280
2. Formulation of the problem 286
3. Local stability and conditions for bifurcations: Floquét theory 287
4. Lyapunov–Floquét transformation 289
5. Nonlinear analysis 291
6. The codimension one bifurcations 301
7. Applications 313
8. Summary and conclusions 332
References 335

Modal Interactions in Asymmetric Vibrations of Circular Plates 339
Won Kyoung Lee

1. Introduction 339
2. Governing equations 341
3. Solution 343
4. Steady-state responses and numerical examples 348
Appendix A 370
Appendix B 371
Appendix C 376
References 376

List of Contributors 377
Author index 379
Subject Index 385

Chapter 1

Bifurcation, Limit Cycle and Chaos of Nonlinear Dynamical Systems

Pei Yu

Department of Applied Mathematics, The University of Western Ontario, London, Ontario, N6A 5B7, Canada

E-mail: *pyu@uwo.ca*

Abstract

In this chapter, we shall consider bifurcations of dynamical systems which are described by ordinary differential equations, difference equations and time delayed differential equations. Particular attention is given to limit cycles and chaos, Hopf bifurcation control, chaos control and chaos synchronization. Both mathematical and practical engineering problems are considered. The materials presented in this chapter are based on the research results obtained recently by the author and his co-workers.

Contents

1. Introduction 2
2. Bifurcation of limit cycles 5
   2.1. Lifting 2-D model with delayed feedback control 6
      2.1.1. Aeroelastic model 7
      2.1.2. Linear stability analysis 9
      2.1.3. Center manifold reduction 11
      2.1.4. Normal form and bifurcation analysis 14
      2.1.5. Numerical simulation results 18
   2.2. Internet congestion model 24
      2.2.1. Hopf bifurcation 26
      2.2.2. Hopf bifurcation control 28
      2.2.3. Numerical results 32
   2.3. Hilbert's 16th problem 38
      2.3.1. Nine small limit cycles around an isolated singular point 38
      2.3.2. Bifurcation to 80 limit cycles in a perturbed Hamiltonian system 50

Edited Series on Advances in Nonlinear Science and Complexity
Volume 1 ISSN: 1574-6909 DOI: 10.1016/S1574-6909(06)01001-X

3. Bifurcation control and chaos synchronization 73
3.1. Global ultimate boundedness of chaotic systems 74
3.2. Hopf bifurcation control 77
3.2.1. Feedback controller using polynomial function 77
3.2.2. Controlling Hopf bifurcation in the Lorenz system 79
3.3. Tracking and chaos synchronization 84
3.3.1. Tracking 85
3.3.2. Synchronization of two Lorenz systems 89
4. Competitive modes 92
4.1. Definition of CM 92
4.2. Application of CM: estimating chaotic parameter regimes 97
4.2.1. Example 1: A PDE model 97
4.2.2. Example 2: A psychological model 101
4.2.3. Example 3: The Oregonator model 104
4.2.4. Example 4: The smooth Chua system 105
4.3. Application of CM: constructing new chaotic systems 109
4.3.1. Example 1 113
4.3.2. Example 2 115
4.3.3. Example 3 117
5. Conclusions 120
Acknowledgement 121
References 121

1. Introduction

Nonlinear dynamics, more grandly called "nonlinear science" or "chaos theory", is a rapidly-growing area, which plays an important role in almost all disciplines of science and engineering including mechanics, aeronautics, electrical circuits, control systems, population problems, economics, financial systems, stock markets, ecological systems, etc. In general, dynamical systems contain certain parameters (e.g., control parameters) and thus it is important to study the behavior of such systems when the parameters are varied. These dynamical behaviors include complex phenomena such as instability, bifurcation and chaos (e.g., see Arnold, 1977; Hassard, Kazarinoff and Wan, 1981; Guckenheimer and Holmes, 1993; Wiggins, 1990; Chen and Dong, 1998; Yu, 2003b). Studies of nonlinear dynamical systems may be roughly divided into two main categories: local analysis and global analysis. For instance, post-critical behaviors such as saddle-node bifurcation and Hopf bifurcation can be studied locally in the vicinity of a critical point, while heteroclinic, homoclinic orbits and chaos are essentially global behaviors and have to be studied globally.

These two categories need to be treated with different theories and methodologies.

In the study of local behavior of dynamical systems, in particular for qualitative properties, the first step is usually to simplify a given system. Such a simplification should keep the dynamical behavior of the system unchanged. Many methodologies have been developed in analyzing local dynamics, such as center manifold theory, normal form theory, averaging method, multiple time scales, Lyapunov–Schmidt reduction, the method of succession functions, the intrinsic harmonic balancing technique, etc. These methods can be used to obtain "simplified" governing differential equations in describing the dynamics of the system in the vicinity of a point of interest. The "simplified" system is topologically equivalent to the original system, and thus it greatly simplifies the analysis of the original system. Usually, center manifold theory is applied first to reduce a system to a low-dimensional center manifold, and then the method of normal forms is employed to further simplify the system (Guckenheimer and Holmes, 1993; Nayfeh, 1993). However, approaches combining the two theories into one unified procedure have been developed (Yu, 1998, 2000, 2001; Yu, Zhang and Bi, 2001). In this chapter, study of the local dynamical analysis is mainly based on this unified procedure of normal forms. A normal form is not uniquely defined and computing the explicit formula of a normal form in terms of the coefficients of the original system is not easy. In the past few years, efficient methodologies and software based on symbolic computations using Maple and Mathematica have been successfully employed in normal form computations.

The phenomenon of limit cycle was first discovered and introduced by Poincaré (1892–1899) who presented the break through qualitative theory of differential equations. In order to determine the existence of limit cycles for a given differential equation and the properties of limit cycles, Poincaré introduced the method of topographical system, the well-known Poincaré Map, which is still the most basic tool for studying the stability and bifurcations of periodic orbits. The driving force behind the study of limit cycle theory was the invention of triode vacuum tube which was able to produce stable self-excited oscillations of constant amplitude. It was noted that such kind of oscillation phenomenon could not be described by linear differential equations. At the end of the 1920s Van der Pol (1926) developed a differential equation to describe the oscillations of constant amplitude of a triode vacuum tube:

$$\ddot{x} + \mu(x^2 - 1)\dot{x} + x = 0, \quad \mu \neq 0, \tag{1}$$

which is now called Van der Pol's equation, where the dot denotes differentiation with respect to time t and μ is a parameter. Later a more general equation, called Liénard equation (Liénard, 1928), was developed, of which Van der Pol's equation is a special case.

Limit cycles are generated through bifurcations, among which the most popular and important one is Hopf bifurcation (Hopf, 1942). Consider the following general nonlinear system:

$$\dot{\mathbf{x}} = \mathbf{f}(\mathbf{x}, \mu), \quad \mathbf{x} \in \mathbb{R}^n, \ \mu \in \mathbb{R}, \ \mathbf{f}: \mathbb{R}^{n+1} \to \mathbb{R}^n, \tag{2}$$

where $\mathbf{x}$ is an n-dimensional state vector while μ is a scalar parameter, called bifurcation parameter. (Note that in general one may assume that μ is an m-dimensional vector for $m \geqslant 1$.) The function $\mathbf{f}$ is assumed analytic with respect to both $\mathbf{x}$ and μ. Equilibrium solutions of system (2) can be found by solving the nonlinear algebraic equation $\mathbf{f}(\mathbf{x}, \mu) = \mathbf{0}$ for an arbitrary μ. Let $\mathbf{x}^*$ be an equilibrium (or fixed point) of the system, i.e., $\mathbf{f}(\mathbf{x}^*, \mu) \equiv \mathbf{0}$ for any real values of μ. Further, suppose that the Jacobian of the system evaluated at the equilibrium $\mathbf{x}^*$ has one pair of complex conjugates, denoted by $\lambda_{1,2}(\mu)$ with $\lambda_1 = \bar{\lambda}_2 = \alpha(\mu) + \mathrm{i}\omega(\mu)$ such that $\alpha(\mu^*) = 0$ and $\frac{\mathrm{d}\alpha(\mu^*)}{\mathrm{d}\mu} \neq 0$. The second condition is usually called the transversality condition, implying that the crossing of the complex conjugate pair on the imaginary axis is not tangent to the imaginary axis. Then Hopf bifurcation occurs at the critical point $\mu = \mu^*$ giving rise to bifurcation of a family of limit cycles. Other local bifurcations include double-zero, Hopf-zero, double Hopf and even more complex type of bifurcations. In this chapter, Hopf and double Hopf bifurcations will be particularly studied.

Global bifurcation analysis, on the other hand, is more difficult than local analysis. Besides homoclinic and heteroclinic bifurcations (Guckenheimer and Holmes, 1993), the most exciting discovery in nonlinear dynamics is chaos, which has lead to a new era in the study of nonlinear dynamical systems. Since the discovery of the Lorenz attractor (Lorenz, 1963), many researchers from different disciplines such as mathematics, physics, etc. extensively investigated the dynamical property of chaotic systems. For a quite long period, people thought that chaos was not predictable nor controllable. However, the OGY method (Ott, Grebogi and Yorke, 1990) developed in the 90's of the last century has completely changed the situation, and the study of bifurcation and chaos control begun. The general goal of bifurcation control is to design a controller such that the bifurcation characteristics of a nonlinear system undergoing bifurcation can be modified to achieve certain desirable dynamical behavior, such as changing a Hopf bifurcation from subcritical to supercritical (Chen, 2000; Chen, Moiola and Wang, 2000; Yu, 2003a, 2003b). The main goal of chaos control is to simply eliminate chaotic motion so that the system's trajectories eventually converge to an equilibrium point or a periodic orbit (Chen and Dong, 1998).

In the study of chaos, chaos control and chaos synchronization, the first step is to identify the parameter values at which a system may exhibit chaos. It is well known that chaotic motion is bounded and sensitive to initial conditions. Such well-accepted characteristic of chaos is good enough to be used to determine whether a motion is chaotic or not, but is not helpful in predicting the location in parameter space of a chaotic regime for a given nonlinear system. Usually numerical approaches are used to identify these chaotic parameter values. However, the numerical approach is often so time consuming as to border on hopeless, especially when the system has many parameters. It is thus necessary to develop efficient methods to find the parameter values leading to chaos, which do not heavily depend on numerical simulations. A method based on the concept of modes and "mode competition" has been recently developed (Yao, 2002; Yao, Yu and Essex, 2002; Yao, Yu, Essex and Davison, 2006), which has been shown very useful in practical applications, though it only provides necessary conditions for the existence of chaos. Moreover, this method can be employed to systematically generate chaotic systems.

The rest of the chapter is organized as follows. In the next section, bifurcations of limit cycles are particularly studied, including systems described by ordinary differential equations, time delayed differential equations and discrete equations. Both Hopf and double Hopf bifurcations are considered. Hopf bifurcation control for discrete systems is also discussed. Then Hilbert's 16th problem is presented in this section, and both local and global bifurcation analyzes are applied to produce maximal number of limit cycles. Section 3 is devoted to the studies of Hopf bifurcation control, chaos control and chaos synchronization. In particular, with the Lorenz system as an example, a feedback controller using polynomial functions are used to control Hopf bifurcation. The result of global ultimate boundedness of the Lorenz system is presented, and then tracking to periodic motion (chaos control) and chaos synchronization are discussed. In Section 4, the new concept "competitive modes" is introduced, and then employed to estimate chaotic parameter regimes of nonlinear chaotic systems and to construct new chaotic systems. Finally, conclusions are given in Section 5.

2. Bifurcation of limit cycles

Limit cycles are common solutions for almost all nonlinear dynamical systems. They model systems that exhibit self-sustained oscillations. In other words, these systems oscillate even in the absence of external periodic forcing. For example, the Holling–Tanner predator–prey model (Lynch, 2004) describes the population of prey and predators, and shows that all positive solutions (trajectories) are drawn to a closed periodic cycle. This model appears to match very well with what happens for many predator–prey species in the natural world. For example, house sparrows and sparrow hawks in Europe, muskrat and mink in Central North America and white-tailed deer and wolf in Ontario, Canada. Other examples of self-excited oscillation are beating of a heart, rhythms in body temperature, hormone secretion, chemical reactions that oscillate spontaneously and vibrations in bridges and airplane wings. Due to the wide occurrence of limit cycles in science and technology, limit cycle theory has also been extensively studied by physicists, and more recently by chemists, biologists and economists.

Limit cycles are generated through bifurcations (perturbations). From the view point of dynamical system theory, there are four principal bifurcations in producing limit cycles: (i) multiple Hopf bifurcations from a center or focus, (ii) separatrix cycle bifurcations from homoclinic or heteroclinic orbits, (iii) global center bifurcation from a periodic annulus and (iv) limit cycle bifurcations from multiple limit cycles. Limit cycles bifurcated from a focus, center or limit cycles are called local bifurcations of limit cycles or small limit cycles, which are usually studied by normal form and other local bifurcation theories. The limit cycles generated from separatrix cycles or global period annuli are called global bifurcations of limit cycles, which are usually investigated by global bifurcation theories, such as Poincaré–Pontryagin–Andronov theorem or higher-order Melnikov function analysis (Melnikov, 1963; Perko, 2001).

To demonstrate bifurcations of limit cycles, in the following we first consider an aeroelastic model with time delayed feedback controls. Then we investigate an Internet congestion model, and finally study the well-known Hilbert's 16th problem.

2.1. Lifting 2-D model with delayed feedback control

The study of the aeroelastic behavior of flight vehicles in the pre- and post-flutter regimes is of a crucial importance towards increasing their operational life and the avoidance of catastrophic failures. A nonlinear model of a wing section of the high speed aircraft incorporating active control has been proposed (Marzocca, Librescu and Silva, 2002) and further studied recently using linear and nonlinear feedback controls (Yuan, Yu, Librescu and Marzocca, 2004), in which particularly the effect of the time delayed proportional feedback control on flutter instability boundary and its character (benign/catastrophic) was discussed. Bifurcations into limit cycles (Hopf bifurcation) were investigated with respect to system parameters as well as the time delay. It has been shown that incorporation of a linear feedback control is always beneficial in controlling both the initiation of Hopf bifurcation and the stability of motions, regardless whether the time delay is added or not. Introducing a time delay into the feedback control could have a profound effect on the stability of the bifurcating motions. However, it has been found that larger time delay is not beneficial in delaying Hopf bifurcation. When nonlinear feedback control is applied, the situation becomes even more complicated. It may destabilize the bifurcating motions if the nonlinear control is combined with larger time delay. Therefore, based on the study in Yuan, Yu, Librescu and Marzocca (2004), it was suggested that both linear and nonlinear controls with small time delay should be applied in order to obtain the best control design. However, further studies are necessary to get better understanding of dynamic behavior of the model with other controls. Complex phenomena such as quasiperiodic solutions (Yu, 2001) and chaotic motions (Yu, Yuan and Xu, 2002) need to be investigated. Hopf bifurcation has been extensively studied using many different methods (Guckenheimer and Holmes, 1993; Nayfeh, 1993), for example, Liapunov's quantity used in Yuan, Yu, Librescu and Marzocca (2004) and Librescu (1975). In Marzocca, Librescu and Silva (2002), the dynamic behavior of the system without time delay in the control was studied in the vicinity of a Hopf bifurcation critical point. In particular, the effect of the active control on the character of the flutter boundary (where the Jacobian has a purely imaginary pair) is investigated. It is shown that for different flight speeds, stable (unstable) equilibrium and stable (unstable) limit cycles exist. Also, the effect of structural nonlinearities on the character of the flutter boundary has been considered.

In this subsection, the approach developed in Yu (2001), Yu, Yuan and Xu (2002) and Yuan, Yu, Librescu and Marzocca (2004) will be applied to further study the effect of the time delay involved in the feedback control. The main attention will be focused on Hopf bifurcation. First, linearization is applied to find the critical point at which a Hopf or double Hopf bifurcation may occur. Linear stability conditions determining the critical points are derived. Then center manifold reduction and normal form theory are employed to analyze Hopf and double Hopf bifurcations. Finally, numerical simulation results are presented to verify the analytical predictions.

2.1.1. Aeroelastic model

In this subsection, we shall extend the methodology presented in Yuan, Yu, Librescu and Marzocca (2004) to study a more realistic model with both proportional and velocity feedback controls. Attention will be focused on double Hopf bifurcation which gives more complex dynamical behavior than single Hopf bifurcation. The schematic diagram for a wing section of the high speed aircraft is shown in Figure 1. The governing equations of the nonlinear model can be described by two coupled second-order delay differential equations

$$\begin{aligned} m\ddot{h}(\hat{t}) + S_\alpha\ddot{\alpha}(\hat{t}) + c_h\dot{h}(\hat{t}) + K_h h(\hat{t}) &= L(\hat{t}) - L_c(\hat{t} - \hat{\tau}), \\ S_\alpha\ddot{h}(\hat{t}) + I_\alpha\ddot{\alpha}(\hat{t}) + c_\alpha\dot{h}(\hat{t}) + M_\alpha &= M(\hat{t}) - M_c(\hat{t} - \hat{\tau}), \end{aligned} \tag{3}$$

where $L(\hat{t})$ and $M(\hat{t})$ are the aerodynamic lift and moment, respectively; the dot denotes the differentiation with respect to time $\hat{t}$, $\hat{\tau}$ represents time delay. M_α, $L(\hat{t})$, $M(\hat{t})$, $L_c(\hat{t} - \hat{\tau})$ and $M_c(\hat{t} - \hat{\tau})$ are given as

$$\begin{aligned} M_\alpha &= K_\alpha\alpha(\hat{t}) + \delta_S\hat{K}_\alpha\alpha^3(\hat{t}), \\ L &= -\frac{\gamma}{12\mu M_\infty}\left\{12\alpha + \delta_A M_\infty^2(1+\kappa)\gamma^2\alpha^3 + 12\left[\dot{\xi} + \dot{\alpha}\frac{b - x_{ea}}{b}\right]\right\}, \end{aligned}$$

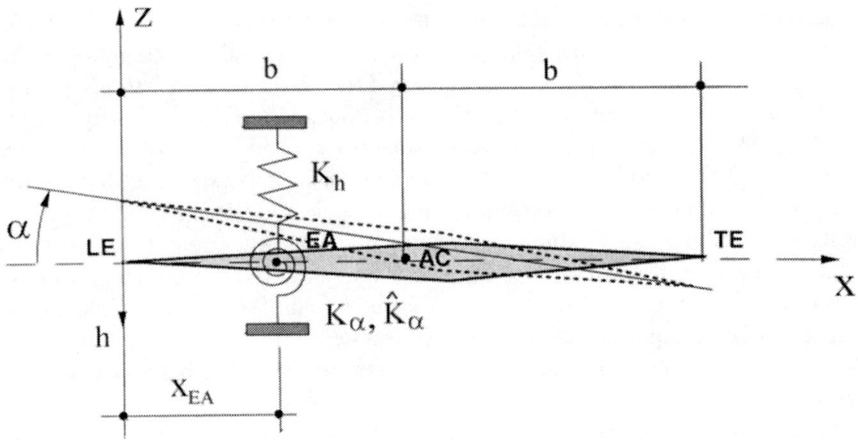

Figure 1. Geometry of the cross-section of lifting surface.

$$M = -\frac{\gamma}{12\mu M_\infty}\frac{1}{r_\alpha^2 b}\left\{12(b - x_{ea})\alpha + \delta_A M_\infty^2 (b - x_{ea})(1+\kappa)\gamma^2\alpha^3 + 4\left[3(b - x_{ea})\dot{\xi} + \dot{\alpha}\frac{4b^2 - 6bx_{ea} + 3x_{ea}^2}{b}\right]\right\},$$

$$L_c = g_1 h(\hat{t} - \hat{\tau}) + g_2 h^3(\hat{t} - \hat{\tau}) + g_3 \dot{h}(\hat{t} - \hat{\tau}),$$

$$M_c = f_1\alpha(\hat{t} - \hat{\tau}) + f_2\alpha^3(\hat{t} - \hat{\tau}) + f_3\dot{\alpha}(\hat{t} - \hat{\tau}), \tag{4}$$

in which g_i, f_i, $i = 1, 2, 3$, are the control gains and are time-independent. Note that the controls include both proportional (linear and cubic order) and velocity (linear order) terms.

To obtain the dimensionless form, assume the following control gains:

$$\Theta_1 = \frac{g_1}{K_h}, \qquad \Theta_2 = \frac{g_2}{K_h}, \qquad \Theta_3 = \frac{g_3}{c_h};$$

$$\Psi_1 = \frac{f_1}{K_\alpha}, \qquad \Psi_2 = \frac{f_2}{K_\alpha}, \qquad \Psi_3 = \frac{f_3}{c_\alpha},$$

then the aeroelastic governing equations are obtained as

$$\ddot{\xi}(t) + \chi_\alpha\ddot{\alpha}(t) + 2\zeta_h\left(\frac{\bar{\omega}}{V}\right)\dot{\xi}(t) + \left(\frac{\bar{\omega}}{V}\right)^2\xi(t) = l_a(t) - l_c(t - \tau),$$

$$\frac{\chi_\alpha}{r_\alpha^2}\ddot{\xi}(t) + \ddot{\alpha}(t) + \frac{2\zeta_\alpha}{V}\dot{\alpha}(t) + \frac{1}{V^2}\alpha(t) + \frac{B}{V^2}\alpha^3(t) = m_a(t) - m_c(t - \tau), \tag{5}$$

where

$$l_c = \left(\frac{\bar{\omega}}{V}\right)^2 \Theta_1 \xi + \left(\frac{\bar{\omega}}{V}\right)^2 \Theta_2 \xi^3 + 2\zeta_h \left(\frac{\bar{\omega}}{V}\right) \Theta_3 \dot{\xi},$$

$$m_c = \left(\frac{\Psi_1}{V^2}\right)\alpha + \left(\frac{\Psi_2}{V^2}\right)\alpha^3 + 2\zeta_\alpha \left(\frac{\Psi_3}{V}\right)\dot{\alpha}, \tag{6}$$

whereas l_a and m_a are the dimensionless terms counterparts of the aerodynamic lift and moment, L and M, respectively (see equation (2)). Here, we still use the dot to denote the differentiation with respect to the normalized time $t = U_\infty \hat{t}/b$. τ is a dimensionless time-delay $\tau = \hat{\tau}\omega_\alpha$ and $\xi = h/b$. The meaning of other parameters can be found in the nomenclature of Yuan, Yu, Librescu and Marzocca (2004).

2.1.2. Linear stability analysis

In order to capture the effect of time delay, τ, related to the various feedback gains Θ_i and Ψ_i, $i = 1, 2, 3$, let

$$\xi = x_1, \qquad \alpha = x_2, \qquad \dot{\xi} = x_3, \qquad \dot{\alpha} = x_4,$$
$$x_{it} = x_i(t - \tau), \quad i = 1, 2, 3, 4.$$

Then, one can transfer equation (5) as a set of four first-order differential equations:

$$\begin{aligned}
\dot{x}_1 &= x_3, \\
\dot{x}_2 &= x_4, \\
\dot{x}_3 &= a_1 x_1 + a_2 x_2 + a_3 x_3 + a_4 x_4 + a_5 x_{1t} + a_6 x_{2t} \\
&\quad + a_7 x_{3t} + a_8 x_{4t} + a_9 x_2^3 + a_{10} x_{1t}^3 + a_{11} x_{2t}^3, \\
\dot{x}_4 &= b_1 x_1 + b_2 x_2 + b_3 x_3 + b_4 x_4 + b_5 x_{1t} + b_6 x_{2t} \\
&\quad + b_6 x_{3t} + b_8 x_{4t} + b_9 x_2^3 + b_{10} x_{1t}^3 + b_{11} x_{2t}^3,
\end{aligned} \tag{7}$$

where all the coefficients a_i's and b_i's are explicitly expressed in terms of the parameters of equation (5).

For convenience in the following analysis, rewrite equation (7) in the vector form,

$$\dot{\mathbf{x}}(t) = A_1 \mathbf{x}(t) + A_2 \mathbf{x}(t - \tau) + \mathbf{F}\big(\mathbf{x}(t), \mathbf{x}(t - \tau)\big), \tag{8}$$

where $\mathbf{x}, \mathbf{F} \in \mathbb{R}^4$, A_1 and A_2 are 4×4 matrices. A_1, A_2 and $\mathbf{F}$ are given by

$$A_1 = \begin{bmatrix} 0 & 0 & 1 & 0 \\ 0 & 0 & 0 & 1 \\ a_1 & a_2 & a_3 & a_4 \\ b_1 & b_2 & b_3 & b_4 \end{bmatrix}, \qquad A_2 = \begin{bmatrix} 0 & 0 & 0 & 0 \\ 0 & 0 & 0 & 0 \\ a_5 & a_6 & a_7 & a_8 \\ b_5 & b_6 & b_7 & b_8 \end{bmatrix}$$

and

$$\mathbf{F} = \begin{pmatrix} 0 \\ 0 \\ a_9 x_2(t)^3 + a_{10} x_1^3(t-\tau) + a_{11} x_2^3(t-\tau) \\ b_9 x_2(t)^3 + b_{10} x_1^3(t-\tau) + b_{11} x_2^3(t-\tau) \end{pmatrix},$$

respectively.

As the first step, we analyze the stability of the trivial solution of the linearized system of (5), which is given by

$$\dot{\mathbf{x}}(t) = A_1 \mathbf{x}(t) + A_2 \mathbf{x}(t-\tau), \quad \mathbf{x} \in \mathbb{R}^4. \tag{9}$$

The characteristic function can be obtained by substituting the trivial solution, $\mathbf{x}(t) = \mathbf{c}e^{\lambda\tau}$, where $\mathbf{c}$ is a constant vector, into the linear part to find (I represents the identify matrix in the following equation)

$$\begin{aligned} D(\lambda) &= \det\big(\lambda I - A_1 - A_2 e^{-\lambda\tau}\big) \\ &= \lambda^4 - (a_3 + b_3)\lambda^3 + (a_3 b_4 - a_4 b_3 - b_2 - a_1)\lambda^2 \\ &\quad + (b_2 a_3 - b_3 a_2 + a_1 b_4 - b_1 a_4)\lambda + a_1 b_2 - a_2 b_1 \\ &\quad + \big[(a_7 b_8 - a_8 b_7)\lambda^2 + (a_5 b_8 - a_8 b_5 + a_7 b_6 - a_6 b_7)\lambda \\ &\qquad + (a_5 b_6 - a_6 b_5)\big] e^{-2\lambda\tau} \\ &\quad - \big[(a_7 + b_8)\lambda^3 + (a_5 + b_6 + a_4 b_7 - a_7 b_4 + a_8 b_3 - a_3 b_8)\lambda^2 \\ &\qquad + (a_8 b_1 - a_1 b_8 + a_6 b_3 - a_3 b_6 + a_4 b_5 - a_5 b_4 \\ &\qquad\quad + a_2 b_7 - a_7 b_2)\lambda + (a_6 b_1 - a_1 b_6 + a_2 b_5 - a_5 b_2)\big] e^{-\lambda\tau}. \end{aligned} \tag{10}$$

Based on equation (10), it can be shown that (Yu, Yuan and Xu, 2002) *the number of the eigenvalues of the characteristic equation* (10) *with negative real parts, counting multiplicities, can change only when the eigenvalues become pure imaginary pairs as the time delay τ and the components of A_1 and A_2 are varied.*

It is seen from equation (10) that when

$$a_1 b_2 - a_2 b_1 + a_2 b_5 - a_5 b_2 + a_5 b_6 - a_6 b_5 + a_6 b_1 - a_1 b_6 \neq 0,$$

none of the roots of $D(\lambda)$ is zero. Thus, the trivial equilibrium $\mathbf{x} = \mathbf{0}$ becomes unstable only when equation (10) has at least one pair of purely imaginary roots: one purely imaginary pair corresponds to Hopf bifurcation, while two purely imaginary pairs lead to double Hopf bifurcation. The critical values for a Hopf or double Hopf bifurcation to occur can be found from the equation: $D(i\omega)$ from which setting real and imaginary parts zero yields

$$\begin{aligned} &P_0 + P_1 \cos(\omega\tau) + P_2 \sin(\omega\tau) + P_3 \cos(2\omega\tau) + P_4 \sin(2\omega\tau) = 0, \\ &Q_0 + Q_1 \cos(\omega\tau) + Q_1 \sin(\omega\tau) + Q_3 \cos(2\omega\tau) + Q_4 \cos(2\omega\tau) = 0, \end{aligned} \tag{11}$$

where P_i and Q_i are functions of ω, given by

$$
\begin{aligned}
P_0 &= -(a_3+b_4)\omega^3 + (a_4b_1 - a_1b_4 + a_2b_3 - a_3b_2)\omega,\\
P_1 &= (a_7+b_8)\omega^3\\
&\quad + (a_1b_8 - a_8b_1 + a_3b_6 - a_6b_3 + a_5b_4 - a_4b_5 + a_7b_2 - a_2b_7)\omega,\\
P_2 &= -(a_5+b_6+a_4b_7 - a_7b_4 + a_8b_3 - a_3b_8)\omega^2\\
&\quad - (a_1b_6 - a_6b_1 + a_5b_2 - a_2b_5),\\
P_3 &= (a_5b_8 - a_8b_5 + a_7b_6 - a_6b_7)\omega,\\
P_4 &= (a_7b_8 - a_8b_7)\omega^2 + a_6b_5 - a_5b_6;\\
Q_0 &= -\omega^4 - (a_1+b_2+a_4b_3 - a_3b_4)\omega^2 - a_1b_2 + a_2b_1,\\
Q_1 &= (a_5+b_6+a_3b_4 - a_4b_3 + a_8b_3 - a_3b_8)\omega\\
&\quad + a_1b_6 - a_6b_1 + a_5b_2 - a_2b_5,\\
Q_2 &= (a_7+b_8)\omega^3\\
&\quad + (a_1b_8 - a_8b_1 + a_3b_6 - a_6b_3 + a_5b_4 - a_4b_5 + a_7b_2 - a_2b_7)\omega,\\
Q_3 &= (a_8b_7 - a_7b_8)\omega^2 + a_5b_6 - a_6b_5,\\
Q_4 &= (a_5b_8 - a_8b_5 + a_7b_6 - a_6b_7)\omega.
\end{aligned}
$$

One can manipulate the two equations given in equation (11) to eliminate the trigonometric functions, resulting in the following 42nd-degree polynomial equation of ω:

$$\omega^{42} + q_1\omega^{41} + q_2\omega^{40} + \cdots + q_{41}\omega + q_{42} = 0, \tag{12}$$

where q_i's are known expressions in terms of the systems coefficients. If equation (12) has no positive real roots, then system (7) does not contain center manifold, but stable and unstable manifolds. On the other hand, if equation (12) has at least one positive solution for ω, one may substitute the solution(s) into equation (10) with $\lambda = \mathrm{i}\omega$ to find the smallest $\tau_{\min}$, at which the system undergoes a Hopf bifurcation. If equation (12) simultaneously has two positive solutions for ω, then double Hopf bifurcation occurs. Similarly, the smallest $\tau_{\min}$ corresponding to double Hopf bifurcation can be found.

Since equation (12) is a high degree polynomial equation, one can only use numerical approach to determine the relations at the critical point among the flutter speed V_{F}, Mach number M_∞, time delay τ and control gains Θ_i and Ψ_i, $i = 1, 2, 3$. More computation results will be given in Section 2.1.5.

2.1.3. Center manifold reduction

In order to obtain the explicit analytical expressions for the stability conditions of Hopf and double Hopf bifurcation solutions, we need to reduce system (5) or (7) to its center manifold. Since the center manifold reduction for Hopf bifurcation has been given in a recent paper (Yuan, Yu, Librescu and Marzocca, 2004) we shall only consider double Hopf bifurcation in this subsection. To achieve this, we express the infinite dimensional problem described by delay equations to an abstract evolution equation on Banach space H of continuously differentiable function $\mathbf{u}: [-\tau, 0] \to \mathbb{R}^4$ as (Hale and Lunel, 1993)

$$\dot{\mathbf{x}} = A\mathbf{x}_t + \mathbf{F}(t, \mathbf{x}_t), \tag{13}$$

where $\mathbf{x}_t(\theta) = \mathbf{x}(t+\theta)$ for $-\tau \leqslant \theta \leqslant 0$ and A is a linear operator for the critical case, expressed by

$$A\mathbf{u}(\theta) = \begin{cases} \frac{\mathrm{d}\mathbf{u}(\theta)}{\mathrm{d}\theta} & \text{for } \theta \in [-\tau, 0), \\ A_1\mathbf{u}(0) + A_2\mathbf{u}(-\tau) & \text{for } \theta = 0. \end{cases} \tag{14}$$

The nonlinear operator F is in the form of

$$\mathbf{F}(\mathbf{u})(\theta) = \begin{cases} 0 & \text{for } \theta \in [-\tau, 0), \\ \mathbf{F}\big(\mathbf{u}(0), \mathbf{u}(-\tau)\big) & \text{for } \theta = 0. \end{cases} \tag{15}$$

Similarly, we can define the dual/adjoint space H^* of continuously differentiable function $\mathbf{v}: [0, \tau] \to \mathbb{R}^4$ with the dual operator

$$A^*\mathbf{v}(\sigma) = \begin{cases} -\frac{\mathrm{d}\mathbf{v}(\sigma)}{\mathrm{d}\sigma} & \text{for } \sigma \in (0, \tau], \\ A_1^*\mathbf{v}(0) + A_2^*\mathbf{v}(\tau) & \text{for } \sigma = 0. \end{cases} \tag{16}$$

From the discussion given in the previous subsection, we know that the characteristic equation (12) has two pairs of purely imaginary eigenvalues $\lambda_1 = \pm \mathrm{i}\omega_{1,2}$ and $\lambda_{3,4} = \pm \mathrm{i}\omega_2$ at the double Hopf critical point. Therefore, H can be split into two subspace as $H = P_\Lambda \oplus Q_\Lambda$, where P_Λ is a four-dimensional space spanned by the eigenvectors of the operator A associated with the eigenvalues $\lambda_{1,2}$ and $\lambda_{3,4}$ and Q_Λ is the complementary space of P_Λ. Here, we shall focus on *nonresonant case*, i.e., the ratio ω_1/ω_2 is an irrational number.

Further, for $\mathbf{u} \in H$ and $\mathbf{v} \in H^*$, we can define a bilinear operator

$$\begin{aligned} \langle \mathbf{v}, \mathbf{u} \rangle &= \bar{\mathbf{v}}^{\mathrm{T}}(0)\mathbf{u}(0) - \int_{-\tau}^{0}\int_{0}^{\theta} \bar{\mathbf{v}}^{\mathrm{T}}(\xi - \theta)\big[\mathrm{d}\eta(\theta)\big]\mathbf{u}(\xi)\,\mathrm{d}\xi \\ &= \bar{\mathbf{v}}^{\mathrm{T}}(0)\mathbf{u}(0) + \int_{-\tau}^{0} \bar{\mathbf{v}}^{\mathrm{T}}(\xi + \theta)A_2(\xi)\mathbf{u}(\xi)\,\mathrm{d}\xi. \end{aligned} \tag{17}$$

Corresponding to the critical characteristic root $\mathrm{i}\omega_1$, the complex eigenvector $\mathbf{q}(\theta) \in H$ satisfies

$$\begin{aligned} &\frac{\mathrm{d}\mathbf{q}_1(\theta)}{\mathrm{d}\theta} = \mathrm{i}\omega_1\mathbf{q}_1(\theta) \quad \text{for } \theta \in [-\tau, 0), \\ &A_1\mathbf{q}_1(0) + A_2\mathbf{q}_1(-\tau) = \mathrm{i}\omega_1\mathbf{q}_1(0) \quad \text{for } \theta = 0. \end{aligned} \tag{18}$$

The general solution of equation (18) is $\mathbf{q}_1(\theta) = \mathbf{C}_1 e^{i\omega_1\theta}$. Then from the boundary conditions we find the matrix equation $[A_1 + A_2 e^{-i\omega_1\tau} - i\omega_1 I_4]\mathbf{C}_1 = \mathbf{0}$, where I_4 denotes 4×4 identity matrix. By letting $\mathbf{C}_1 = (C_{11}, C_{12}, C_{13}, C_{14})^T$ and choosing $C_{11} = 1$, we can uniquely determine C_{12}, C_{13} and C_{14}. Thus, the eigenvector $\mathbf{q}_1 = \mathbf{C}_1 e^{i\omega_1\theta}$ is found.

Similarly, for the root $i\omega_2$, one obtains the characteristic equation $[A_1 + A_2 e^{-i\omega_2\tau} - i\omega_2 I_4]\mathbf{C}_2 = \mathbf{0}$, and then finds the eigenvector $\mathbf{q}_2 = \mathbf{C}_2 e^{i\omega_2\theta}$. Thus, the real basis for P_Λ is obtained as $\Phi(\theta) = (\varphi_1, \varphi_2, \varphi_3, \varphi_4) = (\mathrm{Re}(\mathbf{q}_1(\theta)), \mathrm{Im}(\mathbf{q}_1(\theta)), \mathrm{Re}(\mathbf{q}_2(\theta)), \mathrm{Im}(\mathbf{q}_2(\theta)))$.

Next, from the equations

$$
\begin{aligned}
&-\frac{d\mathbf{q}_1^*(\sigma)}{d\sigma} = -i\omega_1 \mathbf{q}_1^*(\sigma) \quad \text{for } \sigma \in [0, \tau), \\
&A_1^* \mathbf{q}_1^*(0) + A_2^* \mathbf{q}_1^*(\tau) = -i\omega_1 \mathbf{q}_1^*(0) \quad \text{for } \sigma = 0,
\end{aligned}
\tag{19}
$$

one obtains the general solution, given by $\mathbf{q}_1^*(\sigma) = \mathbf{D}_1 e^{i\omega_1\sigma}$ which, with the aid of the boundary condition, yields $[A_1 + A_2 e^{-i\omega_2\tau} + i\omega_1 I_4]\mathbf{D}_1 = \mathbf{0}$. Let $\mathbf{D}_1 = (D_{11}, D_{12}, D_{13}, D_{14})^T$, we may choose $D_{14} = N_{11} + iN_{12}$, where N_{11} and N_{12} are real numbers, and then uniquely determine the other components of $\mathbf{D}_1$.

Similarly, for the root $-i\omega_2$, one obtains the characteristic equation $[A_1 + A_2 e^{-i\omega_2\tau} + i\omega_2 I_4]\mathbf{D}_2 = \mathbf{0}$, and then finds the eigenvector $\mathbf{q}_2^* = \mathbf{D}_2 e^{i\omega_2\sigma}$. Thus, the real basis for the dual space Q_Λ are $\Psi(\sigma) = (\psi_1, \psi_2, \psi_3, \psi_4) = (\mathrm{Re}(\mathbf{q}_1^*(\sigma)), \mathrm{Im}(\mathbf{q}_1^*(\sigma)), \mathrm{Re}(\mathbf{q}_2^*(\sigma)), \mathrm{Im}(\mathbf{q}_2^*(\sigma)))$, and N_{11}, N_{12}, N_{21}, N_{22} can be obtained from the relation $\langle \Psi, \Phi \rangle = I$ which are expressed in terms of ω, τ as well as the coefficients a_i's and b_i's. The lengthy expressions of N_{ij} are omitted here. Now, by defining $\mathbf{w} \equiv (w_1, w_2, w_3, w_4)^T = \langle \Psi, \mathbf{u}_t \rangle$ (which actually represents the local coordinate system on the 4-D center manifold, induced by the basis Ψ), then with the aid of transformations $\Phi(\theta)$ and $\Psi(\sigma)$, one can decompose $\mathbf{u}_t$ into two parts to obtain

$$
\mathbf{u}_t = \mathbf{u}_t^{P_\Lambda} + \mathbf{u}_t^{Q_\Lambda} = \Phi \langle \Psi, \mathbf{u}_t \rangle + \mathbf{u}_t^{Q_\Lambda} = \Phi \mathbf{w} + \mathbf{u}_t^{Q_\Lambda},
\tag{20}
$$

which implies that the projection of $\mathbf{u}_t$ on the center manifold is $\Phi\mathbf{w}$. Next, applying equations (13) and (20) results in

$$
\langle \Psi, \Phi\dot{\mathbf{w}} + \dot{\mathbf{u}}_t^Q \rangle = \langle \Psi, A(\Phi\mathbf{w} + \mathbf{u}_t^Q) \rangle + \langle \Psi, F(\mathbf{t}, \Phi\mathbf{w} + \mathbf{u}_t^Q) \rangle,
\tag{21}
$$

and therefore,

$$
\langle \Psi, \Phi \rangle \dot{\mathbf{w}} = \langle \Psi, A\Phi \rangle \mathbf{w} + \langle \Psi, \mathbf{F}(\mathbf{t}, \Phi\mathbf{w} + \mathbf{u}_t^Q) \rangle
$$

which can be written as $I\dot{\mathbf{w}} = D_\Lambda \mathbf{w} + \mathcal{N}(\mathbf{w})$, and finally we obtain the equation of the center manifold:

$$
\dot{\mathbf{w}} = \begin{bmatrix} 0 & \omega_1 & 0 & 0 \\ -\omega_1 & 0 & 0 & 0 \\ 0 & 0 & 0 & \omega_2 \\ 0 & 0 & -\omega_2 & 0 \end{bmatrix} \mathbf{w} + \mathcal{N}(\mathbf{w}),
\tag{22}
$$

where $\mathcal{N}(\mathbf{w})$ represents the nonlinear terms contributed from the original system to the center manifold.

The lowest-order nonlinear terms of the center manifold, needed for determining stability, are

$$\mathcal{N}_3(\mathbf{w}) = \Psi^{\mathrm{T}}(0)F(\Phi\mathbf{w}) = \Psi^{\mathrm{T}}(0)\,(0 \quad 0 \quad (\mathbf{F}_3)_2^3 \quad (\mathbf{F}_4)_2^3)^{\mathrm{T}}$$
$$= \begin{pmatrix} \sum_{i+j+k+l=3} C^1_{ijkl} w_1^i w_2^j w_3^k w_4^l \\ \sum_{i+j+k+l=3} C^2_{ijkl} w_1^i w_2^j w_3^k w_4^l \\ \sum_{i+j+k+l=3} C^3_{ijkl} w_1^i w_2^j w_3^k w_4^l \\ \sum_{i+j+k+l=3} C^4_{ijkl} w_1^i w_2^j w_3^k w_4^l \end{pmatrix}, \tag{23}$$

where

$$(\mathbf{F}_3)_2^3 = a_9\big(\Phi(0)\mathbf{w}\big)_2^3 + a_{10}\big(\Phi(-\tau)\mathbf{w}\big)_2^3 + a_{11}\big(\Phi(-\tau)\mathbf{w}\big)_2^3,$$
$$(\mathbf{F}_4)_2^3 = b_9\big(\Phi(0)\mathbf{w}\big)_2^3 + b_{10}\big(\Phi(-\tau)\mathbf{w}\big)_2^3 + b_{11}\big(\Phi(-\tau)\mathbf{w}\big)_2^3,$$

and "$(\cdots)_2^3$" denotes the cubic order terms taking from the second component of the vector $(\cdots)$. In fact, since Φ is a 4×4 matrix and $\mathbf{w}$ is a 4×1 vector, $\Phi\mathbf{w}$ is a 4×1 vector which may include higher-order terms in the components, we just intercept the third-order terms.

2.1.4. Normal form and bifurcation analysis

For Hopf critical point, the normal form up to third order can be obtained (Yu, 1998):

$$\dot{\rho} = \rho\big(\alpha\mu + a_{20}\rho^2\big),$$
$$\dot{\theta} = \omega_1 + \beta\mu + c_{20}\rho^2. \tag{24}$$

Then, the approximate bifurcation solutions can be found from the equation $\dot{\rho} = 0$. The stability for the limit cycles arising from a Hopf bifurcation is determined by the sign of the coefficient a_{20} (referred to as the first Lyapunov quantity): the Hopf bifurcation is supercritical (subcritical) if $a_{20} < 0$ ($a_{20} > 0$).

For the double Hopf critical point, the formulas given in Yu (2001) can be applied to the center manifold (22) to yield the normal form for double Hopf bifurcation:

$$\dot{\rho}_1 = \rho_1\big(\alpha_{11}\mu_1 + \alpha_{12}\mu_2 + a_{20}\rho_1^2 + a_{02}\rho_2^2\big),$$
$$\dot{\rho}_2 = \rho_2\big(\alpha_{21}\mu_1 + \alpha_{22}\mu_2 + b_{20}\rho_1^2 + b_{02}\rho_2^2\big), \tag{25}$$

$$\dot{\theta}_2 = \omega_1 + \beta_{11}\mu_1 + \beta_{12}\mu_2 + c_{20}\rho_1^2 + c_{02}\rho_2^2,$$
$$\dot{\theta}_2 = \omega_2 + \beta_{21}\mu_1 + \beta_{22}\mu_2 + d_{20}\rho_1^2 + d_{02}\rho_2^2. \tag{26}$$

Then, the bifurcation solutions can be readily found from the equations $\dot{\rho}_1 = \dot{\rho}_2 = 0$. The stability for the periodic solutions (limit cycles) and quasiperiodic motions (tori) can be determined from the Jacobian of equation (25) evaluated on the bifurcation solutions. It has been shown (Yu, 2001; Yu, Yuan and Xu, 2002) that in the case of double Hopf bifurcation, stable 2-D and 3-D tori may exist, and even chaotic motions can be found.

Since all the coefficients given in equations (25) and (26) are expressed in terms of the original system parameters such as Φ_i, Θ_i and τ, parametric studies can be carried out to consider the effect of the control parameters and the time delay on the stability of the system, and on the characteristics of the instability flutter boundary, that is of benign or catastrophic nature:

(I) the initial equilibrium solution (E.S.):

$$\rho_1 = \rho_2 = 0 \quad (\text{i.e., } w_i = 0 \text{ or } x_i = 0); \tag{27}$$

(II) Hopf bifurcation solution (H.B.(I) with frequency ω_1):

$$\begin{aligned} \rho_1^2 &= -\frac{1}{a_{20}}(\alpha_{11}\mu_1 + \alpha_{12}\mu_2), \qquad \rho_2 = 0, \\ \Omega_1 &= \omega_1 + \beta_{11}\mu_1 + \beta_{12}\mu_2 + c_{20}\rho_1^2; \end{aligned} \tag{28}$$

(III) Hopf bifurcation solution (H.B.(II) with frequency ω_2):

$$\begin{aligned} \rho_1 &= 0, \qquad \rho_2^2 = -\frac{1}{b_{02}}(\alpha_{21}\mu_1 + \alpha_{22}\mu_2), \\ \Omega_2 &= \omega_2 + \beta_{21}\mu_1 + \beta_{22}\mu_2 + d_{02}\rho_2^2; \end{aligned} \tag{29}$$

(IV) quasiperiodic solution (2-D tori with frequencies ω_1, ω_2):

$$\begin{aligned} \rho_1^2 &= \frac{1}{a_{20}b_{02} - a_{02}b_{20}}\big[a_{02}(\alpha_{21}\mu_1 + \alpha_{22}\mu_2) - b_{02}(\alpha_{11}\mu_1 + \alpha_{12}\mu_2)\big], \\ \rho_2^2 &= \frac{1}{a_{20}b_{02} - a_{02}b_{20}}\big[b_{20}(\alpha_{11}\mu_1 + \alpha_{12}\mu_2) - a_{20}(\alpha_{21}\mu_1 + \alpha_{22}\mu_2)\big], \\ \Omega_1 &= \omega_1 + \beta_{11}\mu_1 + \beta_{12}\mu_2 + c_{20}\rho_1^2 + c_{02}\rho_2^2, \\ \Omega_2 &= \omega_2 + \beta_{21}\mu_1 + \beta_{22}\mu_2 + d_{20}\rho_1^2 + d_{02}\rho_2^2. \end{aligned} \tag{30}$$

The Jacobian matrix of equation (25) takes the form

$$J = \begin{bmatrix} \alpha_{11}\mu_1 + \alpha_{12}\mu_2 + 3a_{20}\rho_1^2 + a_{02}\rho_2^2 & 2a_{02}\rho_1\rho_2 \\ 2b_{20}\rho_1\rho_2 & \alpha_{21}\mu_1 + \alpha_{22}\mu_2 + b_{20}\rho_1^2 + 3b_{02}\rho_2^2 \end{bmatrix}. \tag{31}$$

Equation (31) can be used for the stability analysis of the above steady-state solutions (27)–(30).

Evaluating the Jacobian (31) on the solution (27) results in the stability conditions for the E.S.:

$$\alpha_{11}\mu_1 + \alpha_{12}\mu_2 < 0 \quad \text{and} \quad \alpha_{21}\mu_1 + \alpha_{22}\mu_2 < 0,$$

from which two critical lines are obtained: one of them is described by

$$L_1: \quad \alpha_{11}\mu_1 + \alpha_{12}\mu_2 = 0 \quad (\alpha_{21}\mu_1 + \alpha_{22}\mu_2 < 0),$$

where a family of limit cycles bifurcates from the E.S. with the approximate solution H.B.(I) given by equation (28). The second critical line is

$$L_2: \quad \alpha_{21}\mu_1 + \alpha_{22}\mu_2 = 0 \quad (\alpha_{11}\mu_1 + \alpha_{12}\mu_2 < 0),$$

from which another family of limit cycles, given by H.B.(II) solution (29), may occur.

Next, evaluating the Jacobian (31) on the Hopf bifurcation solution (29), results in the stability conditions:

$$\begin{aligned} &\alpha_{11}\mu_1 + \alpha_{12}\mu_2 > 0, \\ &\alpha_{21}\mu_1 + \alpha_{22}\mu_2 - \frac{b_{20}}{a_{20}}(\alpha_{11}\mu_1 + \alpha_{12}\mu_2) < 0. \end{aligned} \tag{32}$$

However, we should check if the H.B.(I) solution exists under the above conditions. Comparing the solution ρ_1^2 given in equation (28) and the first inequality of (32) reveals that the existence of a stable H.B.(I) periodic solution requires $a_{20} < 0$.

The second inequality given in equation (32) implies another critical line

$$\begin{aligned} L_3: \quad &\left(\alpha_{21} - \frac{b_{20}}{a_{20}}\alpha_{11}\right)\mu_1 + \left(\alpha_{22} - \frac{b_{20}}{a_{20}}\alpha_{12}\right)\mu_2 = 0 \\ &(\alpha_{11}\mu_1 + \alpha_{12}\mu_2 > 0) \end{aligned}$$

along which a secondary Hopf bifurcation with frequency ω_2 takes place from the first bifurcating limit cycle H.B.(I), leading to a 2-D torus described by solution (30).

Similarly, a set of stability conditions associated with the H.B.(II) solution (29) can be obtained by evaluating the Jacobian (31) on this solution. Thus, for the H.B.(II) solution to be stable, the inequalities

$$\begin{aligned} &\alpha_{21}\mu_1 + \alpha_{22}\mu_2 > 0, \\ &\alpha_{11}\mu_1 + \alpha_{12}\mu_2 - \frac{a_{02}}{b_{02}}(\alpha_{21}\mu_1 + \alpha_{22}\mu_2) < 0 \end{aligned} \tag{33}$$

must be satisfied. Again, it is easy to show that the existence of a stable H.B.(II) solution requires that $b_{02} < 0$.

The second inequality of (33) defines a critical line

$$L_4: \quad \left(\alpha_{11} - \frac{a_{02}}{b_{02}}\alpha_{21}\right)\mu_1 + \left(\alpha_{12} - \frac{a_{02}}{b_{02}}\alpha_{22}\right)\mu_2 = 0$$
$$(\alpha_{21}\mu_1 + \alpha_{22}\mu_2 > 0),$$

where a secondary Hopf bifurcation with frequency ω_1 takes place from the first bifurcating limit cycle H.B.(II), leading to the same family of 2-D tori described by solution (30).

To find the stability of the family of 2-D tori expressed by equation (30), evaluating the Jacobian (31) on solution (30) yields

$$J_{\text{2-D tori}} = \begin{bmatrix} 2a_{20}\rho_1^2 & 2a_{02}\rho_1\rho_2 \\ 2b_{20}\rho_1\rho_2 & 2b_{02}\rho_2^2 \end{bmatrix}.$$

The stability conditions for the quasiperiodic motion are then obtained from the trace and determinant of the Jacobian, given by

$$\begin{aligned} \mathrm{Tr} &= 2\big(a_{20}\rho_1^2 + b_{02}\rho_2^2\big) < 0, \\ \mathrm{Det} &= 4(a_{20}b_{02} - a_{02}b_{20})\rho_1^2\rho_2^2 > 0. \end{aligned} \tag{34}$$

The condition Det > 0 implies that $a_{20}b_{02} - a_{02}b_{20} > 0$. Note that the above stability conditions must be considered together with the existence conditions of the quasiperiodic solution (see equation (30)):

$$\begin{aligned} &a_{02}(\alpha_{21}\mu_1 + \alpha_{22}\mu_2) - b_{02}(\alpha_{11}\mu_1 + \alpha_{12}\mu_2) > 0, \\ &b_{20}(\alpha_{11}\mu_1 + \alpha_{12}\mu_2) - a_{20}(\alpha_{21}\mu_1 + \alpha_{22}\mu_2) > 0, \end{aligned} \tag{35}$$

due to $a_{20}b_{02} - a_{02}b_{20} > 0$. It is interesting to see from equation (35) that the boundaries of the existence region for the quasiperiodic solution (IV) are actually defined by the critical lines L_4 and L_3. In other words, the periodic solution H.B.(I) (H.B.(II)) bifurcates from the critical line L_3 (L_4) into a quasiperiodic solution which has stability boundary L_4 (L_3).

Under the condition $a_{20}b_{02} - a_{02}b_{20} > 0$ (i.e., Det > 0), the first inequality of (34) becomes

$$a_{20}(a_{02} - b_{02})(\alpha_{21}\mu_1 + \alpha_{22}\mu_2) - b_{02}(a_{22} - b_{20})(\alpha_{11}\mu_1 + \alpha_{12}\mu_2) < 0.$$

Finally, the inequality given in the above equation suggests a critical line

$$\begin{aligned} L_5: \quad &\big[a_{20}(a_{02} - b_{02})\alpha_{21} - b_{02}(a_{22} - b_{20})\alpha_{11}\big]\mu_1 \\ &+ \big[a_{20}(a_{02} - b_{02})\alpha_{22} - b_{02}(a_{22} - b_{20})\alpha_{12}\big]\mu_2 = 0, \end{aligned}$$

from which a quasiperiodic solution loses stability and may bifurcate into a motion on a 3-D torus $(\omega_1, \omega_2, \omega_3)$. However, in order for such a complex motion

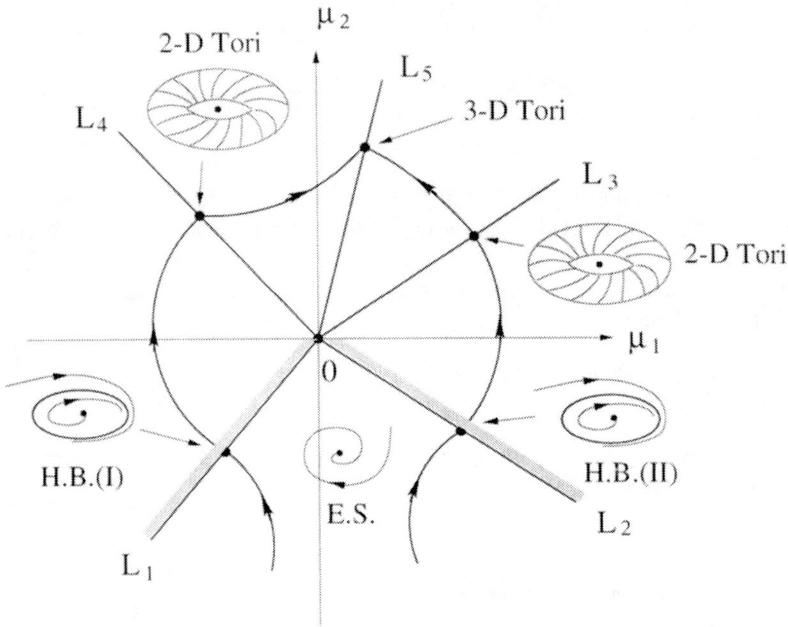

Figure 2. General bifurcation diagram for double Hopf singularity.

to occur, the critical line L_5 must be located in the regime defined by the existence conditions, i.e., the critical line L_5 must be located between the critical lines L_3 and L_4. The bifurcation diagram showing the above results is schematically depicted in Figure 2, where critical lines and bifurcation solutions are illustrated.

The symbolic program using Maple for computing the normal form, the bifurcation solutions and their stabilities can be found in Yu (2001).

2.1.5. Numerical simulation results

As we have shown in the previous subsection, the first equation in the normal form (24) can be used to find the approximation of Hopf bifurcation solutions and determine their stabilities. The normal form given in equation (25) can be employed to study the complex bifurcations and stability behavior of the system near the double Hopf critical point. In this subsection, numerical results are presented to show how periodic solutions bifurcating from the Hopf critical point under the variations of the system parameters such as Θ_1, V and τ, as well as the dampings ζ_h and ζ_α.

In order to compare the results with that given in Marzocca, Librescu and Silva (2002), where no time delay is presented, and in Yuan, Yu, Librescu and

Marzocca (2004), where time delay is involved, we shall take the same parameter values used in these two references. The main chosen varying parameters are Θ_1, V and τ, as well as the dampings ζ_h and ζ_α. Other parameters given in equation (3) take the following fixed values:

$$\begin{array}{llll} b = 1.5, & \mu = 50, & \bar{\omega} = 1.0, & r_\alpha = 0.5, \\ \chi_\alpha = 0.25, & \gamma = 1, & \kappa = 1.4, & \delta_A = 1, \\ B = 1, & x_0 = 0.5, & \omega_\alpha = 60, & \Psi_1 = 0.3, \\ \Psi_2 = \Psi_3 = 0.2, & \Theta_2 = 0.2, & \Theta_3 = 0.4. & \end{array}$$

Further, we choose

$$\Theta_1 = 0.5, \qquad V = 8.0, \qquad \tau = 1.0. \tag{36}$$

Then, for the above chosen parameter values, we use a numerical approach to find the Hopf critical points for both cases with and without damping. A Hopf critical point is obtained for zero damping:

$$\zeta_h = \zeta_\alpha = 0, \qquad M_0 = 3.063142, \qquad \omega = 0.206257.$$

When we take some positive damping values, then the Hopf critical point is changed. For example, when $\zeta_h = \zeta_\alpha = 0.005$, we find the critical point as

$$\zeta_h = \zeta_\alpha = 0.005, \qquad M_0 = 3.786728, \qquad \omega = 0.207129.$$

Now, we perform small perturbations to investigate the effect of the critical parameters Θ_1, V, τ, and the dampings ζ_h and ζ_α. Here we pay particular attention to the dampings. When we vary one of the three parameters, Θ_1, V and τ, we keep the other two parameters unchanged, and then change the damping values. For example, when Θ_1 is changed from 0.5 to, say, 0.7, we shall fix $V = 8.0$ and $\tau = 1.0$ (and $M_0 = 3.063142$ for the case of no damping and $M_0 = 3.786728$ for the case with damping), then we slightly increase the dampings from zero to small positive values. It is shown that zero damping results in diverging motions; intermediate suitable damping leads to stable limit cycles; and relative larger damping yields convergence to a fixed point. For example, when $\Theta_1 = 0.7$ (critical value $\Theta_1 = 0.5$), $V = 8.0$ and $\tau = 1.0$, the system exhibits different motions. The fixed point is unstable (trajectory diverging to infinity) if no damping is involved, the case with a small damping $\zeta_h = 0.005$ and $\zeta_\alpha = 0.05$ shows a similar situation (see Figure 3). Then, for $\zeta_h = 0.015$ and $\zeta_\alpha = 0.005$, Figure 4 depicts stable limit cycles. Further increasing the damping results in the fixed point to become stable. Figure 5 shows that when $\zeta_h = 0.05$ and $\zeta_\alpha = 0.005$, the trajectories converge to the origin (the fixed point).

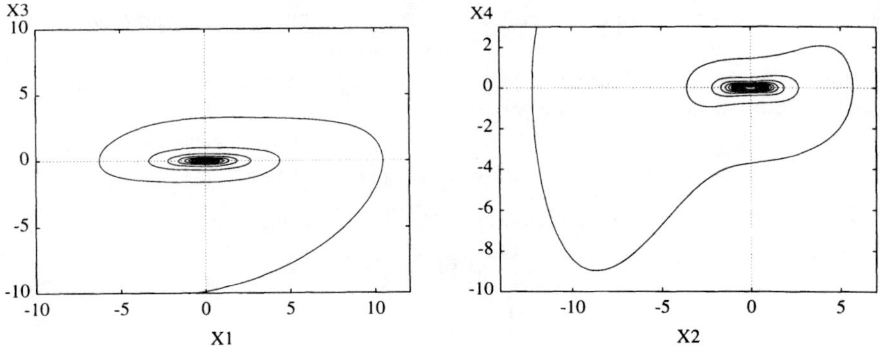

Figure 3. Diverging to infinity when $\Theta_1 = 0.7$, $V = 8.0$, $\tau = 1.0$, with $\zeta_h = 0.005$, $\zeta_\alpha = 0.05$.

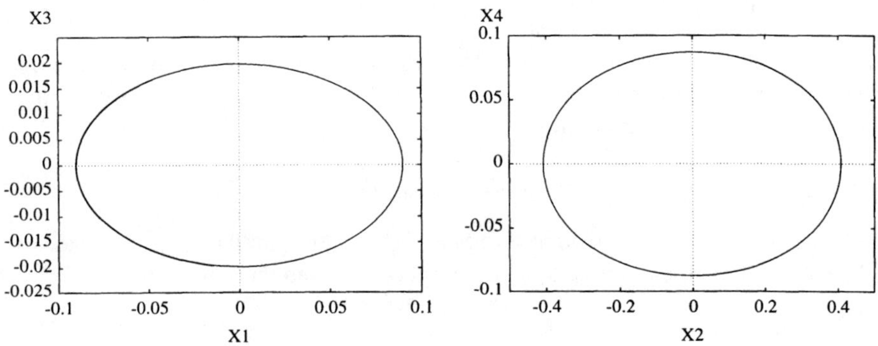

Figure 4. Stable limit cycles for $\Theta_1 = 0.7$, $V = 8.0$, $\tau = 1.0$, with $\zeta_h = 0.015$, $\zeta_\alpha = 0.005$.

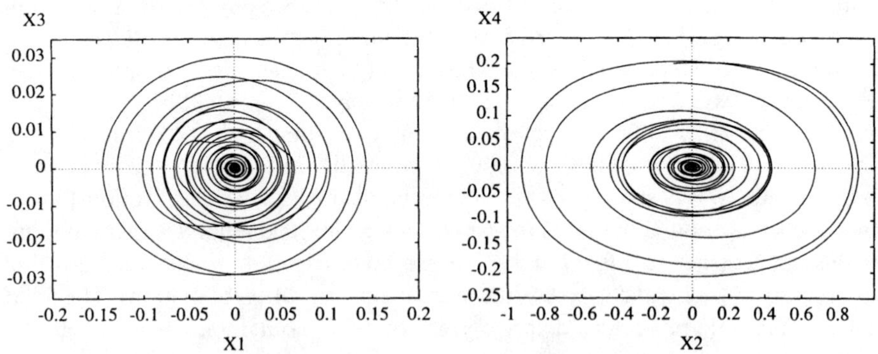

Figure 5. Converging to the origin when $\Theta_1 = 0.7$, $V = 8.0$, $\tau = 1.0$, with $\zeta_h = 0.05$, $\zeta_\alpha = 0.005$.

Figures 6–8 show the case when V is reduced from 8.0 to 7.0, with the variation of the damping. Similar situations to that shown in Figures 3–5 are observed.

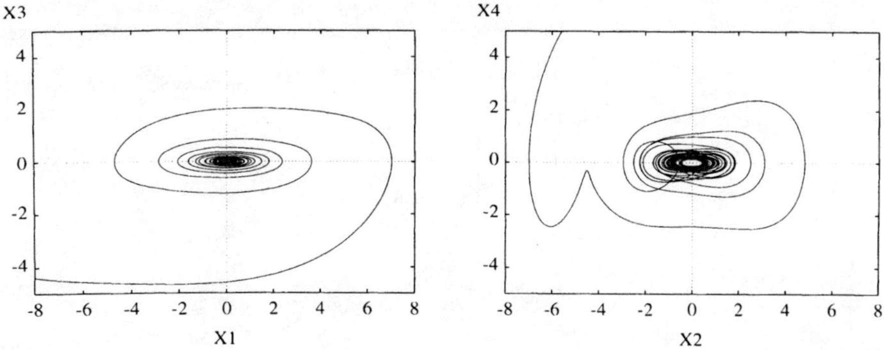

Figure 6. Diverging to infinity when $V = 7.0$, $\Theta_1 = 0.5$, $\tau = 1.0$, with $\zeta_h = \zeta_\alpha = 0.005$.

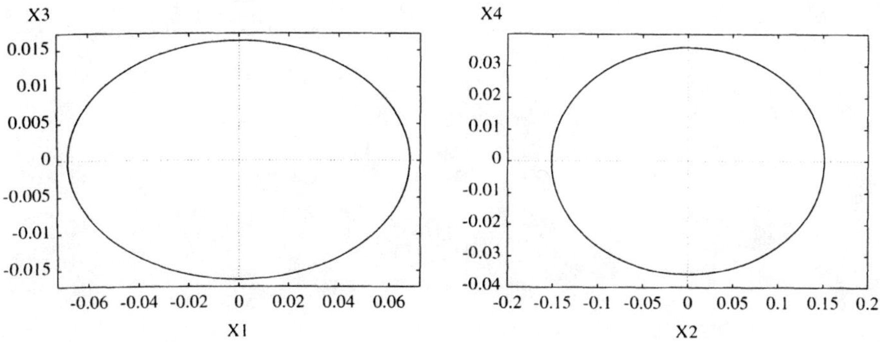

Figure 7. Stable limit cycles for $V = 7.0$, $\Theta_1 = 0.5$, $\tau = 1.0$, with $\zeta_h = 0.018$, $\zeta_\alpha = 0.005$.

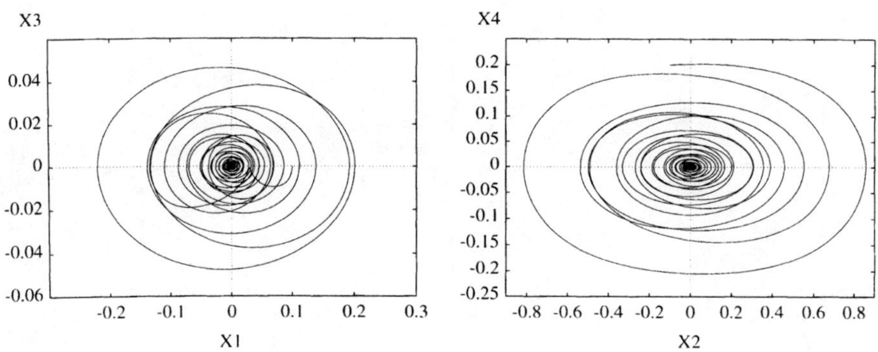

Figure 8. Converging to the origin when $V = 7.0$, $\Theta_1 = 0.5$, $\tau = 1.0$, with $\zeta_h = 0.05$, $\zeta_\alpha = 0.005$.

That is, zero damping or small damping reveal a unstable fixed point, intermediate damping leads to stable limit cycles, and larger damping result in convergence to

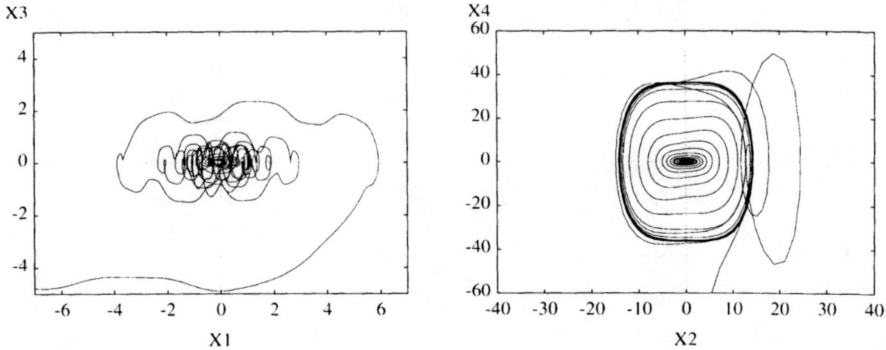

Figure 9. Diverging to infinity when $V = 5.0$, $\Theta_1 = 0.5$, $\tau = 1.0$, with $\zeta_h = \zeta_\alpha = 0.005$.

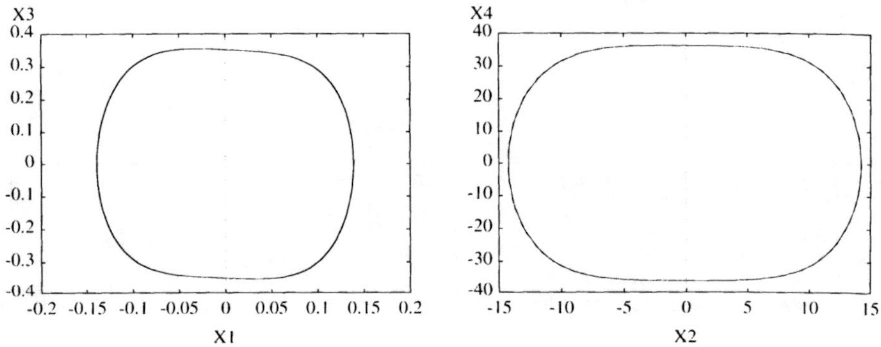

Figure 10. Stable limit cycles for $V = 5.0$, $\Theta_1 = 0.5$, $\tau = 1.0$, with $\zeta_h = 0.03$, $\zeta_\alpha = 0.005$.

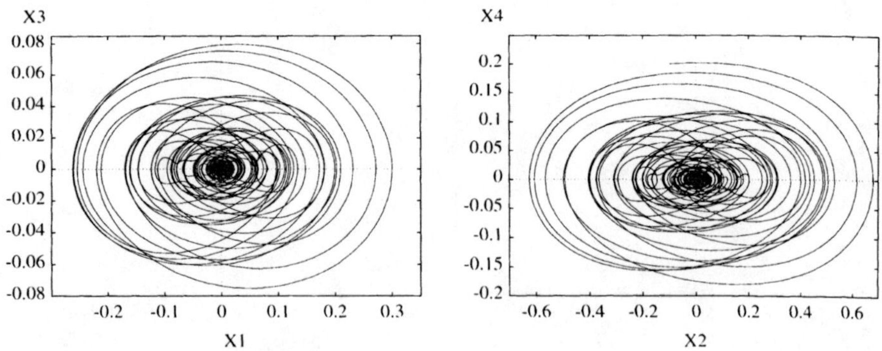

Figure 11. Converging to the origin when $V = 5.0$, $\Theta_1 = 0.5$, $\tau = 1.0$, with $\zeta_h = 0.005$, $\zeta_\alpha = 0.05$.

the fixed point. Figures 9–11 demonstrate the case when $V = 5$, which shows a similar trend as that seen in Figures 6–8. Finally, Figures 12–14 depict the

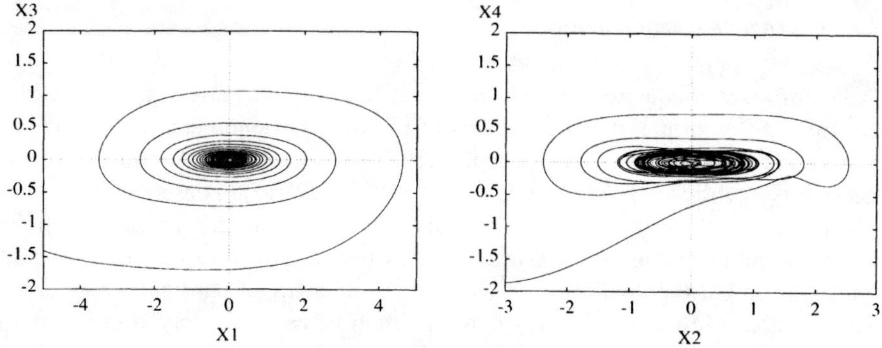

Figure 12. Diverging to infinity when $\tau = 1.1$, $\Theta_1 = 0.5$, $V = 8.0$, with $\zeta_h = \zeta_\alpha = 0.005$.

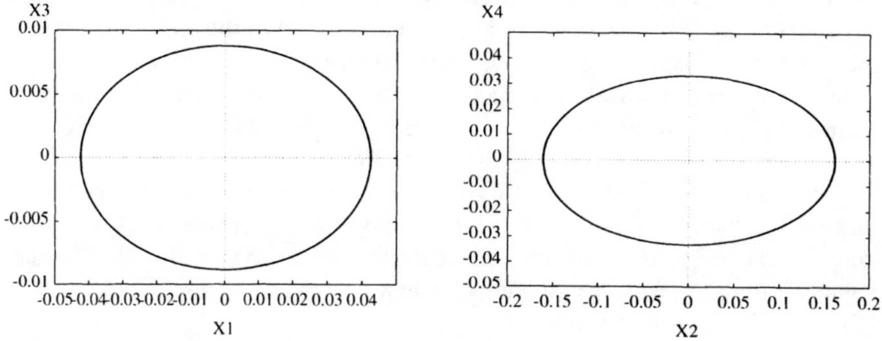

Figure 13. Stable limit cycles for $\tau = 1.1$, $\Theta_1 = 0.5$, $V = 8.0$, with $\zeta_h = 0.01$, $\zeta_\alpha = 0.05$.

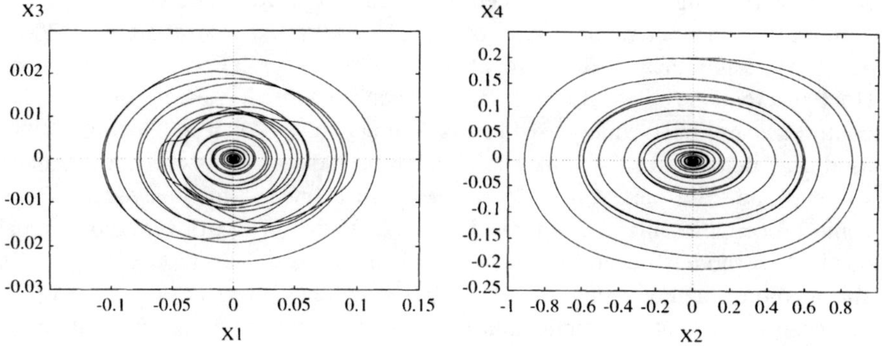

Figure 14. Converging to the origin when $\tau = 1.1$, $\Theta_1 = 0.5$, $V = 8.0$, with $\zeta_h = 0.05$, $\zeta_\alpha = 0.005$.

case when the time delay τ is increased from 1.0 to 1.1. Again, similar trend is found.

2.2. *Internet congestion model*

Delay differential equations have been widely used to study physical models arising from different areas. Many models of biological phenomena involve delays between interactions. An early example is the Hutchinson–Wright equation introduced by Hutchinson (Hassard, Kazarinoff and Wan, 1981) to describe the growth of single species. Many other applications of using delay differential equations can be found in biological neural networks, where the delay occurs in the signal transmission between neurons (Wu, 2001). Another important field where delay exists is control theory, since any control action takes effect only after a certain delay. In some cases, delay may be negligible, but often it is large enough to have impact on the dynamics of the controlled system (Sieber and Krauskopf, 2004). Stimulated by the work of Kelly, Maulloo and Tan (1998), delay differential equation has become an important tool in the study of the congestion control mechanisms and active queue management schemes (AQM) for the Internet.

In recent years, controlling and anticontrolling bifurcation and chaos have attracted many researchers from various disciplines. The aim of bifurcation control is to design a controller to modify the bifurcation properties of a given nonlinear system, thereby to achieve some desirable dynamical behaviors (Chen, Moiola and Wang, 2000; Yu, 2003b). Typical objectives of bifurcation control include delaying the onset of an inherent bifurcation, stabilizing an unstable bifurcated solution or branch, and changing the critical points of an existing bifurcation, etc. (Chen, Moiola and Wang, 2000).

Various methods have been used to control chaos and bifurcations. For example, Abed and Fu (1986) proposed a static state feedback approach; while Chen, Moiola and Wang (2000) developed a dynamic state feedback control law, incorporating washout filter to control the Lorenz system. Yu and Chen (2004) developed a nonlinear feedback controller with polynomial functions to control Hopf bifurcations in the Lorenz and Rössler systems. Time-delayed feedback control has also been widely used in controlling chaos. Bleich and Socolar (1996) used time-delayed feedback to obtain stable periodic orbits in a chaotic system, while Brandt, Shih and Chen (1997) designed a linear, time-delayed feedback control for suppressing a pathological period-2 rhythm in an atrioventricular nodal conduction model. Recently, Song, Yu, Chen, Xu and Tian (2002) proposed a chaos control method using time delay based on repetitive learning. However, it has been noted that the application of time delay in controlling bifurcations is not so popular, especially for controlling bifurcations arising from time-delayed systems.

In this subsection, bifurcation control using a time-delayed feedback controller for an Internet congestion will be considered. This system is used to model the Internet congestion control with a single route and single source (Li, Chen, Liao and Yu, 2004; Kelly, 2000), which can be described as

$$\dot{x}(t) = k\big[w - x(t - D)p\big(x(t - D)\big)\big], \tag{37}$$

where k is a positive gain parameter and $x(t)$ is the rate at which a source sends packets at time t. In the Internet, the communication delay is comprised of propagation delay and queuing delay. As the router hardware and network capacity continue to improve rapidly, the queuing delay becomes smaller compared to the propagation delay. D is the sum of the forward and returning delays, that is, the time during which the packet makes a round trip from a sender to a receiver and back to the sender. As a result, the sum of the forward and returning delays is fixed for resources on a given route. w is a target (set-point) and $p(\cdot)$ is the congestion indication function. When a resource within the network becomes overloaded, one or more packets are lost, and the lost of a packet is taken as an indication of congestion. The congestion indication function is assumed to be increasing, nonnegative and not identically zero (Johari and Tan, 2001).

There mainly exist two types of implementations for the congestion control: by resource or by users. For example, in an ATM network, ensuring quality-of-service is handled by the links in the network; while for the TCP/IP model of Internet, resource-centered congestion controllers are not used, since it is a very difficult design problem in network on the scale of the Internet. In today's Internet, this rate is controlled by the Transmission Control Protocol (TCP), implemented as software on end-systems (Kelly, 2000). The rate control algorithm comprises two components: a steady increase at a rate proportional to w, and a steady decrease as a rate proportional to the stream of congestion-indication signal received (Kelly, 2000). This control model can be used to adjust the sending rate $x(t)$ so that the expected number of marks received by the user will tend to the target w.

Model (37) has been extensively studied by many researchers in the past few years. Kelly (2000) obtained a local stability condition for the equilibrium point of model (37), while the global stability was discussed by Deb and Srikant (2003). Li, Chen, Liao and Yu (2004), on the other hand, have shown that Hopf bifurcation may occur as the positive gain parameter, k, passes through a critical point, where a family of periodic solutions bifurcates from an equilibrium point. Thus, the stationary sending rate is not guaranteed, which is not desirable. We will apply an effective delay feedback control to the Internet model to delay the onset of Hopf bifurcation (Yu and Chen, 2004; Chen and Yu, 2005a, 2005b). It should be noted that for the end-to-end congestion control in the Internet, the congestion control is implemented by end-user software, so technically, designing a controller for the Internet congestion model does not increase the cost or the complexity of the system. We will show, with a Hopf bifurcation controller, that one can increase the critical value of positive gain parameter, thereby, guarantee a stationary sending rate for large parameter values, which benefits congestion controls. Furthermore, we will demonstrate that the delayed feedback controller designed for the Internet congestion model can be used for controlling bifurcations arising from general delayed differential equations.

In the following, first the main results for the Hopf bifurcation of the Internet congestion model, obtained in Li, Chen, Liao and Yu (2004), are summarized for completeness and convenience. Then we design a time-delayed feedback controller using polynomial function to control Hopf bifurcation for the Internet congestion model. Finally, numerical results are presented to verify the analytical predictions.

2.2.1. Hopf bifurcation

In this subsection, the results of Hopf bifurcation for the Internet congestion model, obtained in Li, Chen, Liao and Yu (2004), are summarized here for completeness and convenience.

THEOREM 2.1. *For system* (37), *a Hopf bifurcation occurs from its equilibrium*, x^*, *when the positive gain parameter*, k, *passes through the critical value*, $k^* = \pi/(2D[p(x^*) + x^* p'(x^*)])$, *where* x^* *satisfies* $x^* p(x^*) = w$.

THEOREM 2.2. *The Hopf bifurcation for the Internet congestion model* (37) *is determined by the parameters* μ_2, β_2 *and* τ_2, *where* μ_2 *determines the direction of the Hopf bifurcation*: *the Hopf bifurcation is supercritical* (*subcritical*) *when* $\mu_2 > 0$ ($\mu_2 < 0$), *and the bifurcating periodic solutions exist* (*do not exist*) *if* $k > k^*$ ($k < k^*$); β_2 *determines the stability of the bifurcating periodic solutions*: *the solutions are orbitally stable* (*unstable*) *if* $\beta_2 < 0$ ($\beta_2 > 0$); *and* τ_2 *determines the period of the bifurcating periodic solutions*: *the period increases* (*decreases*) *if* $\tau_2 > 0$ ($\tau_2 < 0$).

The parameters μ_2, β_2 *and* τ_2 *are given by*

$$\mu_2 = -\frac{\operatorname{Re} C_1(0)}{\operatorname{Re} \lambda'(0)}, \qquad \beta_2 = 2 \operatorname{Re} C_1(0),$$
$$\tau_2 = -\frac{\operatorname{Im} C_1(0) + \mu_2 \operatorname{Im} \lambda'(0)}{\omega_0}, \tag{38}$$

where

$$\operatorname{Re} \lambda'(0) = \frac{-b_1^* D[p(x^*) + x^* p'(x^*)]}{1 + (b_1^* D)^2},$$
$$\operatorname{Im} \lambda'(0) = -\frac{1}{b_1^* D} \frac{\mathrm{d}\alpha}{\mathrm{d}k} = \frac{p(x^*) + x^* p'(x^*)}{1 + (b_1^* D)^2},$$
$$C_1(0) = \frac{\mathrm{i}}{2\omega_0}\left(g_{20} g_{11} - 2|g_{11}|^2 - \frac{1}{3}|g_{02}|^2\right) + \frac{g_{21}}{2}, \tag{39}$$

in which

$$
\begin{aligned}
\omega_0 &= \frac{\pi}{2D},\\
g_{20} = g_{02} = -g_{11} &= \frac{-2b_2^*}{1 + b_1^* D e^{-i\omega_0 D}},\\
g_{21} &= \frac{2i}{1 + b_1^* D e^{-i\omega_0 D}}\\
&\quad \times \left[\frac{2(2b_2^* - g_{11} - \bar{g}_{11})b_2^*}{b_1^*} + \frac{(g_{20} - \bar{g}_{02} + 2b_2^*)b_2^*}{b_1^* - 2i\omega_0} - 3b_3^*\right],\\
b_1^* &= -k^*\left[p(x^*) + x^* p'(x^*)\right],\\
b_2^* &= -\frac{k^*}{2}\left[2p'(x^*) + x^* p''(x^*)\right],\\
b_3^* &= -\frac{k^*}{6}\left[3p''(x^*) + x^* p'''(x^*)\right].
\end{aligned}
\tag{40}
$$

The detailed derivation of the above formulas can be found in Li, Chen, Liao and Yu (2004). For convenience in analyzing Hopf bifurcation, we may, based on the above formulas, write the normal form up to third order as

$$
\dot{\rho} = \rho\left[\operatorname{Re}\lambda'(0)(k - k^*) + 2\operatorname{Re}C_1(0)\rho^2 + \cdots\right], \tag{41}
$$

$$
\dot{\theta} = \omega_0 + \operatorname{Im}\lambda'(0)(k - k^*) + 2\operatorname{Im}C_1(0)\rho^2 + \cdots, \tag{42}
$$

where ρ and θ represent the amplitude and phase of periodic solutions, respectively. Thus, the approximate Hopf bifurcation solutions are given by

$$
\rho^2 = -\frac{\operatorname{Re}\lambda'(0)}{2\operatorname{Re}C_1(0)}(k - k^*) = -\frac{\operatorname{Re}\lambda'(0)}{\beta_2}(k - k^*) = \frac{1}{2\mu_2}(k - k^*), \tag{43}
$$

and the approximate frequency for the family of limit cycles (periodic solutions) is

$$
\begin{aligned}
\omega &= \omega_0 + \operatorname{Im}\lambda'(0)(k - k^*) + 2\operatorname{Im}C_1(0)\rho^2\\
&= \omega_0 + \operatorname{Im}\lambda'(0)(k - k^*) + \frac{\operatorname{Im}C_1(0)}{\mu_2}(k - k^*)\\
&= \omega_0\left[1 + \frac{\mu_2 \operatorname{Im}\lambda'(0) + \operatorname{Im}C_1(0)}{\omega_2\mu_2}(k - k^*)\right]\\
&= \omega_0\left[1 - \frac{\tau_2}{\mu_2}(k - k^*)\right].
\end{aligned}
\tag{44}
$$

Note that $\operatorname{Re}\lambda'(0)$ and $\operatorname{Im}\lambda'(0)$ are, respectively, the derivatives of the real and imaginary parts of the complex conjugate eigenvalues which cross the imaginary axis, giving rise to Hopf bifurcation. These two coefficients depend upon only linear terms of the system, while $\operatorname{Re} C_1(0)$ is determined from nonlinear terms of the system. All the three coefficients are fixed (evaluated at the critical point $k = k^*$) regardless the variation of k.

Then all the conclusions given in Theorem 2.2 can be easily obtained from equations (41)–(44). Letting $\dot{\rho} = 0$ yields two steady-state solutions: equilibrium solution $\rho = 0$ (i.e., $x = x^*$) and Hopf bifurcation solution given by equation (43). Suppose k is varied from $k < k^*$ to $k > k^*$. Thus, the equilibrium point $\rho = 0$ is stable (unstable) when $k < k^*$ if $\operatorname{Re} \lambda'(0) > 0$ ($\operatorname{Re} \lambda'(0) < 0$), as expected from linear analysis. (Note that the Jacobian matrix of equation (41) is $J = \operatorname{Re} \lambda'(0)(k - k^*)$ at $\rho = 0$.) Now, assume $\operatorname{Re} \lambda'(0) > 0$, then the equilibrium point loses its stability at the critical point $k = k^*$ and a Hopf bifurcation occurs, giving rise to a family of limit cycles when $k > k^*$. It is easy to see from equation (43) that the Hopf bifurcation is supercritical (subcritical) if $\mu_2 > 0$ ($\mu_2 < 0$), since $\rho^2 = \frac{1}{2\mu_2(k-k^*)} > 0$ for $k > k^*$ when $\mu_2 > 0$. The bifurcating limit cycles are stable (unstable) when $\beta_2 < 0$ ($\beta_2 > 0$). (Note that the Jacobian of equation (41) evaluated on the Hopf bifurcation solution (43) is $J = 2\beta_2\rho^2$.) Further, for a stable periodic solution (for which $\mu_2 > 0$), it can be observed from equation (44) that the frequency of the periodic solution, ω, decreases (increases) as k increases if $\tau_2 > 0$ ($\tau_2 < 0$). This implies that the period of the solution increases (decreases) as k increases if $\tau_2 > 0$ ($\tau_2 < 0$).

2.2.2. *Hopf bifurcation control*

We now turn to design a time-delayed feedback controller in order to control the Hopf bifurcation arising from the Internet congestion model (37). Following the general idea of polynomial function controller (Yu and Chen, 2004), we propose a time-delayed feedback controller as

$$u(x) = \alpha_1\big(x(t-D) - x^*\big) + \alpha_2\big(x(t-D) - x^*\big)^2 + \alpha_3\big(x(t-D) - x^*\big)^3. \tag{45}$$

It is easy to see that this time-delayed feedback controller preserves the equilibrium point of the original Internet congestion model (37). The form of the controller is quite simple, which only involves the first-, second- and third-order terms around the equilibrium point x^*. α_1, α_2 and α_3 are parameters, which can be used to control the Hopf bifurcation to achieve desirable behaviors, such as delaying the onset of a Hopf bifurcation, changing the period of a Hopf bifurcation. Actually, these three terms are sufficient to determine the parameters μ_2, β_2 and τ_2, for the direction, stability and period of a Hopf bifurcation.

With the time-delayed feedback controller (45), the controlled Internet congestion model (37) becomes

$$\frac{\mathrm{d}x(t)}{\mathrm{d}t} = k\big[w - x(t-D)p\big(x(t-D)\big)\big] + \alpha_1\big(x(t-D) - x^*\big) + \alpha_2\big(x(t-D) - x^*\big)^2 + \alpha_3\big(x(t-D) - x^*\big)^3, \tag{46}$$

where x^* denotes an equilibrium point, determined from the equation $x^* \times p(x^*) = w$.

Using Taylor expansion, we can expand the right-hand side of equation (46) around x^*, resulting in the following linearized equation:

$$\dot{x}(t) = b_1\big(x(t-D) - x^*\big), \tag{47}$$

where $b_1 = \alpha_1 - k(p(x^*) + x^* p'(x^*))$. Let $\bar{x}(t) = x(t) - x^*$, then equation (47) becomes $\dot{\bar{x}}(t) = b_1 u(t-D)$, which has the characteristic equation

$$\lambda - b_1 \mathrm{e}^{-\lambda D} = 0. \tag{48}$$

When equation (48) has a pair of purely imaginary roots $\lambda_\pm = \pm \mathrm{i}\omega$ ($\omega > 0$), it is straightforward to obtain that

$$b_1 \cos(\omega D) = 0, \qquad \omega + b_1 \sin(\omega D) = 0,$$

which, in turn, yields

$$\omega D = \frac{(2n+1)\pi}{2},$$
$$\frac{(2n+1)\pi}{2D} + b_1(-1)^n = 0 \quad \text{for } n = 0, 2, 4, \ldots. \tag{49}$$

Then, it follows from equation (49) and $b_1 = \alpha_1 - k(p(x^*) + x^* p'(x^*))$ that

$$k = \frac{\alpha_1}{p(x^*) + x^* p'(x^*)} - \frac{m_1}{p(x^*) + x^* p'(x^*)} = \frac{\alpha_1}{p(x^*) + x^* p'(x^*)} + \frac{(2n+1)\pi}{2D[p(x^*) + x^* p'(x^*)]}.$$

It has been shown by Li, Chen, Liao and Yu (2004) that the characteristic equation does not have roots with positive real parts unless if $\omega_0 = \pi/(2D)$. Thus, we obtain

$$k^* = \frac{\alpha_1}{p(x^*) + x^* p'(x^*)} + \frac{\pi}{2D[p(x^*) + x^* p'(x^*)]}.$$

Letting $\lambda = \alpha + \mathrm{i}\omega$, and then substituting λ into the characteristic equation (48) yields

$$(\alpha + \mathrm{i}\omega) - b_1\big\{\mathrm{e}^{-\alpha D}\big[\cos(-\omega D) + \mathrm{i}\sin(-\omega D)\big]\big\} = 0,$$

from which one can easily obtain

$$\frac{\mathrm{d}\alpha(k^*, \alpha_0, \omega_0)}{\mathrm{d}k} = \frac{-b_1^* D[p(x^*) + x^* p'(x^*)]}{1 + (b_1^* D)^2},$$
$$\frac{\mathrm{d}\omega(k^*, \alpha_0, \omega_0)}{\mathrm{d}k} = -\frac{1}{b_1^* D}\frac{\mathrm{d}\alpha}{\mathrm{d}k}.$$

Summarizing the above results gives the following theorem.

THEOREM 2.3. *For the controlled system* (46), *there exists a Hopf bifurcation emerging from its equilibrium x^*, when the positive gain parameter, k, passes through the critical value*

$$k^* = \frac{\alpha_1}{p(x^*) + x^* p'(x^*)} + \frac{\pi}{2D[p(x^*) + x^* p'(x^*)]},$$

where the equilibrium point x^* is kept unchanged, satisfying $x^* p(x^*) = w$.

Theorem 2.3 indicates that one can delay or advance the onset of a Hopf bifurcation without changing the original equilibrium points by choosing an appropriate value of α_1. If one only needs to delay the onset of the Hopf bifurcation, a linear, time-delayed feedback control with parameter α_1 is sufficient. The next theorem tells us that one can also change the stability and direction of bifurcating periodic solutions by choosing appropriate values of α_1, α_2 and α_3, thereby change the three control parameter μ_2, β_2 and τ_2 of the Hopf bifurcation.

To find the control parameters μ_2, β_2 and τ_2, one can use Taylor expansion to expand the right-hand side of equation (46) at the equilibrium solution, x^*. Thus, we have

$$\dot{x}(t) = b_1 u(t-D) + b_2 u^2(t-D) + b_3 u^3(t-D), \tag{50}$$

where

$$b_1 = \alpha_1 - k\big[p(x^*) + x^* p'(x^*)\big],$$
$$b_2 = \alpha_2 - \frac{k}{2}\big[2p'(x^*) + x^* p''(x^*)\big],$$
$$b_3 = \alpha_3 - \frac{k}{6}\big[3p''(x^*) + x^* p'''(x^*)\big]. \tag{51}$$

Note that the standard procedure for reducing a general one-dimensional delay differential equation (50) to a center manifold and calculating the parameters μ_2, β_2 and τ_2 are described in Li, Chen, Liao and Yu (2004), as well as in the textbook of Hassard, Kazarinoff and Wan (1981). For different one-dimensional systems, we may have different b_1, b_2 and b_3 in the Taylor expansion. Thus, to obtain the formulas of μ_2, β_2 and τ_2 for the controlled Internet congestion model (46), one can simply substitute b_1, b_2 and b_3 given in equation (51) to the formulas (38)–(40) of Theorem 2.2, thereby we have the following theorem for the controlled Internet congestion model.

THEOREM 2.4. *The Hopf bifurcation exhibited by the controlled Internet congestion model* (46) *is determined by the parameters* μ_2, β_2 *and* τ_2, *where* μ_2 *determines the direction of the Hopf bifurcation: the Hopf bifurcation is supercritical (subcritical) when* $\mu_2 > 0$ $(\mu_2 < 0)$ *and the bifurcating periodic solutions exist (do not exist) if* $k > k^*$ $(k < k^*)$; β_2 *determines the stability of the bifurcating periodic solutions: the solutions are orbitally stable (unstable) if* $\beta_2 < 0$ $(\beta_2 > 0)$; *and* τ_2 *determines the period of the bifurcating periodic solutions: the period increases (decreases) if* $\tau_2 > 0$ $(\tau_2 < 0)$.

The parameters μ_2, β_2 *and* τ_2 *can be found using the following formulas*:

$$\mu_2 = -\frac{\operatorname{Re} C_1(0)}{\operatorname{Re}\lambda'(0)}, \qquad \beta_2 = 2\operatorname{Re} C_1(0),$$

$$\tau_2 = -\frac{\operatorname{Im} C_1(0) + \mu_2 \operatorname{Im}\lambda'(0)}{\omega_0},$$

where

$$\operatorname{Re}\lambda'(0) = \frac{-b_1(k^*)D[p(x^*) + x^* p'(x^*)]}{1 + [b_1(k^*)D]^2},$$

$$\operatorname{Im}\lambda'(0) = -\frac{1}{b_1(k^*)D}\operatorname{Re}\lambda'(0),$$

$$C_1(0) = \frac{\mathrm{i}}{2\omega_0}\left(g_{20}g_{11} - 2|g_{11}|^2 - \frac{1}{3}|g_{02}|^2\right) + \frac{g_{21}}{2},$$

in which

$$\omega_0 = \frac{\pi}{2D},$$

$$g_{20} = g_{02} = -g_{11} = \frac{-2b_2^*}{1 + b_1^* D e^{-\mathrm{i}\omega_0 D}},$$

$$g_{21} = \frac{2\mathrm{i}}{1 + b_1^* D e^{-\mathrm{i}\omega_0 D}} \times \left[\frac{2(2b_2^* - g_{11} - \bar{g}_{11})b_2^*}{b_1^*} + \frac{(g_{20} - \bar{g}_{02} + 2b_2^*)b_2^*}{b_1^* - 2\mathrm{i}\omega_0} - 3b_3^*\right],$$

$$b_1^* = \alpha_1 - k^*\left[p(x^*) + x^* p'(x^*)\right],$$

$$b_2^* = \alpha_2 - \frac{k^*}{2}\left[2p'(x^*) + x^* p''(x^*)\right],$$

$$b_3^* = \alpha_3 - \frac{k^*}{6}\left[3p''(x^*) + x^* p'''(x^*)\right].$$

It is easy to see from Theorem 2.4 that one may choose appropriate values of the parameters α_1, α_2 and α_3 to change the values of μ_2, β_2 and τ_2 in order to control the direction, stability and period of the bifurcating periodic solutions of the Internet congestion model (37). The normal form for the controlled system (46) is in the same form of equations (41) and (42). The discussions based on equations (41)–(44), given at the end of the previous subsection for system (37) without the feedback control, can also be applied here for the controlled system (46).

2.2.3. Numerical results

Now, we present numerical results to verify the analytical predictions obtained in the previous subsection, using the time-delayed feedback controller (45) to control the Hopf bifurcation of Internet congestion model (37). The numerical approach is based on a fourth-order Runge–Kutta integration scheme.

For a consistent comparison, we choose the same function, $p(x) = x/(20 - 3x)$, used in Li, Chen, Liao and Yu (2004) to study model (37). For the uncontrolled model, it follows from Theorems 2.1 and 2.2 that

$$x^* = 3.2170, \qquad k^* = 1.7231, \qquad \omega_0 = 1.5708$$

and

$$\mu_2 = 0.4484, \qquad \tau_2 = 0.0769, \qquad \beta_2 = -0.4579.$$

The dynamical behavior of this uncontrolled Internet congestion model is illustrated in Figures 15–17. It is shown that when $k < k^*$, trajectories converge to the equilibrium point (see Figure 15), while as k is increased to pass k^*, x^* loses its stability and a Hopf bifurcation occurs (see Figures 16 and 17). Note that the period of the Hopf bifurcation increases due to $\tau_2 > 0$.

Now we choose appropriate values of α_1, α_2 and α_3 to control Hopf bifurcation. It is easy to see from Theorem 2.3 that for a positive value of α_1 ($\alpha_1 > 0$), we can delay the onset of the Hopf bifurcation. For example, by choosing

$$\alpha_1 = 0.5, \qquad \alpha_2 = 0, \qquad \alpha_3 = 0,$$

we can apply Theorems 2.3 and 2.4 to obtain

$$x^* = 3.2170, \qquad k^* = 2.2717, \qquad \omega_0 = 1.5708$$

and

$$\mu_2 = 0.1139, \qquad \tau_2 = 0.1336, \qquad \beta_2 = -1.0015.$$

Note that the controlled Internet congestion model (46) has the same equilibrium point as that of the original Internet congestion model (37), but the critical value

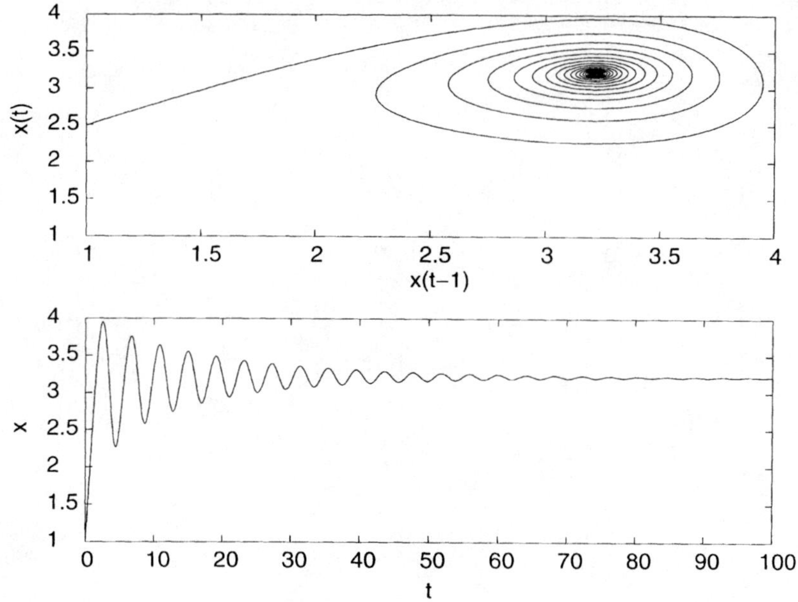

Figure 15. Phase portrait and time history of model (37) for $k = 1.6$.

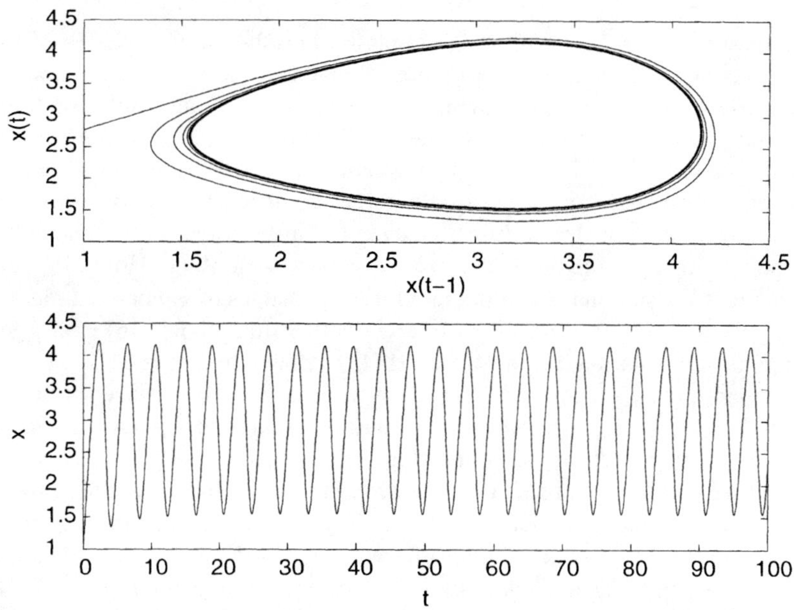

Figure 16. Phase portrait and time history of model (37) for $k = 1.9$.

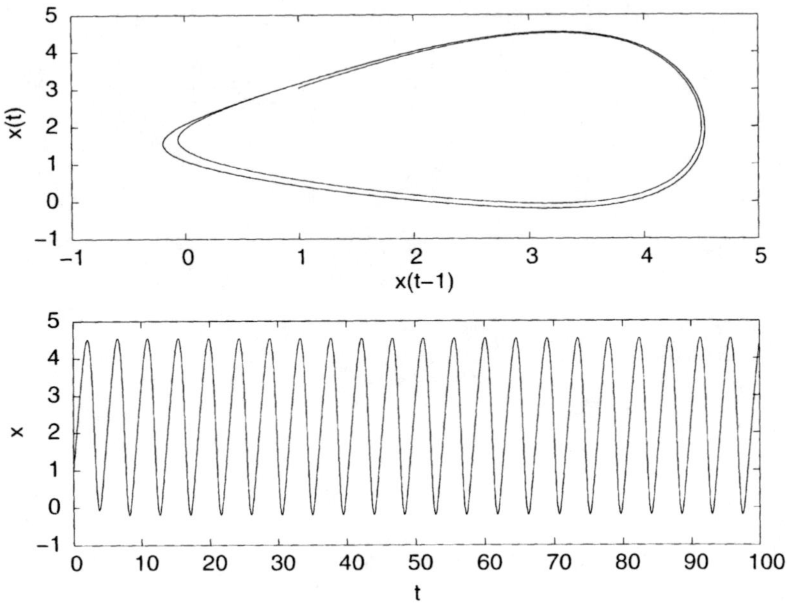

Figure 17. Phase portrait and time history of model (37) for $k = 2.2$.

k^* increases from 1.7231 to 2.2717, implying that the onset of the Hopf bifurcation is delayed. It is seen from Figure 18 that when $k = 2.2$ (the same value used in Figure 17), instead of having a Hopf bifurcation, the controlled Internet congestion model (46) converges to the equilibrium point x^*. When k passes the critical value $k^* = 2.2717$, a Hopf bifurcation occurs (see Figures 19 and 20). It can be shown that if we choose a larger value of α_1, the Internet congestion model may not have a Hopf bifurcation even for larger values of k. This indicates that the time-delayed feedback controller can delay the onset of Hopf bifurcation, thus guarantee a stationary sending rate for larger values of k. For example, when choosing $\alpha_1 = 5.0$, the controlled Internet congestion model (46) converges to the equilibrium solution if $k < k^* = 5.0141$, as shown in Figure 21.

In general, the parameters α_2, α_3, as well as α_1, can be all used to tune the parameters μ_2, τ_2 and β_2. Although it may be enough to use only α_1 for system (37) in delaying the onset of the Hopf bifurcation, it is more effective to use all the three parameters in changing the property of the Hopf bifurcation for general one-dimensional time-delayed system. Note that, under the three parameters, only the linearized systems will be different (i.e., b_1, b_2 and b_3 are adjusted for different systems). The derivation of the formulas for this general case can follow the same procedure described in previous subsections. Choosing different values of α_1, α_2 and α_3, one can efficiently change the stability, direction and period of

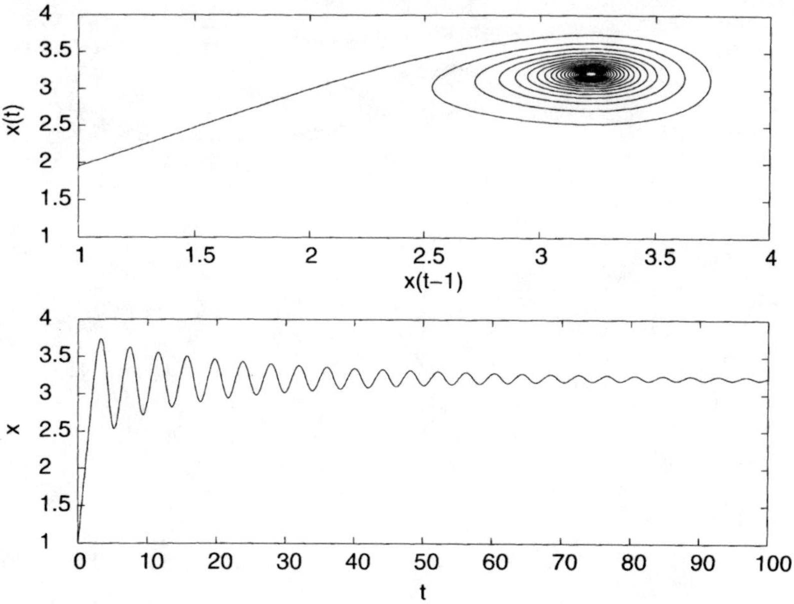

Figure 18. Phase portrait and time history of model (46) for $k = 2.2$ and $\alpha_1 = 0.5$.

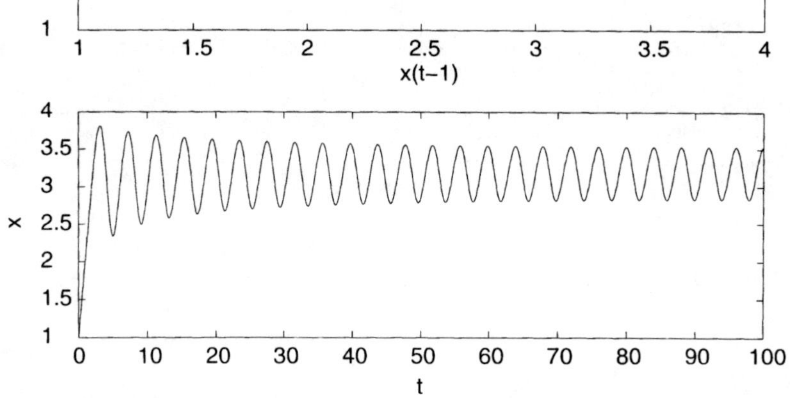

Figure 19. Phase portrait and time history of model (46) for $k = 2.3$ and $\alpha_1 = 0.5$.

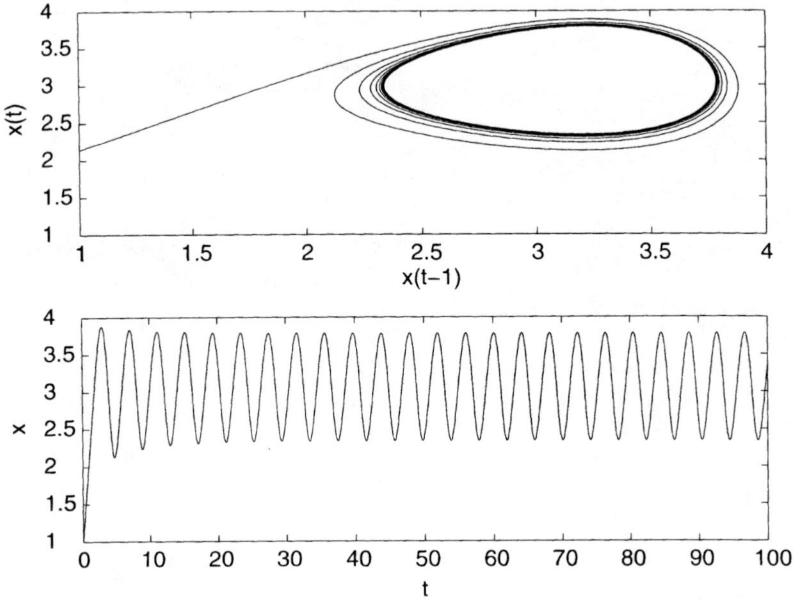

Figure 20. Phase portrait and time history of model (46) for $k = 2.4$ and $\alpha_1 = 0.5$.

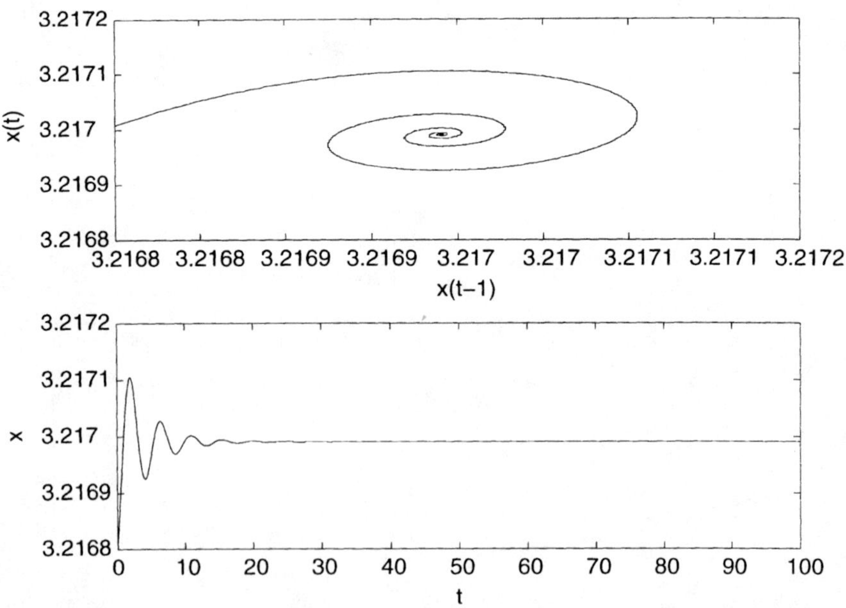

Figure 21. Phase portrait and time history of model (46) for $k = 4.5$ and $\alpha_1 = 5.0$.

a Hopf bifurcation. For example, when

$$\alpha_1 = 0.5, \qquad \alpha_2 = 0.4, \qquad \alpha_3 = 0.1,$$

we obtain

$$\mu_2 = 0.0066, \qquad \tau_2 = 0.0350, \qquad \beta_2 = -0.5785.$$

The critical value $k^* = 2.2717$, and the equilibrium point is the same as that in the case when $\alpha_1 = 0.5$, $\alpha_2 = 0$, $\alpha_3 = 0$. Taking $k = 2.3$ yields the results shown in Figure 22, where k takes the same value as that of the case shown in Figure 19. It is noted that the behavior shown in Figure 22 is quite different from that of Figure 19 even a same value of k is used in the two cases. The amplitude of the limit cycle shown in Figure 22 is larger than that depicted in Figure 19. This can be easily explained by using the normal form (41) and the bifurcation solution (43), because the absolute value of β_2 given for Figure 22 ($\beta_2 = -0.5785$) is smaller than that for Figure 19 ($\beta_2 = -1.0015$), but the linear coefficient $\operatorname{Re}\lambda'(0)$ is same for the two cases. (This result can also be found in Berns, Moiola and Chen (2000), Chen (2000), Nayfeh, Harb and Chin (1996), Yao, Yu and Essex (2001).) This suggests that one may choose appropriate values of α_2 and α_3, in addition to α_1, to obtain the desired behavior of a Hopf bifurcation.

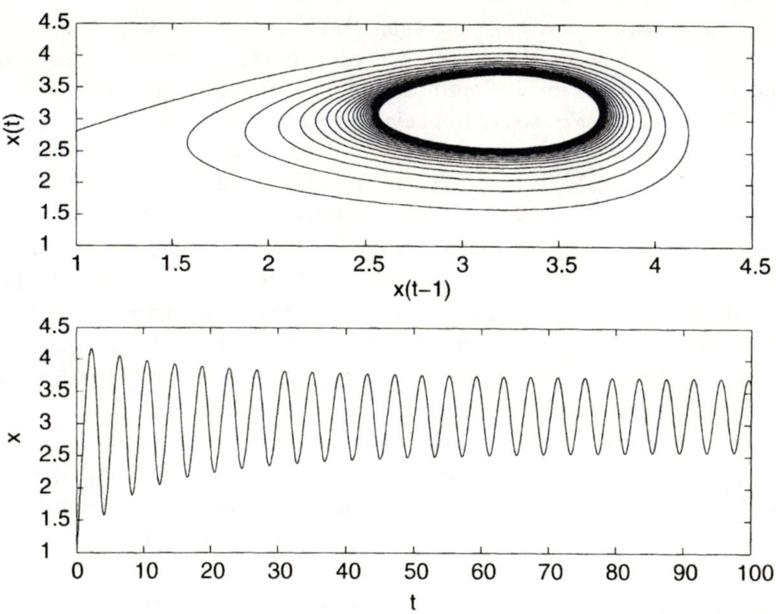

Figure 22. Phase portrait and time history of model (46) for $k = 2.3$, $\alpha_1 = 0.5$, $\alpha_2 = 0.4$ and $\alpha_3 = 0.1$.

2.3. Hilbert's 16th problem

One well-known problem, closely related to limit cycle theory, is Hilbert's 16th problem, one of the 23 mathematical problems proposed by Hilbert at the Second International Congress of Mathematics in 1900 (Hilbert, 1902). Recently, a modern version of the second part of Hilbert's 16th problem was formulated by Smale, and chosen as one of his 18 most challenging mathematical problems for the 21st century (Smale, 1998). To be more specific, consider the following planar system:

$$\dot{x} = P_n(x, y), \qquad \dot{y} = Q_n(x, y), \tag{52}$$

where $P_n(x, y)$ and $Q_n(x, y)$ represent nth-degree polynomials of x and y. The second part of Hilbert's 16th problem is to find an upper bound, denoted by $H(n)$, which only depends upon n.

In general, this is a very difficult problem: in particular, for determining large (global) limit cycles. If Hilbert's 16th problem is restricted to the vicinity of isolated fixed points, the problem is equivalent to studying degenerate Hopf bifurcations, and the main tasks become computing the so-called *focus values* of the point and determining center conditions. In the past half-century, many researchers have investigated the local problem and obtained many results (e.g., see Bautin, 1952; Kukles, 1944; Li and Liu, 1991; Liu and Li, 1989; Malkin, 1964; Han, 1999). For a quadratic system, it is now known that the maximal number of small limit cycles is three (Bautin, 1952). For cubic order systems, the best results published so far are twelve limit cycles (Yu, 2002, 2003a; Yu and Han, 2004, 2005a, 2005b).

In order to find the number of limit cycles of a system in the neighborhood of a fixed point (which is a linear center), one must compute the focus values of the point with efficient computational method using computer algebra systems such as Maple (Maple 10, 2005) or Mathematica (Mathematica, 2004). In particular, a perturbation approach (Yu, 1998) has been proved computationally efficient in calculating focus values. This perturbation method does not require application of center manifold theory, but combines normal form theory and center manifold theory into a unified approach.

In this subsection, we will use Hilbert's 16th problem to demonstrate the efficient computation of limit cycles. First, we will consider small limit cycles around a Hopf-type singular point, and then study the limit cycles generated from a perturbed 9th-order Hamiltonian systems to exhibit 80 limit cycles.

2.3.1. Nine small limit cycles around an isolated singular point

First we give sufficient conditions for the existence of small limit cycles. Suppose that the normal form of system (52) has been obtained in the polar coordinates up to the $(2k+1)$th-order term (interested readers can find the details of normal form computation in Yu (1998)):

$$\dot{r} = r\big(v_0 + v_1 r^2 + v_2 r^4 + \cdots + v_k r^{2k}\big), \tag{53}$$

$$\dot{\theta} = \omega + t_1 r^2 + t_2 r^4 + \cdots + t_k r^{2k}, \tag{54}$$

where r and θ denote the amplitude and phase of a periodic solution (limit cycle), respectively. Both v_k and t_k are explicitly expressed in terms of the original system's coefficients. v_k is called the kth-order focus value of the Hopf-type critical point (the origin). Note that here v_0 is the term obtained from linear perturbation.

The basic idea of finding k small limit cycles of system (52) around the origin is as follows: First, find the conditions such that $v_1 = v_2 = \cdots = v_{k-1} = 0$ (note that $v_0 = 0$ is automatically satisfied at the critical point), but $v_k \neq 0$, and then perform appropriate small perturbations to prove the existence of k limit cycles. This indicates that the procedure for finding multiple limit cycles involves two steps: Computing the focus values (i.e., computing the normal form) and solving the coupled nonlinear equations: $v_1 = v_2 = \cdots = v_{k-1} = 0$. Without loss of generality we may assume $v_k > 0$. Then, finding the solution of r^2 from the equation $\dot{r} = 0$ is equivalent to finding the real roots of the equation

$$y^k + c_{k-1} y^{k-1} + \cdots + c_1 y + c_0 = 0, \tag{55}$$

where $y = r^2$ and $c_{i-1} = v_{i-1}/v_k$, for $i = 1, 2, \ldots, k$. Since we are interested in small amplitude solutions, we may assume that the roots can be put in the form $y = r^2 = \mathrm{O}(\varepsilon)$ $(0 < \varepsilon \ll 1)$. In the following, we give theorems for proving the existence of small limit cycles. The proofs can be found in Yu and Han (2004, 2005a, 2005b).

THEOREM 2.5. *Let $\beta_0, \beta_1, \ldots, \beta_{k-1}$ be real constants such that the following equation*

$$w^k + \beta_{k-1} w^{k-1} + \cdots + \beta_1 w + \beta_0 = 0$$

has k simple positive roots w_i, $i = 1, 2, \ldots, k$. Then for any continuous functions c_i satisfying

$$c_i(\varepsilon) = \beta_i \varepsilon^{k-i} + \mathrm{o}\big(\varepsilon^{k-i}\big), \quad i = 0, 1, \ldots, k-1,$$

equation (55) *has exactly k simple positive roots in the form of $y_i = \varepsilon w_i + \mathrm{o}(\varepsilon)$ for sufficiently small $\varepsilon > 0$. Therefore, if*

$$v_i = \beta_i \varepsilon^{k-i} + \mathrm{o}\big(\varepsilon^{k-i}\big), \quad i = 0, 1, \ldots, k,$$

with $\beta_k \neq 0$, then equation (55) *has exactly k real positive roots* (*i.e., system* (53) *has exactly k limit cycles*) *in a neighborhood of the origin for sufficiently small $\varepsilon > 0$.*

Note that the roots w_j and the coefficients β_j are mutually determined. However, in many cases, v_{2j+1} depends on k parameters:

$$v_j = c_j(\varepsilon_1, \varepsilon_2, \ldots, \varepsilon_k), \quad j = 0, 1, \ldots, k. \tag{56}$$

In this case, the following theorem is more convenient in applications.

THEOREM 2.6. *Suppose that condition* (56) *holds, and further assume that*

$$c_k(0) \neq 0,$$
$$c_j(0) = 0, \quad j = 0, 1, \ldots, k-1,$$

and

$$\det\left[\frac{\partial(c_0, c_1, \ldots, c_{k-1})}{\partial(\varepsilon_1, \varepsilon_2, \ldots, \varepsilon_k)}(0)\right] \neq 0.$$

Then for any given $\varepsilon_0 > 0$, *there exist* $\varepsilon_1, \varepsilon_2, \ldots, \varepsilon_k$ *and* $\delta > 0$ *with* $|\varepsilon_j| < \varepsilon_0$, $j = 1, 2, \ldots, k$, *such that equation* (55) *has exactly k real positive roots* (*i.e., system* (53) *has exactly k limit cycles*) *in a* δ*-ball with the center at the origin.*

An alternate and simple version of the above two theorem can be stated as another theorem.

THEOREM 2.7. *If the focus values* v_i *in equation* (53) *satisfy the following conditions*:

$$v_i v_{i+1} < 0 \quad \text{and} \quad |v_i| \ll |v_{i+1}| \ll 1 \quad \text{for } i = 0, 1, 2, \ldots, k-1,$$

the polynomial equation given by $\dot{r} = 0$ *in equation* (55) *has k positive real roots of* r^2, *and thus the original system* (53) *has k limit cycles in the vicinity of the origin.*

In the following, we will use a cubic system to show the existence of nine small limit cycles in the vicinity of the origin. A general cubic system with a fixed point at the origin (Yu and Corless, 2006) can be written as

$$\begin{aligned}
\dot{x} &= a_{10}x + a_{01}y + a_{20}x^2 + a_{11}xy + a_{02}y^2 \\
&\quad + a_{30}x^3 + a_{21}x^2y + a_{12}xy^2 + a_{03}y^3, \\
\dot{y} &= b_{10}x + b_{01}y + b_{20}x^2 + b_{11}xy + b_{02}y^2 \\
&\quad + b_{30}x^3 + b_{21}x^2y + b_{12}xy^2 + b_{03}y^3,
\end{aligned} \tag{57}$$

where a_{ij}'s and b_{ij}'s are real constant coefficients (parameters). It is obvious that the origin $(x, y) = (0, 0)$ is a fixed point. The system has a total of eighteen parameters. However, not all of them are independent. First, note that we may use a linear transformation such that system (57) can be rewritten as

$$\begin{aligned}\dot{x} &= \alpha x + \beta y + a_{20}x^2 + a_{11}xy + a_{02}y^2 \\ &\quad + a_{30}x^3 + a_{21}x^2y + a_{12}xy^2 + a_{03}y^3, \\ \dot{y} &= \pm\beta x + \alpha y + b_{20}x^2 + b_{11}xy + b_{02}y^2 \\ &\quad + b_{30}x^3 + b_{21}x^2y + b_{12}xy^2 + b_{03}y^3, \end{aligned} \tag{58}$$

where λ and $\beta > 0$ are used to represent the eigenvalues of the linearized system of (57). Note that the other coefficients in (58) should be different from that of system (57), but we use the same notation for convenience. Here, when the negative sign is taken, the origin is a focus point or a center (if $\alpha = 0$); otherwise, it is a saddle point or node.

Now, suppose we are interested in the small limit cycles in the neighborhood of the origin. So the negative sign is taken in (58), and the eigenvalues are now given by $\lambda_{1,2} = \alpha \pm \beta \mathrm{i}$, where i is the imaginary unit, satisfying $\mathrm{i}^2 = -1$. Then we can apply a time scale, $\tau = \beta t$, into system (58) to obtain

$$\begin{aligned}\frac{\mathrm{d}x}{\mathrm{d}\tau} &= \alpha x + y + a_{20}x^2 + a_{11}xy + a_{02}y^2 \\ &\quad + a_{30}x^3 + a_{21}x^2y + a_{12}xy^2 + a_{03}y^3, \\ \frac{\mathrm{d}y}{\mathrm{d}\tau} &= -x + \alpha y + b_{20}x^2 + b_{11}xy + b_{02}y^2 \\ &\quad + b_{30}x^3 + b_{21}x^2y + b_{12}xy^2 + b_{03}y^3, \end{aligned} \tag{59}$$

where again the same notations for the parameters are used. Henceforth we assume that the leading β has been scaled to 1, and rename $\tau = t$. Now, system (59) has only fifteen parameters. Further, by a rotation we can remove one parameter (Bautin, 1952; Lloyd, Blows and Kalenge, 1988) from system (59), which can be written in the general form

$$\begin{aligned}\dot{x} &= \alpha x + y + Ax^2 + (B + 2D)xy + Cy^2 + Fx^3 + Gx^2y \\ &\quad + (H - 3P)xy^2 + Ky^3, \\ \dot{y} &= -x + \alpha y + Dx^2 + (E - 2A)xy - Dy^2 + Lx^3 \\ &\quad + (M - H - 3F)x^2y + (N - G)xy^2 + Py^3. \end{aligned} \tag{60}$$

This form is perhaps the simplest form for cubic systems in the literature (Lloyd, Blows and Kalenge, 1988). The system has fourteen parameters. However, since the same order terms on the right-hand side of (60) are homogeneous, we can remove one more parameter. Suppose $A \neq 0$ (in case $A = 0$ one may use another nonzero parameter in the scaling), we let

$$\begin{array}{llll} B = bA, & C = cA, & D = dA, & E = eA, \\ F = fA^2, & G = gA^2, & H = hA^2, & K = kA^2, \\ L = \ell A^2, & M = mA^2, & N = nA^2, & P = pA^2, \end{array}$$

and apply a spatial scaling $x \to x/A$, $y \to y/A$ to system (60) to obtain the final system

$$\begin{aligned} \dot{x} &= \alpha x + y + x^2 + (b + 2d)xy + cy^2 + fx^3 + gx^2y \\ &\quad + (h - 3p)xy^2 + ky^3, \\ \dot{y} &= -x + \alpha y + dx^2 + (e - 2)xy - dy^2 + \ell x^3 + (m - h - 3f)x^2y \\ &\quad + (n - g)xy^2 + py^3, \end{aligned} \tag{61}$$

which has only thirteen independent parameters. It is easy to see that the zeroth-order focus value is $v_0 = \alpha$. Other focus values are given in terms of the remaining twelve parameters. Let

$$\mathcal{S} = \{b, c, d, e, f, g, h, k, \ell, m, n, p\}.$$

Then

$$v_i = v_i(\mathcal{S}).$$

In general, the maximal number of small limit cycles which exist in the vicinity of the origin is not greater than the number of independent parameters. Here it is 13. In other words, the best possibility one can have is

$$v_i = 0, \quad i = 0, 1, \ldots, 12, \quad \text{but} \quad v_{13} \neq 0.$$

Then according to the theorems given above, the maximal number of small limit cycles which can be obtained by appropriate perturbations is 13. Of course, this conclusion is obtained under the assumption that the origin is a linear center (i.e., the origin is a Hopf-type critical point). If the origin is a saddle point or a node, then the situation is different (Yu and Han, 2004, 2005a, 2005b).

Before we show the existence of 9 limit cycles for system (61), we briefly introduce methods of calculating focus values. There are many approaches which can be used to compute the focus values. Here, we briefly describe two efficient methods for computing focus values.

(A) *A perturbation technique.* This perturbation technique is based on the normal form theory associated with Hopf singularity (Yu, 1998). The basic idea of normal form theory is to apply successive coordinate transformations to obtain a simplified form which is qualitatively equivalent to the original system in the vicinity of an equilibrium. Normal form theory is usually employed after the application of center manifold theory which uses a similar idea to reduce the original equation to a lower-dimensional system. A perturbation technique has been used to develop a unified approach to directly compute the normal forms of Hopf and degenerate Hopf bifurcations for general n-dimensional systems without the application of center manifold theory (Yu, 1998, 2000). In the following, we briefly describe this perturbation approach.

Consider the general n-dimensional differential equation

$$\dot{\mathbf{x}} = J\mathbf{x} + \mathbf{f}(\mathbf{x}), \quad \mathbf{x} \in \mathbb{R}^n,\ \mathbf{f}: \mathbb{R}^n \to \mathbb{R}^n, \tag{62}$$

where $J\mathbf{x}$ represents the linear terms of the system, and the nonlinear function $\mathbf{f}$ is assumed to be analytic; and $\mathbf{x} = \mathbf{0}$ is an equilibrium of the system, i.e., $\mathbf{f}(\mathbf{0}) = \mathbf{0}$. Further, assume that the Jacobian of system (62) evaluated at the equilibrium $\mathbf{0}$ contains a pair of purely imaginary eigenvalues $\pm\mathrm{i}$, and thus the Jacobian of system (62) is in the Jordan canonical form

$$J = \begin{bmatrix} 0 & 1 & 0 \\ -1 & 0 & 0 \\ 0 & 0 & A \end{bmatrix}, \quad A \in \mathbb{R}^{(n-2)\times(n-2)},$$

where A is *stable* (i.e. all of its eigenvalues have negative real parts).

The basic idea of the perturbation technique based on multiple scales is as follows: Instead of a single time variable, multiple independent variables or scales are used in the expansion of the system response. To achieve this, introducing the new independent time variables $T_k = \varepsilon^k t$, $k = 0, 1, 2, \ldots$, yields partial derivatives with respect to T_k as

$$\begin{aligned} \frac{\mathrm{d}}{\mathrm{d}t} &= \frac{\partial T_0}{\partial t}\frac{\partial}{\partial T_0} + \frac{\partial T_1}{\partial t}\frac{\partial}{\partial T_1} + \frac{\partial T_2}{\partial t}\frac{\partial}{\partial T_2} + \cdots \\ &= D_0 + \varepsilon D_1 + \varepsilon^2 D_2 + \cdots, \end{aligned} \tag{63}$$

where $D_k = \frac{\partial}{\partial T_k}$ denotes the differentiation operator. Then, assume that the solutions of system (62) in the neighborhood of $\mathbf{x} = \mathbf{0}$ are expanded in the series,

$$\mathbf{x}(t;\varepsilon) = \varepsilon\mathbf{x}_1(T_0, T_1, \ldots) + \varepsilon^2\mathbf{x}_2(T_0, T_1, \ldots) + \cdots. \tag{64}$$

Note that the same perturbation parameter, ε, is used in both the time and space scalings (see equations (63) and (64)). In other words, this perturbation approach treats time and space in a uniform scaling.

Applying formulas (63) and (64) into system (62), and solving the resulting ordered linear differential equations finally yields the normal form, given in polar coordinates (the detailed procedure can be found in Yu (1998)):

$$\begin{aligned} \frac{\mathrm{d}r}{\mathrm{d}t} &= \frac{\partial r}{\partial T_0}\frac{\partial T_0}{\partial t} + \frac{\partial r}{\partial T_1}\frac{\partial T_1}{\partial t} + \frac{\partial r}{\partial T_2}\frac{\partial T_2}{\partial t} + \cdots \\ &= D_0 r + D_1 r + D_2 r + \cdots \end{aligned} \tag{65}$$

and

$$\begin{aligned}\frac{\mathrm{d}\phi}{\mathrm{d}t} &= \frac{\partial\phi}{\partial T_0}\frac{\partial T_0}{\partial t} + \frac{\partial\phi}{\partial T_1}\frac{\partial T_1}{\partial t} + \frac{\partial\phi}{\partial T_2}\frac{\partial T_2}{\partial t} + \cdots \\ &= D_0\phi + D_1\phi + D_2\phi + \cdots,\end{aligned} \tag{66}$$

where $D_i r$ and $D_i\phi$ are uniquely determined, implying that the normal form given by equations (65) and (66) is unique. Further, it has been shown that (Yu, 1998) the derivatives $D_i r$ and $D_i\phi$ are functions of r only, and only $D_{2k}r$ and $D_{2k}\phi$ are nonzero, which can be expressed as $D_{2k}r = v_k r^{2k+1}$ and $D_{2k}\phi = t_k r^{2k}$, where both v_k and t_k are expressed in terms of the original system's coefficients. The coefficient v_k is called the kth-order focus value of the Hopf-type critical point (the origin).

THEOREM 2.8. *For the general n-dimensional system* (62), *which has a Hopf-type singular point at the origin, i.e., the linearized system of* (62) *has a pair of purely imaginary eigenvalues and the remaining eigenvalues have negative real parts, the normal form for the Hopf or generalized Hopf bifurcations up to* $(2k+1)$*th-order term is given by*

$$\begin{aligned}\dot{r} &= r\left(v_0 + v_1 r^2 + v_2 r^4 + \cdots + v_k r^{2k}\right),\\ \dot{\theta} &= 1 + \dot{\phi} = 1 + t_1 r^2 + t_2 r^4 + \cdots + t_k r^{2k},\end{aligned}$$

where the constants $v_k = D_{2k}r/r^{2k+1}$ *and* $t_k = D_{2k}\phi/r^{2k+1}$ *are explicitly expressed in terms of the original system parameters, and* $D_{2k}r$ *and* $D_{2k}\phi$ *are obtained recursively using multiple time scales.*

Note that the above normal form is exactly the same as that given in equations (53) and (54) with $\omega = 1$.

(B) *Singular point quantity method.* This iterative method computes the focus values via the computation of singular point quantities (see Chen and Liu, 2004; Liu and Huang, 2005; Chen, Liu and Yu, 2006 for details).

To introduce this method, consider the planar polynomial differential system

$$\dot{x} = \delta x - y + \sum_{k=2}^{\infty} X_k(x,y), \qquad \dot{y} = x + \delta y + \sum_{k=2}^{\infty} X_k(x,y), \tag{67}$$

where $X_k(x,y)$ and $Y_k(x,y)$ are homogeneous polynomials of x, y with degree k. The origin $(x,y) = (0,0)$ is a singular point of system (67), which is either a focus or a center (when $\delta = 0$). Since we are interested in the computation of focus values, we assume $\delta = 0$ in the following analysis. Introducing the transformations, given by

$$z = x + \mathrm{i}y, \qquad w = x - \mathrm{i}y, \qquad T = \mathrm{i}t, \quad \mathrm{i} = \sqrt{-1},$$

into system (67) results in

$$\frac{dz}{dT} = z + \sum_{k=2}^{\infty} Z_k(z, w) = Z(z, w),$$

$$\frac{dw}{dT} = -w - \sum_{k=2}^{\infty} W_k(z, w) = -W(z, w), \tag{68}$$

where z, w and T are complex variables and

$$Z_k(z, w) = \sum_{\alpha+\beta=k} a_{\alpha\beta} z^\alpha w^\beta, \qquad W_k = \sum_{\alpha+\beta=k} b_{\alpha\beta} w^\alpha z^\beta.$$

Systems (67) and (68) are said to be concomitant.

If system (67) is a real planar, differential system, the coefficients of system (68) must satisfy the conjugate conditions

$$\overline{a_{\alpha\beta}} = b_{\alpha\beta}, \quad \alpha \geqslant 0,\ \beta \geqslant 0,\ \alpha + \beta \geqslant 2.$$

By the following transformations:

$$z = re^{i\theta}, \qquad w = re^{-i\theta}, \qquad T = it,$$

system (68) can be transformed into

$$\frac{dr}{dT} = \frac{ir}{2} \sum_{m=1}^{\infty} \sum_{\alpha+\beta=m+2} (a_{\alpha,\beta-1} - b_{\beta,\alpha-1}) e^{i(\alpha-\beta)\theta} r^m,$$

$$\frac{d\theta}{dT} = 1 + \frac{1}{2} \sum_{m=1}^{\infty} \sum_{\alpha+\beta=m+2} (a_{\alpha,\beta-1} + b_{\beta,\alpha-1}) e^{i(\alpha-\beta)\theta} r^m. \tag{69}$$

For a complex constant h, $|h| \ll 1$, we may write the solution of (69) satisfying the initial condition $r|_{\theta=0} = h$ as

$$r = \tilde{r}(\theta, h) = h + \sum_{k=2}^{\infty} v_k(\theta) h^k.$$

Evidently, if system (67) is a real system, then $v_k(2\pi)$, $k = 1, 2, \ldots$, is the kth-order focal (or focus) value of the origin.

For system (67), we can uniquely derive the following formal series:

$$\varphi(z, w) = z + \sum_{k+j=2}^{\infty} c_{kj} z^k w^j, \qquad \psi(z, w) = w + \sum_{k+j=2}^{\infty} d_{k,j} w^k z^j,$$

where

$$c_{k+1,k} = d_{k+1,k} = 0, \quad k = 1, 2, \ldots,$$

such that

$$\frac{\mathrm{d}\varphi}{\mathrm{d}T} = \varphi + \sum_{j=1}^{\infty} p_j \varphi^{j+1} \psi^j, \qquad \frac{\mathrm{d}\psi}{\mathrm{d}T} = -\psi - \sum_{j=1}^{\infty} q_j \psi^{j+1} \varphi^j.$$

Let $\mu_0 = 0$, $\mu_k = p_k - q_k$, $k = 1, 2, \ldots$. μ_k is called the kth-order singular point quantity of the origin of system (68) (Chen and Liu, 2004). Based on the singular quantities, we can define the order of the singular critical point and extended center as follows.

If $\mu_0 = \mu_1 = \cdots = \mu_{k-1} = 0$ and $\mu_k \neq 0$, then the origin of system (68) is called the *kth-order weak critical singular point*. In other words, k is the multiplicity of the origin of system (68).

If $\mu_k = 0$ for $k = 1, 2, \ldots$, then the origin of system (68) is called an *extended center* (*complex center*).

If system (67) is a real autonomous differential system with the concomitant system (68), then for the origin, the kth-order focus quantity v_k of system (67) and the kth-order quantity of the singular point of system (68) have the relation (Liu and Huang, 2005; Liu and Li, 1989), given in the following theorem.

THEOREM 2.9. *Given system* (67) ($\delta = 0$) *or* (68), *for any positive integer* m, *the following assertion holds*:

$$v_k(2\pi) = \mathrm{i}\pi\left(\mu_k + \sum_{j=1}^{k-1} \xi_m^{(j)} \mu_j\right), \quad k = 1, 2, \ldots, \tag{70}$$

where $\xi_m^{(j)}$, $j = 1, 2, \ldots, k-1$, *are polynomial functions of coefficients of system* (68).

We have the following recursive formulas for computing the singular point quantities of system (68) (Chen and Liu, 2004): $c_{11} = 1$, $c_{20} = c_{02} = c_{kk} = 0$, $k = 2, 3, \ldots$, and $\forall(\alpha, \beta)$, $\alpha \neq \beta$, and $m \geqslant 1$,

$$C_{\alpha\beta} = \frac{1}{\beta - \alpha} \sum_{k+j=3}^{\alpha+\beta+2} \left[(\alpha - k + 1)a_{k,j-1} - (\beta - j + 1)b_{j,k-1}\right] \times C_{\alpha-k+1,\beta-j+1}$$

and

$$\mu_m = \sum_{k+j=3}^{2m+4} \left[(m - k + 2)a_{k,j-1} - (m - j + 2)b_{j,k-1}\right] C_{m-k+2,m-j+2},$$

where $a_{kj} = b_{kj} = C_{kj} = 0$ for $k < 0$ or $j < 0$.

It is clearly seen from equation (70) that

$$\mu_1 = \mu_2 = \cdots = \mu_{k-1} \quad \Longleftrightarrow \quad v_1 = v_2 = \cdots = v_{k-1}.$$

Therefore, when determining the conditions such that $v_1 = v_2 = \cdots = v_{k-1} = 0$, one can instead use the equations: $\mu_1 = \mu_2 = \cdots = \mu_{k-1} = 0$. If the μ_k's are simpler than the v_k's then this method is better than the method of directly computing v_k. However, in general such μ_k are not necessarily simpler than v_k.

It should be pointed out that since the normal form (focus value) is not unique, the focus values obtained using different methods are not necessarily the same. However, the first nonzero focus value must be identical (ignoring a constant multiplier). This implies that for different focus values obtained using different approaches, the solution to the equations $v_1 = v_2 = \cdots = v_{k-1} = 0$ (or $\mu_1 = \mu_3 = \cdots = \mu_{k-1} = 0$) must be identical.

Now we return to the cubic system (61). So far, in the literature, the maximal number of limit cycles in the neighborhood of one singular point obtained using symbolic computation is 8 (James and Lloyd, 1991). Although it was reported that 11 small limit cycles might exist around one singular point (Zoladek, 1995), the result has not been verified by computation. The difficulty in computing higher-order focus values are obvious. Also, solving coupled higher degree multivariate polynomials is difficult.

It is easy to see that $v_0 = \alpha$, and other focus values are given in terms of the remaining twelve parameters. It has been shown that the origin is a center if

$$\alpha = b = e = h = n = m = 0.$$

We let

$$\alpha = d = e = h = 0, \tag{71}$$

and then compute the focus values. Here, we use the recursion formulas (Chen and Liu, 2004; Chen, Liu and Yu, 2006) to compute the singular point quantities. The first-order quantity is given by $\mu_1 = \frac{1}{4}\mathrm{i}[b(c+1) - m]$. Setting $\mu_1 = 0$ results in

$$m = b(c+1). \tag{72}$$

For simplicity, letting $b = 0$, and so $m = 0$. Therefore, n must be nonzero, otherwise it is a center. Thus, seven free parameters are remained: c, f, g, k, ℓ, n and p.

Under the above choices of parameters, the second-order singular point quantity becomes $\mu_2 = -\frac{1}{4}\mathrm{i}n(p - f)$. In order to have $\mu_2 = 0$, the only choice is

$$p = f, \tag{73}$$

since $n \neq 0$. Then $\mu_3 = \frac{1}{96}\mathrm{i} f n(45 - 30c - 35c^2 + 15\ell + 15k + 3n)$. One can choose n to set $\mu_3 = 0$, yielding

$$n = \frac{1}{3}\big(35c^2 + 30c - 15\ell - 15k - 45\big). \tag{74}$$

The next singular point quantity can be found as

$$\begin{aligned}\mu_4 = \frac{\mathrm{i} f n}{192}\big[&g\big(7c^2 + 30c + 6k + 6\ell - 45\big) - 648 + 162c \\ &- 81k - 72\ell + 30k\ell - 60c\ell - 54ck + 24k^2 \\ &+ 6\ell^2 + 516c^2 - 56c^2\ell - 21c^2k + 434c^3 + 168c^4\big],\end{aligned}$$

from which we solve for g to obtain $g = g(c, k, l)$. Then μ_5 is given by $\mu_5 = -(\mathrm{i} f n)/(10368(30c + 7c^2 + 6k + 6\ell - 45)^2)\bar{\mu}_5(f, c, k, \ell)$, where $\bar{\mu}_5$ is a polynomial of f, c, k, ℓ, from which we can solve for f^2 to get $f^2 = f^2(c, k, l)$. Finally, the next three singular point quantities are

$$\mu_6 = \mu_6(c, k, \ell), \qquad \mu_7 = \mu_7(c, k, \ell), \qquad \mu_8 = \mu_8(c, k, \ell),$$

which are polynomials of c, k and ℓ. The three singular point quantities are coupled, we thus have to simultaneously solve the three polynomial equations: $\mu_6 = \mu_7 = \mu_8 = 0$. To find the solutions of these equations, we might first eliminate one parameter from the three equations to obtain two resultant equations, and then further eliminate one more parameter from the two equation to get a final resultant equation which is a univariate polynomial and thus we could find all possible solutions. This elimination method has been used in our other publications (e.g., see Yu and Han, 2004, 2005a, 2005b). The method may increase the degree of the final resultant polynomial substantially.

Here, we use a numerical approach to simultaneously solve the three equations. The solution of the system of equations may be approached in several ways. A simple numerical approach, using *fsolve* in Maple directly, is successful in many cases. By examining its source code, we find that this routine uses a variant of damped multivariate Newton iteration (there appears to be no published paper describing the routine), and by judicious choices of our initial guess we obtained a result (with $f_2 > 0$) after only two tries.

Now, for system (61), by taking $a = 0.3$, we have used the numerical method to obtain the critical parameter values (results from computer output using Maple up to 1000 digits, but list only 50 digits):

$$\begin{aligned}&b = d = e = h = 0,\\ &m^* = 0,\\ &p^* = f^* = 0.57205513738312947531208972107913988960903196760 78,\\ &n^* = 1.5587366234496833705957910715181908410140471444 1598,\end{aligned}$$

$$
\begin{aligned}
g^* &= 0.32875002652319114041354789797311142571533487466423, \\
f^* &= 0.57205513738312947531208972107913988960903196760780, \\
\ell^* &= -0.28703189754662589381676063046084638581078384981160, \\
c^* &= -0.52179929787663453432829806398960426486646485537068, \\
k^* &= 0.02751217774798897933976121289219741452562430620627.
\end{aligned} \tag{75}
$$

Under the above critical parameter values and conditions, we execute the Maple program (Yu, 1998) to find the following focus values (again up to 1000 decimal digits, but v_9 is only given up to 50 digits here):

$$
\begin{aligned}
& v_i \approx 10^{-999}, \quad i = 1, 2, \ldots, 8, \\
& v_9 = -0.0014315972268236725754647008537844355295621179315106.
\end{aligned}
$$

The above result indeed indicates that an excellent approximate solution has been obtained, and the values of v_i, $i = 1, 2, \ldots, 8$, can be considered as being very close to a true real zero. Thus, the maximal number of small limit cycles which can be obtained in the vicinity of the origin of system (61) is nine. In order to prove that the nine small limit cycles indeed exist, we need to check the determinant obtained from the three real-value equations:

$$
\mathrm{i}\mu_6(\ell, c, k) = \mathrm{i}\mu_7(\ell, c, k) = \mathrm{i}\mu_8(\ell, c, k) = 0, \quad \mathrm{i} = \sqrt{-1},
$$

which yields

$$
\begin{aligned}
\det(J_c) &= \det \begin{bmatrix} \frac{\partial \mathrm{i}\mu_6}{\partial \ell} & \frac{\partial \mathrm{i}\mu_6}{\partial c} & \frac{\partial \mathrm{i}\mu_6}{\partial k} \\ \frac{\partial \mathrm{i}\mu_7}{\partial \ell} & \frac{\partial \mathrm{i}\mu_7}{\partial c} & \frac{\partial \mathrm{i}\mu_7}{\partial k} \\ \frac{\partial \mathrm{i}\mu_8}{\partial \ell} & \frac{\partial \mathrm{i}\mu_8}{\partial c} & \frac{\partial \mathrm{i}\mu_8}{\partial k} \end{bmatrix}_{(\ell,c,k)=(\ell^*,c^*,k^*)} \\
&= \det \begin{bmatrix} 0.0451319561 & 0.2439914595 & 0.0220645269 \\ 0.1223727073 & -0.0022077660 & 0.1580040454 \\ 0.0055179038 & 0.6737398583 & -0.0271869307 \end{bmatrix} \\
&= -0.0019578515 \neq 0.
\end{aligned}
$$

Thus, according to Theorem 2.6 we know that nine small-amplitude limit cycles bifurcating from the origin of system (61) can be obtained by properly perturbing the critical values given in equation (75). The above results are summarized in the following theorem.

THEOREM 2.10. *For the cubic system* (61), *suppose* $b = d = e = h = 0$. *When the remaining parameters are properly perturbed to the critical values*: $\alpha^* = 0$, $m^* = 0$, $p^* = f$, $n^* = \frac{1}{3}(35c^2 + 30c - 15\ell - 15k - 45)$, $g^* = g^*(c^*, k^*, \ell^*)$, $f^{*2} = f^{*2}(c^*, k^*, \ell^*)$, *while* ℓ^*, c^* *and* k^* *are given in* (75), *then system* (61) *has exactly nine small limit cycles around the origin.*

2.3.2. Bifurcation to 80 limit cycles in a perturbed Hamiltonian system

For higher-order system (52), in order to obtain more limit cycles and various configuration patterns of their relative dispositions, we introduce an efficient method (see Li and Li, 1985; Li and Huang, 1987; Li and Zhao, 1989; Li and Liu, 1991; Li, Chan and Chung, 2002) to perturb symmetric Hamiltonian systems to have maximal number of centers, i.e., to study the weakened Hilbert's 16th problem for symmetric planar polynomial Hamiltonian systems, since bifurcation and symmetry are closely connected and symmetric systems play pivotal roles as a bifurcation point in all planar Hamiltonian systems. To investigate perturbed Hamiltonian systems, first we need to know the global behavior of the unperturbed polynomial systems, namely, to determine the global property of the families of real planar algebraic curves defined by the Hamiltonian functions. Then by using proper perturbation techniques, we shall obtain the global information of bifurcations for the perturbed nonintegrable systems. In this sense, we say that our study utilizes both the two parts of Hilbert's 16th problem.

On the basis of the method of detection functions introduced by Li and Li (1985), we have developed a method of control parameters (Wang, 2005; Wang and Yu, 2005; Wang, Yu and Li, 2006) in order to obtain more limit cycles for a Z_{10}-equivariant perturbed polynomial Hamiltonian system of degree 9. With the aid of numerical simulations (using Maple) we show that there exist parameter groups such that our system has at least 80 limit cycles with three different configurations.

(A) *Z_q-equivariant planar vector fields.* First, we present some basic concepts and results which will be used in the following subsections. Let G be a compact Lie group of transformations acting on $\mathbb{R}^n$. A mapping $\Phi : \mathbb{R}^n \to \mathbb{R}^n$ is called G-equivariant if for all $g \in G$ and $x \in \mathbb{R}^n$, $\Phi(gx) = g\Phi(x)$. A function $H : \mathbb{R}^n \to \mathbb{R}$ is called G-invariant function if for all $g \in G$ and $x \in \mathbb{R}^n$, $H(gx) = H(x)$. If Φ is a G-equivariant mapping, the vector field $\dot{x} = \Phi(x)$ is called a G-equivariant vector field.

Let q be an integer. A group Z_q is called a cyclic group of order q if it is generated by a planar counterclockwise rotation of the vector fields through $2\pi/q$ about the origin. Introducing the transformation $z = x + \mathrm{i}y$, $\bar{z} = x - \mathrm{i}y$ into system (52) yields

$$\dot{z} = F(z, \bar{z}), \qquad \dot{\bar{z}} = \overline{F}(z, \bar{z}), \tag{76}$$

where $F(z, \bar{z}) = P(u, v) + \mathrm{i}Q(u, v)$, $u = (z + \bar{z})/2$ and $v = (z - \bar{z})/(2\mathrm{i})$.

THEOREM 2.11 (Li and Zhao, 1989). *A vector field defined by* (76) *is Z_q-equivariant if and only if the function $F(z, \bar{z})$ has the following form:*

$$F(z, \bar{z}) = \sum_{l=1} g_l(|z|^2)\bar{z}^{lq-1} + \sum_{l=0} h_l(|z|^2)z^{lq+1}, \tag{77}$$

where $g_l(|z|^2)$ and $h_l(|z|^2)$ are polynomials with complex coefficients. In addition, equation (76) is a Hamiltonian system having Z_q-equivariance if and only if equation (77) holds and

$$\frac{\partial F}{\partial z} + \frac{\partial \overline{F}}{\partial \bar{z}} \equiv 0. \tag{78}$$

THEOREM 2.12 (Z_q-invariant function (Li, 2003)). *A Z_q-invariant function $I(z, \bar{z})$ has the following form:*

$$I(z, \bar{z}) = \sum_{l=0} g_l(|z|^2) z^{lq} + \sum_{l=1} h_l(|z|^2) \bar{z}^{lq}. \tag{79}$$

COROLLARY 2.13. *For the planar polynomial systems of degree $m-1$ ($m > 7$), the Z_q-equivariant Hamiltonian vector fields defined by equation (76) have the explicit forms, given as*

(1) $q = m = 2k+1$,

$$F(z, \bar{z}) = (b_0 + b_1|z|^2 + \cdots + b_{(m-3)/2}|z|^{m-3})zi + A_{(m-1)/2}\bar{z}^{m-1};$$

$q = m = 2k$,

$$F(z, \bar{z}) = (b_0 + b_1|z|^2 + \cdots + b_{(m-2)/2}|z|^{m-2})zi + A_{m/2}\bar{z}^{m-1};$$

(2) $q = m = 2k, m = 2k+1$,

$$F(z, \bar{z}) = (b_0 + b_1|z|^2 + \cdots + b_{(m-3)/2}|z|^{m-3})zi + A_{(m-1)/2}\bar{z}^{m-2};$$

$q = m-1, m = 2k$,

$$F(z, \bar{z}) = (b_0 + b_1|z|^2 + \cdots + b_{(m-2)/2}|z|^{m-2})zi + A_{m/2}\bar{z}^{m-2};$$

(3) $q = m-2, m = 2k+1$,

$$F(z, \bar{z}) = (b_0 + b_1|z|^2 + \cdots + b_{(m-3)/2}|z|^{m-3})zi + A_{(m-1)/2}z^{m-1} + (A_{(m+1)/2} - (m-1)\overline{A}_{(m-1)/2}|z|^2)\bar{z}^{m-3};$$

$q = m-2, m = 2k$,

$$F(z, \bar{z}) = (b_0 + b_1|z|^2 + \cdots + b_{(m-2)/2}|z|^{m-2})zi + A_{m/2}\bar{z}^{m-1} + (A_{(m+2)/2} - (m-1)\overline{A}_{m/2}|z|^2)\bar{z}^{m-3};$$

(4) $q = m-2, m = 2k+1$,

$$F(z, \bar{z}) = (b_0 + b_1|z|^2 + \cdots + b_{(m-3)/2}|z|^{m-3})zi + A_{(m-1)/2}z^{m-2} + (A_{(m+1)/2} - (m-2)\overline{A}_{(m-1)/2}|z|^2)\bar{z}^{m-4};$$

$q = m-3, m = 2k$,

$$F(z, \bar{z}) = (b_0 + b_1|z|^2 + \cdots + b_{(m-2)/2}|z|^{m-2})zi + A_{m/2}z^{m-2} + (A_{(m+2)/2} - (m-2)\overline{A}_{m/2}|z|^2)\bar{z}^{m-4};$$

where b_i's are real while A_i's are complex.

The orbits of Hamiltonian polynomial systems given in Corollary 2.13 define different families of m algebraic curves having Z_q-equivariance. One of the main questions in real algebraic geometry is: what schemes of the mutual arrangement (schemes or configurations) of ovals can be realized by the curves of a given degree? By using some Z_q-equivariant Hamiltonian systems, we can realize various configurations of ovals for planar algebraic curves of degree m.

A group D_q is called dihedral group of order 2_q which is characterized as the symmetry group of the regular q-gon. In is generated by two elements, the (plane) rotation by an angle $2\pi/q$ and a reflection in $\mathbb{R}^2$.

THEOREM 2.14 (Li, 2003). *A vector field defined by equation* (76) *is D_q-equivariant if and only if all coefficients A_j, $j = 0, 1, \ldots,$ of the functions $g_l(|z|^2)$ and $h_l(|z|^2)$ in equation* (77) *are real numbers.*

(B) *The method of detection functions: perturbations of Z_q-equivariant Hamiltonian systems.* Consider the following perturbed planar Hamiltonian system:

$$\dot{x} = \frac{\partial H}{\partial y} - \varepsilon x\big(p(x, y) - \lambda\big),$$
$$\dot{y} = -\frac{\partial H}{\partial x} - \varepsilon y\big(p(x, y) - \lambda\big), \tag{80}$$

where $H(x, y)$ is the Hamiltonian, $p(0, 0) = 0$, $0 < \varepsilon \ll 1$ and $\lambda \in \mathbb{R}$. It is called rough perturbation because there are two linear terms λx and λy in the perturbation.

Suppose that the origin in the phase plane is a singular point of equation (80) and the following conditions hold:

(A1) The unperturbed system (80) ($\varepsilon = 0$) is a Z_q-equivariant Hamiltonian vector field. There are families of curves $\{\Gamma^h\}$ defined by the Hamiltonian function $H(x, y) = h$ lying in periodic annuli which enclose at least one singular point. As the value of h changes toward the value defined by the singular points, Γ^h approaches a singular point or an inner boundary of the periodic annuli which consist of a heteroclinic (or homoclinic) loop.

(A2) Surrounding the periodic annuli, there exist heteroclinic (or homoclinic) loops which connect some hyperbolic saddle points at the values of h defined by these singular points.

(A3) The divergence $2\varepsilon(\lambda - F(x, y)) \equiv 2\varepsilon(\lambda - \frac{x}{2}\frac{\partial p}{\partial x} - \frac{y}{2}\frac{\partial p}{\partial y} - p(x, y))$ of the perturbed vector field is a Z_q-invariant function.

Therefore, we can define

$$\lambda = \lambda(h) = \left(\iint\limits_{D^h} F(x, y)\,\mathrm{d}x\,\mathrm{d}y\right) \Big/ \left(\iint\limits_{D^h} \mathrm{d}x\,\mathrm{d}y\right) = \frac{\psi(h)}{\phi(h)}, \tag{81}$$

which is called a detection function corresponding to the periodic family $\{\Gamma^h\}$. The graph of $\lambda = \lambda(h)$ in the plane (h, λ) is called a detection curve, where D^h is the area inside Γ^h.

Clearly, if $H(x, y) = h$ is a polynomial, then $\lambda(h)$ is a ratio between two Abelian integrals. In this case, $\lambda(h)$ is a differentiable function with respect to h. However, when the degree of $H(x, y)$ is greater than 4, classical mathematical analyzes cannot be used to calculate $\lambda(h)$. In general, numerical techniques are applied to compute these Abelian integrals. The approach described below is based on numerical computation with the aid of Maple 10 (2005), which is quite efficient to find more limit cycles and their complicated patterns.

With the Poincaré–Pontryagin–Andronov theorem on the global center bifurcations and Melnikov method (Melnikov, 1963), we have the following two theorems (as in Li and Li, 1985; Li and Huang, 1987; Li and Zhao, 1989; Li and Liu, 1991; Li, Chan and Chung, 2002).

THEOREM 2.15 (Bifurcation of limit cycles). *Suppose that the conditions* (A1) *and* (A3) *hold. For a given* $\lambda = \lambda_0$, *considering the set S of the intersection points of the straight line* $\lambda = \lambda_0$ *and the curve* $\lambda = \lambda(h)$ *in the* (h, λ)*-plane, we have the following conclusions*:

(i) *If S consists of exactly one point* (h_0, λ_0) *and* $\lambda'(h) > 0$ (< 0), *then there exists a stable* (*unstable*) *limit cycle of system* (80) *near* Γ^{h_0}.
(ii) *If S consists of two points* (h_0, λ_0) *and* $(\tilde{h}_0, \lambda_0)$ *having* $\tilde{h}_0 > h_0$ *and* $\lambda'(\tilde{h}_0) > 0$, $\lambda'(h_0) < 0$, *then there exist two limit cycles near* $\Gamma^{\tilde{h}_0}$ *and* Γ^{h_0}, *respectively, with the former stable and the latter unstable.*
(iii) *If S contains a point* (h_0, λ_0) *and* $\lambda'(h_0) = \lambda''(h_0) = \cdots = \lambda^{(k-1)}(h_0) = 0$ *but* $\lambda^{(k)}(h_0) \neq 0$, *then system* (80) *has at most k limit cycles near* Γ^{h_0}.
(iv) *If S is empty, then system* (80) *has no limit cycles.*

THEOREM 2.16 (Bifurcation parameter created by a heteroclinic or homoclinic loop). *Suppose that the conditions* (A1), (A2) *and* (A3) *hold. Then for* $0 < \varepsilon \ll 1$, *when* $\lambda = \lambda(h_2) + \mathrm{O}(\varepsilon)$, *system* (80) *has a heteroclinic* (*or homoclinic*) *loop with* Z_q*-equivariance at* $h = h_2$.

The following two propositions describe the properties of the detection function at the boundary values of h.

PROPOSITION 2.17 (Parameter value of Hopf bifurcation). *Suppose that as* $h \to h_1$, *the periodic orbit* Γ^h *of system* (80) *approaches a singular point* (ξ, η), *then at this point the Hopf bifurcation parameter value is given by*

$$b_{\mathrm{H}} = \lambda(h_1) + \mathrm{O}(\varepsilon) = \lim_{h \to h_1} \lambda(h) + \mathrm{O}(\varepsilon) = F(\xi, \eta) + \mathrm{O}(\varepsilon). \tag{82}$$

PROPOSITION 2.18 (Bifurcation direction of heteroclinic or homoclinic loop). *Suppose that as* $h \to h_2$, *the periodic orbit* Γ^h *of system* (80) *approaches a heteroclinic (or homoclinic) loop connecting a hyperbolic saddle point* (α, β), *where the saddle point value satisfies*

$$SQ(\alpha, \beta) = 2\varepsilon\sigma(\alpha, \beta) \equiv 2\varepsilon\big(\lambda(h_2) - F(\alpha, \beta)\big) > 0 \quad (< 0),$$

then we have

$$\lambda'(h_2) = \lim_{h \to h_2} \lambda'(h) = -\infty \quad (+\infty). \tag{83}$$

REMARKS.

(1) If Γ^h contracts inwards as h increases, then the stability of limit cycles defined in Theorem 2.15 and the sign of $\lambda'(h_2)$ in equation (83) change accordingly.
(2) If the curve Γ^h defined by $H(x, y) = h$ $(h \in (h_1, h_2))$ consists of m components of oval families having Z_q-equivariance, then Theorem 2.15 gives rise to simultaneous global bifurcations of limit cycles from all the oval families.
(3) If the unperturbed system (80) ($\varepsilon = 0$) has several different periodic annuli filled by periodic orbit families $\{\Gamma_i^h\}$, then by calculating the detection functions for every oval families, the global information of bifurcations of system (80) can be obtained.

We have noticed that hypothesis (A3) is a very important symmetry condition to guarantee the existence of simultaneous global bifurcations of limit cycles bifurcating from all symmetric ovals of unperturbed systems.

(C) Z_{10}-*equivariant planar vector field.* From the results given above, it is easy to find that for $n = 9$, system (52) is Z_{10}-equivariant, if and only if it has the following form given in the polar coordinates:

$$\begin{aligned}\dot r &= r\big[a_0 + a_1 r^2 + a_2 r^4 + a_3 r^6 + \big(a_4 + a_5\cos(10\theta) + b_5\sin(10\theta)\big)r^8\big],\\ \dot\theta &= b_0 + b_1 r^2 + b_2 r^4 + b_3 r^6 + \big(b_4 + b_5\cos(10\theta) - a_5\sin(10\theta)\big)r^8,\end{aligned}$$

which is a Hamiltonian if $a_0 = a_1 = a_2 = a_3 = a_4 = 0$, reduced to

$$\dot{r} = r^9\big[a_5 \cos(10\theta) + b_5 \sin(10\theta)\big],$$
$$\dot{\theta} = b_0 + b_1 r^2 + b_2 r^4 + b_3 r^6 + b_4 r^8 + \big[b_5 \cos(10\theta) - a_5 \sin(10\theta)\big] r^8.$$

By changing the polar axis and time scale, we can further reduce the above equation, by eliminating two parameters, to the following five-parameter system:

$$\dot{r} = br^9 \sin(10\theta) = R(r, \theta),$$
$$\dot{\theta} = 1 + cr^2 + dr^4 + er^6 + \big[a + b\cos(10\theta)\big] r^8 = \Theta(r, \theta), \tag{84}$$

which has the Hamiltonian

$$H(r, \theta) = -\frac{1}{2} r^2 - \frac{1}{4} cr^4 - \frac{1}{6} dr^6 - \frac{1}{8} er^8 - \frac{1}{10}\big[a + b\cos(10\theta)\big] r^{10}. \tag{85}$$

Without loss of generality, suppose that $(r, \theta) = (1, 0)$ is a singular point of equation (84). Then, from $\Theta(r, 0) = 0$ and $\Theta(r, \pi/10) = 0$, we have $a + b + c + d + e + 1 = 0$ and

$$f_1(r) = (a + b) r^8 + er^6 + dr^4 + cr^2 + 1 = 0,$$
$$f_2(r) = (a - b) r^8 + er^6 + dr^4 + cr^2 + 1 = 0.$$

Further, suppose that the following parametric conditions in group G hold: $a = 6.138618693$, $b = 0.02861869300$, $c = -6.684452552$, $d = 16.14239117$, $e = -16.62517600$, then system (84) has 81 finite singular points at $(0, 0)$ and $(z_1, 0)$, $(z_2, 0)$, $(z_3, 0)$, $(z_4, 0)$, $(z_5, \pi/10)$, $(z_6, \pi/10)$, $(z_7, \pi/10)$, $(z_8, \pi/10)$ and their Z_{10}-equivariant symmetric points. Here, the values of z_i's are defined as $z_1 = 0.65$, $z_2 = 0.7$, $z_3 = 0.885$, $z_4 = 1$, $z_5 = 0.6389658748$, $z_6 = 0.7454834324$, $z_7 = 0.8086840806$, $z_8 = 1.050231017$. As h is varied, the level curves $H(r, \theta) = h$ of the Hamiltonian defined by equation (85) give rise to a family of tenth algebraic curves in the affine real plane (see Figure 24 later).

Now the values of Hamiltonian function at critical points are calculated as

$$h_1 = H(z_1, 0) = -0.05793645019,$$
$$h_2 = H(z_2, 0) = -0.05790831192,$$
$$h_3 = H(z_3, 0) = -0.0588607574,$$
$$h_4 = H(z_4, 0) = -0.0578621301,$$
$$h_5 = H(z_5, \pi/10) = -0.05786629736,$$
$$h_6 = H(z_6, \pi/10) = -0.05769015970,$$
$$h_7 = H(z_7, \pi/10) = -0.05773717706,$$
$$h_8 = H(z_8, \pi/10) = -0.0502778305$$

and

$$-\infty < h_3 < h_1 < h_2 < h_5 < h_4 < h_7 < h_6 < h_8 < 0.$$

As h increases from $-\infty$ to 0, the schemes of ovals of the nine algebraic curves are varied, with details given below.

(1) For $h \in (-\infty, h_3)$: there is a global periodic annulus $\{\Gamma_8^h\}$, enclosing all 81 finite singular points.

(2) For $h \in (h_3, h_1)$: there exist 10 periodic annuli $\{\Gamma_{1i}^h\}$, $i = 1, 2, \ldots, 10$, enclosing the center $(z_3, 0)$ and its Z_{10}-equivariant symmetric points, respectively. Together with $\{\Gamma_8^h\}$, there exist 11 period annuli.

(3) For $h \in (h_1, h_2)$: there exist 20 periodic annuli $\{\Gamma_{1i}^h\}$, $\{\Gamma_{2i}^h\}$, $i = 1, 2, \ldots, 10$, enclosing the centers $(z_1, 0)$, $(z_3, 0)$ and their Z_{10}-equivariant symmetric points, respectively. Together with $\{\Gamma_8^h\}$, there exist 21 period annuli.

(4) For $h = h_2$: there exist 10 8-shaped loops consisting of $\Gamma_{3i}^{h_2}$, $i = 1, 2, \ldots, 10$, which enclose the centers $(z_1, 0)$, $(z_3, 0)$, and are homoclinic to $(z_2, 0)$, and their Z_{10}-equivariant symmetric points, respectively. There is also a closed orbit $\Gamma_8^{h_2}$.

(5) For $h \in (h_2, h_5)$: $\{\Gamma_{3i}^h\}$, $i = 1, 2, \ldots, 10$, expand outwards into vase-shaped loops, enclosing the centers $(z_1, 0)$, $(z_3, 0)$, the saddle point $(z_2, 0)$, and their Z_{10}-equivariant symmetric points, respectively. $\{\Gamma_8^h\}$ still exists.

(6) For $h = h_5$: there exists a 10-cycle heteroclinic orbit $\Gamma_0^{h_5}$ enclosing the origin $(0, 0)$ and 10 heteroclinic orbits $\Gamma_{3i}^{h_5}$, connected to the saddle point $(z_5 \pi/10)$ and its Z_{10}-equivariant symmetric points. There is also a closed orbit $\Gamma_8^{h_5}$.

(7) For $h \in (h_5, h_4)$: there exist 3 periodic annuli: $\{\Gamma_0^h\}$, enclosing the center $(0, 0)$, $\{\Gamma_4^h\}$, enclosing 41 finite singular points, and $\{\Gamma_8^h\}$. Thus, for every fixed $h \in (h_5, h_4)$, there is a nest of ovals with the depth 3 denoted by $C(1, 1, 1)$ (see Li, 2003).

(8) For $h = h_4$: there are 10 heteroclinic orbits denoted by $\pm\Gamma_{5i}^{h_4}$, $i = 1, 2, \ldots, 10$, connected to the saddle point $(z_4, 0)$ and its Z_{10}-equivariant symmetric points, enclosing the centers $(z_6, \pi/10)$, $(z_8, \pi/10)$, the saddle point $(z_7, \pi/10)$ and their Z_{10}-equivariant symmetric points, respectively. There is also a closed orbit $\Gamma_0^{h_4}$.

(9) For $h \in (h_4, h_7)$: there exist 10 periodic annuli $\{\Gamma_{5i}^h\}$, enclosing the centers $(z_6, \pi/10)$, $(z_8, \pi/10)$, the saddle point $(z_7, \pi/10)$ and their Z_{10}-equivariant symmetric points, respectively. There is also a periodic annulus $\{\Gamma_0^h\}$.

(10) For $h = h_7$: there exist 10 8-shaped loops $\Gamma_{5i}^{h_7}$, $i = 1, 2, \ldots, 10$, enclosing the centers $(z_6, \pi/10)$, $(z_8, \pi/10)$, and homoclinic to $(z_7, \pi/10)$ and

their Z_{10}-equivariant symmetric points, respectively. There is also a closed orbit $\Gamma_0^{h_7}$.

(11) For $h \in (h_7, h_6)$: the 8-shaped loops are split into 20 periodic annuli $\{\Gamma_{6i}^h\}$ and $\{\Gamma_{7i}^h\}$, enclosing the centers $(z_6, \pi/10)$, $(z_8, \pi/10)$ and their Z_{10}-equivariant symmetric points, respectively. $\{\Gamma_0^h\}$ still exists.

(12) For $h \in (h_6, h_8)$: $\{\Gamma_{7i}^h\}$ shrink inwards but still enclose the centers $(z_8, \pi/10)$, and $\{\Gamma_0^h\}$ encloses the origin.

(13) For $h \in (h_8, 0)$: $\{\Gamma_{7i}^h\}$ disappear and $\{\Gamma_0^h\}$ keeps shrinking inwards until reaching the origin when $h = 0$.

The above described changing process of schemes of the ovals are shown in Figure 23.

To sum up, by using Gudkov's oval notation (see Hilbert, 1976), we have the following conclusion which is relevant to the first part of Hilbert's 16th problem.

PROPOSITION 2.19. *In the affine real plane, the integral curves of the Hamiltonian system* (84) *can realize the following oval schemes*: 1, 11, 21, $\frac{10}{1}$, $\frac{20}{1}$ *and* $C(1, 1, 1)$.

(D) *Bifurcation parameter values of* Z_{10}*-equivariant perturbed Hamiltonian systems.* Consider the following perturbed Z_{10}-equivariant vector field:

$$\begin{aligned} \dot{r} &= br^9 \sin(10\theta) - \varepsilon r\left(5sr^8 + 4ur^6 + 3vr^4 + 2wr^2 - \lambda\right), \\ \dot{\theta} &= 1 + cr^2 + dr^4 + er^6 + \left[a + b\cos(10\theta)\right]r^8. \end{aligned} \tag{86}$$

Corresponding to equation (86), the function $F(x, y)$ in the divergence of vector field of the hypothesis (A3) has the form

$$F(r, \theta) = 5sr^8 + 4ur^6 + 3vr^4 + 2wr^2.$$

We continue to consider the parametric condition G, i.e., the unperturbed system (84) has the phase portrait as shown in Figure 24.

We compute nine detection functions defined by equation (81), corresponding to the above 9 types of periodic annuli $\{\Gamma_i^h\}$, $i = 0, 1, \ldots, 8$,

$$\begin{aligned} \lambda_i(h) &= \frac{\iint_{D_i^h} F(r, \theta) r \, dr \, d\theta}{\iint_{D_i^h} r \, dr \, d\theta} \\ &= \frac{\psi_i(h)}{\phi_i(h)} = \frac{1}{\phi_i(h)}\left[5sI_{i1}(h) + 4uI_{i2}(h) + 3vI_{i3}(h) + 2wI_{i4}(h)\right] \\ &= 5sJ_{i1}(h) + 4uJ_{i2}(h) + 3vJ_{i3}(h) + 2wJ_{i4}(h), \quad i = 0, 1, \ldots, 8, \end{aligned}$$

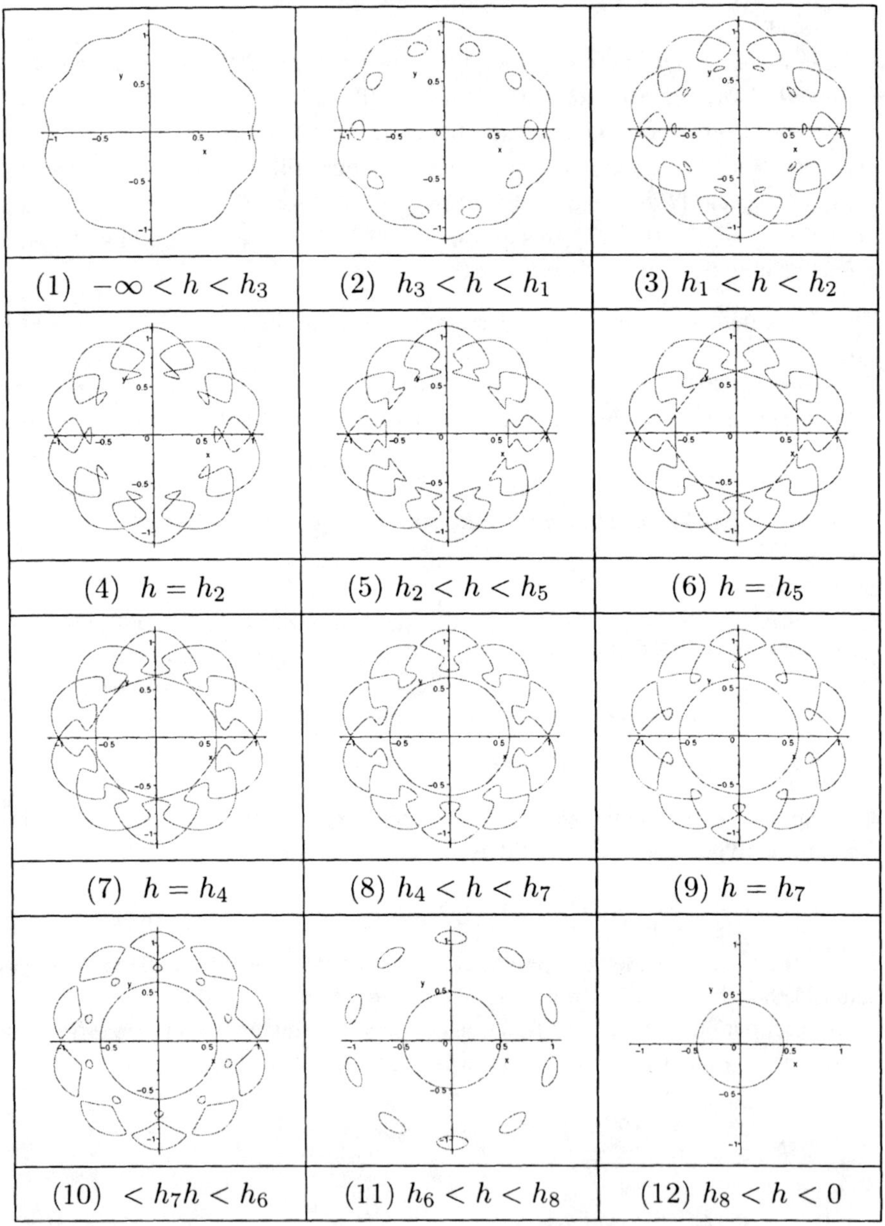

Figure 23. Different schemes of ovals defined by equation (84) under the parameter condition G.

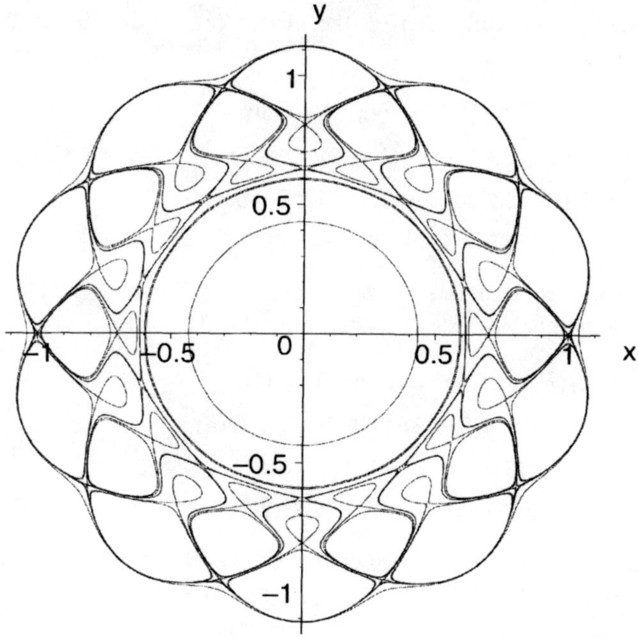

Figure 24. A Z_{10}-equivariant polynomial Hamiltonian vector field of degree 9.

where $J_{ij}(h) = I_{ij}(h)/\phi_i(h)$, $j = 1, 2, 3, 4$, and $\phi_i(h) = \iint_{D_i^h} r\, \mathrm{d}r\, \mathrm{d}\theta$, in which D_i^h is the area inside Γ_i^h, and

$$I_{i1}(h) = \iint\limits_{D_i^h} r^9\, \mathrm{d}r\, \mathrm{d}\theta, \qquad I_{i2}(h) = \iint\limits_{D_i^h} r^7\, \mathrm{d}r\, \mathrm{d}\theta,$$

$$I_{i3}(h) = \iint\limits_{D_i^h} r^5\, \mathrm{d}r\, \mathrm{d}\theta, \qquad I_{i4}(h) = \iint\limits_{D_i^h} r^3\, \mathrm{d}r\, \mathrm{d}\theta.$$

For the given parameter group G, the functions $J_{ij}(h)$ can be numerically calculated to a given degree of accuracy. Here, the accuracy is given up to 10 digit decimal points (as the default of Maple).

By using the theory given above, we immediately obtain the following values of bifurcation parameters and bifurcation direction detections.

Hopf bifurcation parameters. The parameter values for Hopf bifurcations are given below.

(1) Bifurcation from the origin (0, 0):

$$b_0^{\mathrm{H}} = \lambda_0(0) + \mathrm{O}(\varepsilon) = 0 + \mathrm{O}(\varepsilon).$$

(2) Simultaneous bifurcations from the center $(z_1, 0)$ and its Z_{10}-equivariant symmetric points:

$$\begin{aligned}
b_1^{\mathrm{H}} &= \lambda_2(h_1) + \mathrm{O}(\varepsilon) = F(z_1, 0) + \mathrm{O}(\varepsilon)\\
&= 5sz_1^8 + 4uz_1^6 + 3vz_1^4 + 2wz_1^2 + \mathrm{O}(\varepsilon)\\
&= 0.1593224150s + 0.3016755748u\\
&\quad + 0.5355187644v + 0.8450000114w + \mathrm{O}(\varepsilon).
\end{aligned}$$

(3) Simultaneous bifurcations from the center $(z_3, 0)$ and its Z_{10}-equivariant symmetric points:

$$\begin{aligned}
b_3^{\mathrm{H}} &= \lambda_1(h_3) + \mathrm{O}(\varepsilon) = F(z_3, 0) + \mathrm{O}(\varepsilon)\\
&= 5sz_3^8 + 4uz_3^6 + 3vz_3^4 + 2wz_3^2 + \mathrm{O}(\varepsilon)\\
&= 1.881551910s + 1.921850679u\\
&\quad + 1.840324275v + 1.566450031w + \mathrm{O}(\varepsilon).
\end{aligned}$$

(4) Simultaneous bifurcations from the center $(z_6, \pi/10)$ and its Z_{10}-equivariant symmetric points:

$$\begin{aligned}
b_6^{\mathrm{H}} &= \lambda_6(h_6) + \mathrm{O}(\varepsilon) = F(z_6, \pi/10) + \mathrm{O}(\varepsilon)\\
&= 5sz_6^8 + 4uz_6^6 + 3vz_6^4 + 2wz_6^2 + \mathrm{O}(\varepsilon)\\
&= 0.4769512304s + 0.6865749724u\\
&\quad + 0.9265593423v + 1.111491096w + \mathrm{O}(\varepsilon).
\end{aligned}$$

(5) Simultaneous bifurcations from the center $(z_8, \pi/10)$ and its Z_{10}-equivariant symmetric points:

$$\begin{aligned}
b_8^{\mathrm{H}} &= \lambda_7(h_8) + \mathrm{O}(\varepsilon) = F(z_8, \pi/10) + \mathrm{O}(\varepsilon)\\
&= 5sz_8^8 + 4uz_8^6 + 3vz_8^4 + 2wz_8^2 + \mathrm{O}(\varepsilon)\\
&= 7.400289800s + 5.367462680u\\
&\quad + 3.649728981v + 2.205970378w + \mathrm{O}(\varepsilon).
\end{aligned}$$

Bifurcations from heteroclinic or homoclinic loops. The parameter values for the bifurcations of heteroclinic or homoclinic loops can be found as follows.

(1) The homoclinic and heteroclinic bifurcation values from $\Gamma_{1i}^{h_2}$, $\Gamma_{2i}^{h_2}$ and $\Gamma_{3i}^{h_2}$:

$$\begin{aligned}
\lambda_1(h_2) &= 5sJ_{14}(h_2) + 4uJ_{13}(h_2) + 3vJ_{12}(h_2) + 2wJ_{11}(h_2)\\
&= 1.494582740s + 1.567061760u\\
&\quad + 1.571896568v + 1.431780302w,
\end{aligned}$$

$$
\begin{aligned}
\lambda_2(h_2) &= 5sJ_{24}(h_2) + 4uJ_{23}(h_2) + 3vJ_{22}(h_2) + 2wJ_{21}(h_2) \\
&= 0.1736217000s + 0.3204819432u \\
&\quad + 0.5560802067v + 0.8599433548w, \\
\lambda_3(h_2) &= 5sJ_{34}(h_2) + 4uJ_{33}(h_2) + 3vJ_{32}(h_2) + 2wJ_{31}(h_2) \\
&= 1.349290145s + 1.429844341u \\
&\quad + 1.460010416v + 1.368765508w.
\end{aligned}
$$

(2) The homoclinic and heteroclinic bifurcation values from $\Gamma_4^{h_4}$, $\Gamma_{5i}^{h_4}$ and $\Gamma_8^{h_4}$:

$$
\begin{aligned}
\lambda_4(h_4) &= 5sJ_{44}(h_4) + 4uJ_{43}(h_4) + 3vJ_{42}(h_4) + 2wJ_{41}(h_4) \\
&= 0.5141286510s + 0.5748284220u \\
&\quad + 0.6561545736v + 0.7800965728w, \\
\lambda_5(h_4) &= 5sJ_{54}(h_4) + 4uJ_{53}(h_4) + 3vJ_{52}(h_4) + 2wJ_{51}(h_4) \\
&= 3.804153395s + 3.051358988u \\
&\quad + 2.374665417v + 1.722704272w, \\
\lambda_8(h_4) &= 5sJ_{84}(h_4) + 4uJ_{83}(h_4) + 3vJ_{82}(h_4) + 2wJ_{81}(h_4) \\
&= 1.825990245s + 1.562318063u \\
&\quad + 1.341392083v + 1.155951154w.
\end{aligned}
$$

(3) The homoclinic and heteroclinic bifurcation values from $\Gamma_0^{h_5}$, $\Gamma_3^{h_5}$ and $\Gamma_4^{h_5}$:

$$
\begin{aligned}
\lambda_0(h_5) &= 5sJ_{04}(h_5) + 4uJ_{03}(h_5) + 3vJ_{02}(h_5) + 2wJ_{01}(h_5) \\
&= 0.02276719494s + 0.05856372540u \\
&\quad + 0.1507250682v + 0.3881297704w, \\
\lambda_3(h_5) &= 5sJ_{34}(h_5) + 4uJ_{33}(h_5) + 3vJ_{32}(h_5) + 2wJ_{31}(h_5) \\
&= 1.149960387s + 1.242025078u \\
&\quad + 1.307868795v + 1.283886171w, \\
\lambda_4(h_5) &= 5sJ_{44}(h_5) + 4uJ_{43}(h_5) + 3vJ_{42}(h_5) + 2wJ_{41}(h_5) \\
&= 0.5133362970s + 0.5736214608u \\
&\quad + 0.6543290313v + 0.7779745778w.
\end{aligned}
$$

(4) The homoclinic and heteroclinic bifurcation values from $\Gamma_{5i}^{h_7}$, $\Gamma_{6i}^{h_7}$ and $\Gamma_{7i}^{h_7}$:

$$
\begin{aligned}
\lambda_5(h_7) &= 5sJ_{54}(h_7) + 4uJ_{53}(h_7) + 3vJ_{52}(h_7) + 2wJ_{51}(h_7) \\
&= 4.706190119s + 3.691042625u \\
&\quad + 2.768329870v + 1.891552834w,
\end{aligned}
$$

$$\lambda_6(h_7) = 5sJ_{64}(h_7) + 4uJ_{63}(h_7) + 3vJ_{62}(h_7) + 2wJ_{61}(h_7)$$
$$= 0.5035693170s + 0.7106306624u$$
$$+ 0.9441224925v + 1.119645944w,$$
$$\lambda_7(h_7) = 5sJ_{74}(h_7) + 4uJ_{73}(h_7) + 3vJ_{72}(h_7) + 2wJ_{71}(h_7)$$
$$= 5.068765340s + 3.948173491u$$
$$+ 2.925710805v + 1.958148019w.$$

The values of bifurcation direction detections of heteroclinic and homoclinic loops. The parameter values for the bifurcation direction detections of heteroclinic and homoclinic loops are given by

(1) $$\sigma_1(z_2, 0) = \lambda_1(h_2) - F(z_2, 0)$$
$$= 1.206342717s + 1.096465793u$$
$$+ 0.8515966016v + 0.4517803250w,$$

(2) $$\sigma_2(z_2, 0) = \lambda_2(h_2) - F(z_2, 0)$$
$$= -0.1146183230s - 0.1501140236u$$
$$- 0.1642197597v - 0.1200566222w,$$

(3) $$\sigma_3(z_2, 0) = \lambda_3(h_2) - F(z_2, 0)$$
$$= 1.061050122s + 0.9592483742u$$
$$+ 0.7397104496v + 0.3887655310w,$$

(4) $$\sigma_4(z_4, 0) = \lambda_4(h_4) - F(z_4, 0)$$
$$= -4.485871145s - 3.425171456u$$
$$- 2.343845365v - 1.219903407w,$$

(5) $$\sigma_5(z_4, 0) = \lambda_5(h_4) - F(z_4, 0)$$
$$= -3.867579956s + 2.774889880u$$
$$- 1.706166761v - 0.724146345w,$$

(6) $$\sigma_8(z_4, \pi/10) = \lambda_8(h_4) - F(z_4, 0)$$
$$= -3.174009551s - 2.437681815u$$
$$- 1.658607856v - 0.844048826w,$$

(7) $$\sigma_0(z_5, \pi/10) = \lambda_0(h_5) - F(z_5, \pi/10)$$
$$= -0.1161612965s - 0.2136600031u$$
$$- 0.3493462113v - 0.4284250080w,$$

(8) $$\sigma_3(z_5, \pi/10) = \lambda_3(h_5) - F(z_5, \pi/10) = 1.011031896s + 0.9698013495u + 0.8077975155v + 0.4673313926w,$$

(9) $$\sigma_4(z_5, 0) = \lambda_4(h_5) - F(z_5, \pi/10) = 0.3744078056s + 0.3013977323u + 0.1542577518v + 0.0385802006w,$$

(10) $$\sigma_5(z_7, \pi/10) = \lambda_5(h_7) - F(z_7, \pi/10) = 3.791653383s + 2.572291836u + 1.485299814v + 0.583612950w,$$

(11) $$\sigma_6(z_7, \pi/10) = \lambda_6(h_7) - F(z_7, \pi/10) = -0.4109674190s - 0.4081201266u - 0.3389075635v - 0.188293940w,$$

(12) $$\sigma_7(z_7, \pi/10) = \lambda_7(h_7) - F(z_7, \pi/10) = 4.154228604s + 2.829422702u + 1.642680749v + 0.650208135w.$$

Some extra detection function values obtained for parameter control purpose. Now, we consider some extra detection function values which can be used to control parameter values so that we can find exactly 80 limit cycles. These detection function values are given by

(1) $$\lambda_1(h_2 - 0.00001) = 1.516119470s + 1.586538033u + 1.586800516v + 1.439503895w,$$

(2) $$\lambda_1(h_3 + 0.00065) = 1.927815s + 1.759589378u + 1.718028714v + 1.505943701w,$$

(3) $$\lambda_2(h_1 + 0.000005) = 0.1608676140s + 0.3037155236u + 0.5377511325v + 0.8466185292w,$$

(4) $$\lambda_2(h_2 - 0.00001) = 1.835063992s + 1.568650522u + 1.345336723v + 1.157801190w,$$

(5) $$\lambda_5(h_7 + 0.000037) = 0.4796272212s + 0.6887476724u + 0.9278715450v + 1.111893668w.$$

Now we are ready to determine the perturbed parameter group PG: (s, u, v, w) such that our nine detection functions $\lambda_i(h)$, $i = 0, 1, \ldots, 8$, can create extra limit cycles. In other words, we can choose parameters to "control" the number and configuration of limit cycles created by equation (86). To this end, three limit cycle configurations have been obtained with different control groups.

CASE 1. Suppose that the following conditions hold:

C1: $\lambda_1(h_2) - \lambda_2(h_2) = 0$,
C2: $\lambda_6(h_7) - \lambda_2(h_1) = 0$,
C3: $\lambda_3(h_5) - \lambda_7(h_7) = 0$,

which imply that

$$s = s, \qquad u = -3.568366338s,$$
$$v = 4.880734178s, \qquad w = -3.201327275s.$$

As an example, taking $s = 2.8$, then

$$PG_1 = (s, u, v, w) = (2.8, -9.991425746, 13.66605570, -8.963716370),$$

and we obtain the bifurcation values of $\lambda_i(h)$ and the values of direction detections as follows:

$$\begin{aligned}
\lambda_0(h_5) &= -1.940654968, & \lambda_0(h_4) &= -1.935209920,\\
\lambda_0(h_7) &= -1.869002816, & \lambda_0(h_6) &= -1.853292180,\\
\lambda_0(h_8) &= -1.261270431, & \lambda_1(h_3) &= -2.82492274,\\
\lambda_1(h_1) &= -2.82447110, & \lambda_1(h_2) &= -2.82479602,\\
\lambda_2(h_1) &= -2.823977513, & \lambda_2(h_2) &= -2.824796027,\\
\lambda_3(h_2) &= -2.82481327, & \lambda_3(h_5) &= -2.824695946,\\
\lambda_5(h_4) &= -2.82531986, & \lambda_5(h_7) &= -2.82463888,\\
\lambda_6(h_7) &= -2.823977518, & \lambda_6(h_6) &= -2.825078763,\\
\lambda_7(h_7) &= -2.82469595, & \lambda_7(h_6) &= -2.82477103,\\
\lambda_7(h_8) &= -2.80408660, & \lambda_8(h_3) &= -2.53086534,\\
\lambda_8(h_1) &= -2.52776115, & \lambda_8(h_2) &= -2.52755337,\\
\lambda_8(h_5) &= -2.52715451, & \lambda_8(h_4) &= -2.52709159
\end{aligned}$$

and

$$\begin{aligned}
\sigma_1(z_2, 0) &= 0.010838949 > 0, & \sigma_2(z_2, 0) &= 0.010838942 > 0,\\
\sigma_3(z_2, 0) &= 0.010821699 > 0, & \sigma_0\left(z_5, \frac{\pi}{10}\right) &= 0.875611901 > 0,
\end{aligned}$$

$$\sigma_3\left(z_5, \frac{\pi}{10}\right) = -0.008429077 < 0, \qquad \sigma_4\left(z_5, \frac{\pi}{10}\right) = 0.490865799 > 0,$$
$$\sigma_4(z_4, 0) = 0.565653889 > 0, \qquad \sigma_5(z_4, 0) = 0.06964877 > 0,$$
$$\sigma_8(z_4, 0) = 0.36787704 > 0, \qquad \sigma_5(z_7, 0) = -0.017384361 < 0,$$
$$\sigma_6(z_7, 0) = -0.016722999 < 0, \qquad \sigma_7(z_7, 0) = -0.017441431 < 0.$$

It follows from the above discussions that under the parameter conditions of G and PG_1, system (86) has the graphs of detection curves as shown in Figure 25(a) and a zoom-in is depicted in Figure 25(b). It is seen from Figure 25(a) that when

$$\tilde{\lambda} \in \big(\lambda_7(h_6), \lambda_7(h_7)\big) = (-2.82477103, -2.82469595), \tag{87}$$

in the (h, λ)-plane the straight line $\lambda = \tilde{\lambda}$ intersects the curves $\lambda = \lambda_2(h)$, $\lambda_3(h)$, $\lambda_5(h)$ and $\lambda = \lambda_6(h)$ at least once and intersects the curves $\lambda_1(h)$ and $\lambda_7(h)$ at least twice, which makes the total of 8 intersections in one of the ten equivariance sections. Thus, there exist 8 limit cycles in one section. Since system (86) is Z_{10}-equivariance, it has at least 80 limit cycles with configuration shown in Figure 26.

CASE 2. If the control conditions given in Case 1 are changed to

C1: $\lambda_1(h_2) - \lambda_5(h_7) = 0$,
C2: $\lambda_6(h_7) - \lambda_1(h_3 + 0.00065) = 0$,
C3: $\lambda_2(h_1 + 0.00001) - \lambda_1(h_1 + 0.000005) = 0$,

then another parameter group PG_2 is obtained as

$$\begin{aligned} PG_2 &= (s, u, v, w) \\ &= (2.803157747, -10, 13.67228371, -8.962712462), \end{aligned}$$

under which the bifurcation values of $\lambda_i(h)$ and the values of bifurcation direction detections are given as follows:

$$\lambda_0(h_5) = -1.939756850, \qquad \lambda_0(h_4) = -1.934318541,$$
$$\lambda_0(h_7) = -1.868187780, \qquad \lambda_0(h_6) = -1.852494142,$$
$$\lambda_0(h_8) = -1.260916583, \qquad \lambda_1(h_3) = -2.82242559,$$
$$\lambda_1(h_1) = -2.822285732, \qquad \lambda_1(h_2) = -2.82228573,$$
$$\lambda_2(h_1) = -2.821877538, \qquad \lambda_2(h_2) = -2.822669090,$$
$$\lambda_3(h_2) = -2.82234534, \qquad \lambda_3(h_5) = -2.822279772,$$
$$\lambda_4(h_5) = -2.323842278, \qquad \lambda_4(h_4) = -2.327750296,$$
$$\lambda_5(h_4) = -2.82295157, \qquad \lambda_5(h_7) = -2.82228573,$$

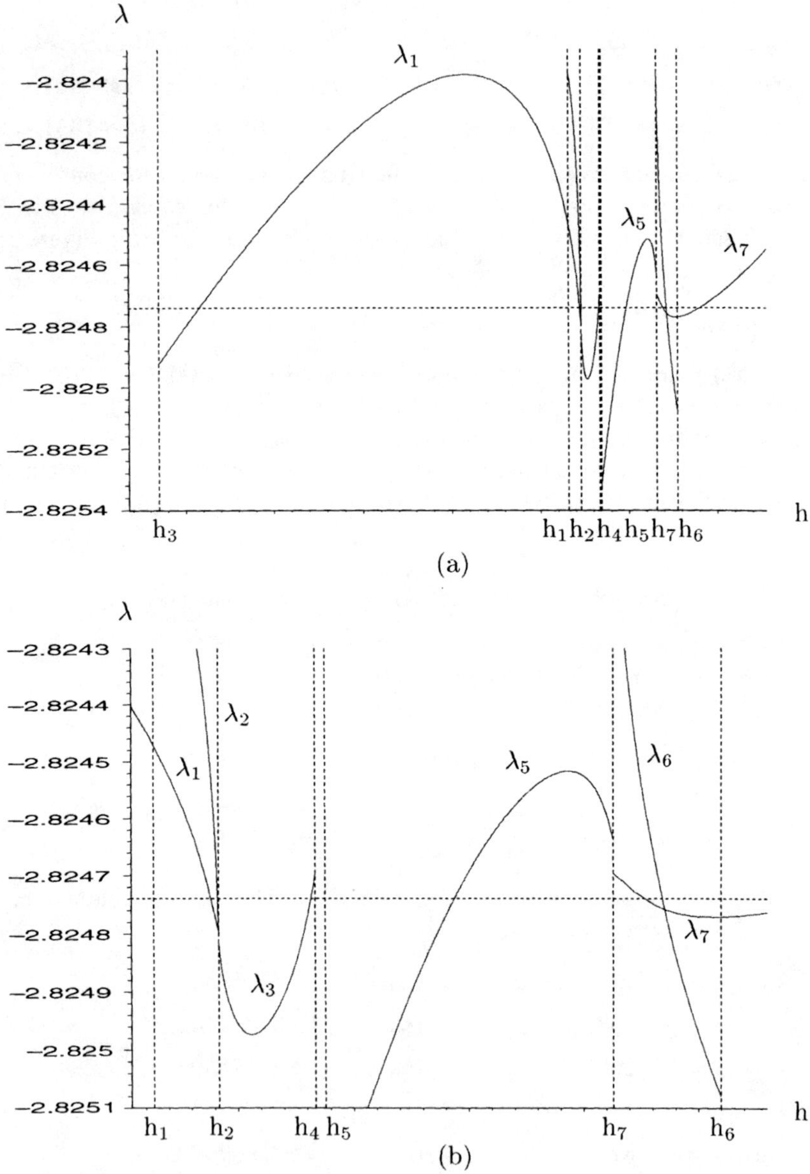

Figure 25. (a) Graphs of detection curves of equation (86) for parameters G and PG_1; and (b) a zoom-in of part (a).

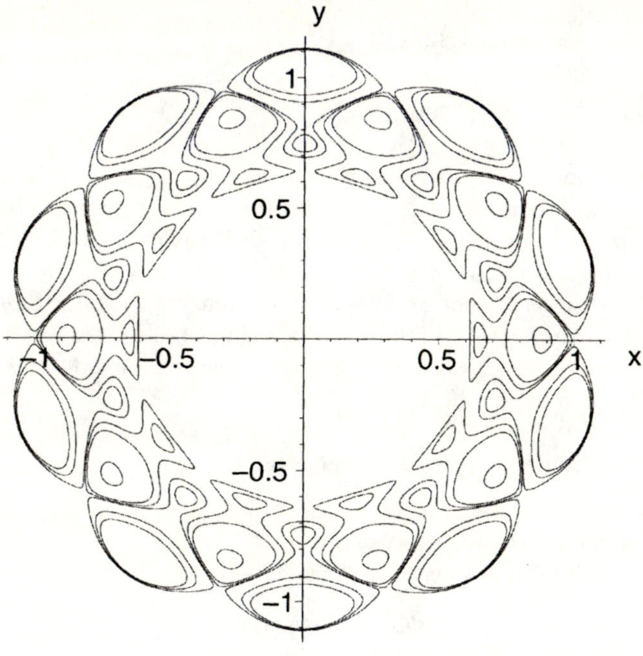

Figure 26. Existence of 80 limit cycles in system (86) with PG_1.

$$\lambda_6(h_7) = -2.821476482, \qquad \lambda_6(h_6) = -2.822573086,$$
$$\lambda_7(h_7) = -2.82235555, \qquad \lambda_7(h_6) = -2.82243550,$$
$$\lambda_7(h_8) = -2.80179522, \qquad \lambda_8(h_3) = -2.52897104,$$
$$\lambda_8(h_1) = -2.52587518, \qquad \lambda_8(h_2) = -2.52566773,$$
$$\lambda_8(h_5) = -2.52526946, \qquad \lambda_8(h_4) = -2.52520661$$

and

$$\sigma_1(z_2, 0) = 0.011004192 > 0, \qquad \sigma_2(z_2, 0) = 0.010620832 > 0,$$
$$\sigma_3(z_2, 0) = 0.010944582 > 0, \qquad \sigma_0\left(z_5, \frac{\pi}{10}\right) = 0.874471237 > 0,$$
$$\sigma_3\left(z_5, \frac{\pi}{10}\right) = -0.008051685 < 0, \qquad \sigma_4\left(z_5, \frac{\pi}{10}\right) = 0.490385809 > 0,$$
$$\sigma_4(z_4, 0) = 0.565034774 > 0, \qquad \sigma_5(z_4, 0) = 0.06983350 > 0,$$
$$\sigma_8(z_4, 0) = 0.36757846 > 0, \qquad \sigma_5\left(z_7, \frac{\pi}{10}\right) = -0.017630406 < 0,$$

$$\sigma_6\left(z_7, \frac{\pi}{10}\right) = -0.016821158 < 0,$$

$$\sigma_7\left(z_7, \frac{\pi}{10}\right) = -0.017700226 < 0.$$

Based on the above parameter values, system (86) has the graphs of detection curves shown in Figure 27(a) and a zoom-in is given in Figure 27(b). It is observed that when $\tilde{\lambda} \in (\lambda_6(h_6), \lambda_6(h_7)) = (-2.82243550, -2.82235555)$, in the (h, λ)-plane the straight line $\lambda = \tilde{\lambda}$ intersects the curves $\lambda = \lambda_1(h), \lambda_2(h), \lambda_5(h)$ and $\lambda = \lambda_6(h)$ at least once and intersects the curves $\lambda_3(h)$ and $\lambda_7(h)$ at least twice, which makes the total of 8 intersections in one of the ten equivariance sections. Therefore, it comes out 8 limit cycles in one section and 80 for the whole plane, as shown in Figure 28.

CASE 3. If the control conditions are changed to

C1: $\lambda_1(h_3) - \lambda_4(h_7) = 0$,
C2: $\lambda_5(h_7 + 0.000037) - \lambda_4(h_7) = 0$,
C3: $\lambda_2(h_1 + 0.00001) - \lambda_1(h_1 + 0.000005) = 0$,

then another parameter group PG_3 is obtained as

$$PG_3 = (s, u, v, w) = (2.8, -9.988759290, 13.65655815, -8.951989483),$$

under which the bifurcation values of $\lambda_i(h)$ and the values of bifurcation direction detections are listed as follows:

$$\lambda_0(h_5) = -1.939756850, \qquad \lambda_0(h_4) = -1.934318541,$$
$$\lambda_0(h_7) = -1.868187780, \qquad \lambda_0(h_6) = -1.852494142,$$
$$\lambda_0(h_8) = -1.259390380, \qquad \lambda_1(h_3) = -2.82242558,$$
$$\lambda_1(h_1) = -2.82195797, \qquad \lambda_1(h_2) = -2.82228573,$$
$$\lambda_2(h_1) = -2.821877540, \qquad \lambda_2(h_2) = -2.822669090,$$
$$\lambda_3(h_2) = -2.82234534, \qquad \lambda_3(h_5) = -2.822279772,$$
$$\lambda_4(h_5) = -2.323842278, \qquad \lambda_4(h_4) = -2.327750296,$$
$$\lambda_5(h_4) = -2.822573086, \qquad \lambda_5(h_7) = -2.82228573,$$
$$\lambda_6(h_7) = -2.821476482, \qquad \lambda_6(h_6) = -2.822573086,$$
$$\lambda_7(h_7) = -2.81899242, \qquad \lambda_7(h_6) = -2.81907658,$$
$$\lambda_7(h_8) = -2.79856885, \qquad \lambda_8(h_3) = -2.52587366,$$
$$\lambda_8(h_1) = -2.52277812, \qquad \lambda_8(h_2) = -2.52257080,$$
$$\lambda_8(h_5) = -2.52217277, \qquad \lambda_8(h_4) = -2.52210997$$

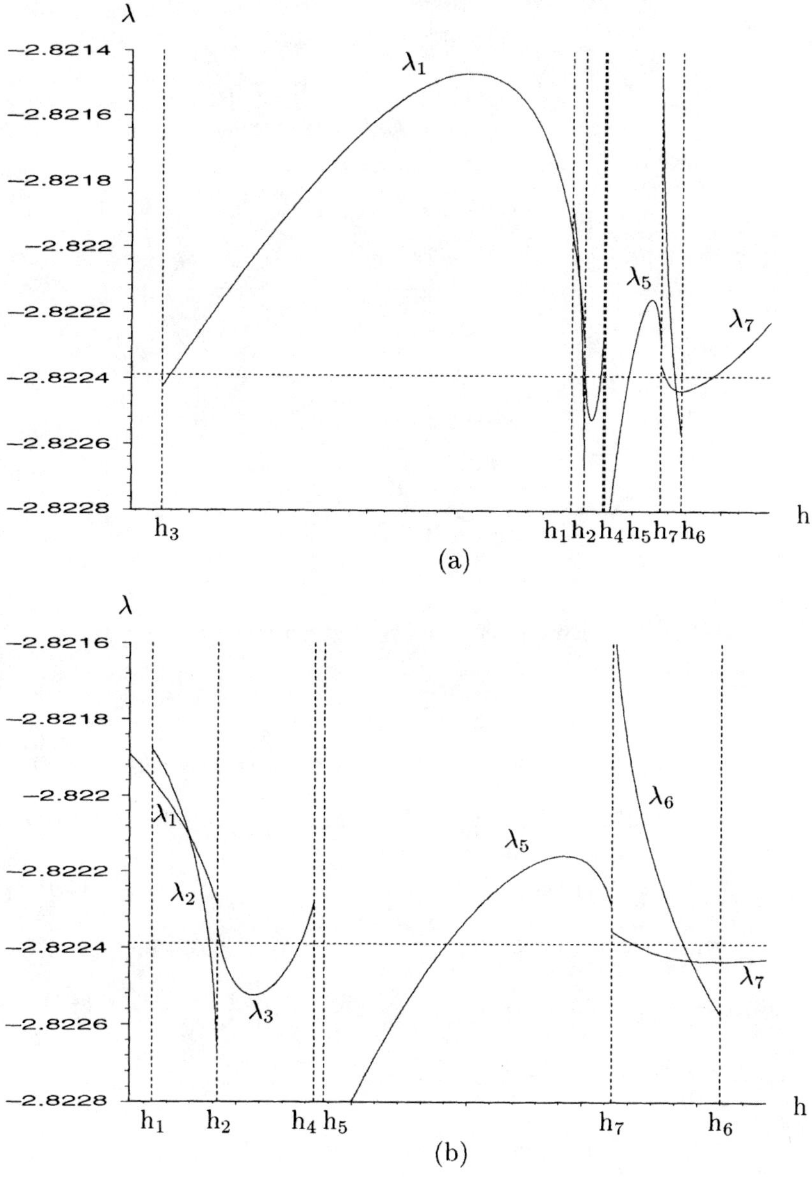

Figure 27. (a) Graphs of detection curves of equation (86) for parameters G and PG_2; and (b) a zoom-in of part (a).

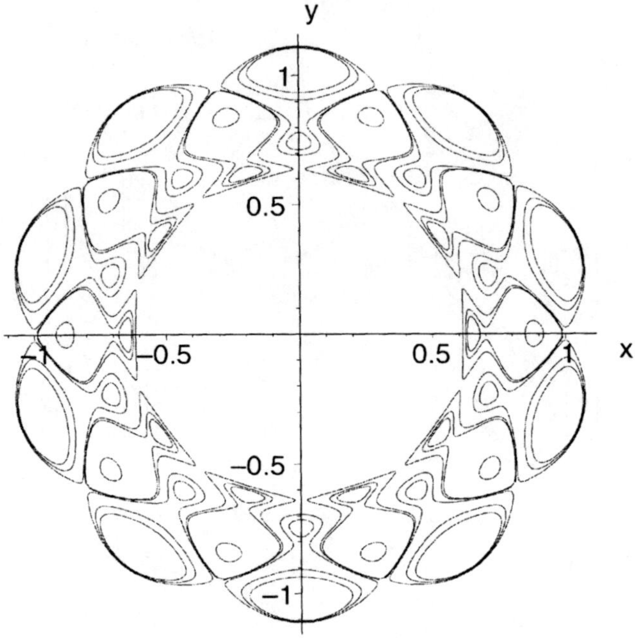

Figure 28. Existence of 80 limit cycles in system (86) with PG_2.

and

$$\sigma_1(z_2, 0) = 0.010972523 > 0, \qquad \sigma_2(z_2, 0) = 0.010590467 > 0,$$

$$\sigma_3(z_2, 0) = 0.010913073 > 0, \qquad \sigma_0\left(z_5, \frac{\pi}{10}\right) = 0.873336026 > 0,$$

$$\sigma_3\left(z_5, \frac{\pi}{10}\right) = -0.008034906 < 0, \qquad \sigma_4\left(z_5, \frac{\pi}{10}\right) = 0.489751962 > 0,$$

$$\sigma_4(z_4, 0) = 0.564475938 > 0, \qquad \sigma_5(z_4, 0) = 0.06980657 > 0,$$

$$\sigma_8(z_4, 0) = 0.36723171 > 0, \qquad \sigma_5\left(z_7, \frac{\pi}{10}\right) = -0.017788211 < 0,$$

$$\sigma_6\left(z_7, \frac{\pi}{10}\right) = -0.016800544 < 0,$$

$$\sigma_7\left(z_7, \frac{\pi}{10}\right) = -0.017873421 < 0.$$

The above parameter values generate another set of detection function curves shown in Figure 29(a) and see a zoom-in in Figure 29(b). When $\tilde{\lambda} \in (\lambda_5(h_7), \lambda_3(h_2)) = (-2.82228573, -2.82234534)$, in the (h, λ)-plane the straight line

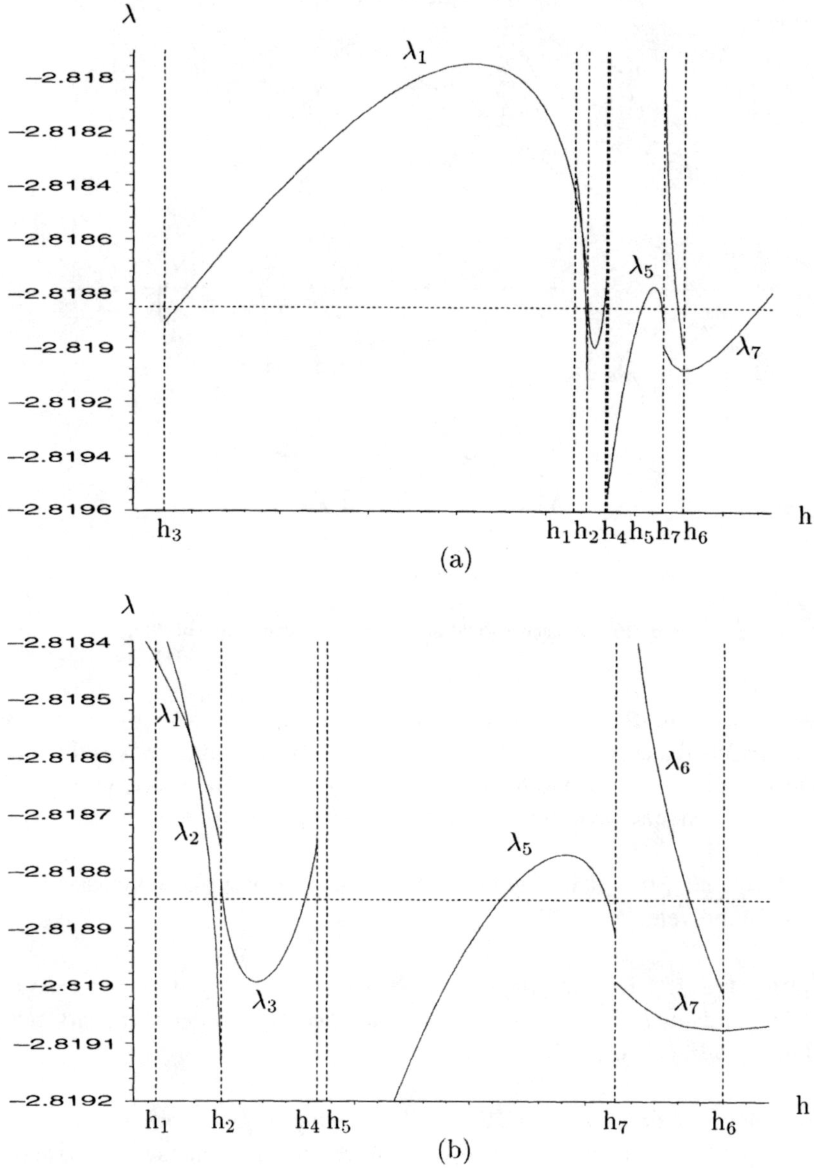

Figure 29. (a) Graphs of detection curves of equation (86) for parameters G and PG_3; and (b) a zoom-in of part (a).

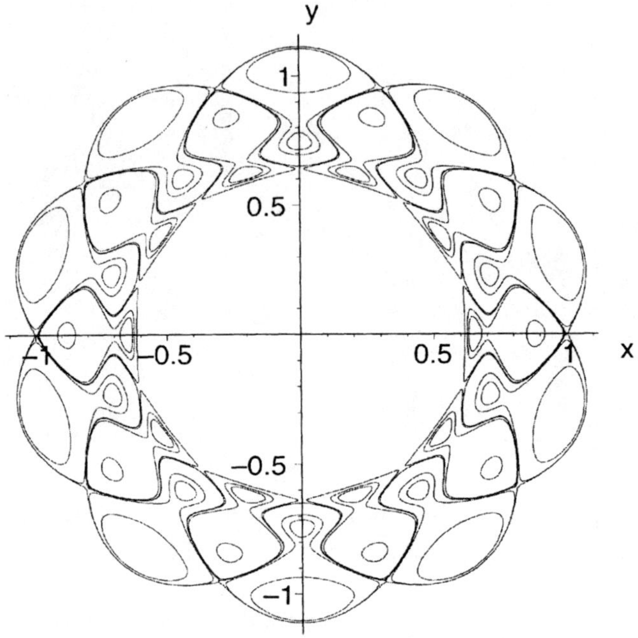

Figure 30. Existence of 80 limit cycles in system (86) with PG_3.

$\lambda = \tilde{\lambda}$ intersects the curves $\lambda = \lambda_1(h)$, $\lambda_2(h)$, $\lambda_6(h)$ and $\lambda_7(h)$ at least once and intersects the curves $\lambda_3(h)$ and $\lambda_5(h)$ at least twice, which makes the total of 8 intersections in one of the ten equivariance sections. Therefore, it again gives 8 limit cycles in one section and 80 for the whole plane, as shown in Figure 30.

Consequently, the main results of this section is proved, summarized in the following theorem.

THEOREM 2.20. *For the unperturbed and perturbed parameter groups G and PG, and small* $\varepsilon > 0$, *when* $\lambda = \tilde{\lambda}$ *satisfies* (87), *system* (86) *has at least* 80 *limit cycles, i.e.,* $H(9) \geqslant 9^2 - 1$.

For the vector fields of degree 9 with Z_{10}-symmetry, global bifurcation analysis with the detection function method is employed to show that such a system can have at least 80 limit cycles, i.e., $H(9) \geqslant 9^2 - 1$. Very recently, it has also been shown that $H(11) \geqslant 11^2$ and $H(6) \geqslant 6^2 - 1$. Combining these results with previously published results $H(5) \geqslant 5^2 - 1$, $H(7) \geqslant 7^2$ and $H(4) \geqslant 4^2 - 1$, we propose a conjecture as follows (Wang and Yu, 2005).

CONJECTURE 2.21. *The lower bound of the Hilbert number, $H(n)$, of planar polynomial systems is either $H(n) \geqslant n^2$ or $H(n) \geqslant n^2 - 1$.*

3. Bifurcation control and chaos synchronization

The discover of the Lorenz chaotic system (Lorenz, 1963, 1993; Stewart, 2002) in 60's of the last century began a new era of chaos, as the well-known physicist, Ford, said: "The discover of chaos is the third revolution of physics in 20th century". As the first milestone in the chaos history, the Lorenz system revealed the fundamental characteristics of nonlinear systems. Since then, based on the Lorenz system, physicists, mathematicians and many researchers from other disciplines extensively studied the characteristics and applications of Lorenz systems. However, there still exist many fundamental problems in this area to be solved (Chen and Lü, 2003).

Due to that chaos is externally sensitive to initial conditions, it has been a long time that people consider it is impossible to control chaos, and there is no way for two chaotic systems being synchronized. However, the developments in this area in 1990s have completely changed the situation. These developments include the OGY method (Ott, Grebogi and Yorke, 1990), and in particular, the concept of chaos synchronization and its application in secure communication, proposed by Pecora and Carroll (1990, 1991) and later realized in circuit (Cuomo and Oppenheim, 1993). These studies have indeed generated a high tide in the research of chaos in the past two decades and many good results have been obtained (e.g., see Chen, 2000; Chen and Dong, 1998; Sprott, 1977; Sprott and Ling, 2000; Sanchez, Peren, Martizez and Chen, 2002; Lü, Zhou and Zhang, 2002; Gua, Shieh, Chen and Lin, 2000; Rössler, 1976, 1979; Yu and Chen, 2004; Liao and Chen, 2003a, 2003b; Liao, 2001; Liao and Yu, 2005).

In general, chaos control has two objectives: one is stabilizing and the other is tracking. The purpose of stabilization is, with an appropriate control, to stabilize an equilibrium point, which is either unstable or locally stable, to reach globally asymptotic stability or even globally exponential stability. This objective is relatively easy to realize and many results have been obtained. The second objective, on the other hand, is much difficult than the first objective and not much has been achieved so far. It has been found that there usually exist rich unstable periodic orbits in a chaotic attractor, which is important in many research areas, in particular, for biological systems (Chen and Lü, 2003). It is of great important and interesting to consider the tracking of periodic orbits. That is, for a given periodic orbit, which is either unstable or only locally stable, one wants to design a controller such that all trajectories would converge to the periodic orbit globally or even with a globally exponential converging rate. The first objective can be considered as a static tracking, while the second one as a dynamic tracking which is still a study of asymptotic behavior between two solutions of the system. Chaos synchronization, on the other hand, is to consider the relative asymptotic behavior of the solutions between two chaotic systems. Due to its potential application in secure communication, many results have been achieved in past few years and many relative reports can be found in the literature.

Although the Lorenz system is a strong nonlinear system and does not satisfy a global Lipschitz condition, it has a very good property, namely, it has a globally attractive compact set. Leonov obtained the estimations of the globally attractive set for all variables (Leonov, Bunin and Koksch, 1987) and partial variables (Leonov and Reitmann, 1987). Recently, we improved the estimations (Yu and Liao, 2006a, 2006b). In this section, this property of the Lorenz system and the Floquet–Lyapunov theory on periodic linear systems are used to construct feedback control laws and Lyapunov functions. The main attention here is focused on the globally exponential tracking of periodic orbits in the Lorenz system. The globally exponential synchronization of two Lorenz systems and globally exponential stabilization of equilibrium points are also considered. Numerical simulations are presented to verify the theoretical results.

3.1. Global ultimate boundedness of chaotic systems

In this subsection, we will consider the Lorenz system and show that it is globally ultimately bounded, and will present some of our new results obtained recently.

Consider the Lorenz equation (Lorenz, 1963):

$$\begin{aligned} \dot{x} &= a(y - x), \\ \dot{y} &= cx - xz - y, \\ \dot{z} &= xy - bz, \end{aligned} \tag{88}$$

where a, b and c are parameters. The typical values of the parameters for the system to exhibit the butterfly chaotic attractor are $a = 10$, $b = 8/3$ and $c = 28$. Most of the results given below do not depend upon these special values. System (88) has three equilibrium points, given by

$$\begin{aligned} E_0 &= (0, 0, 0), \\ E_1 &= \big(\sqrt{b(c-1)}, \sqrt{b(c-1)}, c-1\big), \\ E_2 &= \big(-\sqrt{b(c-1)}, -\sqrt{b(c-1)}, c-1\big). \end{aligned}$$

First, we need to introduce some notations and definitions. Let $X = (x, y, z)$, and $\Omega \subseteq \mathbb{R}^3$ be a compact (bounded and closed) set containing the origin. $X(t, t_0, X_0)$ denotes the solution of system (88) satisfying the initial condition $X(t_0, t_0, X_0) = X_0$. Further, let $\rho(X, \Omega) = \inf_{\widetilde{X} \in \Omega} \|X - \widetilde{X}\|$ denote the distance between X and Ω.

DEFINITION 3.1. If there exists a compact set $Q \subseteq \mathbb{R}^3$ such that

$$\forall X_0 \in \mathbb{R}^3, \qquad \lim_{t \to +\infty} \rho\big(X(t, t_0, X_0), Q\big) \to 0,$$

then we call Q a globally attractive set of system (88). A system having a globally attractive set is considered as dissipative in the sense that any of the system's trajectory is ultimately bounded. If $\forall X_0 \in Q$, $X(t, t_0, X_0) \subseteq Q$ when $t \geqslant t_0$, then Q is called positive invariant set of system (88).

In the following, we summarize the results reported so far for the globally attractive and positive invariant set of the Lorenz system. In 1987, Leonov, Bunin and Koksch and Leonov and Reitmann obtained two estimations for the globally attractive and positive invariant set of the Lorenz system:

$$y^2 + (z - c)^2 \leqslant \frac{b^2c^2}{4(b-1)} \quad \text{and}$$
$$x^2 + y^2 + (z - a - c)^2 \leqslant \frac{b^2(a+c)^2}{4(b-1)}, \tag{89}$$

where the parameters are assumed to be $a > 0$, $b > 1$ and $c > 1$. It is easy to see that the above two estimations become very poor as $b \to 1^+$. Later, Leonov (2001) improved the first estimation to obtain the following new result: Assume that $c > 1$ and $2a > b > 0$, then

$$\begin{cases} y^2 + (z - c)^2 \leqslant R^2, \\ z \geqslant \frac{x^2}{2a}, \end{cases} \quad \text{where } R^2 = \begin{cases} c^2 & \text{for } b \leqslant 2, \\ \frac{b^2c^2}{4(b-1)} & \text{for } b \geqslant 2. \end{cases} \tag{90}$$

As pointed out by Leonov (2001), if one of the conditions $c > 1$ and $2a > b > 0$ fails to hold, then the Lorenz system will be globally, asymptotically stable, that is, any of its solutions (trajectories) tend to some equilibrium state as $t \to +\infty$.

Recently, the second estimation given in (89) was improved by Li, Lu, Wu and Chen (2005) using the approach of Lagrange multiplier. The new estimation is given as follows: For $a > 0$, $b > 0$ and $c > 0$, the following result holds:

$$x^2 + y^2 + (z - a - c)^2 \leqslant \overline{R}^2 = \begin{cases} \frac{(a+c)^2b^2}{4(b-1)} & \text{for } a \geqslant 1,\ b \geqslant 2, \\ (a+c)^2 & \text{for } 2a > b,\ b \leqslant 2, \\ \frac{(a+c)^2b^2}{4a^2(b-a)^2} & \text{for } 0 < a < 1,\ 2a \leqslant b. \end{cases} \tag{91}$$

In the same paper (Li, Lu, Wu and Chen, 2005), it is also shown that $z \geqslant x^2/(2a)$ when $2a > b$, which is identical to that obtained by Leonov (2001).

Since it is known that the origin $(0, 0, 0)$ is globally attractive for $c < 1$, and stable when $c = 1$, it is not necessary to consider $0 < c \leqslant 1$ in equation (91). In addition, according to the Remarks, the third case listed in (91) is not necessary either.

Summarizing the above results given in equations (90) and (91) gives the following theorem.

THEOREM 3.2 (Leonov, 2001; Li, Lu, Wu and Chen, 2005). *Assume that $c > 1$ and $2a > b > 0$. The globally attractive and positive invariant set of the Lorenz system is bounded by*

$$D_1: \quad \begin{cases} y^2 + (z-c)^2 \leqslant R^2 = \begin{cases} c^2 & \text{for } b \leqslant 2, \\ \frac{b^2c^2}{4(b-1)} & \text{for } b \geqslant 2, \end{cases} \\ x^2 + y^2 + (z-a-c)^2 \leqslant \overline{R}^2 = \begin{cases} (a+c)^2 & \text{for } b \leqslant 2, \\ \frac{b^2(a+c)^2}{4(b-1)} & \text{for } b \geqslant 2, \end{cases} \\ z \geqslant \frac{x^2}{2a}. \end{cases} \tag{92}$$

The following corollary is a direct result of Theorem 3.2.

COROLLARY 3.3. *Assume that $c > 1$ and $2a > b > 0$. The maximum values of the ultimate state variables of the Lorenz system are bounded by*

$$\Omega_1: \quad |y| \leqslant R, \qquad |x| \leqslant \overline{R}, \qquad \frac{x^2}{2a} \leqslant z \leqslant R + c,$$

where R and $\overline{R}$ are given in equation (92).

It is obvious to see from equation (92) that $\overline{R} > R$. In other words, the estimation for the variable x is not as good as that for the variable y. In the following, based on the results given in Theorem 3.2, we provide a simple proof to improve the estimation on the variable x. The new estimations on the globally attractive and positive invariant set of the Lorenz system are summarized in the following theorem.

THEOREM 3.4. *Assume that $c > 1$ and $2a > b > 0$. The globally attractive and positive invariant set of the Lorenz system is bounded by*

$$D_2: \quad \begin{cases} |x| \leqslant R, \\ y^2 + (z-c)^2 \leqslant R^2, \\ z \geqslant \frac{x^2}{2a}, \end{cases} \quad \text{where } R = \begin{cases} c & \text{for } b \leqslant 2, \\ \frac{bc}{2\sqrt{b-1}} & \text{for } b \geqslant 2. \end{cases} \tag{93}$$

PROOF. We only need to prove $|x| \leqslant R$. By Corollary 3.3, we may assume $|y| \leqslant R$. Consider the following generalized Lyapunov function:

$$V = |x|.$$

Computing the Dini derivative of V with respect to time t along the trajectory of system (88) results in

$$\begin{aligned} D^+V &= \dot{x}\,\mathrm{sign}(x) = a(y-x)\,\mathrm{sign}(x) = a\big(y\,\mathrm{sign}(x) - |x|\big) \\ &\leqslant a\big(|y| - |x|\big) \leqslant a\big(R - |x|\big), \end{aligned} \tag{94}$$

where R is given in equation (93), and the estimation $|y| \leqslant R$ has been used. Inequality (94) clearly indicates that $\frac{\mathrm{d}V}{\mathrm{d}t} < 0$ when $|x| > R(1+\varepsilon)$, where ε is an arbitrarily small positive number. This implies that the x variable is ultimately bounded by

$$|x| \leqslant R, \quad \text{where } R = \begin{cases} c & \text{for } b \leqslant 2, \\ \frac{bc}{2\sqrt{b-1}} & \text{for } b \geqslant 2. \end{cases}$$

This completes the proof for Theorems 3.4. □

Similarly, we have the following corollary.

COROLLARY 3.5. *Assume that $c > 1$ and $2a > b > 0$. The maximum values of the ultimate state variables of the Lorenz system are bounded by*

$$\Omega_2: \quad |x| \leqslant R, \qquad |y| \leqslant R, \qquad \frac{x^2}{2a} \leqslant z \leqslant R + c,$$
$$R = \begin{cases} c & \textit{for } b \leqslant 2, \\ \frac{bc}{2\sqrt{b-1}} & \textit{for } b \geqslant 2. \end{cases}$$

3.2. Hopf bifurcation control

Now we want to particularly consider Hopf bifurcation since the limit cycles generated by Hopf bifurcation is the most popular phenomenon exhibited in nonlinear dynamical systems. For convenience, we will use the Lorenz system as an example to illustrate the theory and methodology of Hopf bifurcation control. A state feedback controller using polynomial function has been used in Section 2.2.2 for controlling Hopf bifurcation of a discrete system. In the following, we will first briefly describe the state feedback law for continuous systems (Yu and Chen, 2004).

3.2.1. Feedback controller using polynomial function

Consider the general nonlinear system (2):

$$\dot{\mathbf{x}} = \mathbf{f}(\mathbf{x}, \mu), \quad \mathbf{x} \in \mathbb{R}^n, \ \mu \in \mathbb{R}, \ \mathbf{f}: \mathbb{R}^{n+1} \to \mathbb{R}^n, \tag{95}$$

where $\mathbf{x}$ is an n-dimensional state vector while μ is a scalar parameter, called bifurcation parameter. Suppose that at the critical point $\mu = \mu^*$ on an equilibrium solution $\mathbf{x} = \mathbf{x}^*$, the Jacobian of the system has a complex pair of eigenvalues first crossing the imaginary axis. Then Hopf bifurcation occurs at the critical point and a family of limit cycles bifurcates from the equilibrium solution $\mathbf{x}^*$.

Suppose system (95) has k equilibria, given by

$$\mathbf{x}_i^*(\mu) = \left(x_{1i}^*, x_{2i}^*, \ldots, x_{ni}^*\right), \quad i = 1, 2, \ldots, k. \tag{96}$$

A general nonlinear state feedback control is applied so that system (95) becomes

$$\dot{\mathbf{x}} = \mathbf{f}(\mathbf{x}, \mu) + \mathbf{u}(\mathbf{x}, \mu). \tag{97}$$

In order for the controlled system (97) to keep all the original k equilibria unchanged under the control $\mathbf{u}$, it requires that the following conditions be satisfied:

$$\mathbf{u}\left(\mathbf{x}_i^*, \mu\right) \equiv (u_1, u_2, \ldots, u_n)^{\mathrm{T}} = \mathbf{0} \tag{98}$$

for $i = 1, 2, \ldots, k$.

A general formula satisfying condition (98) can be constructed as (Yu and Chen, 2004)

$$\begin{aligned}
&u_q\left(\mathbf{x}, \mathbf{x}_1^*, \mathbf{x}_2^*, \ldots, \mathbf{x}_k^*, \mu\right) \\
&\quad = \sum_{i=1}^{n} A_{qi} \prod_{j=1}^{k} \left(x_i - x_{ij}^*\right) \\
&\qquad + \sum_{i=1}^{n} \sum_{j=1}^{k} B_{qij} \left(x_i - x_{ij}^*\right) \prod_{p=1}^{k} \left(x_i - x_{ip}^*\right) \\
&\qquad + \sum_{i=1}^{n} \sum_{j=1}^{k} C_{qij} \left(x_i - x_{ij}^*\right)^2 \prod_{p=1}^{k} \left(x_i - x_{ip}^*\right) \\
&\qquad + \sum_{i=1}^{n} \sum_{j=1}^{k} D_{qij} \left(x_i - x_{ij}^*\right)^2 \prod_{p=1}^{k} \left(x_i - x_{ip}^*\right)^2 \\
&\qquad + \cdots, \qquad q = 1, 2, \ldots, n.
\end{aligned} \tag{99}$$

It is easy to verify that $u_q(\mathbf{x}_i^*, \mathbf{x}_1^*, \mathbf{x}_2^*, \ldots, \mathbf{x}_k^*, \mu) = 0$ for $i = 1, 2, \ldots, k$. Usually terms given in equation (99) up to D_{qij} are enough for controlling a bifurcation if the singularity of the system is not highly degenerate. The coefficients A_{qi}, B_{qij}, C_{qij} and D_{qij}, which may be functions of μ, are determined from the stabilities of an equilibrium under consideration and that of the associated bifurcation solutions. More precisely, linear terms are determined by the requirement of shifting an existing bifurcation (e.g., delaying an existing Hopf bifurcation). The nonlinear terms, on the other hand, can be used to change the stability of an existing bifurcation or create a new bifurcation (e.g., changing an existing subcritical Hopf bifurcation to supercritical). Note that not just A_{qi} terms may involve linear terms; B_{qij} terms, etc. may also contain linear terms.

It is not necessary to take all the components u_i, $i = 1, 2, \ldots, n$, in the controller. In most cases, using fewer components or just one component may be enough to satisfy the pre-designed control objectives. It is preferable to have a simplest possible design for engineering applications. If $x_{i1}^* = x_{i2}^* = \cdots = x_{ik}^*$ for some i, then one only needs to use these terms and omits the remaining terms in the control law. Moreover, lower-order terms related to these equilibrium components can be added. This greatly simplify the control formula. For example, if $i = 1$, then the general controller can be taken as

$$u_q = \sum_{i=1}^{k-1} a_{qi}\left(x_1 - x_{11}^*\right)^i + A_{q1}\left(x_1 - x_{11}^*\right)^k + B_{q11}\left(x_1 - x_{11}^*\right)^{k+1} + C_{q11}\left(x_1 - x_{11}^*\right)^{k+2},$$

where a_{qi}'s denote the added lower-order terms.

The goals of Hopf bifurcation control are

(i) to move the critical point $(\mathbf{x}^*, \mu^*)$ to a designate position $(\tilde{\mathbf{x}}, \tilde{\mu})$;
(ii) to stabilize all possible Hopf bifurcations.

Goal (i) only requires linear analysis, while goal (ii) must apply nonlinear systems theory. In general, if the purpose of the control is to avoid bifurcations, one should employ linear analysis to maximize the stable interval for the equilibrium. The best result is to completely eliminate possible bifurcations using a feedback control. If this is not feasible, then one may have to consider stabilizing the bifurcating limit cycles using a nonlinear state feedback.

3.2.2. Controlling Hopf bifurcation in the Lorenz system

It is well known that the Lorenz system can exhibit very rich periodic and chaotic motions. In this subsection, we will use a different version of Lorenz equation, which contains only two parameters, given as (Wang and Abed, 1995; Chen, Moiola and Wang, 2000)

$$\begin{aligned} \dot{x} &= -p(x - y), \\ \dot{y} &= -xz - y, \\ \dot{z} &= xy - z - r, \end{aligned} \tag{100}$$

where p and r are positive constants, which are considered as control parameters. One can easily show that system (100) is a special case of system (88) by letting $b = 1$, $p = a$ and $r = c$, with a constant shift.

System (100) has three equilibrium solutions, C_0, C_+ and C_-, given by

$$\begin{aligned} &C_0: \quad x_e^0 = y_e^0 = 0, \qquad z_e^0 = -r, \\ &C_\pm: \quad x_e^\pm = y_e^\pm = \pm\sqrt{r-1}, \qquad z_e^\pm = -1. \end{aligned} \tag{101}$$

Suppose the parameters p and r are positive. Then C_0 is stable for $0 \leqslant r < 1$, and a pitchfork bifurcation occurs at $r = 1$, where the equilibrium C_0 looses its stability and bifurcates into either C_+ or C_-. The two equilibria C_+ and C_- are stable for $1 < r < r_{\mathrm{H}}$, where

$$r_{\mathrm{H}} = \frac{p(p+4)}{p-2}, \qquad p > 2, \tag{102}$$

and at this critical point C_+ and C_- loose their stabilities, giving rise to Hopf bifurcation. We fix $p = 4$, which was used in Wang and Abed (1995) and Chen, Moiola and Wang (2000). Then $r_{\mathrm{H}} = 16$, the Lorenz system (100) exhibits chaotic motions when $r > 16$. In fact, one can employ numerical simulation to show the coexistence of locally stable equilibria $C_\pm$ and (global) chaotic attractors at a same value of r, with different initial conditions (Yu and Chen, 2004).

We first consider system (100) without control. The critical point is $p = 4$, $r_{\mathrm{H}} = 16$. The Jacobian of system (100) evaluated at C_+ and C_- has a real eigenvalue -6 and a purely imaginary pair $\pm 2\sqrt{5}$. Using the shift, given by

$$x = \pm\sqrt{r-1} + \tilde{x}, \qquad y = \pm\sqrt{r-1} + \tilde{y}, \qquad z = -1 + \tilde{z}, \tag{103}$$

to move $C_\pm$ to the origin and then applying an appropriate linear transformation to system (100), we obtain the following system:

$$\begin{aligned}
\dot{\tilde{x}} &= 2\sqrt{5}\tilde{y} + \frac{1}{84}(\tilde{x} + 4\sqrt{5}\tilde{y} - 6\tilde{z})\mu - \frac{\sqrt{15}}{21}(\tilde{x} - 2\sqrt{5}\tilde{y})(\tilde{x} - 2\tilde{z}) + \cdots,\\
\dot{\tilde{y}} &= -2\sqrt{5}\tilde{x} - \frac{\sqrt{5}}{2100}(155\tilde{x} - 10\sqrt{5}\tilde{y} - 6\tilde{z})\mu\\
&\quad - \frac{\sqrt{3}}{105}(55\tilde{x} - 5\sqrt{5}\tilde{y} + 42\tilde{z})(\tilde{x} - 2\tilde{z}) + \cdots,\\
\dot{\tilde{z}} &= -6\tilde{z} + \frac{1}{168}(\tilde{x} + 4\sqrt{5}\tilde{y} - 6\tilde{z})\mu - \frac{\sqrt{15}}{42}(\tilde{x} - 2\sqrt{5}\tilde{y})(\tilde{x} - 2\tilde{z}) + \cdots,
\end{aligned}$$

where $\mu = r - 16$ is a bifurcation parameter.

Employing the Maple programs developed in Yu (1998) for computing the normal forms of Hopf and generalized Hopf bifurcations yields the following normal form:

$$\begin{aligned}
\dot{\rho} &= \rho\left(\frac{1}{56}\mu + \frac{31}{3248}\rho^2\right) + \cdots,\\
\dot{\theta} &= 2\sqrt{5}\left(1 + \frac{17}{560}\mu - \frac{851}{48720}\rho^2\right) + \cdots,
\end{aligned} \tag{104}$$

where ρ and θ represent the amplitude and phase of the motion, respectively. The first equation of (104) clearly shows that the Hopf bifurcation is subcritical since the coefficient of ρ^3 is $\frac{31}{3248} > 0$.

Next, we apply feedback controls to stabilize system (100). A washout filter control has been used by Wang and Abed (1995). The disadvantage of this method is that it increases the dimension of the original system by one, unnecessarily increases the complexity of the system and difficulty in analysis. Here we apply the control formula (102) to controlling the Hopf bifurcation. Due to the symmetry and $z_{\pm} = -1$, we may find a control law with one variable only:

$$u_3 = -k_{31}(z+1) - k_{33}(z+1)^3. \tag{105}$$

The closed-loop system is now given by

$$\begin{aligned} \dot{x} &= -p(x-y), \\ \dot{y} &= -xz - y, \\ \dot{z} &= xy - z - r - k_{31}(z+1) - k_{33}(z+1)^3, \end{aligned} \tag{106}$$

where the negative signs are used for k_{ij}'s for consistence and comparison with that of the controller based on the washout filter. Introducing the transformation (103) into equation (106) results in

$$\begin{aligned} \dot{x} &= -p(\tilde{x} - \tilde{y}), \\ \dot{y} &= -\tilde{x}\tilde{z} + \tilde{x} - \tilde{y} \mp \sqrt{r-1}\,\tilde{z}, \\ \dot{z} &= \tilde{x}\tilde{y} \pm \sqrt{r-1}(\tilde{x} + \tilde{y}) - \tilde{z} - k_{31}\tilde{z} - k_{33}\tilde{z}^3. \end{aligned} \tag{107}$$

Then $O_e = (\tilde{x}, \tilde{y}, \tilde{z}) = (0, 0, 0)$ is an equilibrium of system (107), corresponding to the equilibria C_+ and C_- of the original system (100). The characteristic polynomial of system (107) is

$$P(\lambda) = \lambda^3 + (p + 2 + k_{31})\lambda^2 + (p + r + k_{31} + pk_{31})\lambda + 2p(r-1),$$

which shows that only the linear term of the controller u_3 affects the linear stability. The stability conditions for O_e (under the assumption $p, r > 0$) can be obtained as

$$\begin{aligned} &p + 2 + k_{31} > 0, \\ &p + r + k_{31}(p+1) > 0, \\ &2p(r-1) > 0, \\ &p(p+4) - r(p - 2 - k_c) + k_{31}^2(p+1) + k_{31}\left(p^2 + 4p + 2\right) > 0. \end{aligned}$$

If choosing $k_{31} > 0$, then it only requires $r > 1$. The last condition in above equation implies a critical point at which the controlled system has a Hopf bifurcation emerging from the equilibrium O_e, defined by

$$r_H = \frac{p(p+4) + k_{31}^2(p+1) + k_{31}(p^2 + 4p + 2)}{(p - 2 - k_{31})}, \tag{108}$$

for $0 < k_{31} < p - 2$.

Setting $k_{31} = 0$ yields $r_{\mathrm{H}} = p(p+4)/(p-2)$ $(p > 2)$ which is the condition given in equation (102) for the system without control. It can be seen from equation (108) that the parameter r_{H} for the controlled system can reach very large values as long as k_{31} is chosen close to $p - 2$. For example, when $p = 4$, choosing $k_{31} = 1.5$ gives $r_{\mathrm{H}} = 188.5$, and $r_{\mathrm{H}} = 71$ if $k_{31} = 1$. These values of r_{H} are much larger than $r_{\mathrm{H}} = 16$ for the uncontrolled system. If we choose $r > 1$ and $0 < p - 2 < k_{31}$, then the equilibria C_+ and C_- are always stable, and no Hopf bifurcations can occur from the two equilibria.

Next, we perform a nonlinear analysis to determine the stability of Hopf bifurcation. If $p = 4$, then $k_{31} \in (0, 2)$, and for determination we choose $k_{31} = (2\sqrt{1006} - 58)/5 \approx 1.087$, thus $r_{\mathrm{H}} = 82$. Let $r = r_{\mathrm{H}} + \mu = 82 + \mu$, where μ is a perturbation from the critical point. Then, we have the closed-loop system

$$
\begin{aligned}
\dot{\tilde{x}} &= -8(\tilde{x} - \tilde{y}),\\
\dot{\tilde{y}} &= -\tilde{x}\tilde{z} + \tilde{x} - \tilde{y} \mp \sqrt{81 + \mu}\,\tilde{z},\\
\dot{\tilde{z}} &= \tilde{x}\tilde{y} \pm \sqrt{81 + \mu}(\tilde{x} + \tilde{y}) - \frac{2\sqrt{1006} - 53}{5}\tilde{z} - k_{33}\tilde{z}^3.
\end{aligned}
\tag{109}
$$

The eigenvalues of the Jacobian of system (109), when evaluated at the equilibrium O_{e}, are $\lambda_{1,2} = \pm\sqrt{2\sqrt{1006} + 28}\,\mathrm{i} \approx 9.5621\mathrm{i}$ and $\lambda_3 = -(2\sqrt{1006} - 28)/5 \approx -7.0870$. To apply the method of normal forms (Guckenheimer and Holmes, 1993; Yu, 1998, 2000), we introduce the following transformation:

$$
\begin{aligned}
\tilde{x} &= u - \frac{24 + \sqrt{1006}}{43}w,\\
\tilde{y} &= u + \frac{2\sqrt{1006} + 28}{4}v + w,\\
\tilde{z} &= \pm\frac{\sqrt{1006} + 14}{18}u \mp \frac{5(2\sqrt{1006} + 28)}{36}v \pm \frac{9\sqrt{1006} - 171}{215}w,
\end{aligned}
$$

to equation (109), and then employ the Maple program (Yu, 1998) to obtain an identical normal form for the system associated with the two equilibria C_+ and C_-, given in the polar coordinates up to the third order:

$$
\begin{aligned}
\dot{\rho} = \rho\Bigg[&\frac{1249 - 34\sqrt{1006}}{52942}\mu\\
&+ \bigg(\frac{4646315818 - 102399253\sqrt{1006}}{358010321904}\\
&\qquad - \frac{5746272 + 187233\sqrt{1006}}{4235360}k_{33}\bigg)\rho^2\Bigg]
\end{aligned}
\tag{110}
$$

and

$$\dot{\theta} = \sqrt{2\sqrt{1006} + 28}\left[1 + \frac{122602 - 773\sqrt{1006}}{17153208}\mu - \left(\frac{21706679417 + 211691192\sqrt{1006}}{6444185794272} + \frac{34871 + 1594\sqrt{1006}}{16941440}k_{33}\right)\rho^2\right]. \tag{111}$$

Steady-state solutions and their stabilities can be found from equation (110). The solution $\rho = 0$ represents the initial equilibrium solution O_e (C_+ or C_-), which is stable when $\mu < 0$ (i.e., $r < r_H = 82$) and unstable when $\mu > 0$ ($r > 82$). The supercritical Hopf bifurcation solution can be obtained, if

$$\frac{4646315818 - 102399253\sqrt{1006}}{358010321904} - \frac{5746272 + 187233\sqrt{1006}}{4235360}k_{33} < 0,$$

i.e.,

$$k_{33} > \frac{3672843514\sqrt{1006} - 115816173526}{478327912875} \approx 0.001416.$$

Choosing $k_{33} = 0.01$, we have the controller

$$u = -1.087(z + 1) - 0.01(z + 1)^3. \tag{112}$$

So the controlled system described in the original states is given by

$$\begin{aligned} \dot{x} &= -4(x - y), \\ \dot{y} &= -xz - y, \\ \dot{z} &= xy - z - r - 1.087(z + 1) - 0.01(z + 1)^3. \end{aligned} \tag{113}$$

The corresponding normal form then becomes

$$\begin{aligned} \dot{\rho} &= \rho\big(0.003222\mu - 0.023683\rho^2\big) + \cdots, \\ \dot{\theta} &= 9.562165 + 0.054678\mu - 0.046512\rho^2 + \cdots \end{aligned}$$

and the solution for the family of bifurcating limit cycles is obtained as

$$\rho = 0.136070\sqrt{\mu} = 0.136070\sqrt{r - 82}. \tag{114}$$

Some numerical simulation results, obtained from the controlled system (113), are given in Figures 31 and 32. Figure 31 depicts that the trajectories converge to the equilibria C_+ and C_- for $1 < r < 82$, while Figure 32 demonstrates the stable limit cycles bifurcated from the system when $r > 82$. By using equation (114), one can estimate the amplitudes of the three limit cycles shown in Figure 32 as 0.136, 0.385 and 0.593, respectively. These approximations give a

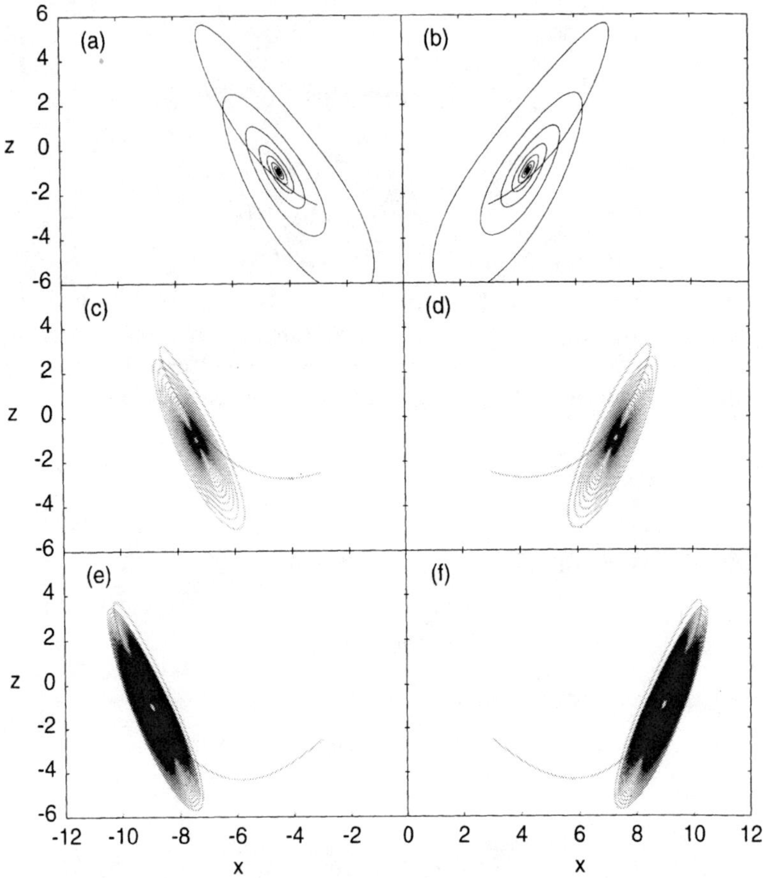

Figure 31. Stable equilibria $C_{\pm}$ of the controlled Lorenz system (113) with new control law for the initial conditions $(x_0, y_0, z_0) = (\pm 3.0, \pm 12.0, -2.5)$ when (a) and (b) $r = 20$; (c) and (d) $r = 55$ and (e) and (f) $r = 81$.

good prediction, confirmed by the numerical simulation results. It can be seen from Figures 31 and 32 that the symmetry of the two equilibria C_+ and C_- remain unchanged before and after the Hopf bifurcation generated by using the simple control (105).

3.3. Tracking and chaos synchronization

In the previous subsection, we presented stabilizing equilibrium points and controlling Hopf bifurcation. Now we turn to discuss tracking of periodic motion and synchronizing two identical chaotic systems.

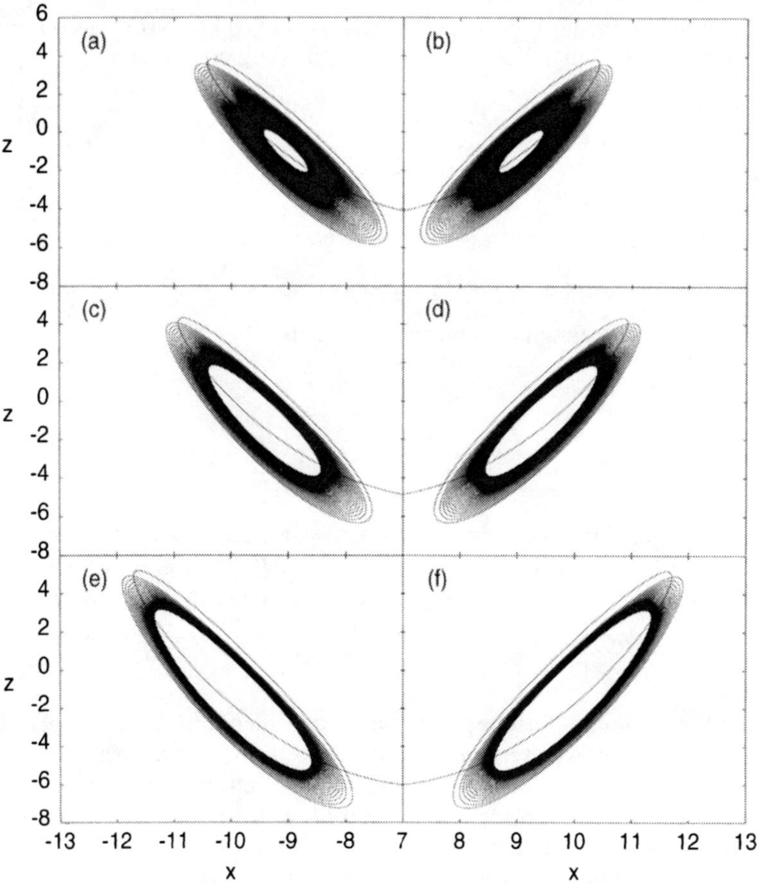

Figure 32. Stable limit cycles around $C_{\pm}$ of the controlled Lorenz system (113) with the control law for the initial conditions $(x_0, y_0, z_0) = (\pm 3.0, \pm 12.0, -2.5)$ when (a) and (b) $r = 83$; (c) and (d) $r = 90$ and (e) and (f) $r = 101$.

3.3.1. Tracking

If there exists a stable or an unstable periodic solution in chaotic attractor, one wishes to employ a control such that any trajectory of the system ultimately converges to the periodic solution. Since the Lorenz system (88) has a globally attractive and positive invariant set, the function on the right-hand side of equation (88) must satisfy the Lipschitz condition on the globally attractive set. This property greatly facilitates the design of feedback control laws in controlling chaos. In this subsection, we consider globally stabilizing a given equilibrium point or globally tracking a given periodic orbit. Assume that $X^*(t) = (x^*(t), y^*(t), z^*(t))$ is a periodic solution or an equilibrium point of system (88). Let $\overline{X}(t) = X(t) - X^*(t)$, then it is easy to find that $\overline{X}(t)$ satisfies the following error system:

$$\dot{\bar{x}} = a(\bar{y} - \bar{x}),$$
$$\dot{\bar{y}} = c\bar{x} - xz + x^*z^* - \bar{y},$$
$$\dot{\bar{z}} = xy - x^*y^* - b\bar{z}. \tag{115}$$

Now, add the negative linear feedback controls

$$u_1 = k_1\bar{x}, \qquad u_2 = k_2\bar{y}, \qquad u_3 = k_3\bar{z} \tag{116}$$

to equation (115) to obtain the controlled system

$$\dot{\bar{x}} = a(\bar{y} - \bar{x}) - u_1 = -(a + k_1)\bar{x} + a\bar{y},$$
$$\dot{\bar{y}} = c\bar{x} - xz + x^*z^* - \bar{y} - u_2 = c\bar{x} - x\bar{z} - z^*\bar{x} - (1 + k_2)\bar{y},$$
$$\dot{\bar{z}} = xy - x^*y^* - b\bar{z} - u_3 = x\bar{y} + y^*\bar{x} - (b + k_3)\bar{z}. \tag{117}$$

THEOREM 3.6. *There always exist $k_i \geqslant 0, i = 1, 2, 3$, such that the zero solution of system* (117) *is globally, exponentially stable. If $X^*(t)$ is a periodic solution, then all trajectories of system* (88) *globally, exponentially converge to $X^*(t)$ under the control; if X^* is an equilibrium point, then X^* is globally, exponentially stabilized.*

PROOF. Since an equilibrium point can be considered as a special periodic solution with an arbitrary constant period, we will not distinguish the two cases. From Theorems 3.4, we know that $X(t)$ and $X^*(t)$ ultimately must enter into D_2. Thus, we may assume that for $c > 1$, $2a > b > 0$, $|x(t)| \leqslant R$, $|y(t)| \leqslant R$ and $0 < z(t) \leqslant R + c$, where R is given in equation (93). Let $f(x, z) = xz$ and $f(x, y) = xy$. By the intermediate value theory, we have $f(x, z) - f(x^*, z^*) = \tilde{z}(x - x^*) + \tilde{x}(z - z^*) = \tilde{z}\bar{x} + \tilde{x}\bar{z}$, where $\tilde{z}$ is between z and $\bar{z}$ and $\tilde{x}$ is between x and $\bar{x}$. Hence, $|f(x, z) - f(x^*, z^*)| \leqslant |\tilde{z}||\bar{x}| + |\tilde{x}||\bar{z}| \leqslant (R + c)|\bar{x}| + R|\bar{z}|$. Similarly, we have $|f(x, y) - f(x^*, y^*)| \leqslant |\tilde{y}||\bar{x}| + |\tilde{x}||\bar{y}| \leqslant R|\bar{x}| + R|\bar{y}|$. We may choose $k_i \geqslant 0, i = 1, 2, 3$, such that the matrix

$$W = \begin{bmatrix} a + k_1 & -a & 0 \\ -(R + c) & 1 + k_2 & -R \\ -R & -R & b + k_3 \end{bmatrix}$$

is an M matrix (Liao, 1993). Therefore, $\forall \eta_i > 0, i = 1, 2, 3$, let $\eta = (\eta_1, \eta_2, \eta_3)^{\mathrm{T}}$ and $\xi = (\xi_1, \xi_2, \xi_3)^{\mathrm{T}}$, the linear equation $W^{\mathrm{T}}\xi = \eta$ has a positive solution $\xi = (W^{\mathrm{T}})^{-1}\eta$, i.e., $\xi_i > 0, i = 1, 2, 3$.

Now for equation (117), construct the positive definite and radially unbounded Lyapunov function

$$V = \xi_1|x| + \xi_2|y| + \xi_3|z|,$$

and then computing the Dini derivative of V with respect to time t along the trajectory of system (117) yields

$$\begin{aligned}
D^+V &= \xi_1\dot{\bar{x}}\,\mathrm{sign}\,\bar{x} + \xi_2\dot{\bar{y}}\,\mathrm{sign}\,\bar{y} + \xi_3\dot{\bar{z}}\,\mathrm{sign}\,\bar{z} \\
&\leqslant -\xi_1(a+k_1)|\bar{x}| + \xi_1 a|\bar{y}| - \xi_2(1+k_2)|\bar{y}| + \xi_2(c+R)|\bar{x}| \\
&\quad + \xi_2 R|\bar{z}| + \xi_3 R|\bar{x}| + \xi_3 R|\bar{y}| - \xi_3(b+k_3)|\bar{z}| \\
&= \big[-(a+k_1)\xi_1 + (c+R)\xi_2 + R\xi_3\big]|\bar{x}| \\
&\quad + \big[a\xi_1 - (1+k_2)\xi_2 + R\xi_3\big]|\bar{y}| + \big[R\xi_2 - (b+k_3)\xi_3\big]|\bar{z}| \\
&= -\eta_1|\bar{x}| - \eta_2|\bar{y}| - \eta_3|\bar{z}| \\
&\leqslant -\min_{1\leqslant i\leqslant 3}\eta_i\big(|\bar{x}| + |\bar{y}| + |\bar{z}|\big) \\
&\leqslant \frac{-\min(\eta_i)}{\max(\xi_i)}\big(\xi_1|\bar{x}| + \xi_2|\bar{y}| + \xi_3|\bar{z}|\big).
\end{aligned} \tag{118}$$

Equation (118) yields the estimation

$$\min(\xi_i)\big(|\bar{x}(t)| + |\bar{y}(t)| + |\bar{z}(t)|\big) \leqslant V(t) \leqslant V(t_0)\mathrm{e}^{-\min(\eta_i)/\max(\xi_i)(t-t_0)},$$

from which we finally obtain

$$|\bar{x}(t)| + |\bar{y}(t)| + |\bar{z}(t)| \leqslant \frac{V(t_0)}{\min(\xi_i)}\mathrm{e}^{-\min(\eta_i)/\max(\xi_i)(t-t_0)}. \tag{119}$$

□

To illustrate the above analytical results, we present two examples: one for tracking a periodic solution and the other for stabilizing an equilibrium point. We fix $a = 10$, $b = 8/3 > 2$, so $R = 4c/\sqrt{15}$. First, we consider tracking periodic orbits. It is known that the zero equilibrium point E_1 (the origin) of the Lorenz system is globally stable when $0 < c < 1$. It becomes unstable when c is crossing $c = 1$, and bifurcates into two nonzero equilibrium points, E_2 and E_3, which are stable for $1 < c < 24.74$ and become unstable at $c = 24.74$, giving rise to Hopf bifurcation. The Lorenz system exhibits chaotic attractors when $c \in (24.74, 148.4)$ and periodic solutions for $c \in (148.8, 166.07)$. When $c = 149.7$, the system exhibits a stable limit cycle $\widetilde{X}(t)$, its phase portrait is shown in Figure 33(a) and the time history of the component $\tilde{x}(t)$ is depicted in Figure 33(c). At $c = 130.0$, the motion is chaotic, see Figure 33(b). The initial conditions for the simulation are taken as $\tilde{x}(0) = \tilde{y}(0) = \tilde{z}(0) = 10$ for the periodic motion, but $x(0) = -20$, $y(0) = -10$, $z(0) = 30$ for the chaotic motion. The feedback control law (116) is taken for this example, where the control parameters must satisfy $k_2 > 397.53$ and $k_3 > 265.86$. Together with $k_1 > 0$, we choose

$$k_1 = 1, \qquad k_2 = 400, \qquad k_3 = 270. \tag{120}$$

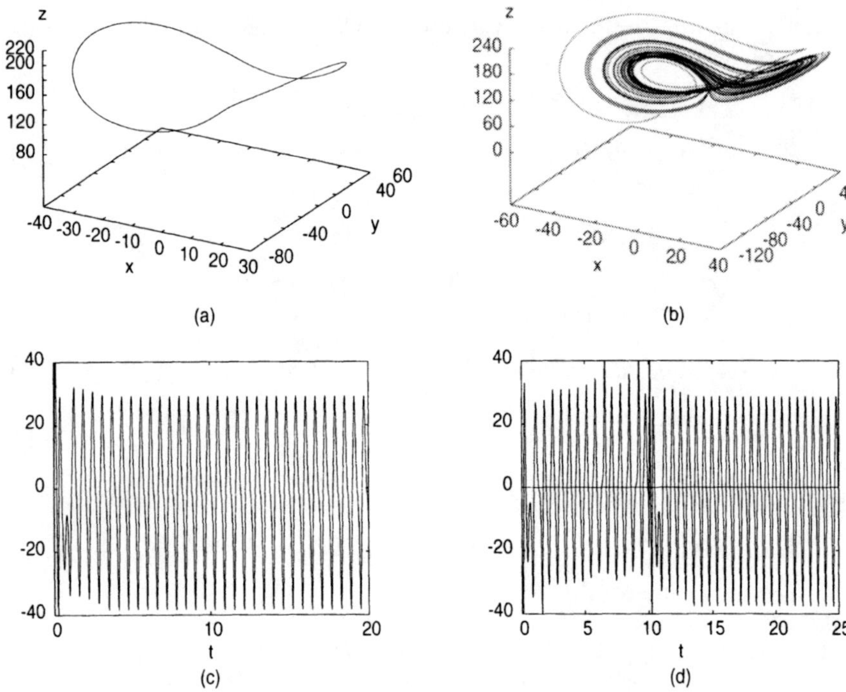

Figure 33. Tracking periodic orbits when $a = 10$, $b = 8/3$: (a) the 3-D phase portrait of a periodic orbit when $c = 149.7$; (b) the 3-D phase portrait of a chaos when $c = 130$; (c) the time history of the periodic solution $\tilde{x}(t)$; (d) the time history of $x(t)$ before and after applying the control law (116) with $k_1 = 1$, $k_2 = 400$, $k_3 = 270$. The initial conditions are $\tilde{x}(0) = \tilde{y}(0) = \tilde{z}(0) = 10$ for (a) and (c), but $x(0) = -20$, $y(0) = -10$, $z(0) = 10$ for (b) and (d).

The time history of $x(t)$ before and after applying the feedback control (116) is shown in Figure 33(d), where the motion shown for $t \in [0, 10]$ is chaotic, while the remaining part (for $t \in [10, 25]$) shows a short transient period when the control is added, and then the chaotic motion becomes purely periodic.

Next, we consider the global stabilization of equilibrium points for two cases, $c = 28$ and $c = 130$, both of them show chaos without controls. We apply the control law (116) to force the system to converge to the equilibrium point E_2, with the control parameter values: $k_1 = 1$, $k_2 = 90$, $k_3 = 58$ for $c = 28$, and that given in equation (120) for $c = 130$. Under the above choices of parameter values with the designed equilibrium point E_2 (which is unstable without control), the matrix W is an M matrix. According to Theorem 3.6, the controlled trajectories converge to the designed equilibrium point exponentially. The simulating trajectories are shown in Figures 34.

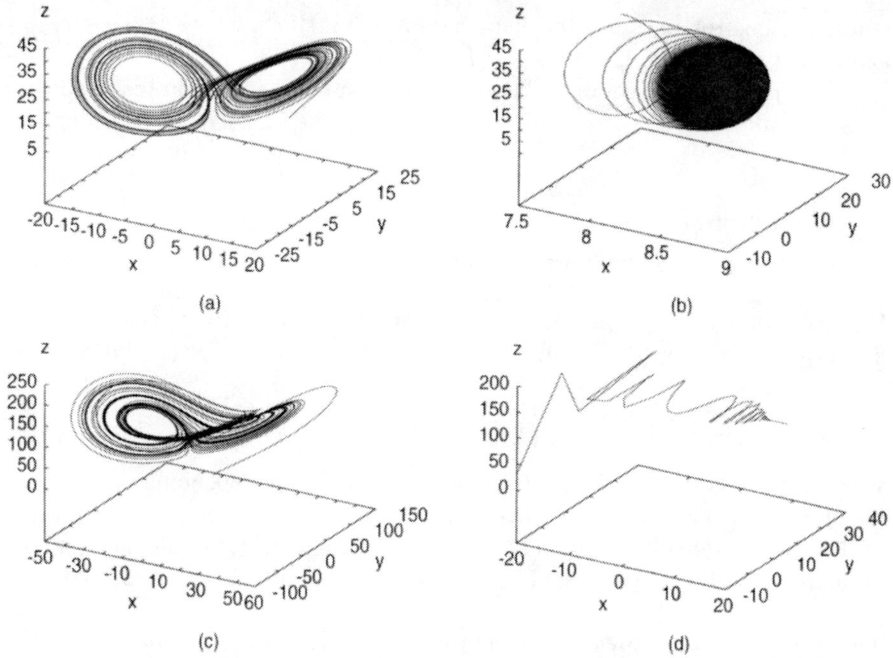

Figure 34. Stabilizing the chaotic Lorenz system: (a) a chaos for $a = 10$, $b = 8/3$, $c = 28$; (b) globally stabilizing the chaos shown in part (a) to the equilibrium point E_2 using control law (116) with $k_1 = 1$, $k_2 = 90$, $k_3 = 58$; (c) a chaos for $a = 10$, $b = 8/3$, $c = 130$; (d) globally stabilizing the chaos shown in part (c) to the equilibrium point E_2 using control law (116) with $k_1 = 1$, $k_2 = 400$, $k_3 = 270$.

3.3.2. Synchronization of two Lorenz systems

In this subsection, we study globally exponential synchronization between two Lorenz systems. Consider the transmitter system, given by

$$\begin{aligned}\dot{x}_d &= a(y_d - x_d),\\ \dot{y}_d &= cx_d - x_d z_d - y_d,\\ \dot{z}_d &= x_d y_d - bz_d,\end{aligned} \tag{121}$$

and the receiver (response) system with feedback controls:

$$\begin{aligned}\dot{x}_r &= a(y_r - x_r) - u_1(x_r - x_d, y_r - y_d, z_r - z_d),\\ \dot{y}_r &= cx_r - x_r z_r - y_r - u_2(x_r - x_d, y_r - y_d, z_r - z_d),\\ \dot{z}_r &= x_r y_r - bz_r - u_3(x_r - x_d, y_r - y_d, z_r - z_d),\end{aligned} \tag{122}$$

where the controls u_i's are linear or nonlinear functions of x_d, x_r, y_d, y_r, z_d and z_r, satisfying $u_i(0,0,0) = 0$, $i = 1, 2, 3$. Let $e_x = x_r - x_d$, $e_y = y_r - y_d$, $e_z = z_r - z_d$, subtracting equation (121) from equation (122) results in the following error system:

$$\begin{aligned}
\dot{e}_x &= a(e_y - e_x) - u_1(e_z, e_y, e_z),\\
\dot{e}_y &= ce_x - x_r z_r + x_d z_d - e_y - u_2(e_z, e_y, e_z),\\
\dot{e}_z &= x_r y_r - x_d y_d - be_z - u_3(e_z, e_y, e_z).
\end{aligned} \tag{123}$$

DEFINITION 3.7. If for the transmitter system, $\forall (x_d(0), y_d(0), z_d(0)) \in \mathbb{R}^3$, and the corresponding $(x_r(0), y_r(0), z_r(0)) \in \mathbb{R}^3$ for the receiver system, the solution of the error system (123) has the following estimation:

$$e_x^2(t) + e_y^2(t) + e_z^2(t) \leqslant K\big(E(t_0)\big)\mathrm{e}^{-\alpha(t-t_0)}, \quad \alpha < 0, \tag{124}$$

where $K(E(t_0)) = K(e_x(t_0), e_y(t_0), e_z(t_0))$ is a constant depending upon $E(t_0)$, while $\alpha > 0$ is positive real number, independent of $E(t_0)$, then the zero solution of the error system (123) is said to be globally, exponentially stable, and the two systems (121) and (122) are said to be globally, exponentially synchronized.

THEOREM 3.8. *For system* (123), *there always exist the following linear feedback controls*:

$$u_1 = k_1 e_x, \qquad u_2 = k_2 e_y, \qquad u_3 = k_3 e_z, \tag{125}$$

under which one can choose appropriate large $k_i > 0$, $i = 1, 2, 3$, *such that the zero solution of system* (123) *is globally, exponentially stable, and thus the receiver system* (122) *is globally, exponentially synchronized with the transmitter system* (121).

PROOF. We can always choose enough large $k_i > 0$, $i = 1, 2, 3$, such that the matrix

$$Q = \begin{bmatrix} -2(a+k_1) & a+R & R \\ a+R & -2(1+k_2) & 0 \\ R & 0 & -2(b+k_3) \end{bmatrix}$$

is negative definite. Construct the Lyapunov function in the positive definite, quadratic form $V = e_x^2 + e_y^2 + e_z^2$. Then, computing $\frac{\mathrm{d}V}{\mathrm{d}t}$ along the trajectory of system (123) yields

$$\begin{aligned}
\frac{\mathrm{d}V}{\mathrm{d}t} &= 2e_x\dot{e}_x + 2e_y\dot{e}_y + 2e_z\dot{e}_z\\
&= \begin{pmatrix} e_x \\ e_y \\ e_z \end{pmatrix}^{\mathrm{T}} \begin{bmatrix} -2(a+k_1) & a+c-z_d & y_d \\ a+c-z_d & -2(1+k_2) & 0 \\ y_d & 0 & -2(b+k_3) \end{bmatrix} \begin{pmatrix} e_x \\ e_y \\ e_z \end{pmatrix}
\end{aligned}$$

$$
\leqslant \begin{pmatrix} |e_x| \\ |e_y| \\ |e_z| \end{pmatrix}^{\mathrm{T}} \begin{bmatrix} -2(a+k_1) & a+R & R \\ a+R & -2(1+k_2) & 0 \\ R & 0 & -2(b+k_3) \end{bmatrix} \begin{pmatrix} |e_x| \\ |e_y| \\ |e_z| \end{pmatrix}
$$

$$
\leqslant \lambda_{\max}(Q)\left[e_x^2(t) + e_y^2(t) + e_z^2(t)\right],
$$

from which we obtain

$$
e_x^2(t) + e_y^2(t) + e_z^2(t) \leqslant \left[e_x^2(t_0) + e_y^2(t_0) + e_z^2(t_0)\right] \mathrm{e}^{\lambda_{\max}(Q)(t-t_0)}, \tag{126}
$$

where $\lambda_{\max}(Q)$ denotes the largest eigenvalue of matrix Q. Equation (126) indicates that the conclusion of Theorem 3.8 is true. □

Finally, we present two numerical examples to illustrate the synchronization. We again choose the two typical sets of parameter values used for the simulation of stabilization in the previous subsection: $a = 10$, $b = 8/3$, $c = 28$ or $c = 130$. With the control law given in equation (125), we choose the following control parameter values: $k_1 = 25$, $k_2 = 19$, $k_3 = 13$ for the first set of parameter values, and $k_1 = 130$, $k_2 = 72$, $k_3 = 65$ for the second set of parameter values. Under these two sets of parameter values, the matrix Q is negative definite. The simulation results show that the two Lorenz systems (one is transmitter and the other receiver) are perfectly synchronized. The time history of the error signal $e_x(t)$ is depicted in Figure 35(a) for the first case ($c = 28$) and in Figure 35(b) for the second case ($c = 130$). They indeed indicate that the trajectories converge to zero exponentially.

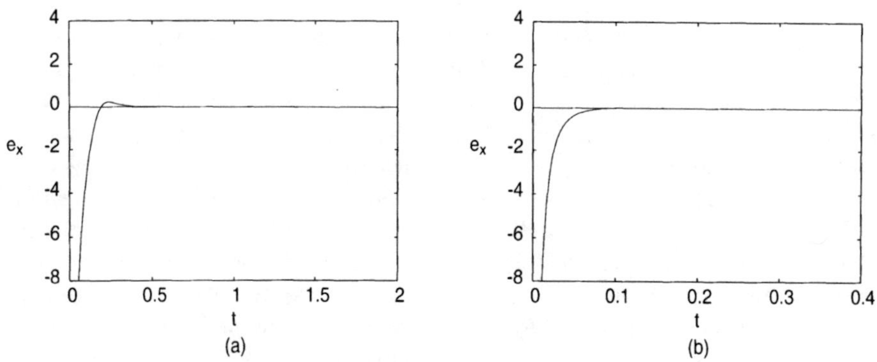

Figure 35. Synchronizing two chaotic Lorenz systems: (a) error signal e_x for the synchronization when $a = 10$, $b = 8/3$, $c = 28$ using control law (125), with $k_1 = 25$, $k_2 = 19$, $k_3 = 13$; and (b) error signal e_x for the synchronization when $a = 10$, $b = 8/3$, $c = 130$ using control law (125), with $k_1 = 130$, $k_2 = 72$, $k_3 = 65$.

4. Competitive modes

Recently, we have developed a new concept for *mode*, and use mode competition to study chaos (Yao, 2002; Yao, Essex and Yu, 2002; Yao, Yu and Essex, 2002; Yao, Yu, Essex and Davison, 2006). It is conjectured that a chaotic system has at least two competitive modes (CMs). For a general nonlinear dynamical system, we give a simple, dynamically motivated definition of mode suitable for this concept. Since for most chaotic systems it is difficult to obtain the form of a CM, we focus on the competition between the corresponding modulated frequency components of the CMs. Some direct applications result from the explicit form of the frequency functions. One application is to estimate parameter regimes which may lead to chaos. It is shown that chaos may be found by analyzing the frequency function of the CMs without applying a numerical integration scheme. Another application is to create new chaotic systems using *custom-designed* CMs.

4.1. Definition of CM

The word "mode" comes from *modus*, which is a Latin word for "measure" or "size". In physics, a mode often denotes an oscillation of single frequency. A related usage is the term, normal mode, which refers to an oscillation in which all particles move with the same frequency and phase. Recent studies consider a mode as the solution of a *model system*. The model system captures the main dynamical behavior of the original system. Thus, the solution of the original system could be well approximated by the solution of the model system. For a complex dynamical system, there can be more than one model system, and the solution of the original system may be obtained by combining the solutions of the model systems.

Every field of science and engineering seems to have its own definition of mode. For example, when a system is described by a partial differential equation (PDE), in order to simplify the computation and analysis, one often reduces the PDE into some ordinary differential equations (ODEs). Modes are often considered as the key to constructing a Fourier series decomposition. To reflect the complexity of the PDE, more than one mode should be included in the decomposition.

However, the above mode definitions either cannot be directly used to study chaos and complexity, or have not been used to study the dynamical behavior from the mode competition point of view. In the sense of Haken (1983), Nicolis and Prigogine (1977), modes may be considered to be competing with each other in some way, for example, to occupy the maximal resource of a system. A fundamental problem in the study of modes is how to mathematically define such a competitive mode (CM). First of all, certainly we need to know what form a CM should have. For a simple sinusoidal mode, all components of the mode, including the frequency, phase and amplitude, are constants. The result of competition between such modes is straightforward, and therefore such a mode cannot be expected to imply complex dynamical phenomena such as chaos, since it is expected that the results of competition in a chaotic system are complicated. In fact, it has been observed that the power spectrum of a chaotic system is broad band, which strongly suggests that frequencies associated with CM might be expected to be functions of evolution variables.

Thus, we assume that the modes determine the dynamical behavior of a system. If one mode dominates all others, the system's behavior is controlled by the dominant mode, which seizes the most efficient way of occupying the system's resources. If, on the other hand, no mode dominates the others for all time, the competition amongst the modes results in complicated behavior. It should be noted that the classical definition of "mode shape" used in engineering emphasizes the shape/solution of the motion, while in this section by "mode competition" we emphasize the characteristics of the motion.

Let us start by considering the linear system

$$\ddot{x} + \omega^2 x = A \sin \Omega t, \tag{127}$$

where $A \sin \Omega t$ is the external force ($A \neq 0$) and ω is a parameter not equal to Ω. In system (127), there are two modes $m_1(t)$ and $m_2(t)$ which satisfy

$$\begin{aligned} &\ddot{m}_1 + \omega^2 m_1 = 0, \\ &\ddot{m}_2 + \Omega^2 m_2 = 0. \end{aligned} \tag{128}$$

The solution of the system can be expressed as $x = c_1 m_1 + c_2 m_2$, where c_1 and c_2 are constants determined by the initial condition of system (127). Thus, the mode m_1 corresponds to the frequency ω and m_2 to the frequency Ω. We may call m_1 the *internal mode* and m_2 the *external mode*. Clearly, the system's dynamical behavior (periodic or divergent) is determined by these two modes.

Now we rewrite system (127), by introducing $y = \sin \Omega t$, as

$$\begin{aligned} &\ddot{x} + \omega^2 x = Ay, \\ &\ddot{y} + \Omega^2 y = 0, \end{aligned} \tag{129}$$

which can be further put in the vector form:

$$\ddot{\mathbf{x}} + \mathbf{g}\mathbf{x} = \mathbf{h},$$

where in general $\mathbf{g}$ and $\mathbf{h}$ are functions of t. (For system (129) $\mathbf{g}$ is a constant vector.) As we have mentioned, ω corresponds to the internal mode and Ω to the external mode. For system (129), we restate that ω is associated with the mode in the x direction and Ω with the mode in the y direction.

Now, we define our competitive mode. Consider the general nonlinear autonomous system, given by

$$\dot{x}_i = f_i(x_1, x_2, \ldots, x_n), \quad i = 1, 2, \ldots, n, \tag{130}$$

where $f_i \in C^1(\mathbf{R})$ and for any $j \in [1, n]$, $|\frac{\partial f_i}{\partial x_j}|$ is bounded.

To study the general system (130) using modes, one first constructs some modes $m_k(t)$, $k \leqslant n$, based on some properties of the system such as the invariant manifold (Shaw and Pierre, 1993). In general, these modes satisfy the following model equations:

$$\ddot{\mathbf{M}} + \mathbf{A}\dot{\mathbf{M}} + \mathbf{B}\mathbf{M} = \mathbf{0}, \tag{131}$$

where $\mathbf{M} = [m_1, m_2, \ldots, m_k]^{\mathrm{T}}$, both $\mathbf{A}$ and $\mathbf{B}$ are $k \times k$ matrices which can be functions of $\mathbf{M}$ and t.

The solution of system (131) may be expressed as a linear or nonlinear combination of these modes. In the linear case, one has

$$\mathbf{x} = \mathbf{C}\mathbf{M},$$

where $\mathbf{x} = [x_1, x_2, \ldots, x_n]^{\mathrm{T}}$ and $\mathbf{C}$ is an $n \times k$ coefficient matrix, which can be determined using a method such as the Galerkin approach (Pesheck, Pierre and Shaw, 2002).

The main difficulties of the above mode procedure are to find the appropriate manifold, namely, to construct the mode equation (131) based on the manifold, and to obtain the modes by solving the equation, especially when the system is chaotic. To overcome the difficulties, we construct our mode equation below.

Differentiating equation (130) with respect to t yields

$$\begin{aligned} \ddot{x}_i &= \sum_{j=1}^{n} f_j \frac{\partial f_i}{\partial x_j} \\ &= -x_i g_i(x_1, x_2, \ldots, x_i, \ldots, x_n) \\ &\quad + h_i(x_1, x_2, \ldots, x_{i-1}, x_{i+1}, \ldots, x_n), \end{aligned} \tag{132}$$

where g_i and h_i are functions of $\mathbf{x}$ and g_i is bounded. Comparing equation (132) to system (129) shows that h_i should not contain the variable x_i.

It can be seen from equation (132) that the dynamical behavior of x_i is determined by the functions g_i and h_i, as well as the system's initial condition. Furthermore, the dynamical behaviors of systems (130) and (132) are related. If the motion of system (132) is periodic, i.e., $f_j(t + T) = f_j(t)$, $j \in [1, n]$, for some $T > 0$, then system (130) must be periodic too; suppose system (132) is chaotic, implying that f_j, $j \in [1, n]$, must be chaotic, and thus system (130) is chaotic too. Hence, system (132) can be taken as a model system of system (130).

DEFINITION 4.1. For system (130), the competitive modes are defined to be the solution, x_i, $i \in [1, n]$, of the model system (132).

The above mode definition is simple but may not be perfect. For some specific systems, one can have a better mode definition. Even with such a simple definition, we still, in general, cannot solve the mode equations except when the g_i's and h_i's take very simple forms. However, if we focus on the mode competition via their functional frequency components g_i's, namely, g_i's are not constants but variables (i.e., explicitly or implicitly functions of time), we shall show that the dynamical behavior of the system is relative to the competition. In this sense with a comparison to system (128), we have the following definition.

DEFINITION 4.2. For system (132), if $g_i > 0$ at some t, we call g_i the frequency component of the competitive mode x_i. If $g_i \leqslant 0$ at any time, we shall simply state that g_i does not exist.

Now, we conjecture necessary conditions for chaos based on the competition between CMs via their frequency components, g's.

CONJECTURE 4.3. *The conditions for a dynamical system to be chaotic are given below*:

(1) *there exist at least two* g*'s in the system*;
(2) *at least two* g*'s are competitive or nearly competitive, that is, there are* $g_i \simeq g_j > 0$ *at some* t;
(3) *at least one of* g*'s is the function of evolution variables such as* t; *and*
(4) *at least one of* h*'s is the function of the system variables.*

In the following, we use the well-known Lorenz system and Rössler system to show by example how g's determine the dynamical behavior of nonlinear systems.

For the Lorenz system (88), we fix $a = 10$, then the system has three fixed points, given by $(x_0, y_0, z_0) = (0, 0, 0)$ and $(\pm\sqrt{b(c-1)}, \pm\sqrt{b(c-1)}, c-1)$. Differentiating the first equation of (88) with respect to t and then substituting the other two equations of (88) into the resulting equation yields

$$\ddot{x} = -10\dot{x} + 10\dot{y} = (100 + 10\alpha - 10z)x - 110y,$$

which indicates that

$$g_x = -100 - 10\alpha + 10z. \tag{133}$$

Similarly, one can obtain g_y and g_z as

$$g_y = -1 - 10\alpha + 10z + x^2, \tag{134}$$

$$g_z = -\beta^2 + x^2. \tag{135}$$

The competition may happen in three pairs of g's, namely, (g_x, g_y), (g_x, g_z) and (g_y, g_z). Obviously, $g_x < g_y$ for any t. Thus, we see that the frequency component of the CM in the x direction is "dominated" by that in the y direction, and we need only consider the competition in the other two pairs. It should be

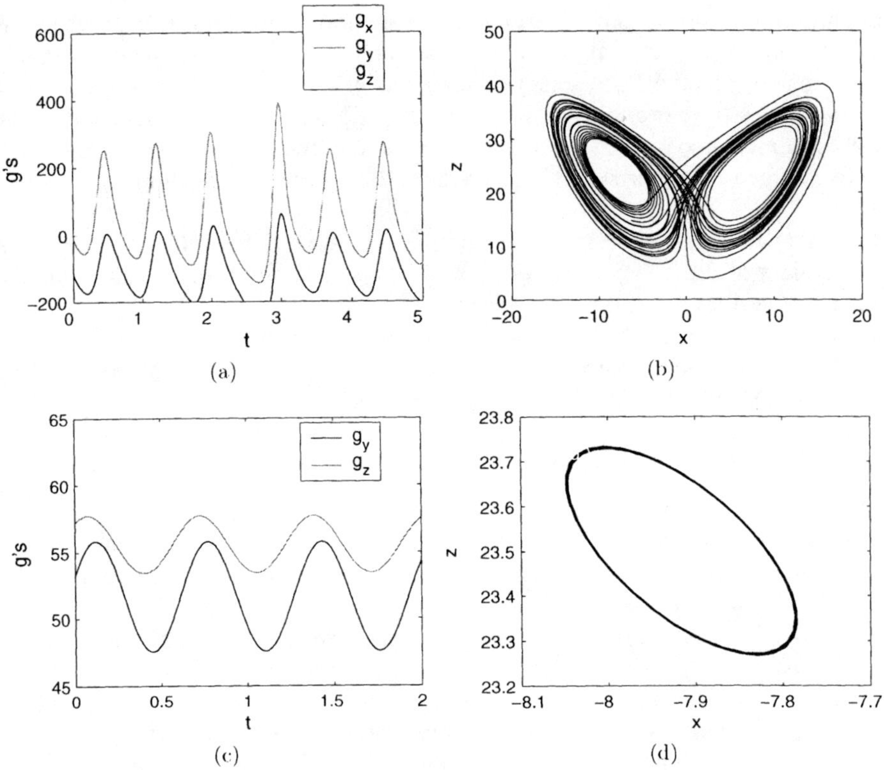

Figure 36. The CM's components, g_y and g_z, and phase portraits of the Lorenz system (88): (a) and (b) chaos when $c = 25$, $b = 2.6667$; (c) and (d) periodic motion when $c = 24.5$, $b = 2.6667$.

pointed out that hereafter by "dominated" or "dominating" we mean the two corresponding modes are neither interactive (i.e., separated) nor very close to each other since, in general, a mode with larger g need not always "dominate" the one with smaller g.

Figures 36(a) and 36(c) show the functions g_x, g_y and g_z when the Lorenz system exhibits chaotic and period-1 motions, respectively (g_x (< 0) is not depicted in Figure 36(c)). The values of g_i displayed in Figure 36 are obtained by first numerically integrating the Lorenz system under the initial condition *taken near the fixed point*, for the parameter values $\beta = 2.6667$ and $\alpha = 25$ and 24.5, respectively, to get $x(t)$, $y(t)$ and $z(t)$, and then using $x(t)$, $y(t)$ and $z(t)$ to calculate g_x, g_y and g_z given in equations (133)–(135). It is seen that (i) $g_x < g_z$ in both cases, which indicates that there is no competition between the modes in the x and z directions; (ii) $g_z > 0$ in both cases, while $g_y > 0$ for some times in Figure 36(a) and all time in Figure 36(c), which means that CMs exist in the y and z directions;

(iii) in the chaotic case (see Figure 36(a)), g_y and g_z are predominant over each other alternatively, while in the periodic case (see Figure 36(c)), g_z dominates g_y so that only one of g's is in effect. Therefore, we may conclude that *the competition between g_y and g_z results in chaotic motion* (see Figure 36(b)), *while the lack of competition leads to the simple dynamical behavior* (see Figure 36(d)).

The Rössler system (Rössler, 1976) is given by

$$\begin{aligned}
\dot{x} &= -y - z,\\
\dot{y} &= x + 0.2y,\\
\dot{z} &= \beta + z(x - \alpha),
\end{aligned} \tag{136}$$

which has two fixed points $(x_0, y_0, z_0) = (0.5\gamma, -2.5\gamma, 2.5\gamma)$, $\gamma = \alpha \pm \sqrt{\alpha^2 - 0.8\beta}$, where α and β are parameters.

From system (136), one obtains

$$\begin{aligned}
g_x &= 1 + z,\\
g_y &= 0.96,\\
g_z &= -(x - \alpha)^2 + y + z.
\end{aligned} \tag{137}$$

The functions g_x, g_y and g_z vs t are depicted in Figures 37(a) and 37(c) when system (136) is in a chaotic state and a period-1 state, respectively. It is seen from Figure 37(a) that g_y is dominated by g_x but competition exists in the both pairs, (g_x, g_z) and (g_y, g_z), and the motion is chaotic (see the phase portrait given in Figure 37(b)); while in Figure 37(c) that g's are separated at any time for the periodic motion, as shown in Figure 37(d), indicating no mode competition.

The Lorenz and Rössler systems are good examples for demonstrating how g's interact in chaotic or periodic motions. Applications of the CM defined above to more complicated chaotic systems are given in the next subsections.

4.2. Application of CM: estimating chaotic parameter regimes

4.2.1. Example 1: A PDE model

Modes are often considered as the key to constructing a Fourier series decomposition. To reflect the complexity of the PDE, more than one mode should be included in the decomposition. As an example, consider the following PDE describing a thin plate in a flow (Yu, Zhang and Bi, 2001):

$$\begin{aligned}
&D\nabla^4 w + \rho h\frac{\partial^2 w}{\partial t^2} - \frac{\partial^2 w}{\partial x^2}\frac{\partial^2 \phi}{\partial y^2} - \frac{\partial^2 w}{\partial y^2}\frac{\partial^2 \phi}{\partial x^2} + 2\frac{\partial^2 w}{\partial x\,\partial y}\frac{\partial^2 \phi}{\partial x\,\partial y} - \mu\frac{\partial w}{\partial t}\\
&\quad = F(x, y)\cos \Omega_1 t
\end{aligned} \tag{138}$$

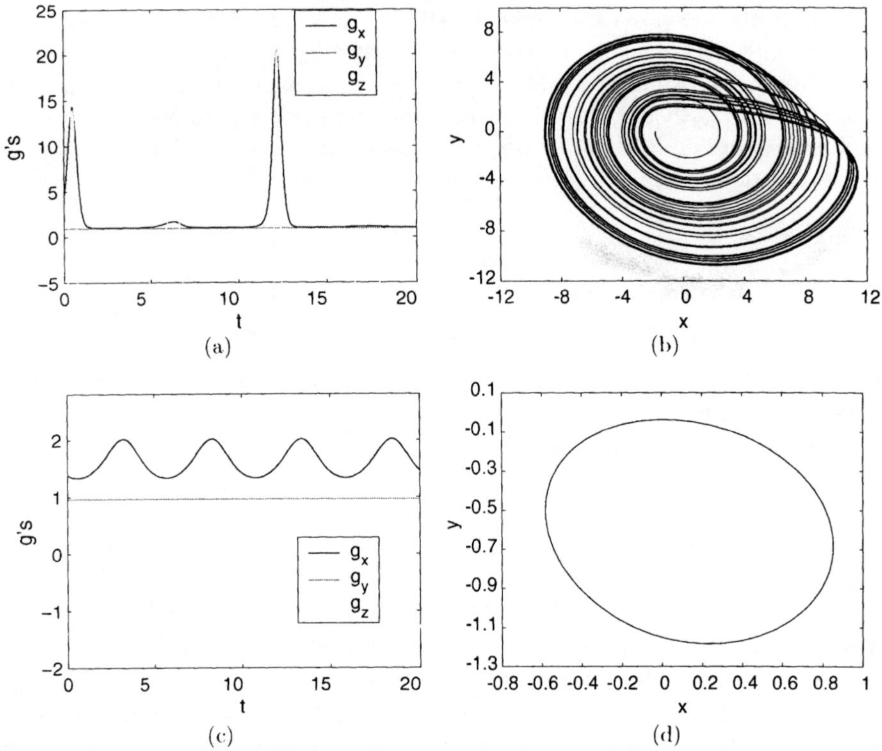

Figure 37. The CM's components, g_x, g_y and g_z of the Rössler system (136) when $\beta = 0.2$ for (a) and (b) chaos when $\alpha = 0.5$; (c) and (d) periodic motion when $\alpha = 5.7$.

and

$$\nabla^4 \phi = Eh\left[\left(\frac{\partial^2 w}{\partial x\,\partial y}\right)^2 - \frac{\partial^2 w}{\partial x^2}\frac{\partial^2 w}{\partial y^2}\right], \tag{139}$$

where w is the displacement of a point in the center plane of the thin plate in the z direction, ϕ is the stress function, ρ, h, μ, E are parameters and $F(x, y)\cos \Omega_1 t$ is the transverse force of the flow on the plate.

By considering the boundary conditions and the first *two modes*,

$$m_1(x, y) = \sin\frac{\pi x}{a}\sin\frac{3\pi y}{b},$$
$$m_2(x, y) = \sin\frac{3\pi x}{a}\sin\frac{\pi y}{b}, \tag{140}$$

one may construct an approximate solution of w in the form of

$$w(x, y, t) = u_1(t)m_1(x, y) + u_2(t)m_2(x, y), \tag{141}$$

where $u_i(t)$, $i = 1, 2$, are, respectively, the amplitudes of the two modes (Yu, Zhang and Bi, 2001).

Correspondingly, the transverse force can also be expressed approximately by

$$F(x, y) = F_1 m_1(x, y) + F_2 m_2(x, y), \tag{142}$$

where F_1 and F_2 are parameters. By substituting equations (141) and (142) into equations (138) and (139), and then using a technique like the Galerkin method (Pesheck, Pierre and Shaw, 2002) to determine the expression of $u_1(t)$ and $u_2(t)$, one finally obtains the following dimensionless equations:

$$\ddot{x}_1 + \varepsilon\mu\dot{x}_1 + (\omega_1^2 + 2\varepsilon F_1 \cos\Omega_2 t)x_1 + \varepsilon(\alpha_1 x_1^3 + \alpha_2 x_1 x_2^2) \\ = \varepsilon F_1 \cos\Omega_1 t, \tag{143}$$

$$\ddot{x}_2 + \varepsilon\mu\dot{x}_2 + (\omega_2^2 + 2\varepsilon F_2 \cos\Omega_2 t)x_2 + \varepsilon(\beta_1 x_2^3 + \beta_2 x_2 x_1^2) \\ = \varepsilon F_2 \cos\Omega_1 t, \tag{144}$$

where $x_i = (\sqrt{ab}/h^2)u_i$, $i = 1, 2$, ω_1 and ω_2 are the two linear natural frequencies of the thin plate, while Ω_1 and Ω_2 are the frequencies of the external forces. $\mu > 0$ is the damping coefficient and ε is a small perturbation parameter. For a more detailed description of the equations and parameters, see Yu, Zhang and Bi (2001).

Now, instead of the original system described by equations (138) and (139), one wants to use equations (143) and (144) to determine the parameter regimes where chaos appears, which agrees with the result obtained from the original system. By introducing

$$\begin{aligned} y_1 &= x_1, & y_2 &= \dot{x}_1, \\ y_3 &= x_2, & y_4 &= \dot{x}_2, \\ y_5 &= \cos(\Omega_1 t), & y_6 &= \cos(\Omega_2 t), \end{aligned}$$

we can rewrite equations (143) and (144) as

$$\begin{aligned} \dot{y}_1 &= y_2, \\ \dot{y}_2 &= \varepsilon F_1 y_5 - \varepsilon\mu y_2 - (\omega_1^2 + 2\varepsilon f_1 y_6)y_1 - \varepsilon(\alpha_1 y_1^3 + \alpha_2 y_1 y_3^2), \\ \dot{y}_3 &= y_4, \\ \dot{y}_4 &= \varepsilon F_2 y_5 - \varepsilon\mu y_4 - (\omega_2^2 + 2\varepsilon f_2 y_6)y_3 - \varepsilon(\beta_1 y_3^3 + \beta_2 y_3 y_1^2), \\ \ddot{y}_5 &= -\Omega_1^2 y_5, \\ \ddot{y}_6 &= -\Omega_2^2 y_6. \end{aligned}$$

The g-functions are easily found to be

$$
\begin{aligned}
g_{y_1} &= \omega_1^2 + \varepsilon\bigl(2f_1 y_6 + \alpha_1 y_1^2 + \alpha_2 y_3^2\bigr),\\
g_{y_2} &= \omega_1^2 + \varepsilon\bigl(2f_1 y_6 + 3\alpha_1 y_1^2 + \alpha_2 y_3^2\bigr) - \varepsilon^2\mu^2,\\
g_{y_3} &= \omega_2^2 + \varepsilon\bigl(2f_2 y_6 + \beta_1 y_3^2 + \beta_2 y_1^2\bigr),\\
g_{y_4} &= \omega_2^2 + \varepsilon\bigl(2f_2 y_6 + 3\beta_1 y_3^2 + \beta_2 y_1^2\bigr) - \varepsilon^2\mu^2,\\
g_{y_5} &= \Omega_1^2,\\
g_{y_6} &= \Omega_2^2.
\end{aligned}
$$

When, as it should be, the term $\varepsilon^2\mu^2$ is very small, one of g_{y_1} or g_{y_2} will dominate the other, depending upon the sign of α_1. Similarly g_{y_3} or g_{y_4} dominates the other depending upon the sign of β_1. For g_{y_5} and g_{y_6}, only the larger of Ω_1 and Ω_2 need be considered because the smaller will be dominated by the larger. Without loss of generality, assume that $\alpha_1 > 0$, $\beta_1 > 0$ and $\Omega_2 \geqslant \Omega_1$, then only the competition among g_{y_2}, g_{y_4} and g_{y_6} need be studied because if $g_{y_i} < 0$, then $g_{y_{i-1}} < 0$ too, where $i = 2, 4, 6$.

For instance, when $\varepsilon \sim O(1)$ and $\Omega_2 \neq 0$, but ω_1 or $\omega_2 \gg \Omega_2$, then g_{y_6} is out of competition. Only g_{y_2} and g_{y_4} need to be considered. Further, if $\omega_1 = \omega_2$, then the difference between f_1 and f_2 should be large so that g_{y_2} and g_{y_4} can predominate over each other alternatively. For example, when $\varepsilon = 0.1$, $\mu = 0.01$, $\omega_1 = \omega_2 = 1.0$, $f_1 = 0.9$, $f_2 = 6.0$, $\Omega_1 = \Omega_2 = 0.1$, $\alpha_1 = \alpha_2 = 1.0$, $\beta_1 = 0.1$, $\beta_2 = 1.0$, $F_1 = 0.5$, $F_2 = 0.8$, the system exhibits chaotic motions, as shown in Figures 38(a) and 38(b). Figure 38(c) depicts the functions g_{y_2} and g_{y_4}.

For most nonlinear systems the relationship between g's cannot be directly analyzed. But since g's are given in analytic form, we may roughly analyze their behavior from the averaged form over the evolution variable t where t is taken from t_0 to ∞. We propose the following heuristic: *When there exist at least two positive averaged g's, the system may exhibit chaotic motions. Otherwise, chaos is not expected.*

For example, consider the Lorenz system again. We may define $\langle g_y\rangle = -1 - 10\alpha + 10\langle z\rangle + \langle x^2\rangle$, where $\langle\cdots\rangle$ denotes an average of "$\cdots$" over t from t_0 to ∞. For instance, $\langle z\rangle = \lim_{t\to\infty} \frac{1}{t}\int_0^t z(t')\,dt'$ is an average value – the mean. If we restrict attention to the local behavior of the system, then $x \approx x_0 + A\cos(\omega_1 t + \phi_1)$, $z \approx z_0 + B\cos(\omega_2 t + \phi_2)$, where A, B, ω_1, ω_2, ϕ_1 and ϕ_2 are constants, and x_0 and y_0 represent the fixed points. Around the fixed points, we have $\langle z\rangle \approx z_0$, $\langle x^2\rangle \gtrapprox x_0^2$. Therefore,

$$\langle g_y\rangle \gtrapprox \beta - \beta - 11,$$

when the nontrivial fixed points are considered. A similar discussion gives

$$\langle g_z\rangle \gtrapprox \beta(\alpha - 1 - \beta).$$

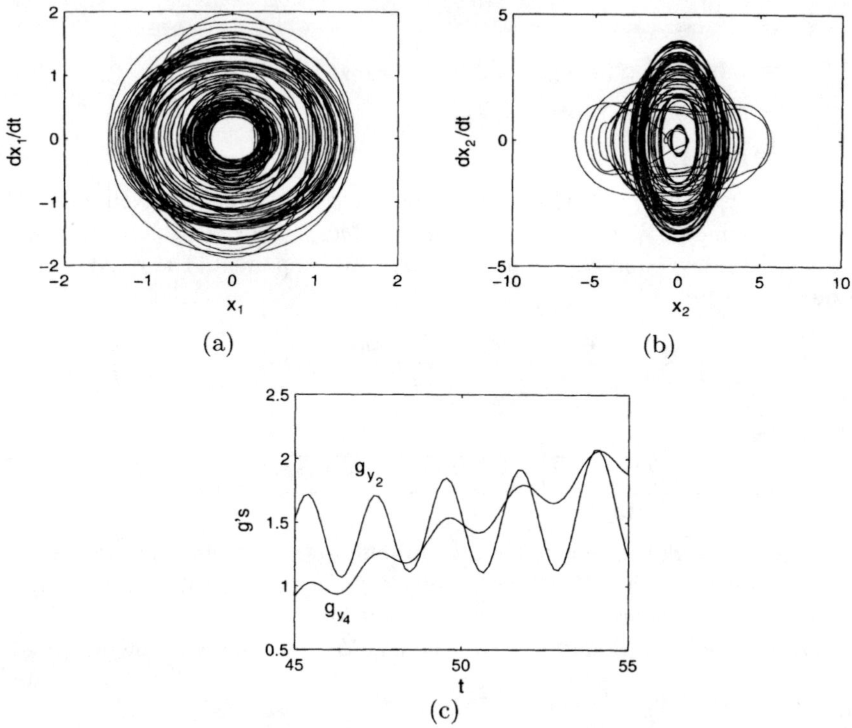

Figure 38. The phase portraits and function g's of system (143)–(144). (a) x_1 vs $\dot{x}_1$; (b) x_2 vs $\dot{x}_2$; (c) g_{y_2} and g_{y_4} vs t.

If both $\langle g_y \rangle$ and $\langle g_z \rangle$ are positive, chaotic motion may exist. For example, when $\alpha = 25$, $\beta = 2.6667$, both $\langle g_y \rangle$ and $\langle g_z \rangle$ are greater than zero, and the system is truly chaotic. However, when $\alpha = 24.5$, $\beta = 2.6667$, both $\langle g_y \rangle$ and $\langle g_z \rangle$ are greater than zero, but the system is periodic. This indicates that the averaged g's condition for predicting chaos is at most necessary. Even though the condition is not sufficient, it is still very useful in estimating chaotic parameter regimes easily and efficiently, in particular for high dimensional systems involving a large number of parameters. We demonstrate this using the following three examples.

4.2.2. Example 2: A psychological model

Our next example is a 6-dimensional psychological model for stress and coping (Neufeld, 1999). The 6-dimensional equation is given by

$$\dot{y}_1 = a - by_3y_4 - cy_1,$$
$$\dot{y}_2 = y_5(a - by_3y_4 - cy_1)(1 + y_6) - ey_2 - fy_3y_4y_1 + g,$$

$$\begin{aligned}
\dot{y}_3 &= h - iy_2,\\
\dot{y}_4 &= y_6(a - by_3y_4 - cy_1) - jy_4 + k(fy_3y_4y_1 - g) + d,\\
\dot{y}_5 &= 1 - y_5(y_1 + y_4) - y_5,\\
\dot{y}_6 &= 1 - y_6y_1 - y_6,
\end{aligned} \tag{145}$$

where $a, b, c, d, e, f, g, h, i, j, k$ are nonnegative parameters, see Neufeld (1999) for the meaning of y_i, $i = 1, \ldots, 6$. The variables y_2 and y_4, which describe the stress arousal level and coping activity, are most important. The fixed points of system (145) are

$$\left(\frac{1-r}{r}, \frac{h}{i}, \frac{ij(-c+ar+cr)}{b(id-hek)r}, \frac{id-hek}{ij}, \frac{ijr}{idr-hekr+ij}, r\right),$$

where

$$r = \frac{afi + 2cfi \pm \sqrt{a^2f^2i^2 + 4bcfieh - 4bcfgi^2}}{2(afi + cfi - beh + bgi)}.$$

Numerical work has shown that it is difficult to find chaotic and even periodic motions of the system (145). It is also very complicated to analytically determine the parameter boundary where the fixed points lose stability.

Let us apply our CM approach to system (145) and find the following six g's:

$$\begin{aligned}
g_{y_1} &= bkfy_3^2y_4 - bcy_3y_6 - c^2,\\
g_{y_2} &= -biy_4y_5(1 + y_6) - e^2 - fiy_1y_4,\\
g_{y_3} &= -biy_4y_5(1 + y_6) - fiy_1y_4,\\
g_{y_4} &= by_6\dot{y}_3 + by_3\dot{y}_6 - kfy_3\dot{y}_1 - kfy_1\dot{y}_3\\
&\quad - bcy_3y_6 - (j + by_3y_6 - kfy_1y_3)^2,\\
g_{y_5} &= \dot{y}_1 + \dot{y}_4 - (1 + y_1 + y_4)^2,\\
g_{y_6} &= \dot{y}_1 - (1 + y_1)^2.
\end{aligned}$$

Note that $g_{y_2} < g_{y_3}$. To estimate the chaotic parameter regimes, we further consider the averaged g's. It is always true that for a bounded variable $x(t)$, the average of $\dot{x}$ over $t \to \infty$ is 0 (otherwise, $x(t)$ will diverge). Using $\langle \dot{y}_1 \rangle = \langle \dot{y}_4 \rangle = 0$, we have $\langle g_{y_5} \rangle < 0$ and $\langle g_{y_6} \rangle < 0$. Therefore, we may only need to consider $\langle g_{y_1} \rangle$, $\langle g_{y_2} \rangle$, $\langle g_{y_3} \rangle$ and $\langle g_{y_4} \rangle$ expressed by the parameters. We use a simple program to search the values of the parameters for which at least two of $\langle g_{y_1} \rangle$, $\langle g_{y_2} \rangle$, $\langle g_{y_3} \rangle$ and $\langle g_{y_4} \rangle$ are greater than zero. Executing the program gives the results almost instantaneously because the program does not need to numerically integrate system (145). The output contains many groups of the values of the parameters, for instance, the group $(a, b, c, d, e, f, g, h, i, j, k) = (0.2, 0.1, 0.1, 0.1, 0.1, 0.2, 0.1, 2.1, 1.1, 2.4, 4.93)$. Figure 39 shows some phase

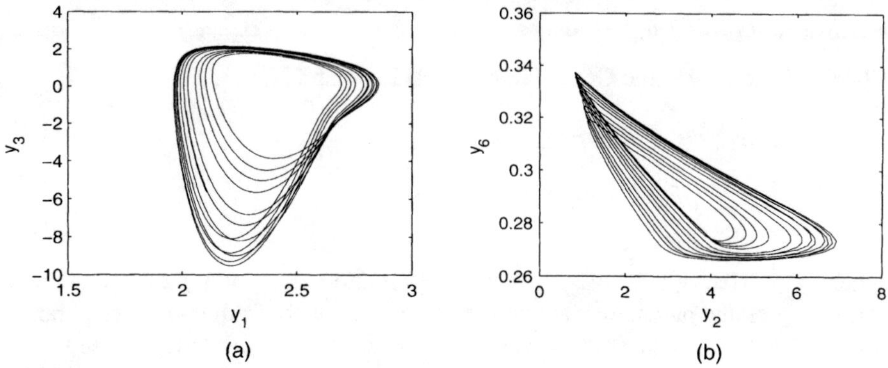

Figure 39. The phase portraits of system (145): (a) y_1 vs y_3; (b) y_2 vs y_6.

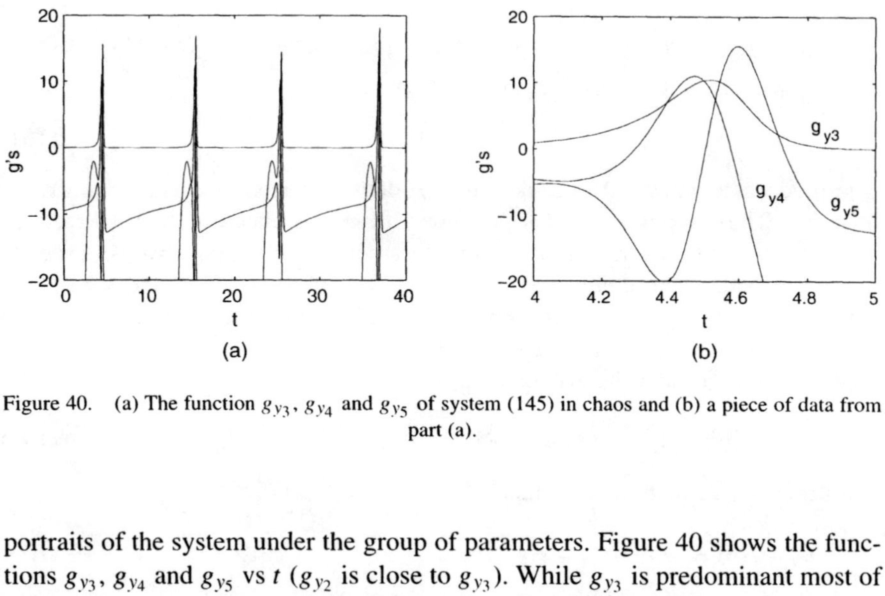

Figure 40. (a) The function g_{y_3}, g_{y_4} and g_{y_5} of system (145) in chaos and (b) a piece of data from part (a).

portraits of the system under the group of parameters. Figure 40 shows the functions g_{y_3}, g_{y_4} and g_{y_5} vs t (g_{y_2} is close to g_{y_3}). While g_{y_3} is predominant most of the time, g_{y_4} and either g_{y_5} dominate the others in some intervals of t. g_{y_1} and g_{y_6} are not shown in Figure 40 because they are always less than zero.

We have also applied the CM approach to study a number of chaotic systems such as the Rössler hyperchaotic system (Rössler, 1991) and the disk dynamo system (Hardy and Steeb, 1999). The results are very encouraging: Only the parameter regimes where the nonlinear system has at least two positive averaged g's may exhibit chaotic behavior. This suggests that one may, instead of g's, use the averaged g's to estimate chaotic parameter regimes.

4.2.3. Example 3: The Oregonator model

The third example is the Oregonator model (Field and Noyes, 1974):

$$\dot{\alpha} = s\big(\eta - \eta\alpha + \alpha - q\alpha^2\big), \tag{146}$$

$$\dot{\eta} = s^{-1}(-\eta - \eta\alpha + f\rho), \tag{147}$$

$$\dot{\rho} = w(\alpha - \rho), \tag{148}$$

where $\alpha \propto [HB_rO_2]$, $\eta \propto [B_r^-]$, $\rho \propto [C_e(IV)]$ and s, w, q and f are parameters. All variables and parameters are nonnegative. A group of frequently used parameters is $s = 77.27$, $w = 0.1610$, $q = 8.375 \times 10^{-6}$, $f = 1.0$. Under these values of the parameters, the system's fixed point is $\alpha_0 = \rho_0 = 488.68$, $\eta_0 = 0.99796$. Because the scales of these parameters are very different, it is extremely difficult to find chaos via numerical simulation. Here we consider the general case.

From equations (146)–(148), we can find

$$g_\alpha = f\rho - \alpha\eta - s^2\big(2q^2\alpha^2 + 3q\alpha\eta - 3q\alpha + \eta^2 - 2\eta - 2q\eta + 1\big), \tag{149}$$

$$g_\eta = \eta - \eta\alpha + \alpha - q\alpha^2 - (1+\alpha)^2/s^2, \tag{150}$$

$$g_\rho = -w^2. \tag{151}$$

The mode associated with the direction ρ does not exist because g_ρ is strictly negative. Thus, we only need to consider those associated with the directions α and η. Because all the variables and parameters are nonnegative, in order to have g_η greater than 0, equation (150) suggests that q be small and s be large. This conclusion agrees with the selection of parameter values often used in the literature. However, such a choice may cause $g_\alpha < 0$.

We further consider the averaged g_η,

$$\langle g_\eta \rangle = \big\langle \eta - \eta\alpha + \alpha - q\alpha^2 - (1+\alpha)^2/s^2 \big\rangle. \tag{152}$$

It follows from equation (146) that

$$\langle \dot{\alpha} \rangle = s\big\langle \eta - \eta\alpha + \alpha - q\alpha^2 \big\rangle = 0. \tag{153}$$

Substituting equation (153) into equation (152) yields

$$\langle g_\eta \rangle = -\big\langle (1+\alpha)^2/s^2 \big\rangle < 0. \tag{154}$$

Because both $\langle g_\eta \rangle$ and $\langle g_\rho \rangle$ are less than zero, the system has at most one positive averaged g function, $\langle g_\alpha \rangle$. Based on this analysis, we conclude that chaos in the Oregonator model is rare though we cannot absolutely exclude the existence of chaos from this model. This agrees with the results from numerical experiment (Field and Noyes, 1974).

4.2.4. Example 4: The smooth Chua system

The Chua circuit (Matsumoto, 1984; Chua, Komuro and Matsumoto, 1986) is the first real electronic system used to exhibit chaos. Unlike the Lorenz and Rössler systems which have nonlinear coupled terms, the function used in the Chua system is piecewise-linear. This means that our CM approach cannot be applied because the piecewise-linear term is not differentiable at the turning points. Our CM approach can however be applied to the recently proposed the smooth Chua system. See Tsuneda (2005) for the details of how the original Chua system is transformed to the smooth Chua system, given by

$$
\begin{aligned}
\dot{x} &= k_1 y + k_2 x + k_3 x^3,\\
\dot{y} &= k_4(x - y + z),\\
\dot{z} &= k_5 y + k_6 z,
\end{aligned}
\tag{155}
$$

where k_i, $i = 1, \ldots, 6$, are parameters. We note that the system has only one nonlinear term, k_3x^3, and the system equations are not nonlinearly but linearly coupled. This simple Chua system still displays very interesting and complicated dynamical behavior, as well as many chaotic attractors, even when the nonlinear term is very weak, namely, $|k_3/k_1| \ll 1$, $|k_3/k_2| \ll 1$ (Tsuneda, 2005).

Performing the CM on the system, we obtain

$$
\begin{aligned}
g_x &= -\big[k_1k_4 + k_2^2 + k_2k_3x^2 + 3k_3x\big(k_1y + k_2x + k_3x^3\big)\big],\\
g_y &= -k_4(k_1 + k_4 + k_5),\\
g_z &= -\big(k_4k_5 + k_6^2\big).
\end{aligned}
$$

Since g_y and g_z are constant, they are not competitive. For the system to be chaotic, g_x must exist and compete with at least one of g_y and g_z. Since the last term of g_x can be rewritten as $3k_3x\dot{x}$, this term oscillates around zero no matter what values of k_3 are chosen. Thus, we may interpret the behavior of g_x as an oscillation around the value of $-(k_1k_4+k_2^2+k_2k_3x^2)$. If $-(k_1k_4+k_2^2+k_2k_3x^2) > 0$, g_x may exist, and if $-(k_1k_4+k_2^2+k_2k_3x^2)$ is close to g_y or g_z, chaos may appear. Based on the forms of g_x, g_y and g_z, we have the following results.

CASE 1. If g_x and g_y are competitive, then $g_x \simeq g_y > 0$.

(1a) If $k_4 < 0$ and $k_1 < 0$, then $k_5 > -k_1 - k_4 > 0$ and $k_2k_3 < 0$.
(1b) If $k_4 < 0$ and $k_1 > 0$, then $k_5 > -k_1 - k_4$, and further, if $k_1k_4 + k_2^2 > 0$, we have $k_2k_3 < 0$.
(1c) If $k_4 > 0$ and $k_1 < 0$, then $k_5 < -k_1 - k_4$, and further, if $k_1k_4 + k_2^2 > 0$, we have $k_2k_3 < 0$.
(1d) If $k_4 > 0$ and $k_1 > 0$, then $k_5 < -k_1 - k_4 < 0$ and $k_2k_3 < 0$.

CASE 2. If g_x and g_z are competitive, then $g_x \simeq g_z > 0$ and $k_4 k_5 < -k_6^2 < 0$.

(2a) If $k_4 < 0$ and $k_1 < 0$, then $k_5 > 0$ and $k_2 k_3 < 0$.
(2b) If $k_4 > 0$ and $k_1 > 0$, then $k_5 < 0$ and $k_2 k_3 < 0$.

The smooth Chua system has been extensively studied by Tsuneda (2005) using numerical simulations. The 23 parameter regions, denoted by C-1–C-20 and C-7′, C-13′, C-17′, for which the system is chaotic or periodic, are identified in Tsuneda (2005). For convenience, Table 2 in Tsuneda (2005) is copied here as Table 1. Comparing these 23 regions with the above CM analysis, we have found that case (1a) contains C-4, 8, 12, 13, 15, 16, 13′, case (1b) contains C-18, case (1c) contains C-3, 5, 9, 10, 14, 20 and case (1d) contains C-1, 2, 6, 7, 11, 17, 19, 7′, 17′. We also have noticed that case (2a) contains C-4, 8, 12, 13, 15, 16, 13′ and case (2b) contains C-1, 2, 6, 7, 11, 17, 19, 7′, 17′. Thus, from the CM point of view, in these 23 parameter regions investigated by Tsuneda (2005) the smooth Chua system could exhibit complicated dynamical behavior such as chaos. Said another way, the regions classified by CM agree with those identified by Tsuneda (2005).

Based on the work of Tsuneda (2005), in some of these 23 parameter regions the smooth Chua system is not chaotic. In order to filter out some nonchaotic parameter regions and to find new chaotic attractors in new parameter regions, we further transform the linear coupled smooth Chua system into a nonlinear coupled smooth Chua system, under which our CM approach may be more useful from the resonance viewpoint. From the first equation of system (155), we have

$$y = \frac{1}{k_2}\left(\dot{x} - k_2 x - k_3 x^3\right). \tag{156}$$

($k_1 \neq 0$, otherwise there is no chaos.) Substituting equation (156) into the other two equations of system (155), and denoting by $x_1 = x$, $x_2 = \dot{x}$, we obtain the following nonlinear coupled smooth Chua system:

$$\begin{aligned}
\dot{x}_1 &= x_2,\\
\dot{x}_2 &= \left(k_2 - k_4 + 3k_3 x_1^2\right)x_2 + k_4\left(k_1 + k_2 + k_3 x_1^2\right)x_1 + k_1 k_4 z,\\
\dot{z} &= \frac{k_5}{k_2}\left(x_2 - k_2 x_1 - k_3 x_1^3\right) + k_6 z.
\end{aligned} \tag{157}$$

Then the g's based on system (157) are found to be

$$\begin{aligned}
g_{x_1} &= -k_4\left(k_1 + k_2 + k_3 x_1^2\right) - 3k_3 x_1 x_2,\\
g_{x_2} &= -k_4\left(k_1 + k_2 + k_5 + 3k_3 x_1^2\right) - 6k_3 x_1 x_2 - \left(k_2 - k_4 + 3k_3 x_1^2\right)^2,\\
g_z &= -\left(k_4 k_5 + k_6^2\right),
\end{aligned} \tag{158}$$

Table 1 Parameter values for the smooth Chua system (157)

No.	k_1	k_2	k_3	k_4	k_5	k_6
C-1	9.3515908493	1.6682747877	−0.6973867753	1.0	−14.7031980540	−0.0160739649
C-2	3.7091002664	7.1642927876	−1.6802567087	1.0	−24.0799705758	0.8592556780
C-3	−6.6919100000	−0.4602892985	0.1835914401	1.0	1.52061	0.0
C-4	−143.1037	14.6570625639	−2.2369405036	−1.0	207.34198	−3.8767721000
C-5	−1.301814	1.4311608333	−0.0695653464	1.0	0.0136073	0.0296996800
C-6	8.4562218418	−1.3714231681	0.1384203393	1.0	−12.0732335925	−0.0051631393
C-7	6.5792294673	1.5321955041	−0.6247272525	1.0	−10.8976626192	0.0447440294
C-8	−4.006	0.6347865975	−0.1504582008	−1.0	54.459671	−0.93435708
C-9	−4.08685	−0.3829118599	0.1411663497	1.0	2.0	0.0
C-10	−75.0	−50.9876254261	1.3092138284	1.0	−31.25	3.125
C-11	15.6	2.6072521221	−1.0283200040	1.0	−28.58	0.0
C-12	−37.195804	3.8096934345	−0.5814301135	−1.0	73.049688	−1.161224
C-13	−13.070921	2.0712045602	−0.4909204336	−1.0	53.612186	−0.7508709600
C-14	−45,012.877058	21.9725901325	−5.0576444486	1.0	14,125.7874586260	0.2326833338
C-15	−3.505	0.5553986581	−0.1316415362	−1.0	66.672752	−0.9477989200
C-16	−12.141414	1.9239158468	−0.4560098118	−1.0	95.721132	−0.8982235
C-17	1800.0	51.2859375	−12.1816406250	1.0	−10,000.0	0.0
C-18	1.7327033212	−1.5599365845	0.0337768366	−1.0	0.0421159445	0.2973436607
C-19	62.3168864230	0.5524169777	−0.0207591313	1.0	−94.7189263	0.6013155140
C-20	−1.424557325	1.3377628122	−0.0281432267	1.0	−0.02944201	−0.3226735790
C-7′	6.62	1.5416902979	−0.6285985969	1.0	−10.8976626192	0.0447440294
C-13′	−19.5	3.2779542798	−0.8536807567	−1.0	53.612186	−0.7508709600
C-17′	2300.0	69.5277041101	−18.1433331473	1.0	−10,000.0	0.0

in which only g_z is constant. Again, we can interpret g_{x_1} to oscillate around $-k_4(k_1 + k_2 + k_3x_1^2)$ and g_{x_2} oscillates around $-k_4(k_1 + k_2 + k_5 + 3k_3x_1^2) - (k_2 - k_4 + 3k_3x_1^2)^2 < -k_4(k_1 + k_2 + k_5 + 3k_3x_1^2)$. Then, for system (157), we have the following results.

CASE 1. If g_{x_1} and g_{x_2} are competitive, then $k_4(k_1+k_2+k_5+3k_3x_1^2) < k_4(k_1+k_2+k_3x_1^2) < 0$, so $k_4k_5 < 0$. Comparing with the 23 parameter regions identified by Tsuneda (2005), we have found that all the regions, except C-3, 5, 9, 14, are contained in this case.

CASE 2. If g_{x_1} and g_z are competitive, then $k_4(k_1+k_2+k_3x_1^2) \simeq k_4k_5+k_6^2 < 0$. Hence, we have $k_4(k_1 + k_2 - k_5 + k_3x_1^2) \simeq k_6^2 > 0$ and

(2a) if $k_4 > 0$ and $k_1 + k_2 - k_5 < 0$, then $k_3 > 0$, $k_5 < 0$, $k_1 + k_2 < 0$;
(2b) if $k_4 < 0$ and $k_1 + k_2 - k_5 > 0$, then $k_3 < 0$, $k_5 > 0$, $k_1 + k_2 > 0$.

Thus, it is found that C-9 and C-10 are contained in case (2a), and none of these 23 regions is contained in case (2b).

CASE 3. If g_{x_2} and g_z are competitive, then $k_4(k_1 + k_2 + k_5 + 3k_3x_1^2) \simeq k_4k_5 + k_6^2 < 0$. Hence, we have $k_4(k_1 + k_2 + 3k_3x_1^2) \simeq k_6^2 > 0$ and

(3a) if $k_4 > 0$ and $k_1 + k_2 < 0$, then $k_3 > 0$ and $k_5 < 0$;
(3b) if $k_4 < 0$ and $k_1 + k_2 > 0$, then $k_3 < 0$ and $k_5 > 0$.

Thus, C-10 is contained in case (3a) and none in case (3b).

One of the advantages of our CM is to find chaos. From the above analysis, we have seen that if the parameter values satisfy case (2b) or (3b), chaos may appear in the smooth Chua system, however these cases were not investigated in Tsuneda (2005). For case (2b), we found a chaotic attractor when $(k_1, k_2, k_3, k_4, k_5, k_6) = (-2, 6.5, -0.15, -1, 4, -1)$ and the initial condition is taken near the origin. (Here, all the initial conditions are chosen near the origin as done in Tsuneda (2005).) The attractor and corresponding g's are displayed in Figures 41(a) and 41(c), respectively. The attractor obtained from the linear coupled smooth Chua system (155), with the same parameter values and initial condition, is depicted in Figure 41(b) for the sake of comparison with the attractors in Tsuneda (2005). For case (3b), another chaotic attractor was found when $(k_1, k_2, k_3, k_4, k_5, k_6) = (-2, 3.7, -0.15, -1, 12, -1)$, as shown in Figures 42(a) and 42(b), and the corresponding g's for Figure 42(a) is given in Figure 42(c). It is observed that all g's are competitive in these two cases and the attractors are somewhat complicated. Since these two attractors exist in different categories of parameter regions from those investigated by Tsuneda (2005) in the viewpoint of CM, the two sets may have some aspects different from the others.

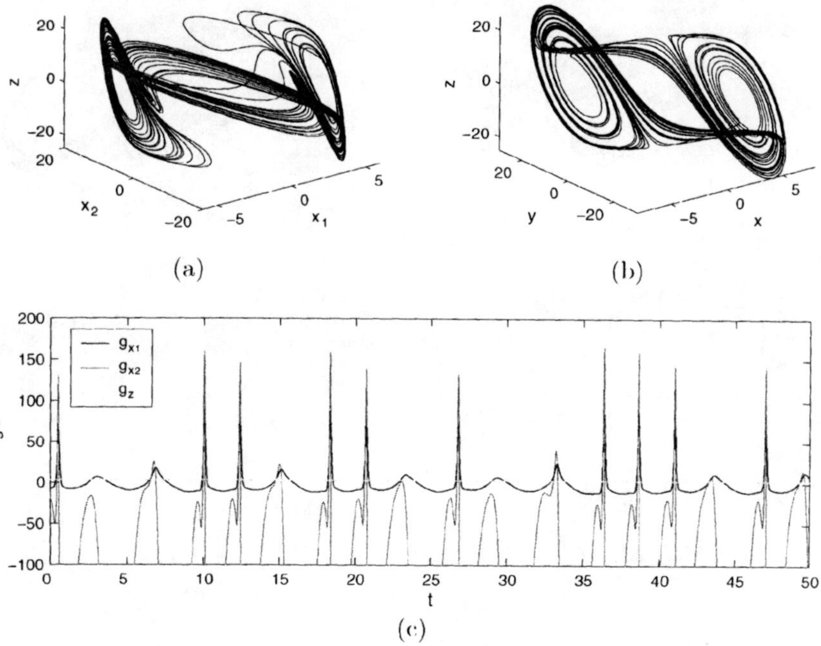

Figure 41. For the smooth Chua system (155), case (2b) when $(k_1, k_2, k_3, k_4, k_5, k_6) = (-2, 6.5, -0.15, -1, 4, -1)$: (a) the strange attractor of system (157); (b) the strange attractor of system (155) and (c) the corresponding g's vs t to the attractor given in part (a).

We repeated numerical simulations for all the 23 cases listed in Tsuneda (2005) and checked the competition of g's in equation (158) obtained from the nonlinear coupled system (157). We have found that for C-6 g_x and g_z are nearly competitive, and for all other chaotic cases, there are at least two g's in competition. Figures 43 and 44 display the attractors and corresponding g's for C-1 and C-6, respectively. Since for C-6 g's do not cross although they are close to each other (near resonance), chaos is not strong. We have also seen that for some periodic cases found in Tsuneda (2005), the g's are not in competition. One of the examples is C-14, as depicted in Figure 45.

4.3. Application of CM: constructing new chaotic systems

In the previous subsections, we have shown that function g plays an important role in the study of chaotic systems. We have estimated chaotic parameter regimes based on the necessary condition that a chaotic system should have at least two CMs. In this subsection, we will apply the same rule to actively create new chaotic systems.

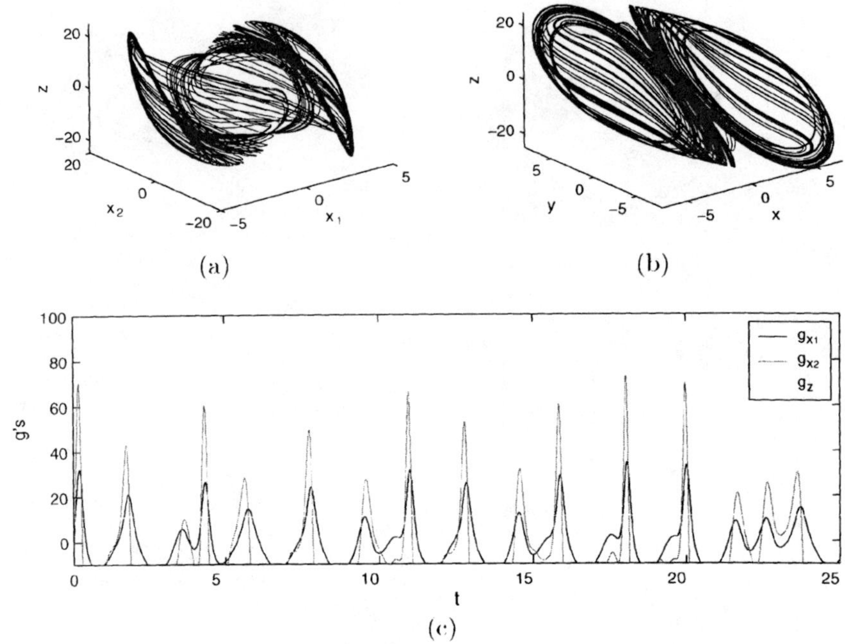

Figure 42. For the smooth Chua system (155), case (3b) when $(k_1, k_2, k_3, k_4, k_5, k_6) = (-2, 3.7, -0.15, -1, 12, -1)$: (a) the strange attractor of system (157); (b) the strange attractor of system (155) and (c) the corresponding g's vs t to the attractor given in part (a).

For constructing a chaotic system, we first design n ($\geqslant 2$) g's so that at least two of the g's are competitive in some parameter regime. The n g's are expressed in n second-order differential equations similar to equation (132). These equations form a system of at least $2n$ dimensions. If we are interested in lower-dimensional systems, or particularly, lower-dimensional first-order differential systems, we may add some constraints to the second-order differential equations. In other words, we seek a subspace of the $2n$-dimensional space so that a lower-dimensional system exists in the subspace. Note that the constraints only help derive a lower-dimensional system from the higher one, but do not affect the designed g's.

For example, to obtain a three-dimensional first-order differential system, a general approach can be described as follows.

First, we design the functions g_x and g_y in the following equations to be competitive in some parameter regime so that the competition may result in chaos:

$$\ddot{x} = -g_x(x, y, z)x + h_1(y, z), \tag{159}$$

$$\ddot{y} = -g_y(x, y, z)y + h_2(x, z). \tag{160}$$

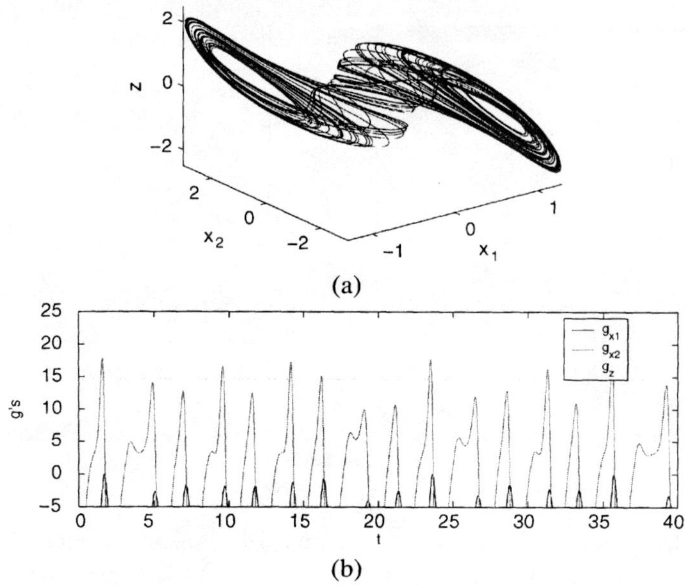

Figure 43. Results for the smooth Chua system (155), case C-1: (a) the strange attractor and (b) the corresponding g's.

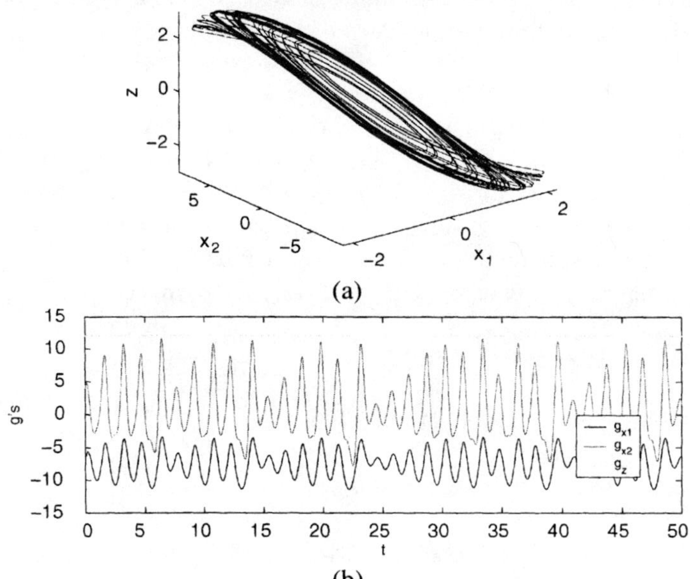

Figure 44. Results for the smooth Chua system (155), case C-6: (a) the strange attractor and (b) the corresponding g's.

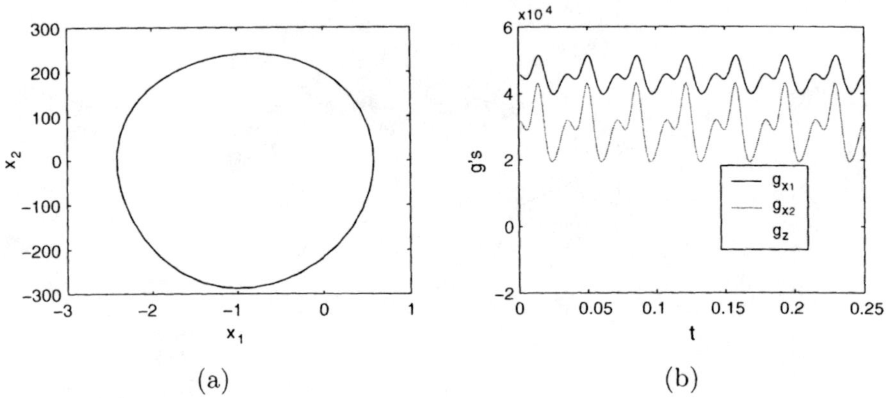

Figure 45. Results for the smooth Chua system (155), case C-14: (a) the strange attractor and (b) the corresponding g's.

In general, equations (159) and (160) construct a dynamical system in $\mathbb{R}^5$ spanned by $\{x, \dot{x}, y, \dot{y}, z\}$ because we are going to use z as a variable. In other words, we need five independent initial values for, respectively, x, $\dot{x}$, y, $\dot{y}$ and z to solve the system. In this space, we seek a three-dimensional subspace in which x, $\dot{x}$, y, $\dot{y}$ and z are not independent but constrained, and the system may be chaotic.

Second, we design the constraint, given by

$$\dot{x} = f(x, y, z). \tag{161}$$

At this stage, the system, which consists of equations (159)–(161), is not five- but, in general, four-dimensional because the state of $\dot{x}$ is determined by x, y and z. Also, the constraint has to be designed in such a form so that the competition between g_x and g_y holds in some part of the parameter space induced by these equations. Actually, in these two steps, we have prepared for obtaining $\dot{y}$ and $\dot{z}$.

Next, differentiating equation (161) with respect to t results in

$$\ddot{x} = \frac{\partial f}{\partial x}\dot{x} + \frac{\partial f}{\partial y}\dot{y} + \frac{\partial f}{\partial z}\dot{z} = f(x, y, z)\frac{\partial f}{\partial x} + \frac{\partial f}{\partial y}\dot{y} + \frac{\partial f}{\partial z}\dot{z}. \tag{162}$$

Then combining equations (159) and (162) yields

$$\frac{\partial f}{\partial y}\dot{y} + \frac{\partial f}{\partial z}\dot{z} = -g_x(x, y, z)x + h_1(y, z) - f(x, y, z)\frac{\partial f}{\partial x}.$$

Now, if $\frac{\partial f}{\partial z} = 0$ and $\frac{\partial f}{\partial y} \neq 0$, then we have

$$\dot{y} = \left[-g_x(x, y, z)x + h_1(y, z) - f(x, y, z)\frac{\partial f}{\partial x}\right] \Big/ \frac{\partial f}{\partial y}. \tag{163}$$

$\dot{y}$ is uniquely determined by x, y, z. In this case, the system which consists of equations (160), (161) and (163) is three-dimensional because we can solve the system by using the initial conditions for x, y and z. That is, our new system no longer includes equation (159).

Similarly, differentiating equation (163) with respect to t and combining with equation (160), we obtain

$$F\left(\dot{z}, x, y, z, h_1, h_2, f, \frac{\partial^2 f}{\partial x\, \partial y}, \ldots\right) = 0. \tag{164}$$

Finally, we may solve $\dot{z}$ from equation (164).

Equations (161), (163) and (164) form the expected system. From this system, we can obtain g_x, g_y and g_z, in which g_x and g_y are the same as that in equations (159) and (160) respectively, while g_z is new. Since we need only two competing g's, the form of g_z will not resist the appearance of chaos.

To have chaos, the forms of g_x and g_y should be designed to be greater than 0 in some parameter regimes. The requirement on the forms of $h_1(y, z)$ and $h_2(x, z)$ by condition 4 given in Conjecture 4.3 is easily satisfied, and we choose the forms as simple as possible. We present three examples in the following, some of which have been used in improving the security of communications via chaos synchronization (Yao, 2002; Yao, Essex and Yu, 2002).

4.3.1. Example 1

We design g_x, g_y and the corresponding second-order differential equations as

$$\ddot{x} = -(\alpha - y)x - z, \tag{165}$$

$$\ddot{y} = -\left(1 - \beta^2\right)y + \beta x - z, \tag{166}$$

where α and β are parameters. $g_x = \alpha - y$ and $g_y = 1 - \beta^2$. When $|\beta| < 1$, g_y exists. Then one can adjust α so that g_x will oscillate around g_y because g_x contains y. The competition of these two g's may result in chaos. The forms of $h_1 = -z$ and $h_2 = \beta x - z$ are very simple and relative to z from which we can deduce $\dot{z}$. Except for these requirements, all of these forms are chosen arbitrarily.

To obtain a three-dimensional first-order system, we add the following constraint:

$$\dot{y} = -\beta y - x, \tag{167}$$

to equations (165) and (166). Differentiating equation (167) with respect to t and then using equation (166), we obtain

$$\dot{x} = y + z. \tag{168}$$

Then differentiating equation (168) with respect to t and then using equation (165) gives

$$\dot{z} = -z + (1-\alpha)x + \beta y + xy. \tag{169}$$

Equations (167)–(169) form the required system.

From this system, we have

$$g_z = -2 + \alpha - y,$$

which is dominated by g_x.

To estimate the chaotic parameter regimes, we further use the averaged g's. The nontrivial fixed point of the system is $(-1-\alpha\beta, \alpha + \frac{1}{\beta}, -\alpha - \frac{1}{\beta})$ near which

$$\langle g_x \rangle \approx -\frac{1}{\beta},$$
$$\langle g_y \rangle = 1 - \beta^2,$$
$$\langle g_z \rangle \approx -\frac{1}{\beta} - 2.$$

We note that α does not appear in $\langle g_x \rangle$, $\langle g_y \rangle$ and $\langle g_z \rangle$. This is because we use first-order approximation of y to get $\langle y \rangle$. When $\beta \in (-1, 0)$, there exist at least two averaged g's. Especially, for $\beta \in (-0.5, 0)$, three averaged g's exist. Because α is free, we can first choose β in the regime, then adjust α to find the chaos, as shown in Figure 46.

This example shows that it is easy to apply CMs to construct a chaotic system which allows much freedom to make parameter changes. When the form of the function g_i is complicated, the result of competition among CMs may become intricate too. Next, we show an example with more complicated forms of g_i.

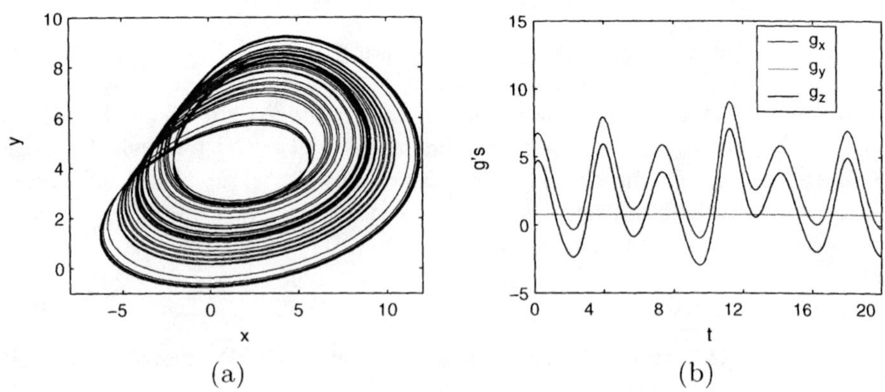

Figure 46. (a) The strange attractors of the system (167)–(169) when $\alpha = 8.3$ and $\beta = -0.48$; and (b) the corresponding g's vs t.

4.3.2. Example 2

In this example, we construct a four-dimensional system. In this case, one may use four-differential equations, for example, two second-order differential equations which define two g's and two first-order differential equations. The two second-order differential equations contains two other free variables. The dimension of the subsystem is then six. However, the two first-order differential equations constrain the second-order equations. Therefore, the whole system is four-dimensional. Thus, let

$$\ddot{x} = -(w + \alpha - 1 - y^2)x + \beta_1 y + \gamma w - (1 + \beta_2)z, \tag{170}$$

$$\ddot{y} = -(w + x - \gamma - \beta_1^2)y - (\beta_1 - \beta_2)(x - \gamma)w - zw - xz + \gamma z, \tag{171}$$

$$\dot{x} = y + z, \tag{172}$$

$$\dot{y} = \beta_1 y - xw + \gamma w, \tag{173}$$

where α, β_1, β_2 and γ are parameters and x, y, z and w are variables. From equations (170), (172) and (173), we find

$$\dot{z} = (1 - \alpha)x - (1 + \beta_2)z + xy^2. \tag{174}$$

From equations (171)–(173), we can similarly obtain

$$\dot{w} = -\beta_2 w + y + z. \tag{175}$$

Equations (172)–(175) form a possibly chaotic system. The function g's of the system are

$$g_x = w + \alpha - 1 - y^2,$$
$$g_y = w + x - \gamma - \beta_1^2,$$
$$g_z = \alpha - 1 - (1 + \beta_2)^2 - y^2,$$
$$g_w = x - \beta_2^2 - \gamma.$$

One may adjust the parameters so that the g's are competitive. Because the only fixed point of the system is (0, 0, 0, 0), one can easily get the averaged g's, given by

$$\langle g_x \rangle \lessapprox \alpha - 1,$$
$$\langle g_y \rangle \approx -\beta_1^2 - \gamma,$$
$$\langle g_z \rangle \lessapprox \alpha - 1 - (1 + \beta_2)^2,$$
$$\langle g_w \rangle \approx -\beta_2^2 - \gamma.$$

Therefore, when $\alpha > 1$ and $\gamma < -\max(\beta_1^2, \beta_2^2)$, there exist at least two averaged g's. The local structure of the chaotic attractors of the system is expected to be complicated. For example, when $\alpha = 20.2$, $\beta_1 = 0.322$, $\beta_2 = 0.2$ and

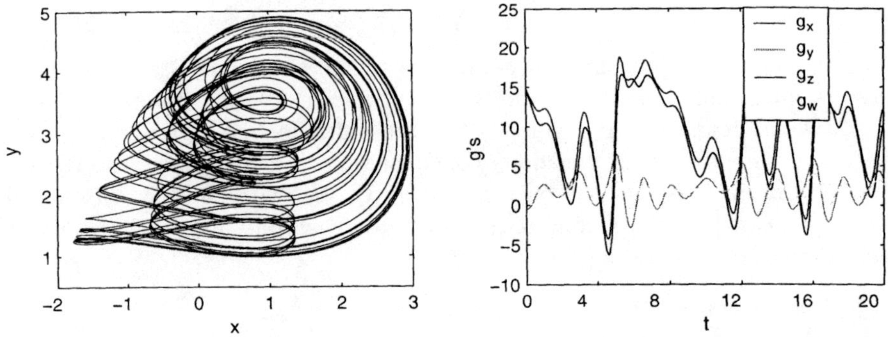

(a) (b)

Figure 47. (a) The chaotic attractor of the system (172)–(175) when $\alpha = 20.2$, $\beta_1 = 0.322$, $\beta_2 = 0.2$ and $\gamma = -1.4$; and (b) g_x, g_y, g_z and g_w vs t in this case.

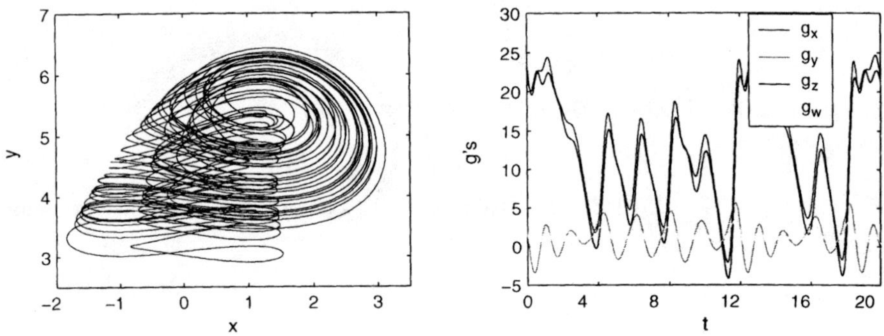

(a) (b)

Figure 48. (a) The chaotic attractor of the system (172)–(175) when $\alpha = 38.2$, $\beta_1 = \beta_2 = 0.2$ and $\gamma = -0.6$; and (b) g_x, g_y, g_z and g_w vs t in this case.

$\gamma = -1.4$, the system is chaotic as shown in Figure 47(a). The attractor is more entangled than that of the system consisting of equations (167)–(169) (see Figure 46(a)). The corresponding g_x, g_y, g_z and g_w vs t are displayed in Figure 47(b). It is seen that all g's are in competition with each other. The competition results in the complicated structure of the chaotic attractor.

When $\beta_1 = \beta_2$, $\langle g_y \rangle \simeq \langle g_w \rangle$, then the competition between g_y and g_w may become stronger. The structure of the chaotic attractor is expected to be more complicated than that when $\beta_1 \neq \beta_2$. Figures 48(a) and 48(b) show, respectively, the chaotic attractor and the functions g_x, g_y, g_z and g_w vs t when $\alpha = 38.2$, $\beta_1 = \beta_2 = 0.2$ and $\gamma = -0.6$. It is observed that the local structure of the attractor is even more entangled than that in Figure 47(a), and that all the g's are in competition.

4.3.3. Example 3

In the last example, we create a class of systems exhibiting complex chaotic attractors using complicated g's. In addition, we want to use this example to further show that CMs reveal the nature of chaos to some extent. Let

$$\ddot{x}_i = -\left(\alpha_i + \sum_{j=1, j\neq i}^{n} x_j^2\right) x_i + \sum_{j=1, j\neq i}^{n} x_j, \tag{176}$$

where α_i are parameters, $i = 1, 2, \ldots, n$ ($n \geqslant 2$). When these parameters are greater than 0, we may have at least n positive g's. As it may not be easy to find a corresponding n-dimensional first-order differential equations, we use this second-order differential equations directly, which turn out to be $2n$-dimensional in phase space. If all the parameters and initial conditions take the same corresponding values, the $2n$-dimensional systems reduce to two-dimensional no matter the value of n (> 1), and the dynamical behavior of the systems is trivial. However, complex motions can be triggered by any small perturbations in the initial conditions or the parameters.

The functions g_i's are

$$g_{x_i} = g_{\dot{x}_i} = \alpha_i + \sum_{j=1, j\neq i}^{n} x_j^2.$$

Therefore, we have $2n$ positive g's when the parameters are greater than 0, but n pairs of them are equal.

These systems can have very complicated attractors. Here, we consider the simplest case of the system when $n = 2$. It has been found that with larger n, the dynamical behavior of the systems is more complicated. For instance, when $n = 3$ we have observed an interesting attractor similar to a three-dimensional poodle.

When $n = 2$, the system is rewritten as

$$\begin{aligned}\ddot{x} &= -(\alpha + y^2)x + y,\\ \ddot{y} &= -(\beta + x^2)y + x,\end{aligned} \tag{177}$$

where α and β are parameters.

Obviously, when α is close to β, and the magnitudes of x and y are approximately equal (which is possible because the system is symmetric), the g's of the system are in full competition. Highly chaotic motions may result from the competition. On the other hand, when the difference between α and β is large, the magnitudes of x and y may be quite different. In this case, there may have no

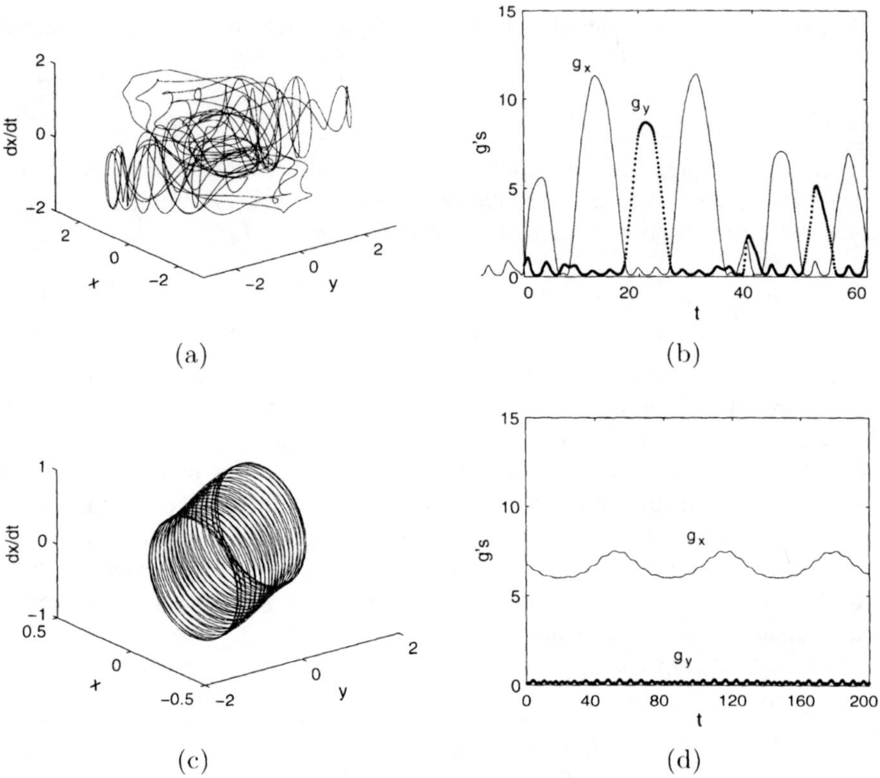

Figure 49. (a) A chaotic attractor of system (177) when $\alpha = \beta = 0.1$ with the initial condition $(x(0), \dot{x}(0), y(0), \dot{y}(0)) = (-0.17, 0.86, -0.97, -0.2)$; (b) the g's corresponding to part (a); (c) a quasiperiodic attractor when $\alpha = 6.0$, $\beta = 0.1$ and the same initial condition used in part (a); and (d) the g's corresponding to part (c).

competition among g's except in the equal pairs $(g_x, g_{\dot{x}})$ and $(g_y, g_{\dot{y}})$, and thus the behavior of the system becomes simpler. Figures 49 and 50 show the attractors and the g's of the system when $\alpha = \beta$ and $\alpha \gg \beta$, respectively. It is seen that the numerical results agree with our analytical predictions based on g's.

To show the complexity of the chaotic attractors obtained above, we further consider the relation between the competition of the g's and the largest Lyapunov exponent (λ) of the system. In general, for a given system, a larger positive λ implies a more complex chaotic attractor. Figure 51 depicts λ when α increases from 0.1 to 10 while $\beta \equiv 0.1$. It clearly indicates that the maximum λ appears when $\alpha = \beta$ for which the g's are in good competition (see Figure 49(b)). As α increases (leaving away from β), λ becomes smaller, which agrees with that observed from Figure 49(d): the competition among the g's is getting weaker.

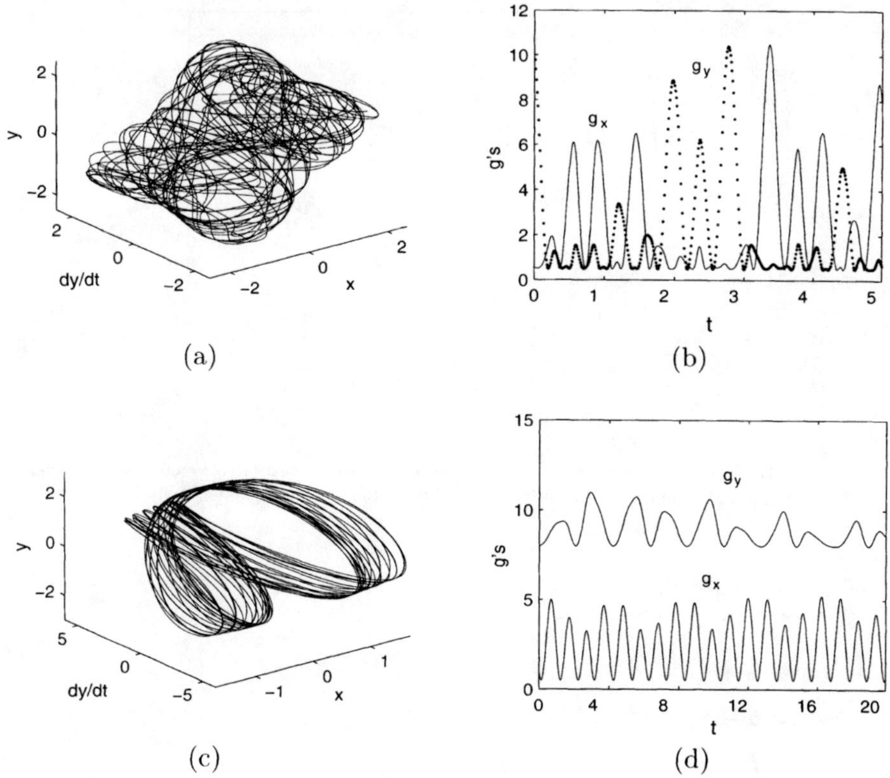

Figure 50. (a) A chaotic attractor of system (177) when $\alpha = \beta = 0.5$ with the initial condition $(x(0), \dot{x}(0), y(0), \dot{y}(0)) = (-0.5, 0.7, -2.1, -1.6)$; (b) the g's corresponding to part (a); (c) a quasiperiodic attractor when $\alpha = 0.5$, $\beta = 8.0$ and the same initial condition used in part (a); and (d) the g's corresponding to part (c).

When $\alpha > 5.8$, $\lambda \to 0^+$, and the state of the system is indeed quasiperiodic. The complete Lyapunov exponent spectrum of the system is $(\lambda, 0, 0, -\lambda)$ because $\nabla \cdot \mathbf{F} = 0$.

Further, to see how g's determine the dynamical behavior of a nonlinear system, we construct a dissipative system which has similar g's as those of system (177). To do this, we modify system (177) as

$$
\begin{aligned}
\dot{x}_1 &= c_1 x_1 + x_2, \\
\dot{x}_2 &= -c_2 x_2 - \left(\alpha + y_1^2\right) x_1 - y_1, \\
\dot{y}_1 &= c_1 y_1 + y_2, \\
\dot{y}_2 &= -c_2 y_2 - \left(\beta + x_1^2\right) y_1 + x_1,
\end{aligned} \tag{178}
$$

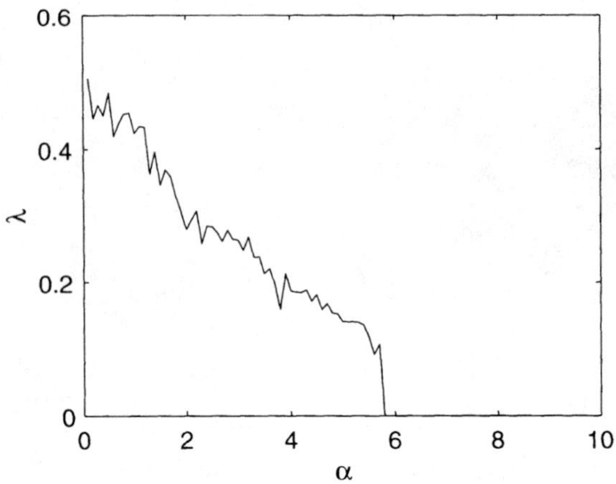

Figure 51. The largest Lyapunov exponent of system (177) with respect to α when $\beta = 0.1$ with the initial condition $(x(0), \dot{x}(0), y(0), \dot{y}(0)) = (-0.17, 0.86, -0.97, 0.20)$.

where α, β, c_1 and c_2 are parameters.

The functions g's now become

$$\begin{aligned} g_{x_1} &= \alpha + y_1^2 - c_1^2, \\ g_{x_2} &= \alpha + y_1^2 - c_2^2, \\ g_{y_1} &= \beta + x_1^2 - c_1^2, \\ g_{y_2} &= \beta + x_1^2 - c_2^2. \end{aligned}$$

When $c_1 < c_2$, system (178) is dissipative. The competition of the g's is the similar to that of system (177) except that $g_{x_1} > g_{x_2}$ and $g_{y_1} > g_{y_2}$. When $\alpha = \beta$, the competition among the g's can be strong, and otherwise may be weak. Figure 52(a) displays a highly chaotic attractor when $\alpha = \beta = 1.5$, while Figure 52(b) shows a quasiperiodic attractor when $\alpha = 1.7$, $\beta = 1.5$. In both cases, we have taken $c_1 = 1.0$, $c_2 = 1.1$. Therefore, we may conclude that the complexity of an attractor depends more on the competition of the CMs than on the dissipation of the system.

5. Conclusions

In this chapter, we have studied bifurcation of limit cycles, chaos, bifurcation control, chaos control and chaos synchronizations. Several recently developed methods are introduced. Illustrative examples chosen from practical problems are presented, and numerical results are given to confirm the analytical predictions.

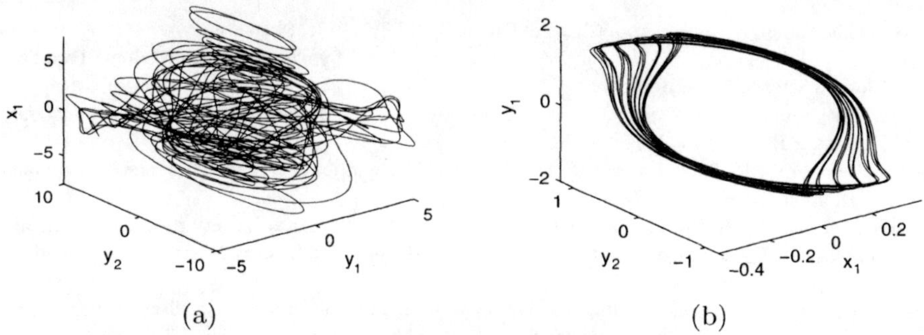

Figure 52. The attractors of system (178) when $c_1 = 1.0$, $c_2 = 1.1$ with the initial condition $(x_1(0), x_2(0), y_1(0), y_2(0)) = (0.13, -0.51, -8.78, 11.73)$ for (a) $\alpha = \beta = 1.5$ and (b) $\alpha = 1.7$, $\beta = 1.5$.

It has been shown that the phenomenon of limit cycle exist in many real problems, and the importance of Hopf bifurcation control is seen from both theoretical development and practical applications. A new concept of mode competition is introduced, which is shown very useful in identifying chaotic parameter regimes and creating new chaotic systems. Chaos control and chaos synchronization, on the other hand, demonstrate that the Lyapunov function method is still a good tool in the study of stability of nonlinear dynamical systems.

Acknowledgement

This work was supported by the Natural Sciences and Engineering Research Council of Canada (NSERC).

References

Abed, E.H., Fu, J.H., 1986. Local feedback stabilization and bifurcation control: I. Hopf bifurcation. Systems Control Lett. 7, 11–17.

Arnold, V.I., 1977. Loss of stability of self-oscillations close to resonance and versal deformations of equivariant vector fields. Funct. Anal. Appl. 11, 85–92.

Bautin, N.N., 1952. On the number of limit cycles which appear with the variation of coefficients from an equilibrium position of focus or center type. Mat. Sb. (N.S.) 30 (72), 181–196.

Berns, D., Moiola, J.L., Chen, G.R., 2000. Controlling oscillation amplitudes via feedback. Internat. J. Bifur. Chaos 10, 2815–2822.

Bleich, M.E., Socolar, J.E.S., 1996. Stability of periodic orbits controlled by time-delay feedback. Phys. Lett. A 210, 87–94.

Brandt, M.E., Shih, H.T., Chen, G.R., 1997. Linear time-delay feedback control of a pathological rhythm in a cardiac conduction model. Phys. Rev. E 56, 1334–1337.

Chen, G.R., 2000. Controlling Chaos and Bifurcation in Engineering and Systems. CRC Press, Boca Raton, FL.

Chen, G.R., Dong, X., 1998. From Chaos to Order. World Scientific, Singapore.
Chen, G.R., Lü, J.H., 2003. Dynamical Analysis, Control and Synchronization of Lorenz Families. Chinese Science Press, Beijing.
Chen, G.R., Moiola, J.L., Wang, H.O., 2000. Bifurcation control: Theories, methods, and applications. Internat. J. Bifur. Chaos 10, 511–548.
Chen, H.B., Liu, Y.R., 2004. Linear recursion formulas of quantities of singular point and applications. Appl. Math. Comput. 148, 163–171.
Chen, H.B., Liu, Y.R., Yu, P., 2006. Center and isochronous center at infinity in a class of planar systems. In: Dynamics of Continuous, Discrete and Impulsive Systems, Series B: Applications and Algorithms, in press.
Chen, Z., Yu, P., 2005a. Controlling and anti-controlling Hopf bifurcations in discrete maps using polynomial functions. Chaos Solitons Fractals 25, 1231–1248.
Chen, Z., Yu, P., 2005b. Hopf bifurcation control for an Internet congestion model. Internat. J. Bifur. Chaos 15 (8), 2643–2651.
Chua, L.O., Komuro, M., Matsumoto, T., 1986. The double scroll family. IEEE Trans. Circuits Syst. 33, 1073–1118.
Cuomo, K.M., Oppenheim, A.V., 1993. Circuit implementation of synchronized chaos with applications to communications. Phys. Rev. Lett. 71, 65–68.
Deb, S., Srikant, R., 2003. Global stability of congestion controller for the Internet. IEEE Trans. Automat. Control 48, 1055–1060.
Field, R.J., Noyes, R.M., 1974. Oscillations in chemical systems. IV. Limit cycle behavior in a model of a real chemical reaction. J. Chem. Phys. 60, 1877–1884.
Gua, S.M., Shieh, L.S., Chen, G.R., Lin, C.F., 2000. Effective chaotic orbit tracker: A perdition-based digital redesign approach. IEEE Trans. Circuits Syst. 47, 1557–1570.
Guckenheimer, J., Holmes, P., 1993. Nonlinear Oscillations, Dynamical Systems, and Bifurcations of Vector Fields, 4th ed. Springer-Verlag, New York.
Haken, H., 1983. Synergetics, An Introduction, Nonequilibrium Phase Transitions and Self-Organization in Physics, Chemistry and Biology, 3rd ed. Springer-Verlag, Berlin.
Hale, J., Lunel, S., 1993. Introduction Functional Differential Equations. Springer-Verlag, New York.
Han, M., 1999. Liapunov constants and Hopf cyclicity of Liénard systems. Ann. Differential Equations 5 (2), 113–126.
Hardy, Y., Steeb, W.H., 1999. The Rikitake two-disk dynamo system and domains with periodic orbits. Internat. J. Theoret. Phys. 38, 2413–2417.
Hassard, B.D., Kazarinoff, N.D., Wan, Y.-H., 1981. Theory and Applications of Hopf Bifurcation. Cambridge Univ. Press, Cambridge.
Hilbert, D., 1902. Mathematical problems. (M. Newton, transl.) Bull. Amer. Math. Soc. 8, 437–479.
Hilbert, D., 1976. Mathematical problems. Proc. Sympos. Pure Math. 28, 1–34.
Hopf, E., 1942. Abzweigung einer periodischen Losung von stationaren Losung einers differential-systems. Ber. Math. Phys. Kl. Sachs Acad. Wiss. Leipzig 94, 1–22; Ber. Math. Phys. Kl. Sachs Acad. Wiss. Leipzig 95 (1942), 3–22.
James, E.M., Lloyd, N.G., 1991. A cubic system with eight small-amplitude limit cycles. IMA J. Appl. Math. 47, 163–171.
Johari, R., Tan, D.K.H., 2001. End-to-end congestion control for the Internet: Delays and stability. IEEE/ACM Trans. Networking 9, 818–832.
Kelly, F.P., 2000. Models for a self-managed Internet. Philos. Trans. R. Soc. London Ser. A 358, 2335–2348.
Kelly, F.P., Maulloo, A., Tan, D., 1998. Rate control in communication networks: Shadow prices, proportional fairness and stability. J. Oper. Res. Soc. 49, 237–252.
Kukles, I.S., 1944. Necessary and sufficient conditions for the existence of center. Dokl. Akad. Nauk 42, 160–163.

Leonov, G., 2001. Bound for attractors and the existence of homoclinic orbits in the Lorenz system. J. Appl. Math. 65, 19–32.

Leonov, G.A., Bunin, A., Koksch, N., 1987. A tractor localization of the Lorenz system. ZAMM Z. Angew. Math. Mech. 67, 649–656.

Leonov, G.A., Reitmann, V., 1987. Attraktoreingrenzung fur Nichtlineare System. Teubner, Leipzig.

Li, C.G., Chen, G.R., Liao, X.F., Yu, J.B., 2004. Hopf bifurcation in an Internet congestion control model. Chaos Solitons Fractals 19, 853–862.

Li, D., Lu, J., Wu, X., Chen, G.R., 2005. Estimating the bounds for the Lorenz family of chaotic systems. Chaos Solitons Fractals 23, 529–534.

Li, J.B., 2003. Hilbert's 16th problem and bifurcations of planar polynomial vector fields. Internat. J. Bifur. Chaos 13, 47–106.

Li, J.B., Chan, H.S.Y., Chung, K.W., 2002. Some lower bounds for $H(n)$ in Hilbert's 16th problem. Qual. Theory Dyn. Syst. 3, 345–360.

Li, J.B., Huang, Q., 1987. Bifurcations of limit cycles forming compound eyes in the cubic system. Chinese Ann. Math. Ser. B 8, 391–403.

Li, J.B., Li, C.F., 1985. Planar cubic Hamiltonian systems and distribution of limit cycles of (E_3). Acta Math. Sin. 28, 509–521.

Li, J.B., Liu, Z., 1991. Bifurcation set and limit cycles forming compound eyes in a perturbed Hamiltonian system. Publ. Mat. 35, 487–506.

Li, J.B., Zhao, X.H., 1989. Rotation symmetry groups of planar Hamiltonian systems. Ann. Differential Equations 5, 25–33.

Liao, X.X., 1993. Absolute Stability of Nonlinear Control Systems. Kluwer Academic–China Science Press, Beijing.

Liao, X.X., 2001. Mathematical Theory and Application of Stability, 2nd ed. Huazhong Normal Univ. Press, Wuhan.

Liao, X.X., Chen, G.R., 2003a. Chaos synchronization of general Lurie system via time-delay feedback control. Internat. J. Bifur. Chaos 13, 207–213.

Liao, X.X., Chen, G.R., 2003b. On feedback-controlled synchronization of chaotic systems. Internat. J. Systems Sci. 43, 453–461.

Liao, X.X., Yu, P., 2005. Analysis on the global exponent synchronization of Chua's circuit using absolute stability theory. Internat. J. Bifur. Chaos 15 (12), 3687–3881.

Librescu, L., 1975. Elastostatics and Kinetics of Anisotropic and Heterogeneous Shell-Type Structures. Noordhoff, Leyden, The Netherlands.

Liénard, A., 1928. Etude des oscillations entretenues. Rev. Générale de lÉlectricité 23, 901–912.

Liu, Y., Li, J., 1989. On the singularity values of complex autonomous differential systems. Sci. China Ser. A 3, 245–255.

Liu, Y.R., Huang, W.T., 2005. A cubic system with twelve small amplitude limit cycles. Bull. Sci. Math. 129, 83–98.

Lloyd, N.G., Blows, T.R., Kalenge, M.C., 1988. Some cubic systems with several limit cycles. Nonlinearity 1, 653–669.

Lorenz, E.N., 1963. Deterministic nonperiodic flow. J. Atmospheric Sci. 20, 130–141.

Lorenz, E.N., 1993. The Essence of Chaos. Univ. of Washington Press.

Lü, J.H., Zhou, T., Zhang, S., 2002. Controlling Chen attractor using feedback function based on parameters identification. Chinese Phys. 11, 12–16.

Lynch, S., 2004. Dynamical Systems with Applications Using Maple. Birkhäuser, Boston.

Malkin, K.E., 1964. Criteria for center of a differential equation. Volg. Mat. Sb. 2, 87–91.

Maple 10, 2005. Maplesoft, Waterloo, Canada.

Marzocca, P., Librescu, L., Silva, W.A., 2002. Flutter, post-flutter and control of a supersonic 2-D lifting surface. J. Guidance Control Dynamics 25, 962–970.

Mathematica, 2004. The Way the World Calculates. Wolfram Research, UK.

Matsumoto, T., 1984. A chaotic attractor from Chua's circuit. IEEE Trans. Circuits Syst. 31, 1055–1058.

Melnikov, V.K., 1963. On the stability of the center for time periodic perturbations. Trans. Moscow Math. Soc. 12, 1–57.

Nayfeh, A.H., 1993. Methods of Normal Forms. Wiley, New York.

Nayfeh, A.H., Harb, A.M., Chin, C.M., 1996. Bifurcations in a power system model. Internat. J. Bifur. Chaos 14, 497–512.

Neufeld, R.W.J., 1999. Dynamic differentials of stress and coping. Psychological Review 106, 385–397.

Nicolis, G., Prigogine, I., 1977. Self-Organization in Nonequilibrium Systems. Wiley, New York.

Ott, E., Grebogi, C., Yorke, J.A., 1990. Controlling chaos. Phys. Rev. Lett. 64, 1096–1199.

Pecora, L.M., Carroll, T.L., 1990. Synchronization in chaotic circuits. Phys. Rev. Lett. 64, 821–824.

Pecora, L.M., Carroll, T.L., 1991. Driving systems with chaotic signals. Phys. Rev. A 44, 2374–2378.

Perko, L.M., 2001. Differential Equations and Dynamical Systems. Springer-Verlag, New York.

Pesheck, E., Pierre, C., Shaw, S.W., 2002. A new Galerkin-based approach for accurate nonlinear normal modes through invariant manifolds. J. Sound Vib. 249, 971–993.

Poincaré, H. (1892–1899). Les Methodes Nouvelles de la Mecanique Celeste.

Rössler, O.E., 1976. An equation for continuous chaos. Phys. Lett. A 57, 397–398.

Rössler, O.E., 1979. Continuous chaos – four prototype equations. Ann. (NY) Acad. Sci. 316, 376–392.

Rössler, O.E., 1991. In: Baier, G., Klein, M. (Eds.), A Chaotic Hierarchy. World Scientific, Singapore.

Sanchez, E.N., Peren, J.P., Martizez, M., Chen, G.R., 2002. Chaos stabilization: An inverse optimal control approach. Latin Amer. Appl. Res.: Int. J. 32, 111–114.

Shaw, S.W., Pierre, C., 1993. Normal modes for non-linear vibratory systems. J. Sound Vib. 164, 85–124.

Sieber, J., Krauskopf, B., 2004. Bifurcation analysis of an inverted pendulum with delayed feedback control near a triple-zero eigenvalue singularity. Nonlinearity 17, 85–103.

Smale, S., 1998. Mathematical problems for the next century. Math. Intelligencer 20, 7–15.

Song, Y.X., Yu, X.H., Chen, G.R., Xu, J.X., Tian, Y.P., 2002. Time delayed repetitive learning control for chaotic systems. Internat. J. Bifur. Chaos 12, 1057–1065.

Sprott, J.C., 1977. Simplest dissipative chaotic flow. Phys. Lett. A 228, 271–274.

Sprott, J.C., Ling, S.J., 2000. Algebraically simple chaotic flows. Int. J. Chaos Theory Appl. 5, 3–22.

Stewart, L., 2002. The Lorenz attractor exists. Nature 406, 948–949.

Tsuneda, A., 2005. A gallery of attractors from smooth Chua's equation. Internat. J. Bifur. Chaos 15 (1), 1–49.

Van der Pol, B., 1926. On relaxation–oscillations. Philos. Mag. 2 (7), 978–992.

Wang, H.O., Abed, E.G., 1995. Bifurcation control of a chaotic system. Automatica 31, 1213–1226.

Wang, S., 2005. Hilbert's 16th Problem and Computation of Limit Cycles. Doctoral thesis, The University of Western Ontario, London, Ontario, Canada.

Wang, S., Yu, P., 2005. Bifurcation of limit cycles in a quintic Hamiltonian system under sixth-order perturbation. Chaos Solitons Fractals 26 (5), 1317–1335.

Wang, S.P., Yu, J., Li, J., 2006. Bifurcation of limit cycles in a Z_{10}-equivariant vector fields of degree 9. Internat. J. Bifur. Chaos 16 (7), in press.

Wiggins, S., 1990. Introduction to Applied Nonlinear Dynamical Systems and Chaos. Spring-Verlag, New York.

Wu, J.H., 2001. Introduction to Neural Dynamics and Signal Transmission Delay. de Gruyter, Berlin.

Yao, W., 2002. Improving Security of Communication via Chaotic Synchronization. Doctoral thesis, The University of Western Ontario, London, Ontario, Canada.

Yao, W., Essex, C., Yu, P., 2002. A new chaotic system for better secure communication. Dyn. Contin. Discrete Impuls. Syst. Ser. B Appl. Algorithms 10, 221–234.

Yao, W., Yu, P., Essex, C., 2001. Delayed stochastic differential model for quiet standing. Phys. Rev. E 63, 021902.

Yao, W., Yu, P., Essex, C., 2002. Estimating chaotic parameter regimes by generalized competition modes approach. Commun. Nonlinear Sci. Numer. Simul. 7, 197–205.

Yao, W., Yu, P., Essex, C., Davison, M., 2006. Competitive modes and their application. Internat. J. Bifur. Chaos 16 (3), in press.

Yu, P., 1998. Computation of normal forms via a perturbation technique. J. Sound Vib. 211, 19–38.

Yu, P., 2000. A method for computing center manifold and normal forms. In: Proc. of Diff. Eqs., vol. 2. World Scientific, Singapore, pp. 832–837.

Yu, P., 2001. Symbolic computation of normal forms for resonant double Hopf bifurcations using multiple time scales. J. Sound Vib. 274, 615–632.

Yu, P., 2002. Limit cycles in 3rd-order planar system. In: International Congress of Mathematicians, Beijing, China, August 20–28.

Yu, P., 2003a. Twelve limit cycles in a cubic case of the 16th Hilbert problem. In: Workshop on Bifurcation Theory and Spatio-Temporal Pattern Formation in PDE, Toronto, Canada, December 11–13.

Yu, P., 2003b. Bifurcation dynamics in control systems. In: Chen, G.R., Hill, D.J., Yu, X. (Eds.), Bifurcation Control: Theory and Applications. Springer-Verlag, Berlin, pp. 99–126.

Yu, P., Chen, G.R., 2004. Hopf bifurcation control using nonlinear feedback with polynomial functions. Internat. J. Bifur. Chaos 14, 1683–1704.

Yu, P., Corless, R., 2006. Symbolic computation of limit cycles associated with Hilbert's 16th problem. J. Symbolic Comput. (revised version submitted for publication).

Yu, P., Han, M., 2004. Twelve limit cycles in a 3rd-order planar system with Z_2 symmetry. Commun. Appl. Pure Anal. 3 (3), 515–526.

Yu, P., Han, M., 2005a. Small limit cycles bifurcating from fine focus points in cubic order Z_2-equivariant vector fields. Chaos Solitons Fractals 24 (1), 329–348.

Yu, P., Han, M., 2005b. Twelve limit cycles in a cubic case of the 16th Hilbert problem. Internat. J. Bifur. Chaos 15 (7), 2191–2205.

Yu, P., Liao, X., 2006a. Globally attractive and positive invariant set of the Lorenz system. Internat. J. Bifur. Chaos 16 (3), in press.

Yu, P., Liao, X., 2006b. New estimations for globally attractive and positive invariant set of the family of the Lorenz systems. Internat. J. Bifur. Chaos, submitted for publication.

Yu, P., Yuan, Y., Xu, J., 2002. Study of double Hopf bifurcation and chaos for an oscillator with time delayed feedback. Commun. Nonlinear Sci. Numer. Simul. 7, 69–91.

Yu, P., Zhang, W., Bi, Q., 2001. Vibration analysis on a thin plate with the aid of computation of normal forms. Internat. J. Non-Linear Mech. 36, 597–627.

Yuan, Y., Yu, P., Librescu, L., Marzocca, P., 2004. Aeroelasticity of time-delayed feedback control of two-dimensional supersonic lifting surface. J. Guidance Control Dynamics 27, 795–803.

Zoladek, H., 1995. Eleven small limit cycles in a cubic vector field. Nonlinearity 8, 843–860.

Chapter 2

Grazing Flows in Discontinuous Dynamic Systems

Albert C.J. Luo

Department of Mechanical and Industrial Engineering,
Southern Illinois University, Edwardsville, IL 62026, USA
E-mail: aluo@siue.edu

Contents

1. Introduction 127
2. Domain accessibility 129
3. Discontinuous dynamic systems 131
4. Oriented boundary and singular sets 132
5. Local singularity and grazing flows 141
6. Piecewise linear systems 150
7. Friction-induced oscillators 158
8. Conclusions 187

Appendix 187
References 189

1. Introduction

Most of the existing theories in dynamics are based on the Lipschitz condition (e.g., Poincare, 1892; Birkhoff, 1927). Indeed, those theories are widely used in science and engineering. However, ones want to develop the expected dynamic behavior to satisfy specified requirements. Hence, discontinuous constraints destroying the Lipschitz conditions are added to dynamic systems. Because of this reason, the established dynamical system theories based on the Lipschitz condition are not adequate for such nonsmooth dynamical systems. For instance, smooth linear dynamical systems with periodic impacting (e.g., Masri and Caughey, 1966; Luo and Han, 1996) have complicated dynamical behaviors which are unpredictable from the traditional dynamical theories. The Lipschitz

Edited Series on Advances in Nonlinear Science and Complexity
Volume 1 ISSN: 1574-6909 DOI: 10.1016/S1574-6909(06)01002-1

condition is very strong for practical dynamical problems, and many dynamical systems cannot satisfy such a condition. To overcome this difficulty, a theory for discontinuous dynamical systems should be developed further.

The early investigation of discontinuous systems in mechanical engineering can be found in the 30's of the last century (e.g., Den Hartog, 1930; Den Hartog and Mikina, 1932). Masri and Caughey (1966) investigated the stability of the symmetrical period-1 motion of a discontinuous oscillator. Masri (1970) gave the further, analytical and experimental investigations on the general motion of impact dampers. The unsymmetrical motion was observed, and the rigorous stability analysis was conducted as well. Since the discontinuity exists widely in engineering and control systems, Utkin (1978) presented sliding modes and the corresponding variable structure systems, and the theory of automatic control systems described with variable structures and sliding motions was also developed in Utkin (1981). Further, Fillippov (1988) developed a geometrical theory of the differential equations with discontinuous right-hand sides, and the local singularity theory of the discontinuous boundary was discussed qualitatively. Ye, Michel and Hou (1998) discussed the stability theory for hybrid systems. From geometrical points of view, Broucke, Pugh and Simic (2001) investigated the structural stability of piecewise smooth systems. So far, an efficient method to model such nonsmooth dynamical systems has not been developed yet. For instance, the linear impacting oscillators cannot be fully understood as one of the simplest discontinuous systems (e.g., Senator, 1970; Bapat, Popplewell and Mclachlan, 1983; Shaw and Holmes, 1983a; Luo, 1995, 2002; Han, Luo and Deng, 1995). Another typical example in engineering is piecewise smooth linear systems. Shaw and Holmes (1983b) used mapping techniques to investigate the chaotic motion of a piecewise linear system with a single discontinuity. Natsiavas (1989) numerically determined the periodic motion and stability for a system with a symmetric, trilinear spring. Nordmark (1991) introduced the grazing mapping to investigate nonperiodic motions. Kleczka, Kreuzer and Schiehlen (1992) investigated the periodic motions and bifurcations of piecewise linear oscillator motion, and numerically observed the grazing motion. Leine and Van Campen (2002) investigated the discontinuous bifurcations of periodic solutions through the Floquet multipliers of periodic solutions. The analytical prediction of periodic responses of piecewise linear systems was presented (e.g., Luo and Menon, 2004; Menon and Luo, 2005). Normal formal mapping for piecewise smooth dynamical systems with/without sliding were discussed (e.g., di Bernardo, Budd and Champney, 2001; di Bernardo, Kowalczyk and Nordmark, 2002). Kunze (2000) presented a mathematical background of a nonsmooth dynamical system with friction. Popp (2000) pointed out: (i) solution methods need to be improved; (ii) efficient methods for stability and bifurcation are required to develop and (iii) the attractor characteristics need to be reconstructed. Luo (2005a) developed a local theory for nonsmooth dynamical systems on connected domains.

This theory was used to piecewise linear systems (e.g., Luo, 2005b, 2006) and friction-inducted oscillator (e.g., Luo and Gegg, 2005, 2006a, 2006b). Luo and Chen (2005, 2006) investigated the periodic and chaotic motions of the piecewise linear systems with impacting. The concept for the strange attractor fragmentation in discontinuous systems was proposed in Luo (2005c).

In this chapter, the grazing flow in the vicinity of the discontinuous boundary will be presented. The accessible and inaccessible subdomains will be introduced. On the accessible domain, the corresponding dynamic systems will be defined. The oriented boundary sets and singular sets will be discussed. The local singularity and tangency of a flow on the separation boundary will be investigated. The necessary and sufficient conditions for such a local singularity and tangency will be presented. The grazing flows in piecewise linear systems and friction-induced oscillators will be investigated, and the grazing conditions of the flows will be determined.

2. Domain accessibility

Before development of a general theory for nonsmooth dynamical systems on a universal domain $\mho \subset \Re^n$ in phase space, the subdomains Ω_i $(i = 1, 2, \ldots)$ of the domain $\mho$ are introduced, and the dynamics on the subdomains are defined differently.

DEFINITION 1. A subdomain in the universal domain $\mho$ is termed the *accessible subdomain* on which a specific, continuous dynamical system can be defined.

DEFINITION 2. A subdomain in a universal domain $\mho$ is termed the *inaccessible subdomain* on which no dynamical system can be defined.

Since the dynamical system can be defined differently on each accessible subdomain, the dynamical behaviors of the system in those accessible subdomains Ω_i can be different from each other in the sense of Newton's mechanics. These different behaviors cause the complexity of motion in the universal domain $\mho$. Owing to the accessible and inaccessible subdomains, the universal domain $\mho$ is classified into the connectable and separable ones. The connectable domain is defined as follows.

DEFINITION 3. A domain $\mho$ in phase space is termed the *connectable domain* if all the accessible subdomains of the universal domain can be connected without any inaccessible subdomain.

Similarly, the separable domain is defined as follows.

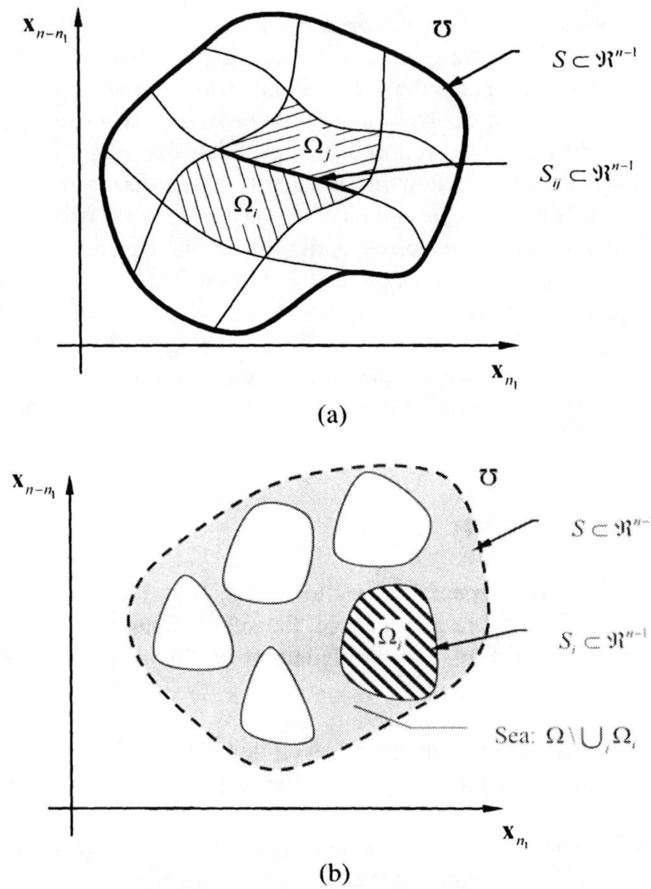

Figure 1. Phase space: (a) connectable and (b) separable domains.

DEFINITION 4. A domain $\mho$ is termed the *separable domain* if the accessible subdomains in the universal domain are separated by inaccessible domains.

The boundary between two adjacent, accessible subdomains is a bridge of dynamical behaviors in two domains for motion continuity. For the connectable domain, it is bounded by the universal boundary surface $S \subset \Re^r$ $(r \leqslant n-1)$, and each subdomain is bounded by the subdomain boundary surface $S_{ij} \subset \Re^r$ $(i, j \in \{1, 2, \ldots\})$ with or without the partial universal boundary. For instance, consider an n-D connectable domain in phase space, as shown in Figure 1(a) through an n_1-dimensional, subvector $\mathbf{x}_{n_1}$ and an $(n-n_1)$-dimensional, subvector $\mathbf{x}_{n-n_1}$. That is, $\mathbf{x}_{n_1} \equiv (x_1, x_2, \ldots, x_{n_1})^{\mathrm{T}}$ and $\mathbf{x}_{n_1} \equiv (x_{n_1+1}, x_{n_1+2}, \ldots, x_n)^{\mathrm{T}}$.

The shaded area Ω_i is a specific subdomain, and other subdomains are white. The dark, solid curve represents the original boundary of the domain $\mho$. In the separable domain, there is at least an inaccessible subdomain to separate the accessible subdomains. The union of inaccessible subdomains is also called the "sea". The sea is the complement of the accessible subdomains to the universal (original) domain $\mho$. That is determined by $\Xi = \mho \backslash \bigcup_i \Omega_i$. The accessible subdomains in the domain $\mho$ are also called the "islands". For illustration of such a definition, an n-D separable domain is shown in Figure 1(b). The dashed surface is the boundary of the universal domain, and the gray area is the sea. The white regions are the accessible domains (or islands). The diagonal-line-shaded region represents a specific accessible subdomain (island). From one island to another, the transport is needed for motion continuity. Due to page limitation, the transport laws will not be discussed. For an accessible domain, grazing flows in discontinuous systems will be presented.

3. Discontinuous dynamic systems

To demonstrate the basic concepts of nonsmooth dynamical system theory, the development of the theory in this chapter is restricted to an n-dimensional, nonsmooth dynamical system. Consider a planar, dynamic system consisting of N subdynamic systems in a universal domain $\mho \subset \Re^n$. The universal domain is divided into N accessible subdomains Ω_i, and the union of all the accessible subdomains $\bigcup_{i=1}^{N} \Omega_i$ and the universal domain $\mho = \bigcup_{i=1}^{N} \Omega_i \cup \Xi$, as shown in Figure 1 through an n_1-dimensional, subvector $\mathbf{x}_{n_1}$ and an $(n - n_1)$-dimensional, subvector $\mathbf{x}_{n-n_1}$. Ξ is the union of the inaccessible domains. For the connectable domain in Figure 1(a), $\Xi = \varnothing$. In Figure 1(b), the union of the inaccessible subdomains is the sea, $\Xi = \mho \backslash \bigcup_{i=1}^{m} \Omega_i$ is the complement of the union of the accessible subdomain. On the ith open subdomain Ω_i, there is a C^r-continuous system ($r \geqslant 1$) in a form of

$$\dot{\mathbf{x}} \equiv \mathbf{F}^{(i)}(\mathbf{x}, t, \boldsymbol{\mu}_i) \in \Re^n, \quad \mathbf{x} = (x_1, x_2, \ldots, x_n)^{\mathrm{T}} \in \Omega_i. \tag{1}$$

The time is t and $\dot{\mathbf{x}} = \mathrm{d}\mathbf{x}/\mathrm{d}t$. In an accessible subdomain Ω_i, the vector field $\mathbf{F}^{(i)}(\mathbf{x}, t, \boldsymbol{\mu}_i)$ with parameter vectors $\boldsymbol{\mu}_i = (\mu_i^{(1)}, \mu_i^{(2)}, \ldots, \mu_i^{(l)})^{\mathrm{T}} \in \Re^l$ is C^r-continuous ($r \geqslant 1$) in $\mathbf{x}$ and for all time t; and the continuous flow in equation (1) $\mathbf{x}^{(i)}(t) = \boldsymbol{\Phi}^{(i)}(\mathbf{x}^{(i)}(t_0), t, \boldsymbol{\mu}_i)$ with $\mathbf{x}^{(i)}(t_0) = \boldsymbol{\Phi}^{(i)}(\mathbf{x}^{(i)}(t_0), t_0, \boldsymbol{\mu}_i)$ is C^{r+1}-continuous for time t.

The nonsmooth dynamic theory developed in this chapter holds for the following conditions:

(A1) The switching between two adjacent subsystems possesses time-continuity.

(A2) For an unbounded, accessible subdomain Ω_i, there is a bounded domain $D_i \subset \Omega_i$ and the corresponding vector field and its flow are bounded, i.e.,

$$\begin{aligned} &\|\mathbf{F}^{(i)}\| \leqslant K_1 \text{ (constant)} \quad \text{and} \\ &\|\boldsymbol{\Phi}^{(i)}\| \leqslant K_2 \text{ (constant)} \quad \text{on } D_i \quad \text{for } t \in [0, \infty). \end{aligned} \tag{2}$$

(A3) For a bounded, accessible domain Ω_i, there is a bounded domain $D_i \subset \Omega_i$ and the corresponding vector field is bounded, but the flow may be unbounded, i.e.,

$$\begin{aligned} &\|\mathbf{F}^{(i)}\| \leqslant K_1 \text{ (constant)} \quad \text{and} \\ &\|\boldsymbol{\Phi}^{(i)}\| < \infty \quad \text{on } D_i \quad \text{for } t \in [0, \infty). \end{aligned} \tag{3}$$

4. Oriented boundary and singular sets

Since dynamical systems on the different accessible subdomains are distinguishing, the relation between flows in the two subdomains should be developed herein for flow continuity. For a subdomain Ω_i, there are k_i-subboundaries ($k_i \leqslant N-1$). Consider a boundary set of any two subdomains, formed by the intersection of the closed subdomains, i.e., $\partial\Omega_{ij} = \overline{\Omega}_i \cap \overline{\Omega}_j$ ($i, j \in \{1, 2, \ldots, N\}$, $j \neq i$), as shown in Figure 2.

DEFINITION 5. The boundary in the n-D phase space is defined as

$$\begin{aligned} S_{ij} \equiv \partial\Omega_{ij} &= \overline{\Omega}_i \cap \overline{\Omega}_j \\ &= \big\{\mathbf{x} \mid \varphi_{ij}(\mathbf{x}, t) = 0 \text{ for specific time } t, \\ &\qquad \text{where } \varphi_{ij} \text{ is } C^r\text{-continuous } (r \geqslant 1)\big\} \subset \Re^{n-1}. \end{aligned} \tag{4}$$

DEFINITION 6. The two subdomains Ω_i and Ω_j are *disjoint* if the boundary $\partial\Omega_{ij}$ is an empty set (i.e., $\partial\Omega_{ij} = \varnothing$).

The boundary values $\mathbf{x}^{(i)}$ and $\mathbf{x}^{(j)}$ are pertaining to the open domains Ω_i and Ω_j, respectively. Based on the boundary definition, we have $\partial\Omega_{ij} = \partial\Omega_{ji}$. From Fillippov (1988) the boundary $\partial\Omega_{ij}$ can be determined by

$$\dot{\mathbf{x}}^{(0)} = \mathbf{F}^{(0)}\big(\mathbf{x}^{(0)}, t\big), \tag{5}$$

where $\mathbf{x}^{(0)} = (x_1^{(0)}, x_2^{(0)}, \ldots, x_n^{(0)})^{\mathrm{T}}$.

DEFINITION 7. If the intersection of the three or more subdomains,

$$\Gamma_{i_1 i_2 \cdots i_k} \equiv \bigcap_{i=i_1}^{i_k} \overline{\Omega}_i \subset \Re^r, \quad r = 0, 1, \ldots, n-2, \tag{6}$$

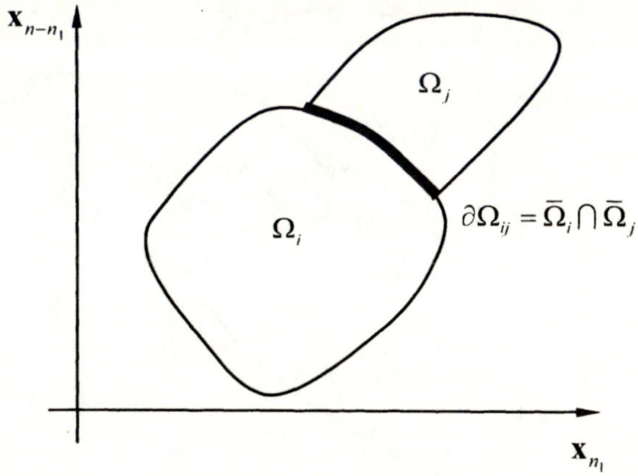

Figure 2. Subdomains Ω_i and Ω_j, the corresponding boundary $\partial\Omega_{ij}$.

where $i_k \in \{1, 2, \ldots, n_2\}$ and $k \geqslant 3$, is nonempty, the subdomain intersection is termed the *singular* set.

The boundary functions relative to the singular sets are C^0-continuous. For $r = 0$, the singular sets will be singular points, which are also termed the *corner points or vertex.* For $r = 1$, the singular sets will be lines, which are termed the singular edges to the $(n-1)$-dimensional discontinuous boundary. For $r \in \{2, 3, \ldots, n-2\}$, the singular sets are the r-dimensional surfaces to the $(n-1)$-dimensional discontinuous boundary. In Figure 3, the singular set for the three closed domains $\{\bar{\Omega}_i, \bar{\Omega}_j, \bar{\Omega}_k\}$ is sketched. The circular symbols represent intersection sets. The largest solid circular symbol stands for the singular set Γ_{ijk}. The corresponding discontinuous boundaries pertaining to the singular set are labeled by $\partial\Omega_{ij}$, $\partial\Omega_{jk}$ and $\partial\Omega_{ik}$. The singular set possesses the parabolic or hyperbolic behavior depending on the properties of the discontinuous boundary set, which can be referred to Luo (2005a).

DEFINITION 8. For a discontinuous dynamical system in equation (1), $\mathbf{x}(t_m) \equiv \mathbf{x}_m \in \partial\Omega_{ij}$ at t_m. For an arbitrarily small $\varepsilon > 0$, there are two time intervals $[t_{m-\varepsilon}, t_m)$ and $(t_m, t_{m+\varepsilon}]$, suppose $\mathbf{x}^{(i)}(t_{m-}) = \mathbf{x}_m = \mathbf{x}^{(j)}(t_{m+})$. The nonempty boundary set $\partial\Omega_{ij}$ to a flow $\mathbf{x}^{(\alpha)}(t)$ $(\alpha \in \{i, j\})$ is *semipassable* from the domain Ω_i to Ω_j (expressed by $\overrightarrow{\partial\Omega}_{ij}$) if the flow $\mathbf{x}^{(\alpha)}(t)$ possesses the following properties either

$$\begin{cases} \mathbf{n}^{\mathrm{T}}_{\partial\Omega_{ij}} \cdot \left[\mathbf{x}^{(0)}(t_{m-\varepsilon}) - \mathbf{x}^{(i)}(t_{m-\varepsilon})\right] > 0 \quad \text{and} \\ \mathbf{n}^{\mathrm{T}}_{\partial\Omega_{ij}} \cdot \left[\mathbf{x}^{(j)}(t_{m+\varepsilon}) - \mathbf{x}^{(0)}(t_{m+\varepsilon})\right] > 0 \quad \text{for } \mathbf{n}_{\partial\Omega_{ij}} \to \Omega_j, \end{cases} \tag{7a}$$

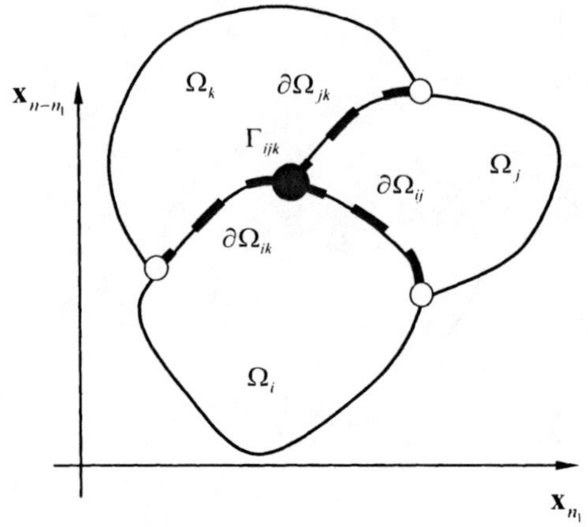

Figure 3. A singular set for the intersection of three closed domains $\{\overline{\Omega}_i, \overline{\Omega}_j, \overline{\Omega}_k\}$. The circular circles represent intersection sets. The largest solid circular symbol stands for the singular set Γ_{ijk}. The corresponding discontinuous boundaries are marked by $\partial\Omega_{ij}$, $\partial\Omega_{jk}$ and $\partial\Omega_{ik}$.

or

$$\begin{cases} \mathbf{n}^{\mathrm{T}}_{\partial\Omega_{ij}} \cdot \left[\mathbf{x}^{(0)}(t_{m-\varepsilon}) - \mathbf{x}^{(i)}(t_{m-\varepsilon})\right] < 0 & \text{and} \\ \mathbf{n}^{\mathrm{T}}_{\partial\Omega_{ij}} \cdot \left[\mathbf{x}^{(j)}(t_{m+\varepsilon}) - \mathbf{x}^{(0)}(t_{m+\varepsilon})\right] < 0 & \text{for } \mathbf{n}_{\partial\Omega_{ij}} \to \Omega_i, \end{cases} \tag{7b}$$

where the normal vector of the boundary $\partial\Omega_{ij}$ is

$$\mathbf{n}_{\partial\Omega_{ij}} = \nabla\varphi_{ij} = \left(\frac{\partial\varphi_{ij}}{\partial x_1}, \frac{\partial\varphi_{ij}}{\partial x_2}, \ldots, \frac{\partial\varphi_{ij}}{\partial x_n}\right)^{\mathrm{T}}_{(\mathbf{x}_m)}. \tag{8}$$

The notations $t_{m\pm\varepsilon} = t_m \pm \varepsilon$ and $t_{m\pm} = t_m \pm 0$ are used. To interpret the geometrical concept of the semipassable boundary sets, consider a flow in equation (1) from the domain Ω_i into the domain Ω_j through the boundary $\partial\Omega_{ij}$. At a time t_m, the flow arrives to the boundary $\partial\Omega_{ij}$, and there is a small neighborhood $(t_{m-\varepsilon}, t_{m+\varepsilon})$ of the time t_m, which is arbitrarily selected. The expression $t_{m\pm\varepsilon} = t_m \pm \varepsilon$. As $\varepsilon \to 0$, the time increment $\Delta t \equiv \varepsilon \to 0$. Before the flow reaches to the boundary, the point $\mathbf{x}^{(i)}(t_{m-\varepsilon}) \equiv (x_1^{(i)}(t_{m-\varepsilon}), x_2^{(i)}(t_{m-\varepsilon}), \ldots, x_n^{(i)}(t_{m-\varepsilon}))^{\mathrm{T}}$ lies in the domain Ω_i. The point $\mathbf{x}_m \equiv (x_1(t_m), x_2(t_m), \ldots, x_n(t_m))^{\mathrm{T}}$ is for the flow on the boundary. After the flow passes through the boundary, the output flow $\mathbf{x}^{(j)}(t_{m+\varepsilon}) \equiv (x_1^{(j)}(t_{m+\varepsilon}), x_2^{(j)}(t_{m+\varepsilon}), \ldots, x_n^{(j)}(t_{m+\varepsilon}))^{\mathrm{T}}$ is a point in the neighborhood of the discontinuous boundary on the side of the domain Ω_j. The input and output flow vectors are $\mathbf{x}^{(i)}(t_m) - \mathbf{x}^{(i)}(t_{m-\varepsilon})$ and $\mathbf{x}^{(j)}(t_{m+\varepsilon}) - \mathbf{x}^{(j)}(t_m)$, respectively. Whether the flow passed through the boundary or not is dependent on the properties of both input and output flow vectors in the neighborhood of

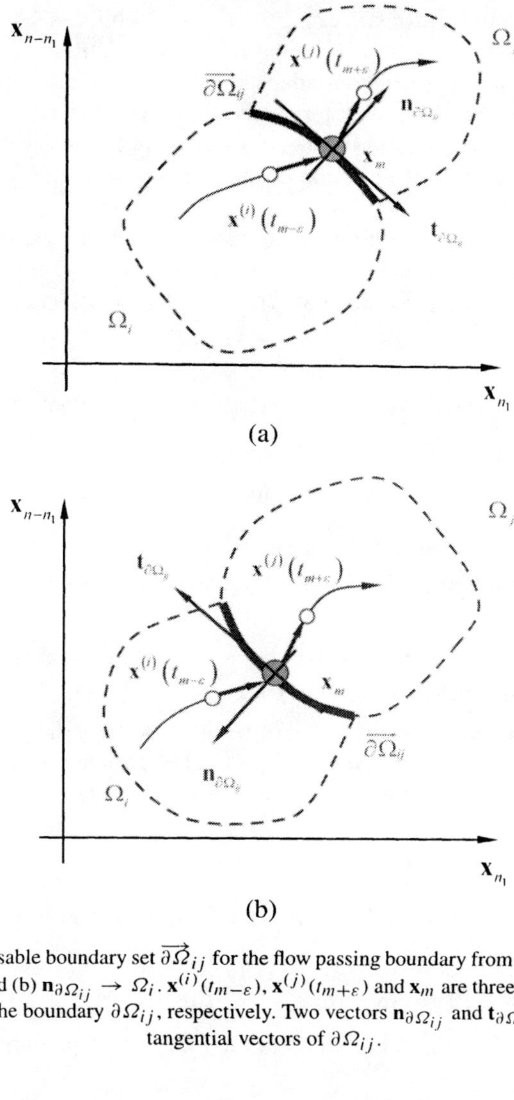

Figure 4. Semipassable boundary set $\overrightarrow{\partial\Omega}_{ij}$ for the flow passing boundary from the domain Ω_i to Ω_j: (a) $\mathbf{n}_{\partial\Omega_{ij}} \to \Omega_j$ and (b) $\mathbf{n}_{\partial\Omega_{ij}} \to \Omega_i$. $\mathbf{x}^{(i)}(t_{m-\varepsilon})$, $\mathbf{x}^{(j)}(t_{m+\varepsilon})$ and $\mathbf{x}_m$ are three points in the domains Ω_i and Ω_j and on the boundary $\partial\Omega_{ij}$, respectively. Two vectors $\mathbf{n}_{\partial\Omega_{ij}}$ and $\mathbf{t}_{\partial\Omega_{ij}}$ are the normal and tangential vectors of $\partial\Omega_{ij}$.

the boundary. The process of the flow passing through the boundary $\partial\Omega_{ij}$ with $\mathbf{n}_{\partial\Omega_{ij}} \to \Omega_j$ and $\mathbf{n}_{\partial\Omega_{ij}} \to \Omega_i$ sets from the domain Ω_i to Ω_j is shown in Figure 4. Two vectors $\mathbf{n}_{\partial\Omega_{ij}}$ and $\mathbf{t}_{\partial\Omega_{ij}}$ are the normal and tangential vectors of the boundary curve $\partial\Omega_{ij}$ determined by $\varphi_{ij}(\mathbf{x}, t) = 0$. When an input flow $\mathbf{x}^{(i)}(t)$ in the domain Ω_i arrives to semipassable boundary $\overrightarrow{\partial\Omega}_{ij}$, the flow can be tangential to, bouncing on semipassable boundary $\overrightarrow{\partial\Omega}_{ij}$, the flow can be tangential to, bouncing on and passing through the semipassable boundary $\overrightarrow{\partial\Omega}_{ij}$. However, once a flow $\mathbf{x}^{(j)}(t)$ in the domain Ω_j arrives to the semipassable boundary $\overrightarrow{\partial\Omega}_{ij}$, the flow cannot pass through the boundary, and either the tangential or bouncing flow $\mathbf{x}^{(j)}(t)$ at the semipassable boundary $\overrightarrow{\partial\Omega}_{ij}$ exists. The tangential (or grazing) flow will be discussed in this chapter. In the following discussion, no any control and transport laws are defined on the semipassable boundary. The direction of $\mathbf{t}_{\partial\Omega_{ij}} \times \mathbf{n}_{\partial\Omega_{ij}}$ is the positive direction of the coordinate by the right-hand rule.

THEOREM 1. *For a discontinuous dynamical system in equation* (1), $\mathbf{x}(t_m) = \mathbf{x}_m \in \partial\Omega_{ij}$ *at* t_m. *For an arbitrarily small* $\varepsilon > 0$, *there are two time intervals* $[t_{m-\varepsilon}, t_m)$ *and* $(t_m, t_{m+\varepsilon}]$. *Suppose* $\mathbf{x}^{(i)}(t_{m-}) = \mathbf{x}_m = \mathbf{x}^{(j)}(t_{m+})$ *and, both flows* $\mathbf{x}^{(i)}(t)$ *and* $\mathbf{x}^{(j)}(t)$ *are* $C^r_{[t_{m-\varepsilon}, t_m)}$*- and* $C^r_{(t_m, t_{m+\varepsilon}]}$*-continuous* $(r \geqslant 2)$ *for time* t, *respectively, and* $\|\mathrm{d}^r\mathbf{x}^{(\alpha)}/\mathrm{d}t^r\| < \infty$ $(\alpha \in \{i, j\})$. *The nonempty boundary set* $\partial\Omega_{ij}$ *for the flows* $\mathbf{x}^{(i)}(t)$ *and* $\mathbf{x}^{(j)}(t)$ *is semipassable from the domain* Ω_i *to* Ω_j *iff*

$$
\begin{aligned}
&\text{either} \quad \begin{cases} \mathbf{n}^{\mathrm{T}}_{\partial\Omega_{ij}} \cdot \dot{\mathbf{x}}^{(i)}(t_{m-}) > 0 & \text{and} \\ \mathbf{n}^{\mathrm{T}}_{\partial\Omega_{ij}} \cdot \dot{\mathbf{x}}^{(j)}(t_{m+}) > 0 & \text{for } \mathbf{n}_{\partial\Omega_{ij}} \to \Omega_j, \end{cases} \\
&\text{or} \quad \begin{cases} \mathbf{n}^{\mathrm{T}}_{\partial\Omega_{ij}} \cdot \dot{\mathbf{x}}^{(i)}(t_{m-}) < 0 & \text{and} \\ \mathbf{n}^{\mathrm{T}}_{\partial\Omega_{ij}} \cdot \dot{\mathbf{x}}^{(j)}(t_{m+}) < 0 & \text{for } \mathbf{n}_{\partial\Omega_{ij}} \to \Omega_i. \end{cases}
\end{aligned} \tag{9}
$$

PROOF. For a point $\mathbf{x}_m \in \partial\Omega_{ij}$ with $\mathbf{n}_{\partial\Omega_{ij}} \to \Omega_j$, suppose $\mathbf{x}^{(i)}(t_{m-}) = \mathbf{x}_m, \mathbf{x}_m = \mathbf{x}^{(j)}(t_{m+})$ and both $\mathbf{x}^{(i)}(t)$ and $\mathbf{x}^{(j)}(t)$ are $C^r_{[t_{m-\varepsilon}, t_m)}$- and $C^r_{(t_m, t_{m+\varepsilon}]}$-continuous $(r \geqslant 2)$ for time t, respectively, and $\|\ddot{\mathbf{x}}^{(\alpha)}(t)\| < \infty$ $(\alpha \in \{i, j\})$ for $0 < \varepsilon \ll 1$. Consider $a \in [t_{m-\varepsilon}, t_{m-})$ and $b \in (t_{m-}, t_{m+\varepsilon}]$. Application of the Taylor series expansion of $\mathbf{x}^{(\alpha)}(t_{m\pm\varepsilon})$ with $t_{m\pm\varepsilon} = t_m \pm \varepsilon$ $(\alpha \in \{i, j\})$ to $\mathbf{x}^{(\alpha)}(a)$ and $\mathbf{x}^{(\alpha)}(b)$ gives

$$
\begin{aligned}
\mathbf{x}^{(i)}(t_{m-\varepsilon}) &\equiv \mathbf{x}^{(i)}(t_{m-} - \varepsilon) \\
&= \mathbf{x}^{(i)}(a) + \dot{\mathbf{x}}^{(i)}(a)(t_{m-} - \varepsilon - a) + \mathrm{o}\big((t_{m-} - \varepsilon - a)\big), \\
\mathbf{x}^{(j)}(t_{m+\varepsilon}) &\equiv \mathbf{x}^{(j)}(t_{m+} + \varepsilon) \\
&= \mathbf{x}^{(j)}(b) + \dot{\mathbf{x}}^{(j)}(t_{m+} + \varepsilon - b)\varepsilon + \mathrm{o}\big((t_{m+} + \varepsilon - b)\big).
\end{aligned}
$$

Let $a \to t_{m-}$ and $b \to t_{m+}$, the limits of the foregoing equations lead to

$$
\begin{cases}
\mathbf{x}^{(i)}(t_{m-\varepsilon}) \equiv \mathbf{x}^{(i)}(t_{m-} - \varepsilon) = \mathbf{x}^{(i)}(t_{m-}) - \dot{\mathbf{x}}^{(i)}(t_{m-})\varepsilon + \mathrm{o}(\varepsilon), \\
\mathbf{x}^{(j)}(t_{m+\varepsilon}) \equiv \mathbf{x}^{(j)}(t_{m+} + \varepsilon) = \mathbf{x}^{(j)}(t_{m+}) + \dot{\mathbf{x}}^{(j)}(t_{m+})\varepsilon + \mathrm{o}(\varepsilon).
\end{cases}
$$

Because of $0 < \varepsilon \ll 1$, the ε^2 and higher-order terms of the foregoing equations can be ignored. Therefore, with the first equation of (9), the following relations exist,

$$
\begin{cases}
\mathbf{n}^{\mathrm{T}}_{\partial\Omega_{ij}} \cdot \big[\mathbf{x}^{(i)}(t_{m-}) - \mathbf{x}^{(i)}(t_{m-\varepsilon})\big] = \mathbf{n}^{\mathrm{T}}_{\partial\Omega_{ij}} \cdot \dot{\mathbf{x}}^{(i)}(t_{m-})\varepsilon > 0, \\
\mathbf{n}^{\mathrm{T}}_{\partial\Omega_{ij}} \cdot \big[\mathbf{x}^{(j)}(t_{m+\varepsilon}) - \mathbf{x}^{(j)}(t_{m+})\big] = \mathbf{n}^{\mathrm{T}}_{\partial\Omega_{ij}} \cdot \dot{\mathbf{x}}^{(i)}(t_{m+})\varepsilon > 0
\end{cases}
$$

and

$$\begin{cases} \mathbf{n}_{\partial\Omega_{ij}}^{\mathrm{T}} \cdot \left[\mathbf{x}^{(0)}(t_{m-}) - \mathbf{x}^{(0)}(t_{m-\varepsilon})\right] = \mathbf{n}_{\partial\Omega_{ij}}^{\mathrm{T}} \cdot \dot{\mathbf{x}}^{(0)}(t_{m-})\varepsilon = 0, \\ \mathbf{n}_{\partial\Omega_{ij}}^{\mathrm{T}} \cdot \left[\mathbf{x}^{(0)}(t_{m+\varepsilon}) - \mathbf{x}^{(0)}(t_{m+})\right] = \mathbf{n}_{\partial\Omega_{ij}}^{\mathrm{T}} \cdot \dot{\mathbf{x}}^{(0)}(t_{m+})\varepsilon = 0. \end{cases}$$

Therefore,

$$\begin{cases} \mathbf{n}_{\partial\Omega_{ij}}^{\mathrm{T}} \cdot \left[\mathbf{x}^{(0)}(t_{m-\varepsilon}) - \mathbf{x}^{(i)}(t_{m-\varepsilon})\right] = \mathbf{n}_{\partial\Omega_{ij}}^{\mathrm{T}} \cdot \dot{\mathbf{x}}^{(i)}(t_{m-})\varepsilon > 0, \\ \mathbf{n}_{\partial\Omega_{ij}}^{\mathrm{T}} \cdot \left[\mathbf{x}^{(j)}(t_{m+\varepsilon}) - \mathbf{x}^{(0)}(t_{m+\varepsilon})\right] = \mathbf{n}_{\partial\Omega_{ij}}^{\mathrm{T}} \cdot \dot{\mathbf{x}}^{(j)}(t_{m+})\varepsilon > 0. \end{cases}$$

From Definition 8, the boundary $\partial\Omega_{ij}$ with $\mathbf{n}_{\partial\Omega_{ij}} \to \Omega_j$ is semipassable under the condition in the first inequality equations of (9). In a similar manner, the boundary $\partial\Omega_{ij}$ with $\mathbf{n}_{\partial\Omega_{ij}} \to \Omega_i$ is semipassable under the conditions in the second inequality equation in (9). □

THEOREM 2. *For a discontinuous dynamical system in equation* (1), $\mathbf{x}(t_m) = \mathbf{x}_m \in \partial\Omega_{ij}$ *for* t_m. *For an arbitrarily small* $\varepsilon > 0$, *three are two time intervals* $[t_{m-\varepsilon}, t_m)$ *and* $(t_m, t_{m+\varepsilon}]$. *Suppose* $\mathbf{x}^{(i)}(t_{m-}) = \mathbf{x}_m = \mathbf{x}^{(j)}(t_{m+})$ *and both* $\mathbf{F}^{(i)}(t)$ *and* $\mathbf{F}^{(j)}(t)$ *are* $C^r_{[t_{m-\varepsilon}, t_m)}$- *and* $C^r_{(t_m, t_{m+\varepsilon}]}$-*continuous* ($r \geqslant 1$) *for time* t, *respectively, and* $\|\mathrm{d}^{r+1}\mathbf{x}^{(\alpha)}/\mathrm{d}t^{r+1}\| < \infty$ ($\alpha \in \{i, j\}$). *The nonempty boundary set* $\partial\Omega_{ij}$ *for the flows* $\mathbf{x}^{(i)}(t)$ *and* $\mathbf{x}^{(j)}(t)$ *is semipassable from the domain* Ω_i *to* Ω_j *iff*

$$\begin{aligned} &\text{either} \quad \begin{cases} \mathbf{n}_{\partial\Omega_{ij}}^{\mathrm{T}} \cdot \mathbf{F}^{(i)}(t_{m-}) > 0 & \text{and} \\ \mathbf{n}_{\partial\Omega_{ij}}^{\mathrm{T}} \cdot \mathbf{F}^{(j)}(t_{m+}) > 0 & \text{for } \mathbf{n}_{\partial\Omega_{ij}} \to \Omega_j, \end{cases} \\ &\text{or} \quad \begin{cases} \mathbf{n}_{\partial\Omega_{ij}}^{\mathrm{T}} \cdot \mathbf{F}^{(i)}(t_{m-}) < 0 & \text{and} \\ \mathbf{n}_{\partial\Omega_{ij}}^{\mathrm{T}} \cdot \mathbf{F}^{(j)}(t_{m+}) < 0 & \text{for } \mathbf{n}_{\partial\Omega_{ij}} \to \Omega_i, \end{cases} \end{aligned} \tag{10}$$

where $\mathbf{F}^{(i)}(t_{m-}) = \mathbf{F}^{(i)}(\mathbf{x}, t_{m-}, \boldsymbol{\mu}_i)$ *and* $\mathbf{F}^{(j)}(t_{m+}) = \mathbf{F}^{(j)}(\mathbf{x}, t_{m+}, \boldsymbol{\mu}_j)$.

PROOF. For a point $\mathbf{x}_m \in \partial\Omega_{ij}$ with $\mathbf{n}_{\partial\Omega_{ij}} \to \Omega_j$, the following relationship $\mathbf{x}^{(i)}(t_{m-}) = \mathbf{x}_m = \mathbf{x}^{(j)}(t_{m+})$ exists. With equation (1), the first inequality equation of (10) gives

$$\begin{cases} \mathbf{n}_{\partial\Omega_{ij}}^{\mathrm{T}} \cdot \dot{\mathbf{x}}^{(i)}(t_{m-}) = \mathbf{n}_{\partial\Omega_{ij}}^{\mathrm{T}} \cdot \mathbf{F}^{(i)}(t_{m-}) > 0 \quad \text{and} \\ \mathbf{n}_{\partial\Omega_{ij}}^{\mathrm{T}} \cdot \dot{\mathbf{x}}^{(j)}(t_{m+}) = \mathbf{n}_{\partial\Omega_{ij}}^{\mathrm{T}} \cdot \mathbf{F}^{(j)}(t_{m+}) > 0. \end{cases}$$

From Theorem 1 and Definition 8, the boundary $\partial\Omega_{ij}$ with $\mathbf{n}_{\partial\Omega_{ij}} \to \Omega_j$ is semipassable. In a similar fashion, the boundary $\partial\Omega_{ij}$ with $\mathbf{n}_{\partial\Omega_{ij}} \to \Omega_i$ is semipassable under the condition in the second inequality equations of (10). □

DEFINITION 9. For a discontinuous dynamical system in equation (1), $\mathbf{x}(t_m) = \mathbf{x}_m \in \partial\Omega_{ij}$ for t_m. For an arbitrarily small $\varepsilon > 0$, there is a time interval $[t_{m-\varepsilon}, t_m)$

and suppose $\mathbf{x}^{(\alpha)}(t_{m-}) = \mathbf{x}_m$ ($\alpha \in \{i, j\}$). So that the nonempty boundary set $\partial\Omega_{ij}$ for the flows $\mathbf{x}^{(i)}(t)$ and $\mathbf{x}^{(j)}(t)$ is the *nonpassable boundary of the first kind*, $\widetilde{\partial\Omega}_{ij}$ (or termed *a sink boundary* between the subdomains Ω_i and Ω_j) if the flows $\mathbf{x}^{(i)}(t)$ and $\mathbf{x}^{(j)}(t)$ in the neighborhood of the boundary $\partial\Omega_{ij}$ possess the following properties,

$$\begin{aligned} &\left\{\mathbf{n}^{\mathrm{T}}_{\partial\Omega_{ij}} \cdot \left[\mathbf{x}^{(i)}(t_{m-}) - \mathbf{x}^{(i)}(t_{m-\varepsilon})\right]\right\} \\ &\quad \times \left\{\mathbf{n}^{\mathrm{T}}_{\partial\Omega_{ij}} \cdot \left[\mathbf{x}^{(j)}(t_{m-}) - \mathbf{x}^{(j)}(t_{m-\varepsilon})\right]\right\} < 0. \end{aligned} \tag{11}$$

DEFINITION 10. For a discontinuous dynamical system in equation (1), $\mathbf{x}(t_m) = \mathbf{x}_m \in \partial\Omega_{ij}$ for t_m. For an arbitrarily small $\varepsilon > 0$, there is a time interval $(t_m, t_{m+\varepsilon}]$ and suppose $\mathbf{x}^{(\alpha)}(t_{m+}) = \mathbf{x}_m$ ($\alpha = \{i, j\}$). So that the nonempty boundary set $\partial\Omega_{ij}$ for the flows $\mathbf{x}^{(i)}(t)$ and $\mathbf{x}^{(j)}(t)$ is *the nonpassable boundary of the second kind* $\widehat{\partial\Omega}_{ij}$ (or termed a *source boundary* between the subdomains Ω_i and Ω_j) if the flows $\mathbf{x}^{(i)}(t)$ and $\mathbf{x}^{(j)}(t)$ in the neighborhood of the boundary $\partial\Omega_{ij}$ possess the following properties,

$$\begin{aligned} &\left\{\mathbf{n}^{\mathrm{T}}_{\partial\Omega_{ij}} \cdot \left[\mathbf{x}^{(i)}(t_{m+\varepsilon}) - \mathbf{x}^{(i)}(t_{m+})\right]\right\} \\ &\quad \times \left\{\mathbf{n}^{\mathrm{T}}_{\partial\Omega_{ij}} \cdot \left[\mathbf{x}^{(j)}(t_{m+\varepsilon}) - \mathbf{x}^{(j)}(t_{m+})\right]\right\} < 0. \end{aligned} \tag{12}$$

The above two concepts for the sink and source boundaries between the two subdomains Ω_i and Ω_j are illustrated in Figures 5(a) and 5(b). The flows in the neighborhood of the boundaries are depicted. When a flow $\mathbf{x}^{(\alpha)}(t)$ ($\alpha \in \{i, j\}$) in the domain Ω_α arrives to the nonpassable boundary of the first kind $\widetilde{\partial\Omega}_{ij}$, the flow can be tangential to or sliding on the nonpassable boundary $\widetilde{\partial\Omega}_{ij}$. For the nonpassable boundary of the second kind $\widehat{\partial\Omega}_{ij}$, a flow $\mathbf{x}^{(\alpha)}(t)$ ($\alpha \in \{i, j\}$) in the domain Ω_α can be tangential to or bouncing on the nonpassable boundary $\widehat{\partial\Omega}_{ij}$. In this chapter, only the tangential motion on the nonpassable boundary will be discussed, which is similar to the one on the semipassable boundary.

THEOREM 3. *For a discontinuous dynamical system in equation* (1), $\mathbf{x}(t_m) = \mathbf{x}_m \in \partial\Omega_{ij}$ *for* t_m. *For an arbitrarily small* $\varepsilon > 0$, *there is a time interval* $[t_{m-\varepsilon}, t_m)$ *and suppose* $\mathbf{x}^{(\alpha)}(t_{m-}) = \mathbf{x}_m$ ($\alpha \in \{i, j\}$). *The flow* $\mathbf{x}^{(\alpha)}(t)$ *is* $C^r_{[t_{m-\varepsilon}, t_m)}$*-continuous* ($r \geqslant 2$) *for time t and* $\|\mathrm{d}^r\mathbf{x}^{(\alpha)}/\mathrm{d}t^r\| < \infty$. *The nonempty boundary set* $\partial\Omega_{ij}$ *for the flows* $\mathbf{x}^{(i)}(t)$ *and* $\mathbf{x}^{(j)}(t)$ *is a nonpassable boundary of the first kind iff*

$$\left[\mathbf{n}^{\mathrm{T}}_{\partial\Omega_{ij}} \cdot \dot{\mathbf{x}}^{(i)}(t_{m-})\right] \times \left[\mathbf{n}^{\mathrm{T}}_{\partial\Omega_{ij}} \cdot \dot{\mathbf{x}}^{(j)}(t_{m-})\right] < 0. \tag{13}$$

PROOF. Following the procedure of the proof of Theorem 1, Theorem 3 can be proved. □

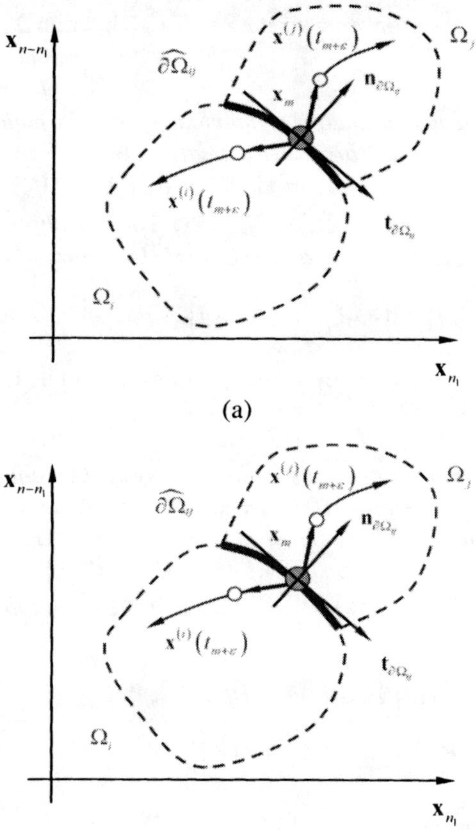

Figure 5. Nonpassable boundary set $\overline{\partial\Omega}_{ij} = \widetilde{\partial\Omega}_{ij} \cup \widehat{\partial\Omega}_{ij}$: (a) the sink boundary (or the nonpassable boundary of the first kind, $\widetilde{\partial\Omega}_{ij}$), (b) the source boundary (or the nonpassable boundary of the second kind, $\widehat{\partial\Omega}_{ij}$). $\mathbf{x}_m \equiv (\mathbf{x}_{n_1}(t_m), \mathbf{x}_{n-n_1}(t_m))^{\mathrm{T}}$, $\mathbf{x}^{(\alpha)}(t_{m\pm\varepsilon}) \equiv (\mathbf{x}_{n_1}^{(\alpha)}(t_{m\pm\varepsilon}), \mathbf{x}_{n-n_1}^{(\alpha)}(t_{m\pm\varepsilon}))^{\mathrm{T}}$ and $\alpha = \{i, j\}$, where $t_{m\pm\varepsilon} = t_m \pm \varepsilon$ for an arbitrary small $\varepsilon > 0$.

THEOREM 4. *For a discontinuous dynamical system in equation* (1), $\mathbf{x}(t_m) = \mathbf{x}_m \in \partial\Omega_{ij}$ *for* t_m. *For an arbitrarily small* $\varepsilon > 0$, *there is a time interval* $[t_{m-\varepsilon}, t_m)$ *and suppose* $\mathbf{x}^{(\alpha)}(t_{m-}) = \mathbf{x}_m$ $(\alpha \in \{i, j\})$. *The vector filed* $\mathbf{F}^{(\alpha)}(t)$ *are* $C^r_{[t_{m-\varepsilon}, t_m)}$*-continuous* $(r \geqslant 1)$ *and* $\|\mathrm{d}^{r+1}\mathbf{x}^{(\alpha)}/\mathrm{d}t^{r+1}\| < \infty$. *The nonempty boundary set* $\partial\Omega_{ij}$ *for the flows* $\mathbf{x}^{(i)}(t)$ *and* $\mathbf{x}^{(j)}(t)$ *is a nonpassable boundary of the first kind iff*

$$\left[\mathbf{n}^{\mathrm{T}}_{\partial\Omega_{ij}} \cdot \mathbf{F}^{(i)}(t_{m-})\right] \times \left[\mathbf{n}^{\mathrm{T}}_{\partial\Omega_{ij}} \cdot \mathbf{F}^{(j)}(t_{m-})\right] < 0, \tag{14}$$

where $\mathbf{F}^{(\alpha)}(t_{m-}) \triangleq \mathbf{F}^{(\alpha)}(\mathbf{x}, t_{m-}, \boldsymbol{\mu}_\alpha)$ $(\alpha \in \{i, j\})$.

PROOF. Following the procedure of the proof of Theorem 2, Theorem 4 can be proved. □

THEOREM 5. *For a discontinuous dynamical system in equation* (1), $\mathbf{x}(t_m) = \mathbf{x}_m \in \partial\Omega_{ij}$ *for* t_m. *For an arbitrarily small* $\varepsilon > 0$, *there is a time interval* $(t_m, t_{m+\varepsilon}]$ *and suppose* $\mathbf{x}^{(\alpha)}(t_{m+}) = \mathbf{x}_m$ $(\alpha = \{i, j\})$. $\mathbf{x}^{(\alpha)}(t)$ *is* $C^r_{(t_m, t_{m+\varepsilon}]}$*-continuous* $(r \geqslant 2)$ *for time t and* $\|\mathrm{d}^r\mathbf{x}^{(\alpha)}/\mathrm{d}t^r\| < \infty$. *The nonempty boundary set* $\partial\Omega_{ij}$ *for the flows* $\mathbf{x}^{(i)}(t)$ *and* $\mathbf{x}^{(j)}(t)$ *is a nonpassable boundary of the second kind iff*

$$\left[\mathbf{n}^{\mathrm{T}}_{\partial\Omega_{ij}} \cdot \dot{\mathbf{x}}^{(i)}(t_{m+})\right] \times \left[\mathbf{n}^{\mathrm{T}}_{\partial\Omega_{ij}} \cdot \dot{\mathbf{x}}^{(j)}(t_{m+})\right] < 0. \tag{15}$$

PROOF. Following the procedure of the proof of Theorem 1, Theorem 5 can be proved. □

THEOREM 6. *For a discontinuous dynamical system in equation* (1), $\mathbf{x}(t_m) = \mathbf{x}_m \in \partial\Omega_{ij}$ *for* t_m. *For an arbitrarily small* $\varepsilon > 0$, *there is a time interval* $(t_m, t_{m+\varepsilon}]$ *and suppose* $\mathbf{x}^{(\alpha)}(t_{m+}) = \mathbf{x}_m$ $(\alpha = \{i, j\})$. *The vector filed* $\mathbf{F}^{(\alpha)}(t)$ *are* $C^r_{[t_{m-\varepsilon}, t_m)}$*-continuous* $(r \geqslant 1)$ *and* $\|\mathrm{d}^{r+1}\mathbf{x}^{(\alpha)}/\mathrm{d}t^{r+1}\| < \infty$. *The nonempty boundary set* $\partial\Omega_{ij}$ *for the flows* $\mathbf{x}^{(i)}(t)$ *and* $\mathbf{x}^{(j)}(t)$ *is a nonpassable boundary of the second kind iff*

$$\left[\mathbf{n}^{\mathrm{T}}_{\partial\Omega_{ij}} \cdot \mathbf{F}^{(i)}(t_{m+})\right] \times \left[\mathbf{n}^{\mathrm{T}}_{\partial\Omega_{ij}} \cdot \mathbf{F}^{(j)}(t_{m+})\right] < 0, \tag{16}$$

where $\mathbf{F}^{(\alpha)}(t_{m+}) \triangleq \mathbf{F}^{(\alpha)}(\mathbf{x}, t_{m+}, \boldsymbol{\mu}_\alpha)$ $(\alpha = \{i, j\})$.

PROOF. Following the procedure of the proof of Theorem 2, Theorem 6 can be proved. □

DEFINITION 11. The nonempty boundary set $\partial\Omega_{ij}$ for the flows $\mathbf{x}^{(i)}(t)$ and $\mathbf{x}^{(j)}(t)$ is passable ($\overleftrightarrow{\partial\Omega_{ij}}$) only if it possesses not only semipassable boundary $\overrightarrow{\partial\Omega}_{ij}$ from the domain Ω_i to Ω_j but $\overleftarrow{\partial\Omega}_{ij}$ from the domain Ω_j to Ω_i.

This definition indicates that the C^0-flow on the boundary set is invertible. The gradients of the flow on both sides of the separation boundary are different in the nonsmooth dynamical systems. If the flow is C^1-smooth on the boundary without effects of sliding motion, the boundary set becomes a trivial boundary set, and the two subdynamical systems become a smooth dynamical system. For illustration of the passable boundary set, the flow passing through the boundary $\partial\Omega_{ij}$ from Ω_i to Ω_j and from Ω_j to Ω_i are presented in Figure 6. The dashed curves are other boundaries for the domains Ω_i and Ω_j. The thicker solid curve represents the boundary $\partial\Omega_{ij}$. The thinner solid curves with arrows are the flow of equation (1) in the two domains.

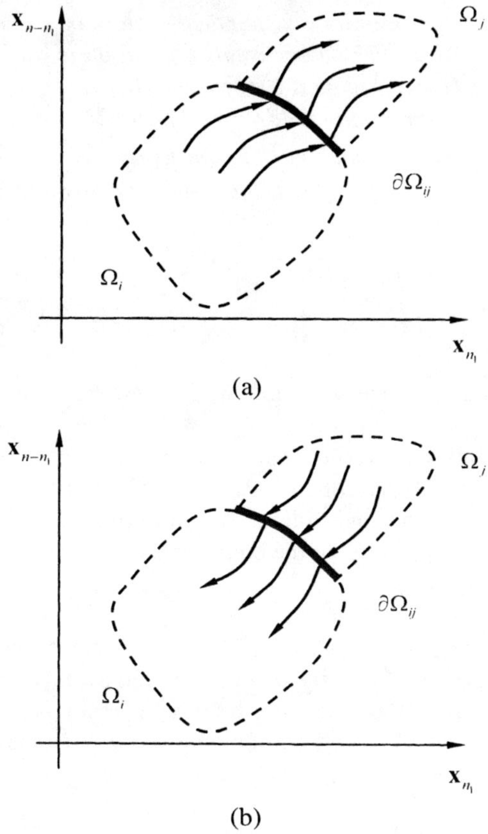

Figure 6. Flow passing through the boundary $\partial\Omega_{ij}$: (a) from Ω_i to Ω_j and (b) from Ω_j to Ω_i.

5. Local singularity and grazing flows

In this section, the flow local singularity and tangential flow will be discussed. The corresponding necessary and sufficient conditions will be presented.

DEFINITION 12. For a discontinuous dynamical system in equation (1), $\mathbf{x}(t_m) = \mathbf{x}_m \in \partial\Omega_{ij}$ for t_m. For an arbitrarily small $\varepsilon > 0$, there are two time intervals (i.e., $[t_{m-\varepsilon}, t_m)$ and $(t_m, t_{m+\varepsilon}]$) and suppose $\mathbf{x}^{(\alpha)}(t_{m-}) = \mathbf{x}_m$, $\mathbf{x}_m = \mathbf{x}^{(\beta)}(t_{m+})$ ($\alpha, \beta \in \{i, j\}$). The two flows $\mathbf{x}^{(\alpha)}(t)$ and $\mathbf{x}^{(\beta)}(t)$ are $C^r_{[t_{m-\varepsilon}, t_m)}$- and $C^r_{(t_m, t_{m+\varepsilon}]}$-continuous ($r \geqslant 2$), respectively. A point $\mathbf{x}_m$ is critical on the nonempty boundary set $\partial\Omega_{ij}$ if the following equation exists

$$\mathbf{n}^{\mathrm{T}}_{\partial\Omega_{ij}} \cdot \dot{\mathbf{x}}^{(\alpha)}(t_{m-}) = 0 \quad \text{and/or} \quad \mathbf{n}^{\mathrm{T}}_{\partial\Omega_{ij}} \cdot \dot{\mathbf{x}}^{(\beta)}(t_{m+}) = 0. \tag{17}$$

THEOREM 7. *For a discontinuous dynamical system in equation* (1), $\mathbf{x}(t_m) = \mathbf{x}_m \in \partial\Omega_{ij}$ *for* t_m. *For an arbitrarily small* $\varepsilon > 0$, *there are two time intervals (i.e.,* $[t_{m-\varepsilon}, t_m)$ *and* $(t_m, t_{m+\varepsilon}]$*) and suppose* $\mathbf{x}^{(\alpha)}(t_{m-}) = \mathbf{x}_m$, $\mathbf{x}_m = \mathbf{x}^{(\beta)}(t_{m+})$ $(\alpha, \beta \in \{i, j\})$. *The two vector fields* $\mathbf{F}^{(\alpha)}(t)$ *and* $\mathbf{F}^{(\beta)}(t)$ *are* $C^r_{[t_{m-\varepsilon}, t_m)}$*- and* $C^r_{(t_m, t_{m+\varepsilon}]}$*-continuous* $(r \geqslant 1)$ *for time* t, *respectively and* $\|\mathrm{d}^{r+1}\mathbf{x}^{(\alpha)}/\mathrm{d}t^{r+1}\| < \infty$ $(\alpha \in \{i, j\})$. *The point* $\mathbf{x}_m \in \partial\Omega_{ij}$ *is critical on the nonempty boundary set* $\partial\Omega_{ij}$ *iff*

$$\mathbf{n}^{\mathrm{T}}_{\partial\Omega_{ij}} \cdot \mathbf{F}^{(\alpha)}(t_{m-}) = 0 \quad \textit{and/or} \quad \mathbf{n}^{\mathrm{T}}_{\partial\Omega_{ij}} \cdot \mathbf{F}^{(\beta)}(t_{m+}) = 0, \tag{18}$$

where $\mathbf{F}^{(\alpha)}(t_{m-}) = \mathbf{F}^{(\alpha)}(\mathbf{x}, t_{m-}, \boldsymbol{\mu}_\alpha)$ *and* $\mathbf{F}^{(\beta)}(t_{m+}) = \mathbf{F}^{(\beta)}(\mathbf{x}, t_{m+}, \boldsymbol{\mu}_\beta)$.

PROOF. Using equation (1) and Definition 12, Theorem 7 can be proved. □

Since the tangential vector of the input and output flows $\mathbf{x}^{(\alpha)}(t_{m-})$ and $\mathbf{x}^{(\alpha)}(t_{m+})$ on the side of the domain Ω_α $(\alpha \in \{i, j\})$ at the boundary $\partial\Omega_{ij}$ is normal to the normal vector of the boundary, it implies that the input and output flow in domain Ω_α is tangential to the boundary. The mathematical description is given as follows.

DEFINITION 13. For a discontinuous dynamical system in equation (1), $\mathbf{x}(t_m) = \mathbf{x}_m \in \partial\Omega_{ij}$ for t_m. For an arbitrarily small $\varepsilon > 0$, there are two time intervals (i.e., $[t_{m-\varepsilon}, t_m)$ and $(t_m, t_{m+\varepsilon}]$) and suppose $\mathbf{x}^{(\alpha)}(t_{m\pm}) = \mathbf{x}_m$ $(\alpha \in \{i, j\})$. The flow $\mathbf{x}^{(\alpha)}(t)$ is $C^r_{[t_{m-\varepsilon}, t_m)}$- and $C^r_{(t_m, t_{m+\varepsilon}]}$-continuous $(r \geqslant 1)$ for time t. The flow $\mathbf{x}^{(\alpha)}(t)$ in Ω_α is tangential to the boundary $\partial\Omega_{ij}$ if the following two conditions hold:

$$\mathbf{n}^{\mathrm{T}}_{\partial\Omega_{ij}} \cdot \dot{\mathbf{x}}^{(\alpha)}(t_{m\pm}) = 0, \tag{19}$$

either

$$\begin{cases} \mathbf{n}^{\mathrm{T}}_{\partial\Omega_{ij}} \cdot \left[\mathbf{x}^{(0)}(t_{m-\varepsilon}) - \mathbf{x}^{(\alpha)}(t_{m-\varepsilon})\right] > 0 & \text{and} \\ \mathbf{n}^{\mathrm{T}}_{\partial\Omega_{ij}} \cdot \left[\mathbf{x}^{(\alpha)}(t_{m+\varepsilon}) - \mathbf{x}^{(0)}(t_{m+\varepsilon})\right] < 0 & \text{for } \mathbf{n}_{\partial\Omega_{ij}} \to \Omega_\beta, \end{cases} \tag{20}$$

where $\beta = \{i, j\}$ but $\alpha \neq \beta$, or

$$\begin{cases} \mathbf{n}^{\mathrm{T}}_{\partial\Omega_{ij}} \cdot \left[\mathbf{x}^{(0)}(t_{m-\varepsilon}) - \mathbf{x}^{(\alpha)}(t_{m-\varepsilon})\right] < 0 & \text{and} \\ \mathbf{n}^{\mathrm{T}}_{\partial\Omega_{ij}} \cdot \left[\mathbf{x}^{(\alpha)}(t_{m+\varepsilon}) - \mathbf{x}^{(0)}(t_{m+\varepsilon})\right] > 0 & \text{for } \mathbf{n}_{\partial\Omega_{ij}} \to \Omega_\alpha. \end{cases} \tag{21}$$

Since $\mathbf{n}^{\mathrm{T}}_{\partial\Omega_{ij}} \cdot \mathbf{t}_{\partial\Omega_{ij}} = 0$ and $\mathbf{t}_{\partial\Omega_{ij}} = \dot{\mathbf{x}}_m$ on the boundary $\partial\Omega_{ij}$, with equation (19), we have

$$\begin{aligned} &\mathbf{n}^{\mathrm{T}}_{\partial\Omega_{ij}} \cdot \dot{\mathbf{x}}^{(\alpha)}(t_{m\pm}) = 0 = \mathbf{n}^{\mathrm{T}}_{\partial\Omega_{ij}} \cdot \dot{\mathbf{x}}_m \quad \text{or} \\ &\dot{\mathbf{x}}^{(\alpha)}(t_{m-}) = \dot{\mathbf{x}}_m = \dot{\mathbf{x}}^{(\alpha)}(t_{m+}). \end{aligned} \tag{22}$$

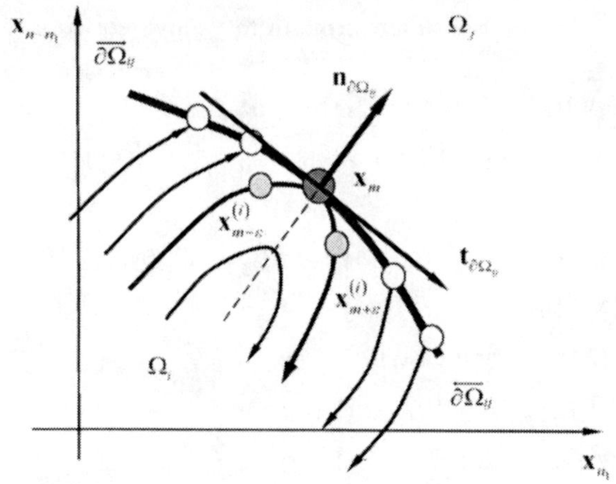

Figure 7. A flow in the domain Ω_i tangential to the boundary $\partial\Omega_{ij}$ with $\mathbf{n}_{\partial\Omega_{ij}} \to \Omega_j$. The grey-filled symbols represent two points ($\mathbf{x}_{m-\varepsilon}^{(i)}$ and $\mathbf{x}_{m+\varepsilon}^{(i)}$) on the flow before and after the tangency. The tangential point $\mathbf{x}_m$ on the boundary $\partial\Omega_{ij}$ is depicted by a large circular symbol.

The above equation implies that the flow $\mathbf{x}^{(\alpha)}$ on the boundary is at least C^1-continuous. To demonstrate the above definition, consider a flow in the domain Ω_i tangential to the boundary $\partial\Omega_{ij}$ with $\mathbf{n}_{\partial\Omega_{ij}} \to \Omega_j$, as shown in Figure 7. The grey-filled symbols represent two points ($\mathbf{x}_{m\pm\varepsilon}^{(i)} = \mathbf{x}^{(i)}(t_m \pm \varepsilon)$) on the flow before and after the tangency. The tangential point $\mathbf{x}_m$ on the boundary $\partial\Omega_{ij}$ is depicted by a large circular symbol. This tangential flow is also termed the *grazing flow*.

THEOREM 8. *For a discontinuous dynamical system in equation* (1), $\mathbf{x}(t_m) = \mathbf{x}_m \in \partial\Omega_{ij}$ *for* t_m. *For an arbitrarily small* $\varepsilon > 0$, *there are two time intervals (i.e.,* $[t_{m-\varepsilon}, t_m)$ *and* $(t_m, t_{m+\varepsilon}]$*) and suppose* $\mathbf{x}^{(\alpha)}(t_{m\pm}) = \mathbf{x}_m$ $(\alpha \in \{i, j\})$. *The flow* $\mathbf{x}^{(\alpha)}(t)$ *is* $C^r_{[t_{m-\varepsilon}, t_m)}$*- and* $C^r_{(t_m, t_{m+\varepsilon}]}$*-continuous* ($r \geqslant 2$) *for time* t *and* $\|\mathrm{d}^r\mathbf{x}^{(\alpha)}/\mathrm{d}t^r\| < \infty$. *The flow* $\mathbf{x}^{(\alpha)}(t)$ *in* Ω_α *is tangential to the boundary* $\partial\Omega_{ij}$ *iff*

$$\mathbf{n}^{\mathrm{T}}_{\partial\Omega_{ij}} \cdot \dot{\mathbf{x}}^{(\alpha)}(t_{m\pm}) = 0, \tag{23}$$

$$\left\{\mathbf{n}^{\mathrm{T}}_{\partial\Omega_{ij}} \cdot \left[\dot{\mathbf{x}}^{(0)}(t_{m-\varepsilon}) - \dot{\mathbf{x}}^{(\alpha)}(t_{m-\varepsilon})\right]\right\} \times \left\{\mathbf{n}^{\mathrm{T}}_{\partial\Omega_{ij}} \cdot \left[\dot{\mathbf{x}}^{(\alpha)}(t_{m+\varepsilon}) - \dot{\mathbf{x}}^{(0)}(t_{m+\varepsilon})\right]\right\} < 0. \tag{24}$$

PROOF. Since equation (23) is identical to equation (19), the first condition in equation (19) is satisfied,

$$\begin{aligned}\mathbf{x}^{(\alpha)}(t_{m\pm}) &\equiv \mathbf{x}^{(\alpha)}(t_{m\pm} \pm \varepsilon \mp \varepsilon) = \mathbf{x}^{(\alpha)}(t_{m\pm} \pm \varepsilon) \mp \varepsilon\dot{\mathbf{x}}^{(\alpha)}(t_{m\pm} \pm \varepsilon) + \mathrm{o}(\varepsilon) \\ &= \mathbf{x}^{(\alpha)}(t_{m\pm\varepsilon}) \mp \varepsilon\dot{\mathbf{x}}^{(\alpha)}(t_{m\pm\varepsilon}) + \mathrm{o}(\varepsilon).\end{aligned}$$

For $0 < \varepsilon \ll 1$, the higher-order terms in the above equation can be ignored. Therefore

$$\begin{cases} \mathbf{n}^{\mathrm{T}}_{\partial\Omega_{ij}} \cdot \left[\mathbf{x}^{(\alpha)}(t_{m+}) - \mathbf{x}^{(\alpha)}(t_{m-\varepsilon})\right] = \varepsilon \mathbf{n}^{\mathrm{T}}_{\partial\Omega_{ij}} \cdot \dot{\mathbf{x}}^{(\alpha)}(t_{m-\varepsilon}), \\ \mathbf{n}^{\mathrm{T}}_{\partial\Omega_{ij}} \cdot \left[\mathbf{x}^{(\alpha)}(t_{m+\varepsilon}) - \mathbf{x}^{(\alpha)}(t_{m+})\right] = \varepsilon \mathbf{n}^{\mathrm{T}}_{\partial\Omega_{ij}} \cdot \dot{\mathbf{x}}^{(\alpha)}(t_{m+\varepsilon}). \end{cases}$$

Similarly

$$\begin{cases} \mathbf{n}^{\mathrm{T}}_{\partial\Omega_{ij}} \cdot \left[\mathbf{x}^{(0)}(t_{m}) - \mathbf{x}^{(0)}(t_{m-\varepsilon})\right] = \varepsilon \mathbf{n}^{\mathrm{T}}_{\partial\Omega_{ij}} \cdot \dot{\mathbf{x}}^{(0)}(t_{m-\varepsilon}), \\ \mathbf{n}^{\mathrm{T}}_{\partial\Omega_{ij}} \cdot \left[\mathbf{x}^{(0)}(t_{m+\varepsilon}) - \mathbf{x}^{(0)}(t_{m})\right] = \varepsilon \mathbf{n}^{\mathrm{T}}_{\partial\Omega_{ij}} \cdot \dot{\mathbf{x}}^{(0)}(t_{m+\varepsilon}). \end{cases}$$

From equation (24), the first case is

$$\mathbf{n}^{\mathrm{T}}_{\partial\Omega_{ij}} \cdot \left[\dot{\mathbf{x}}^{(0)}(t_{m-\varepsilon}) - \dot{\mathbf{x}}^{(\alpha)}(t_{m-\varepsilon})\right] > 0 \quad \text{and}$$

$$\mathbf{n}^{\mathrm{T}}_{\partial\Omega_{ij}} \cdot \left[\dot{\mathbf{x}}^{(\alpha)}(t_{m+\varepsilon}) - \dot{\mathbf{x}}^{(0)}(t_{m+\varepsilon})\right] < 0$$

with which equation (20) holds for $\partial\Omega_{ij}$ with $\mathbf{n}_{\partial\Omega_{ij}} \to \Omega_\beta$ ($\beta \neq \alpha$). However, the second case is

$$\mathbf{n}^{\mathrm{T}}_{\partial\Omega_{ij}} \cdot \left[\dot{\mathbf{x}}^{(0)}(t_{m-\varepsilon}) - \dot{\mathbf{x}}^{(\alpha)}(t_{m-\varepsilon})\right] < 0 \quad \text{and}$$

$$\mathbf{n}^{\mathrm{T}}_{\partial\Omega_{ij}} \cdot \left[\dot{\mathbf{x}}^{(\alpha)}(t_{m+\varepsilon}) - \dot{\mathbf{x}}^{(0)}(t_{m+\varepsilon})\right] > 0$$

from which equation (21) holds for $\partial\Omega_{ij}$ with $\mathbf{n}_{\partial\Omega_{ij}} \to \Omega_\alpha$. Therefore, from Definition 13, the flow $\mathbf{x}^{(\alpha)}(t)$ for $t \in T_m$ in Ω_α is tangential to the boundary $\partial\Omega_{ij}$. □

THEOREM 9. *For a discontinuous dynamical system in equation* (1), $\mathbf{x}(t_m) = \mathbf{x}_m \in \partial\Omega_{ij}$ *for* t_m. *For an arbitrarily small* $\varepsilon > 0$, *there are two time intervals (i.e.,* $[t_{m-\varepsilon}, t_m)$ *and* $(t_m, t_{m+\varepsilon}]$*) and suppose* $\mathbf{x}^{(\alpha)}(t_{m\pm}) = \mathbf{x}_m$ ($\alpha \in \{i, j\}$). *The vector field* $\mathbf{F}^{(\alpha)}(t)$ *are* $C^r_{[t_{m-\varepsilon},t_m)}$*- and* $C^r_{(t_m,t_{m+\varepsilon}]}$*-continuous* ($r \geqslant 1$) *for time* t *and* $\|\mathrm{d}^{r+1}\mathbf{x}^{(\alpha)}/\mathrm{d}t^{r+1}\| < \infty$. *The flow* $\mathbf{x}^{(\alpha)}(t)$ *in* Ω_α *is tangential to the boundary* $\partial\Omega_{ij}$ *iff*

$$\mathbf{n}^{\mathrm{T}}_{\partial\Omega_{ij}} \cdot \mathbf{F}^{(\alpha)}(t_{m\pm}) = 0, \tag{25}$$

$$\left\{\mathbf{n}^{\mathrm{T}}_{\partial\Omega_{ij}} \cdot \left[\mathbf{F}^{(0)}(t_{m-\varepsilon}) - \mathbf{F}^{(\alpha)}(t_{m-\varepsilon})\right]\right\} \times \left\{\mathbf{n}^{\mathrm{T}}_{\partial\Omega_{ij}} \cdot \left[\mathbf{F}^{(\alpha)}(t_{m+\varepsilon}) - \mathbf{F}^{(0)}(t_{m+\varepsilon})\right]\right\} < 0. \tag{26}$$

PROOF. Using equation (1) and Theorem 8, Theorem 9 can be proved. □

THEOREM 10. *For a discontinuous dynamical system in equation* (1), $\mathbf{x}(t_m) = \mathbf{x}_m \in \partial\Omega_{ij}$ *for* t_m. *For an arbitrarily small* $\varepsilon > 0$, *there are two time intervals (i.e.,* $[t_{m-\varepsilon}, t_m)$ *and* $(t_m, t_{m+\varepsilon}]$*) and suppose* $\mathbf{x}^{(\alpha)}(t_{m\pm}) = \mathbf{x}_m$ ($\alpha \in \{i, j\}$). *The flow* $\mathbf{x}^{(\alpha)}(t)$ *is* $C^r_{[t_{m-\varepsilon},t_m)}$*- and* $C^r_{(t_m,t_{m+\varepsilon}]}$*-continuous* ($r \geqslant 3$) *for time* t *and*

$\|\mathrm{d}^r\mathbf{x}^{(\alpha)}/\mathrm{d}t^r\| < \infty$. *The flow* $\mathbf{x}^{(\alpha)}(t)$ *in* Ω_α *is tangential to the boundary* $\partial\Omega_{ij}$ *iff*

$$\mathbf{n}^{\mathrm{T}}_{\partial\Omega_{ij}} \cdot \dot{\mathbf{x}}^{(\alpha)}(t_{m\pm}) = 0, \tag{27}$$

$$\begin{cases} \textit{either} \quad \mathbf{n}^{\mathrm{T}}_{\partial\Omega_{ij}} \cdot \left[\ddot{\mathbf{x}}^{(\alpha)}(t_{m\pm}) - \ddot{\mathbf{x}}^{(0)}(t_m)\right] < 0 \\ \quad \textit{for}\ \mathbf{n}_{\partial\Omega_{ij}} \to \Omega_\beta\ (\beta \in \{i, j\}, \beta \neq \alpha) \\ \textit{or} \quad \mathbf{n}^{\mathrm{T}}_{\partial\Omega_{ij}} \cdot \left[\ddot{\mathbf{x}}^{(\alpha)}(t_{m\pm}) - \ddot{\mathbf{x}}^{(0)}(t_m)\right] > 0 \\ \quad \textit{for}\ \mathbf{n}_{\partial\Omega_{ij}} \to \Omega_\alpha. \end{cases} \tag{28}$$

PROOF. Equation (27) is identical to equation (19), thus the first condition in equation (19) is satisfied. From Definition 13, consider the boundary $\partial\Omega_{ij}$ with $\mathbf{n}_{\partial\Omega_{ij}} \to \Omega_\beta$ ($\beta \neq \alpha$) first. Suppose $\mathbf{x}^{(\alpha)}(t_{m\pm}) = \mathbf{x}_m$ ($\alpha \in \{i, j\}$) and $\mathbf{x}^{(\alpha)}(t)$ are $C^r_{[t_{m-\varepsilon}, t_m)}$- and $C^r_{(t_m, t_{m+\varepsilon}]}$-continuous ($r \geqslant 3$) for time t and $\|\mathrm{d}^r\mathbf{x}^{(\alpha)}/\mathrm{d}t^r\| < \infty$ ($\alpha \in \{i, j\}$). For $a \in [t_{m-\varepsilon}, t_m)$ and $a \in (t_m, t_{m+\varepsilon}]$, the Taylor series expansion of $\mathbf{x}^{(\alpha)}(t_{m\pm\varepsilon})$ to $\mathbf{x}^{(\alpha)}(a)$ up to the third-order term gives

$$\begin{aligned} \mathbf{x}^{(\alpha)}(t_{m\pm\varepsilon}) &\equiv \mathbf{x}^{(\alpha)}(t_{m\pm} - \varepsilon) \\ &= \mathbf{x}^{(\alpha)}(a) + \dot{\mathbf{x}}^{(\alpha)}(a)(t_{m\pm} \pm \varepsilon - a) + \ddot{\mathbf{x}}^{(\alpha)}(a)(t_{m\pm} \pm \varepsilon - a)^2 \\ &\quad + \mathrm{o}\big((t_{m\pm} \pm \varepsilon - a)^2\big). \end{aligned}$$

As $a \to t_{m\pm}$, the limit of the foregoing equation leads to

$$\begin{aligned} \mathbf{x}^{(\alpha)}(t_{m\pm\varepsilon}) &\equiv \mathbf{x}^{(\alpha)}(t_m \pm \varepsilon) \\ &= \mathbf{x}^{(\alpha)}(t_{m\pm}) \pm \dot{\mathbf{x}}^{(\alpha)}(t_{m\pm})\varepsilon + \ddot{\mathbf{x}}^{(\alpha)}(t_{m\pm})\varepsilon^2 + \mathrm{o}\big(\varepsilon^2\big). \end{aligned}$$

The ignorance of the ε^3 and high-order terms, the deformation of the above equation and left multiplication of $\mathbf{n}_{\partial\Omega_{ij}}$ gives

$$\begin{aligned} &\mathbf{n}^{\mathrm{T}}_{\partial\Omega_{ij}} \cdot \left[\mathbf{x}^{(\alpha)}(t_{m+\varepsilon}) - \mathbf{x}^{(\alpha)}(t_{m+})\right] \\ &\quad = \mathbf{n}^{\mathrm{T}}_{\partial\Omega_{ij}} \cdot \dot{\mathbf{x}}^{(\alpha)}(t_{m+})\varepsilon + \mathbf{n}^{\mathrm{T}}_{\partial\Omega_{ij}} \cdot \ddot{\mathbf{x}}^{(\alpha)}(t_{m+})\varepsilon^2, \\ &\mathbf{n}^{\mathrm{T}}_{\partial\Omega_{ij}} \cdot \left[\mathbf{x}^{(\alpha)}(t_{m-}) - \mathbf{x}^{(\alpha)}(t_{m-\varepsilon})\right] \\ &\quad = \mathbf{n}^{\mathrm{T}}_{\partial\Omega_{ij}} \cdot \dot{\mathbf{x}}^{(\alpha)}(t_{m-})\varepsilon - \mathbf{n}^{\mathrm{T}}_{\partial\Omega_{ij}} \cdot \ddot{\mathbf{x}}^{(\alpha)}(t_{m-})\varepsilon^2. \end{aligned}$$

With equation (27), we have

$$\begin{aligned} \mathbf{n}^{\mathrm{T}}_{\partial\Omega_{ij}} \cdot \left[\mathbf{x}^{(\alpha)}(t_{m+\varepsilon}) - \mathbf{x}^{(\alpha)}(t_{m+})\right] &= \mathbf{n}^{\mathrm{T}}_{\partial\Omega_{ij}} \cdot \ddot{\mathbf{x}}^{(\alpha)}(t_{m+})\varepsilon^2, \\ \mathbf{n}^{\mathrm{T}}_{\partial\Omega_{ij}} \cdot \left[\mathbf{x}^{(\alpha)}(t_{m-}) - \mathbf{x}^{(\alpha)}(t_{m-\varepsilon})\right] &= -\mathbf{n}^{\mathrm{T}}_{\partial\Omega_{ij}} \cdot \ddot{\mathbf{x}}^{(\alpha)}(t_{m-})\varepsilon^2, \end{aligned}$$

because $\mathbf{n}^{\mathrm{T}}_{\partial\Omega_{ij}} \cdot \dot{\mathbf{x}}^{(0)}(t_{m\pm}) = 0$

$$\begin{aligned} \mathbf{n}^{\mathrm{T}}_{\partial\Omega_{ij}} \cdot \left[\mathbf{x}^{(0)}(t_{m+\varepsilon}) - \mathbf{x}^{(0)}(t_{m+})\right] &= \mathbf{n}^{\mathrm{T}}_{\partial\Omega_{ij}} \cdot \ddot{\mathbf{x}}^{(0)}(t_{m+})\varepsilon^2, \\ \mathbf{n}^{\mathrm{T}}_{\partial\Omega_{ij}} \cdot \left[\mathbf{x}^{(0)}(t_{m-}) - \mathbf{x}^{(0)}(t_{m-\varepsilon})\right] &= -\mathbf{n}^{\mathrm{T}}_{\partial\Omega_{ij}} \cdot \ddot{\mathbf{x}}^{(0)}(t_{m-})\varepsilon^2, \end{aligned}$$

For the boundary $\partial\Omega_{ij}$ with $\mathbf{n}_{\partial\Omega_{ij}} \to \Omega_\beta$, using the first inequality equation of (28), the foregoing two equations lead to

$$\begin{aligned}
&\mathbf{n}^{\mathrm{T}}_{\partial\Omega_{ij}} \cdot \big[\mathbf{x}^{(0)}(t_{m-\varepsilon}) - \mathbf{x}^{(\alpha)}(t_{m-\varepsilon})\big] \\
&\quad = -\mathbf{n}^{\mathrm{T}}_{\partial\Omega_{ij}} \cdot \big[\ddot{\mathbf{x}}^{(\alpha)}(t_{m-}) - \ddot{\mathbf{x}}^{(0)}(t_{m-})\big]\varepsilon^2 > 0, \\
&\mathbf{n}^{\mathrm{T}}_{\partial\Omega_{ij}} \cdot \big[\mathbf{x}^{(\alpha)}(t_{m+\varepsilon}) - \mathbf{x}^{(\alpha)}(t_{m+})\big] = \mathbf{n}^{\mathrm{T}}_{\partial\Omega_{ij}} \cdot \big[\ddot{\mathbf{x}}^{(\alpha)}(t_{m+}) - \ddot{\mathbf{x}}^{(0)}(t_{m+})\big]\varepsilon^2 < 0.
\end{aligned}$$

Similarly, for the boundary $\partial\Omega_{ij}$ with $\mathbf{n}_{\partial\Omega_{ij}} \to \Omega_\alpha$, using the second inequality equation of (28), the foregoing two equations lead to

$$\begin{aligned}
&\mathbf{n}^{\mathrm{T}}_{\partial\Omega_{ij}} \cdot \big[\mathbf{x}^{(0)}(t_{m-\varepsilon}) - \mathbf{x}^{(\alpha)}(t_{m-\varepsilon})\big] \\
&\quad = -\mathbf{n}^{\mathrm{T}}_{\partial\Omega_{ij}} \cdot \big[\ddot{\mathbf{x}}^{(\alpha)}(t_{m-}) - \ddot{\mathbf{x}}^{(0)}(t_{m-})\big]\varepsilon^2 < 0, \\
&\mathbf{n}^{\mathrm{T}}_{\partial\Omega_{ij}} \cdot \big[\mathbf{x}^{(\alpha)}(t_{m+\varepsilon}) - \mathbf{x}^{(0)}(t_{m+\varepsilon})\big] = \mathbf{n}^{\mathrm{T}}_{\partial\Omega_{ij}} \cdot \big[\ddot{\mathbf{x}}^{(\alpha)}(t_{m+}) - \ddot{\mathbf{x}}^{(0)}(t_{m+})\big]\varepsilon^2 > 0.
\end{aligned}$$

Therefore under condition in equation (28), the flow $\mathbf{x}^{(\alpha)}(t)$ in Ω_α is tangential to the boundary $\partial\Omega_{ij}$. □

THEOREM 11. *For a discontinuous dynamical system in equation* (1), $\mathbf{x}(t_m) = \mathbf{x}_m \in \partial\Omega_{ij}$ *for* t_m. *For an arbitrarily small* $\varepsilon > 0$, *there are two time intervals* (*i.e.*, $[t_{m-\varepsilon}, t_m)$ *and* $(t_m, t_{m+\varepsilon}]$) *and suppose* $\mathbf{x}^{(\alpha)}(t_{m\pm}) = \mathbf{x}_m$ ($\alpha \in \{i, j\}$). *The vector field* $\mathbf{F}^{(\alpha)}(t)$ *are* $C^r_{[t_{m-\varepsilon}, t_m)}$- *and* $C^r_{(t_m, t_{m+\varepsilon}]}$-*continuous* ($r \geqslant 2$) *for time* t *and* $\|\mathrm{d}^{r+1}\mathbf{x}^{(\alpha)}/\mathrm{d}t^{r+1}\| < \infty$. *The flow* $\mathbf{x}^{(\alpha)}(t)$ *in* Ω_α *is tangential to the boundary* $\partial\Omega_{ij}$ *iff*

$$\mathbf{n}^{\mathrm{T}}_{\partial\Omega_{ij}} \cdot \mathbf{F}^{(\alpha)}(t_{m\pm}) = 0 \tag{29}$$

and

$$\begin{cases}
\textit{either} \quad \mathbf{n}^{\mathrm{T}}_{\partial\Omega_{ij}} \cdot \big[D\mathbf{F}^{(\alpha)}(t_{m\pm}) - D\mathbf{F}^{(0)}(t_{m\pm})\big] < 0 \\
\quad \textit{for}\ \mathbf{n}_{\partial\Omega_{ij}} \to \Omega_\beta\ (\beta \in \{i, j\}, \beta \neq \alpha) \\
\textit{or} \quad \mathbf{n}^{\mathrm{T}}_{\partial\Omega_{ij}} \cdot \big[D\mathbf{F}^{(\alpha)}(t_{m\pm}) - D\mathbf{F}^{(0)}(t_{m\pm})\big] > 0 \\
\quad \textit{for}\ \mathbf{n}_{\partial\Omega_{ij}} \to \Omega_\alpha,
\end{cases} \tag{30}$$

where the total differentiation ($p, q \in \{1, 2, \ldots, n\}$) *for* $\lambda = \{0, \alpha\}$,

$$D\mathbf{F}^{(\lambda)}(t_{m\pm}) = \left[\frac{\partial F_p^{(\lambda)}(t_{m\pm})}{\partial x_q}\right]\mathbf{F}^{(\lambda)}(t_{m\pm}) + \frac{\partial \mathbf{F}^{(\lambda)}(t_{m\pm})}{\partial t}.$$

PROOF. Using equations (1), (5) and (29), thus the first condition in equation (19) is satisfied. The derivative of equations (1) and (5) with respect to time gives

$$\ddot{\mathbf{x}}^{(\lambda)} \equiv \left[\frac{\partial F_p^{(\lambda)}(\mathbf{x}, t, \boldsymbol{\mu}_\alpha)}{\partial x_q}\right]\dot{\mathbf{x}} + \frac{\partial}{\partial t}\mathbf{F}^{(\lambda)}(\mathbf{x}, t, \boldsymbol{\mu}_\alpha) = D\mathbf{F}^{(\lambda)}(t_{m\pm}).$$

For $t = t_{m\pm}$ and $\mathbf{x} = \mathbf{x}_m$, the left multiplication of $\mathbf{n}_{\partial\Omega_{ij}}$ to the foregoing equation gives

$$\mathbf{n}^{\mathrm{T}}_{\partial\Omega_{ij}} \cdot \ddot{\mathbf{x}}^{(\lambda)}(t_{m\pm}) = \mathbf{n}^{\mathrm{T}}_{\partial\Omega_{ij}} \cdot D\mathbf{F}^{(\lambda)}(t_{m\pm}),$$

where $\mathbf{F}^{(\alpha)}(\mathbf{x}_m, t_{m\pm}, \boldsymbol{\mu}_\alpha) \triangleq \mathbf{F}^{(\alpha)}(t_{m\pm})$. Using equation (30), the above equation leads to equation (28). From Theorem 10, the flow $\mathbf{x}^{(\alpha)}(t)$ in Ω_α is tangential to the boundary $\partial\Omega_{ij}$. □

DEFINITION 14. For a discontinuous dynamical system in equation (1), $\mathbf{x}(t_m) = \mathbf{x}_m \in \partial\Omega_{ij}$ for t_m. For an arbitrarily small $\varepsilon > 0$, there are two time intervals (i.e., $[t_{m-\varepsilon}, t_m)$ and $(t_m, t_{m+\varepsilon}]$) and suppose $\mathbf{x}^{(\alpha)}(t_{m\pm}) = \mathbf{x}_m$ ($\alpha \in \{i, j\}$). The flow $\mathbf{x}^{(\alpha)}(t)$ is $C^r_{[t_{m-\varepsilon}, t_m)}$- and $C^r_{(t_m, t_{m+\varepsilon}]}$-continuous ($r \geqslant 2n_1$) for time t. The flow $\mathbf{x}^{(\alpha)}(t)$ for $t \in T_m$ in Ω_α is the $(2n_1 - 1)$th-order tangential to the boundary $\partial\Omega_{ij}$ if the three conditions hold:

$$\mathbf{n}^{\mathrm{T}}_{\partial\Omega_{ij}} \cdot \frac{\mathrm{d}^k}{\mathrm{d}t^k}\left[\mathbf{x}^{(\alpha)}(t_{m\pm}) - \mathbf{x}^{(0)}(t_{m\pm})\right] = 0 \quad \text{for } k = 1, 2, \ldots, 2n_1 - 1, \tag{31}$$

$$\mathbf{n}^{\mathrm{T}}_{\partial\Omega_{ij}} \cdot \frac{\mathrm{d}^{2n_1}}{\mathrm{d}t^{2n_1}}\left[\mathbf{x}^{(\alpha)}(t_{m\pm}) - \mathbf{x}^{(0)}(t_{m\pm})\right] \neq 0, \tag{32}$$

either

$$\begin{cases} \mathbf{n}^{\mathrm{T}}_{\partial\Omega_{ij}} \cdot \left[\mathbf{x}^{(0)}(t_{m-\varepsilon}) - \mathbf{x}^{(\alpha)}(t_{m-\varepsilon})\right] > 0 & \text{and} \\ \mathbf{n}^{\mathrm{T}}_{\partial\Omega_{ij}} \cdot \left[\mathbf{x}^{(\alpha)}(t_{m+\varepsilon}) - \mathbf{x}^{(0)}(t_{m+\varepsilon})\right] < 0 & \text{for } \mathbf{n}_{\partial\Omega_{ij}} \to \Omega_\beta, \end{cases} \tag{33}$$

where $\beta = \{i, j\}$ but $\alpha \neq \beta$, or

$$\begin{cases} \mathbf{n}^{\mathrm{T}}_{\partial\Omega_{ij}} \cdot \left[\mathbf{x}^{(0)}(t_{m-\varepsilon}) - \mathbf{x}^{(\alpha)}(t_{m-\varepsilon})\right] < 0 & \text{and} \\ \mathbf{n}^{\mathrm{T}}_{\partial\Omega_{ij}} \cdot \left[\mathbf{x}^{(\alpha)}(t_{m+\varepsilon}) - \mathbf{x}^{(0)}(t_{m+\varepsilon})\right] > 0 & \text{for } \mathbf{n}_{\partial\Omega_{ij}} \to \Omega_\alpha. \end{cases} \tag{34}$$

THEOREM 12. *For a discontinuous dynamical system in equation* (1), $\mathbf{x}(t_m) = \mathbf{x}_m \in \partial\Omega_{ij}$ *for* t_m. *For an arbitrarily small* $\varepsilon > 0$, *there are two time intervals (i.e.,* $[t_{m-\varepsilon}, t_m)$ *and* $(t_m, t_{m+\varepsilon}]$) *and suppose* $\mathbf{x}^{(\alpha)}(t_{m\pm}) = \mathbf{x}_m$ ($\alpha \in \{i, j\}$). *The flow* $\mathbf{x}^{(\alpha)}(t)$ *is* $C^r_{[t_{m-\varepsilon}, t_m)}$*- and* $C^r_{(t_m, t_{m+\varepsilon}]}$*-continuous* ($r \geqslant 2n_1 + 1$) *for time t and* $\|\mathrm{d}^r\mathbf{x}^{(\alpha)}/\mathrm{d}t^r\| < \infty$. *The flow* $\mathbf{x}^{(\alpha)}(t)$ *in* Ω_α *is the* $(2n - 1)$*th-order tangential to the boundary* $\partial\Omega_{ij}$ *iff*

$$\mathbf{n}^{\mathrm{T}}_{\partial\Omega_{ij}} \cdot \frac{\mathrm{d}^k}{\mathrm{d}t^k}\left[\mathbf{x}^{(\alpha)}(t_{m\pm}) - \mathbf{x}^{(0)}(t_{m\pm})\right] = 0 \quad \textit{for } k = 1, 2, \ldots, 2n_1 - 1, \tag{35}$$

$$\mathbf{n}^{\mathrm{T}}_{\partial\Omega_{ij}} \cdot \frac{\mathrm{d}^{2n_1}}{\mathrm{d}t^{2n_1}}\left[\mathbf{x}^{(\alpha)}(t_{m\pm}) - \mathbf{x}^{(0)}(t_{m\pm})\right] \neq 0, \tag{36}$$

and

$$\begin{cases} \textit{either}\ \ \mathbf{n}^{\mathrm{T}}_{\partial\Omega_{ij}} \cdot \frac{\mathrm{d}^{2n_1}}{\mathrm{d}t^{2n_1}}\big[\mathbf{x}^{(\alpha)}(t_{m\pm}) - \mathbf{x}^{(0)}(t_{m\pm})\big] < 0 \\ \quad \textit{for}\ \mathbf{n}_{\partial\Omega_{ij}} \to \Omega_\beta \\ \textit{or}\ \ \mathbf{n}^{\mathrm{T}}_{\partial\Omega_{ij}} \cdot \frac{\mathrm{d}^{2n_1}}{\mathrm{d}t^{2n_1}}\big[\mathbf{x}^{(\alpha)}(t_{m\pm}) - \mathbf{x}^{(0)}(t_{m\pm})\big] > 0 \\ \quad \textit{for}\ \mathbf{n}_{\partial\Omega_{ij}} \to \Omega_\alpha, \end{cases} \tag{37}$$

where $\beta \in \{i, j\}$ *but* $\alpha \neq \beta$.

PROOF. For equations (35) and (36), the first two conditions in Definition 14 are satisfied. Consider the boundary $\partial\Omega_{ij}$ with $\mathbf{n}_{\partial\Omega_{ij}} \to \Omega_\beta$ ($\beta \neq \alpha$) first. Choose $a \in [t_{m-\varepsilon}, t_m)$ or $a \in (t_m, t_{m-\varepsilon}]$, and application of the Taylor series expansion of $\mathbf{x}^{(\alpha)}(t_{m\pm\varepsilon})$ to $\mathbf{x}^{(\alpha)}(a)$ and up to the $2n_1$-order term gives for $\lambda = \{0, \alpha\}$

$$\begin{aligned} \mathbf{x}^{(\lambda)}(t_{m\pm\varepsilon}) &\equiv \mathbf{x}^{(\lambda)}(t_{m\pm} \pm \varepsilon) \\ &= \mathbf{x}^{(\lambda)}(a) + \sum_{k=1}^{2n-1} \frac{\mathrm{d}^k}{\mathrm{d}t^k}\mathbf{x}^{(\lambda)}(a)(t_{m\pm} \pm \varepsilon - a)^k \\ &\quad + \frac{\mathrm{d}^{2n_1}}{\mathrm{d}t^{2n_1}}\mathbf{x}^{(\lambda)}(a)(t_{m\pm} \pm \varepsilon - a)^{2n_1} + \mathrm{o}\big((t_{m\pm} \pm \varepsilon - a)^{2n_1}\big). \end{aligned}$$

As $a \to t_{m\pm}$, we have

$$\begin{aligned} \mathbf{x}^{(\lambda)}(t_{m\pm\varepsilon}) &\equiv \mathbf{x}^{(\lambda)}(t_{m\pm} \pm \varepsilon) \\ &= \mathbf{x}^{(\lambda)}(t_{m\pm}) + \sum_{k=1}^{2n-1} \frac{\mathrm{d}^k}{\mathrm{d}t^k}\mathbf{x}^{(\lambda)}(t_{m\pm})(\pm\varepsilon)^k \\ &\quad + \frac{\mathrm{d}^{2n_1}}{\mathrm{d}t^{2n_1}}\mathbf{x}^{(\lambda)}(t_{m\pm})(\pm\varepsilon)^{2n_1} + \mathrm{o}\big(\pm\varepsilon^{2n_1}\big). \end{aligned}$$

With equations (35) and (36), the deformation of the above equation and left multiplication of $\mathbf{n}_{\partial\Omega_{ij}}$ produces

$$\begin{aligned} &\mathbf{n}^{\mathrm{T}}_{\partial\Omega_{ij}} \cdot \big[\mathbf{x}^{(\alpha)}(t_{m+\varepsilon}) - \mathbf{x}^{(0)}(t_{m+\varepsilon})\big] \\ &\quad = \mathbf{n}^{\mathrm{T}}_{\partial\Omega_{ij}} \cdot \frac{\mathrm{d}^{2n_1}}{\mathrm{d}t^{2n_1}}\big[\mathbf{x}^{(\alpha)}(t_{m+}) - \mathbf{x}^{(0)}(t_{m+})\big]\varepsilon^{2n_1}, \\ &\mathbf{n}^{\mathrm{T}}_{\partial\Omega_{ij}} \cdot \big[\mathbf{x}^{(0)}(t_{m-\varepsilon}) - \mathbf{x}^{(\alpha)}(t_{m-\varepsilon})\big] \\ &\quad = -\mathbf{n}^{\mathrm{T}}_{\partial\Omega_{ij}} \cdot \frac{\mathrm{d}^{2n_1}}{\mathrm{d}t^{2n_1}}\big[\mathbf{x}^{(\alpha)}(t_{m-}) - \mathbf{x}^{(0)}(t_{m-})\big]\varepsilon^{2n_1}. \end{aligned}$$

Under equation (37), the condition in equation (33) is satisfied. Therefore, the flow $\mathbf{x}^{(\alpha)}(t)$ in Ω_α is the $(2n_1 - 1)$th-order tangential to the boundary $\partial\Omega_{\alpha\beta}$ with $\mathbf{n}_{\partial\Omega_{ij}} \to \Omega_\beta$. Similarly, under the condition in equation (37), the flow $\mathbf{x}^{(\alpha)}(t)$

in Ω_α is the $(2n_1-1)$th-order tangential to the boundary $\partial\Omega_{\alpha\beta}$ with $\mathbf{n}_{\partial\Omega_{ij}} \to \Omega_\alpha$. This theorem is proved. □

THEOREM 13. *For a discontinuous dynamical system in* (1), $\mathbf{x}(t_m) = \mathbf{x}_m \in \partial\Omega_{ij}$ *for* t_m. *For an arbitrarily small* $\varepsilon > 0$, *there are two time intervals (i.e.,* $[t_{m-\varepsilon}, t_m)$ *and* $(t_m, t_{m+\varepsilon}]$*) and suppose* $\mathbf{x}^{(\alpha)}(t_{m\pm}) = \mathbf{x}_m$ $(\alpha \in \{i, j\})$. *The vector field* $\mathbf{F}^{(\alpha)}(t)$ *is* $C^r_{[t_{m-\varepsilon},t_m)}$*- and* $C^r_{(t_m,t_{m+\varepsilon}]}$*-continuous* $(r \geqslant 2n_1)$ *for time t and* $\|\mathrm{d}^{r+1}\mathbf{x}^{(\alpha)}/\mathrm{d}t^{r+1}\| < \infty$. *The flow* $\mathbf{x}^{(\alpha)}(t)$ *in* Ω_α *is the* $(2n_1-1)$*th-order tangential to the boundary* $\partial\Omega_{ij}$ *iff*

$$\mathbf{n}^{\mathrm{T}}_{\partial\Omega_{ij}} \cdot D^{k-1}\big[\mathbf{F}^{(\alpha)}(t_{m\pm}) - \mathbf{F}^{(0)}(t_{m\pm})\big] = 0 \quad \text{for } k = 1, 2, \ldots, 2n_1 - 1, \tag{38}$$

$$\mathbf{n}^{\mathrm{T}}_{\partial\Omega_{ij}} \cdot D^{2n_1-1}\big[\mathbf{F}^{(\alpha)}(t_{m\pm}) - \mathbf{F}^{(0)}(t_{m\pm})\big] \neq 0, \tag{39}$$

$$\left\{\begin{array}{l} \mathbf{n}^{\mathrm{T}}_{\partial\Omega_{ij}} \cdot D^{2n_1-1}\big[\mathbf{F}^{(\alpha)}(t_{m\pm}) - \mathbf{F}^{(0)}(t_{m\pm})\big] < 0 \\ \quad \text{for } \mathbf{n}_{\partial\Omega_{ij}} \to \Omega_\beta \quad \text{or} \\ \mathbf{n}^{\mathrm{T}}_{\partial\Omega_{ij}} \cdot D^{2n_1-1}\big[\mathbf{F}^{(\alpha)}(t_{m\pm}) - \mathbf{F}^{(0)}(t_{m\pm})\big] > 0 \\ \quad \text{for } \mathbf{n}_{\partial\Omega_{ij}} \to \Omega_\alpha, \end{array}\right. \tag{40}$$

where the total differentiation for $\lambda = \{0, \alpha\}$

$$D^{k-1}\mathbf{F}^{(\lambda)}(t_{m\pm}) = D^{k-2}\left\{\left[\frac{\partial F_p^{(\lambda)}(t_{m\pm})}{\partial x_q}\right]\mathbf{F}^{(\lambda)}(t_{m\pm}) + \frac{\partial \mathbf{F}^{(\lambda)}(t_{m\pm})}{\partial t}\right\}, \tag{41}$$

$p, q \in \{1, 2, \ldots, n\}$, $k \in \{2, 3, \ldots, 2n_1\}$ *and* $\beta \in \{i, j\}$ *but* $\alpha \neq \beta$.

PROOF. The derivative of equations (1) and (5) with respect to time gives for $\lambda = \{0, \alpha\}$

$$\begin{aligned} \frac{\mathrm{d}^{n_1}\mathbf{x}^{(\lambda)}(t_m)}{\mathrm{d}t^{n_1}} &= \frac{\mathrm{d}^{n_1-1}}{\mathrm{d}t^{n_1-1}}\dot{\mathbf{x}}^{(\lambda)}(t_m) = \frac{\mathrm{d}^{n_1-1}}{\mathrm{d}t^{n_1-1}}\mathbf{F}^{(\lambda)}(\mathbf{x}_m, t_m, \boldsymbol{\mu}_\alpha) \\ &\equiv D^{n_1-1}\mathbf{F}^{(\lambda)}(\mathbf{x}_m, t_m, \boldsymbol{\mu}_\alpha) \\ &= D^{n_1-2}\left\{\left[\frac{\partial F_p^{(\lambda)}(\mathbf{x}_m, t_m, \boldsymbol{\mu}_\alpha)}{\partial x_q}\right]\dot{\mathbf{x}} + \frac{\partial}{\partial t}\mathbf{F}^{(\lambda)}(\mathbf{x}_m, t_m, \boldsymbol{\mu}_\alpha)\right\}. \end{aligned}$$

Using the foregoing equation to the conditions in equations (38)–(41), the flow $\mathbf{x}^{(\alpha)}(t)$ for $t \in T_m$ in Ω_α is the $(2n_1-1)$th-order tangential to the boundary $\partial\Omega_{ij}$ from Theorem 12. Therefore, this theorem is proved. □

DEFINITION 15. A flow $\mathbf{x}^{(\alpha)}(t)$ tangential to $\partial\Omega_{ij}$ $(\alpha \in \{i, j\})$ in Ω_α is termed the local grazing flow if $\mathbf{x}^{(\alpha)}(t)$ starting from $\partial\Omega_{ij}$ in Ω_α is not intersected with another boundary before grazing. Suppose $\mathbf{x}^{(\alpha)}(t)$ has $\mathbf{x}^{(\alpha)}_{m-1}$ and $\mathbf{x}^{(\alpha)}_{m+1}$ on the

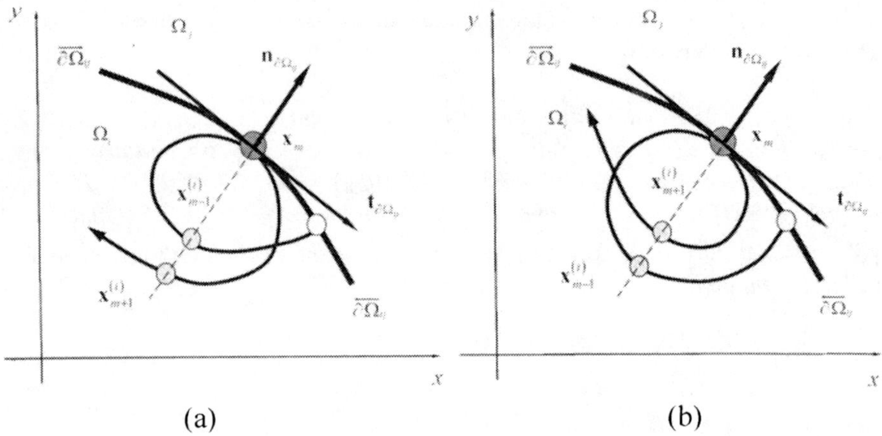

Figure 8. A classification of local grazing flows in Ω_i to $\partial\Omega_{ij}$ with $\mathbf{n}_{\partial\Omega_{ij}} \to \Omega_j$: (a) the first kind of grazing flow, (b) the second kind of grazing flow and (c) the third kind of grazing flow.

$\mathbf{n}_{\partial\Omega_{ij}}$-line relative to $\mathbf{x}_m \in \partial\Omega_{ij}$, then the three grazing flows exist: The tangential flow $\mathbf{x}^{(\alpha)}(t)$ is termed

(a) the *grazing flow of the first kind* if

$$\left\|\mathbf{x}_{m-1}^{(\alpha)} - \mathbf{x}_m\right\| < \left\|\mathbf{x}_{m+1}^{(\alpha)} - \mathbf{x}_m\right\|, \tag{42}$$

(b) the *grazing flow of the second kind* if

$$\left\|\mathbf{x}_{m-1}^{(\alpha)} - \mathbf{x}_m\right\| > \left\|\mathbf{x}_{m+1}^{(\alpha)} - \mathbf{x}_m\right\|, \quad \text{or} \tag{43}$$

(c) the *grazing flow of the third kind* if

$$\left\|\mathbf{x}_{m-1}^{(\alpha)} - \mathbf{x}_m\right\| = \left\|\mathbf{x}_{m+1}^{(\alpha)} - \mathbf{x}_m\right\|. \tag{44}$$

From Definition 15, the local grazing flows in the domain Ω_i to the boundary $\partial\Omega_{ij}$ with $\mathbf{n}_{\partial\Omega_{ij}} \to \Omega_j$ are sketched in Figure 8 for interpretation of the local grazing flows. The first, second and third kinds of grazing flow are arranged in Figures 8(a)–8(c), respectively. The grey-filled symbols represent two points ($\mathbf{x}_{m-1}^{(i)}$ and $\mathbf{x}_{m+1}^{(i)}$) on the normal line relative to tangential point $\mathbf{x}_m$ on the boundary $\partial\Omega_{ij}$ depicted by a large circular symbol.

6. Piecewise linear systems

To demonstrate the local singularity and grazing in nonsmooth dynamical systems, two common examples in mechanical engineering are considered herein.

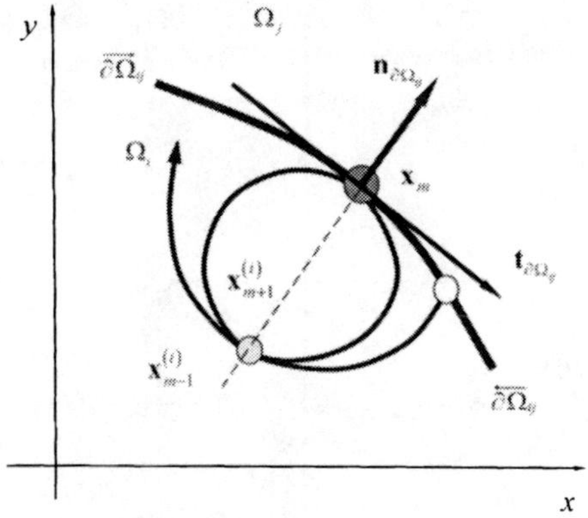

(c)

Figure 8. (*Continued.*)

Consider a periodically excited, piecewise linear system as the first application with

$$\ddot{x} + 2d\dot{x} + k(x) = a\cos\Omega t, \tag{45}$$

where $\dot{x} = \mathrm{d}x/\mathrm{d}t$. The parameters ($\Omega$ and a) are excitation frequency and amplitude, respectively. The restoring force is

$$k(x) = \begin{cases} cx - e & \text{for } x \in [E, \infty), \\ 0 & \text{for } x \in [-E, E], \\ cx + e & \text{for } x \in (-\infty, -E], \end{cases} \tag{46}$$

with $E = e/c$. To describe the motion of the foregoing system, there are three linear regions of the restoring force (Region I: $x \geqslant E$, Region II: $-E \leqslant x \leqslant E$ and Region III: $x \leqslant -E$). The solution for each region can be easily obtained in the Appendix. From equations (45) and (46), the dynamical systems in Regions I and III do not have any singularity. Once all the parameters are finite, the solutions of these motions in Regions I and III are finite. In Region II, the solution of the motion continuous to time t is unbounded, but the displacement boundary of the domain on which the dynamical system is defined is finite. Since the velocity is the derivative of displacement with respect to time, the flows of the system in the displacement-bounded, Region II is bounded. The phase space in equation (45) is divided into three subdomains, and the three subdomains are defined by

$$\begin{cases} \Omega_1 = \{(x, y) \mid x \in [E, \infty), y \in (-\infty, \infty)\}, \\ \Omega_2 = \{(x, y) \mid x \in [-E, E], y \in (-\infty, \infty)\}, \\ \Omega_3 = \{(x, y) \mid x \in (-\infty, -E], y \in (-\infty, \infty)\}. \end{cases} \tag{47}$$

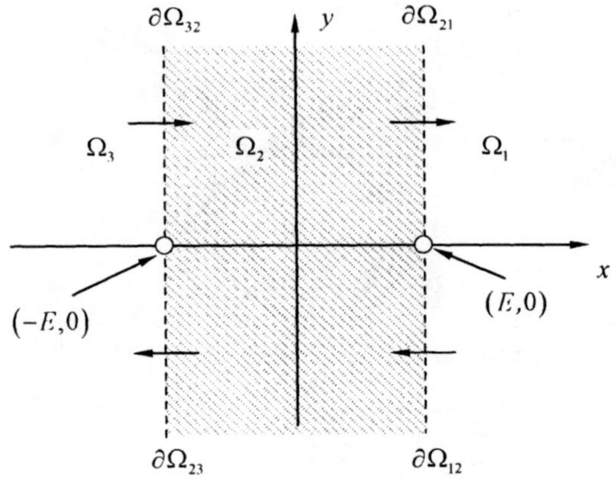

Figure 9. Subdomains, boundaries and equilibriums $(\pm E, 0)$.

The entire phase space is given by

$$\mho = \bigcup_{\alpha=1}^{3} \Omega_\alpha. \tag{48}$$

The corresponding separation boundaries are

$$\begin{cases} \partial\Omega_{12} = \Omega_1 \cap \Omega_2 = \big\{(x, y) \mid \varphi_{12}(x, y) \equiv x - E = 0\big\}, \\ \partial\Omega_{23} = \Omega_2 \cap \Omega_3 = \big\{(x, y) \mid \varphi_{23}(x, y) \equiv x + E = 0\big\}. \end{cases} \tag{49}$$

Such domains and the boundary are sketched in Figure 9. From the above definitions, equations (45) and (46) give

$$\mathbf{x} = \mathbf{F}^{(\alpha)}(\mathbf{x}, t, \boldsymbol{\mu}_\alpha, \boldsymbol{\pi}) \quad \text{for } \alpha \in \{1, 2, 3\}, \tag{50}$$

where

$$\begin{cases} \mathbf{F}^{(1)}(\mathbf{x}, t, \boldsymbol{\mu}_1, \boldsymbol{\pi}) = (y, -2dy - cx + a\cos\omega t)^{\mathrm{T}} & \text{in } \mathbf{x} \in \Omega_1, \\ \mathbf{F}^{(2)}(\mathbf{x}, t, \boldsymbol{\mu}_2, \boldsymbol{\pi}) = (y, -2dy + a\cos\omega t)^{\mathrm{T}} & \text{in } \mathbf{x} \in \Omega_2, \\ \mathbf{F}^{(3)}(\mathbf{x}, t, \boldsymbol{\mu}_3, \boldsymbol{\pi}) = (y, -2dy - cx + a\cos\omega t)^{\mathrm{T}} & \text{in } \mathbf{x} \in \Omega_3. \end{cases} \tag{51}$$

Note that $\boldsymbol{\mu}_1 = \boldsymbol{\mu}_3 = (c, d)^{\mathrm{T}}$, $\boldsymbol{\mu}_2 = (0, d)^{\mathrm{T}}$ and $\boldsymbol{\pi} = (\Omega, a)^{\mathrm{T}}$. To understand the local dynamics of equation (45), it is very important to investigate the sliding dynamics along the separation boundary. In equation (49), using $\mathbf{n}_{\partial\Omega_{ij}} = \nabla\varphi_{ij}$ gives

$$\mathbf{n}_{\partial\Omega_{12}} = \mathbf{n}_{\partial\Omega_{23}} = (1, 0)^{\mathrm{T}}. \tag{52}$$

Therefore, equations (51) and (52) give

$$\begin{cases} \mathbf{n}^{\mathrm{T}}_{\partial\Omega_{12}} \cdot \mathbf{F}^{(1)}(\mathbf{x}, t, \boldsymbol{\mu}_1, \boldsymbol{\pi}) = \mathbf{n}^{\mathrm{T}}_{\partial\Omega_{12}} \cdot \mathbf{F}^{(2)}(\mathbf{x}, t, \boldsymbol{\mu}_2, \boldsymbol{\pi}) = y, \\ \mathbf{n}^{\mathrm{T}}_{\partial\Omega_{23}} \cdot \mathbf{F}^{(2)}(\mathbf{x}, t, \boldsymbol{\mu}_2, \boldsymbol{\pi}) = \mathbf{n}^{\mathrm{T}}_{\partial\Omega_{23}} \cdot \mathbf{F}^{(3)}(\mathbf{x}, t, \boldsymbol{\mu}_3, \boldsymbol{\pi}) = y. \end{cases} \tag{53}$$

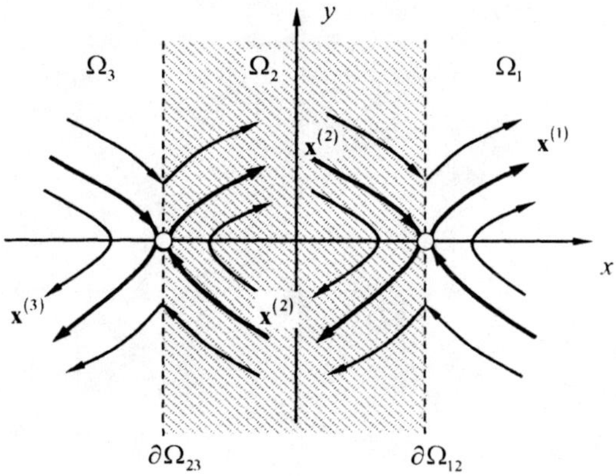

Figure 10. Phase portraits near equilibriums $(\pm E, 0)$ on the boundaries.

From equations (52) and (53) with Theorem 11, the equilibrium points $(\pm E, 0)$ are common tangential points for flows in both subdomains of the two separatrices. Furthermore, the grazing bifurcation conditions on the separatrices for the flows of this discontinuous system in the three subdomains are

$$y = 0 \quad \text{and} \quad \begin{cases} \dot{y} > 0 & \text{for } x = E \text{ in } \Omega_1 \text{ and } x = -E \text{ in } \Omega_2, \\ \dot{y} < 0 & \text{for } x = E \text{ in } \Omega_2 \text{ and } x = -E \text{ in } \Omega_3. \end{cases} \tag{54}$$

Therefore, in the neighborhoods of the two equilibrium points, the local topological structures of the hyperbolic flow for the system in equation (45) are sketched in Figure 10. The detailed discussion can be refereed to Luo (2006). For description of motion in equation (45), two switching sections (or sets) are

$$\begin{cases} \Sigma^{+} = \{(t_i, x_i, y_i) \mid x_i = E, \dot{x}_i = y_i\}, \\ \Sigma^{-} = \{(t_i, x_i, y_i) \mid x_i = -E, \dot{x}_i = y_i\}. \end{cases} \tag{55}$$

The two sets are decomposed as

$$\begin{aligned} &\Sigma^{+} = \Sigma^{+}_{+} \cup \Sigma^{+}_{-} \cup \{t_i, E, 0\} \quad \text{and} \\ &\Sigma^{-} = \Sigma^{-}_{+} \cup \Sigma^{-}_{-} \cup \{t_i, -E, 0\}, \end{aligned} \tag{56}$$

where four subsets are defined as

$$\begin{cases} \Sigma^{+}_{+} = \{(t_i, x_i, y_i) \mid x_i = E, \dot{x}_i = y_i > 0\}, \\ \Sigma^{+}_{-} = \{(t_i, x_i, y_i) \mid x_i = E, \dot{x}_i = y_i < 0\}, \end{cases} \tag{57}$$

$$\begin{cases} \Sigma^{-}_{+} = \{(t_i, x_i, y_i) \mid x_i = -E, \dot{x}_i = y_i > 0\}, \\ \Sigma^{-}_{-} = \{(t_i, x_i, y_i) \mid x_i = -E, \dot{x}_i = y_i < 0\}. \end{cases} \tag{58}$$

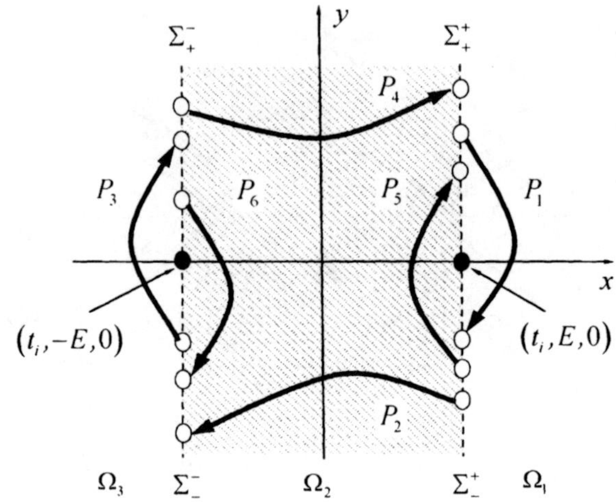

Figure 11. Switching sections and basic mappings in phase plane.

From four subsets, six basic mappings are

$$\begin{cases} P_1 : \Sigma_{+}^{+} \to \Sigma_{-}^{+}, & P_2 : \Sigma_{-}^{+} \to \Sigma_{-}^{-}, & P_3 : \Sigma_{-}^{-} \to \Sigma_{+}^{-}, \\ P_4 : \Sigma_{+}^{-} \to \Sigma_{+}^{+}, & P_5 : \Sigma_{-}^{+} \to \Sigma_{+}^{+}, & P_6 : \Sigma_{+}^{-} \to \Sigma_{-}^{-}. \end{cases} \tag{59}$$

The switching planes and basic mappings are sketched in Figure 11. Consider the initial and final states of $(t, x, \dot{x})$ to be (t_i, x_i, y_i) and $(t_{i+1}, x_{i+1}, y_{i+1})$ in the subdomain Ω_α, $\alpha = 1, 2, 3$, respectively. The local mappings are $\{P_1, P_3, P_5, P_6\}$ and the global mappings are $\{P_2, P_4\}$. The displacement and velocity equations in the Appendix with initial conditions give the governing equations for mapping P_k, $k = 1, 2, \ldots, 6$, i.e.,

$$P_k : \begin{cases} f_1^{(k)}(x_i, y_i, t_i, x_{i+1}, y_{i+1}, t_{i+1}) = 0, \\ f_2^{(k)}(x_i, y_i, t_i, x_{i+1}, y_{i+1}, t_{i+1}) = 0. \end{cases} \tag{60}$$

The necessary and sufficient conditions for the grazing of all the six generic mappings are

$$\begin{cases} y_{i+1} = 0; \\ \dot{y}_{i+1} = a \cos \Omega t_{i+1} > 0 \quad \text{for } P_j,\ j = 1, 2, 6; \\ \dot{y}_{i+1} = a \cos \Omega t_{i+1} < 0 \quad \text{for } P_j,\ j = 3, 4, 5. \end{cases} \tag{61}$$

From the foregoing equation, once one of the initial time or velocity is selected, the grazing bifurcation can be determined through the Newton–Raphson method. In computation, the condition $t_{i+1} > t_i$ should be inserted for fast obtaining solutions. Consider the parameters $a = 20$, $c = 100$, $E = 1$, $d = 0.5$ for illustration of the grazing bifurcation. For given initial velocity y_i, the grazing bifurcations

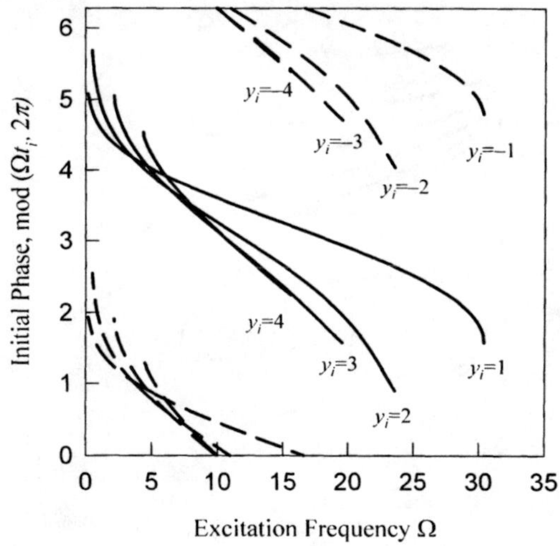

Figure 12. The grazing bifurcation conditions for mappings P_1 ($y_i \geqslant 0$, solid) and P_3 ($y_i \leqslant 0$, dash). ($n = 1, a = 20, c = 100, E = 1, d = 0.5$.)

are computed. In Figure 12, the grazing bifurcation conditions of mapping P_1 ($y_i \geqslant 0$) and P_3 ($y_i \leqslant 0$) are illustrated for $y_i = \pm\{1, 2, 3, 4\}$ and $n = 1$. The grazing bifurcations for both mappings P_2 ($y_i \leqslant 0$) and P_4 ($y_i \geqslant 0$) and mappings P_2 ($y_i \leqslant 0$) and P_4 ($y_i \geqslant 0$) are presented in Figure 13. The mappings $\{P_2, P_5\}$ and $\{P_3, P_6\}$ are represented by the solid and dashed curves, respectively. The initial conditions varying with excitation frequency are presented. Similarly, the initial condition varying with excitation amplitude can be determined. From these initial conditions, based on the six mappings, the final states will be tangential to the corresponding boundaries. From the predictions of grazing bifurcation, the grazing flows are demonstrated for a better understanding of the grazing bifurcation. The same parameters are used as before. The grazing flows for mapping P_1 ($x_i = 1$, $(y_i, \Omega t_i) \approx (1, 3.592), (2, 3.256), (3, 3.125)$) in Ω_1 and mapping P_3 ($x_i = -1$, $(y_i, \Omega t_i) \approx (-1, 0.449), (-2, 0.114), (-3, 6.267)$) in Ω_3 are illustrated in Figures 14(a) and 14(b) for $\Omega = 10$. The grazing flows in Ω_2 are illustrated in Figures 14(c) and 14(d) for mapping P_2 ($x_i = 1$, $(y_i, \Omega t_i) = (0, 2.71), (-2, 3.20), (-4, 3.63), (-6, 4.07)$) and mapping P_4 ($x_i = -1$, $(y_i, \Omega t_i) \approx (0, 5.85), (2, 0.06), (-4, 0.47), (-6, 0.93)$). The grazing flow for $y_i = 0$ seems the half generic separatrix of the pendulum. Finally, the local grazing flows in domain Ω_2 are illustrated in Figures 14(e) and 14(f) for mappings P_5 ($x_i = 1$, $y_i = -2$, $\Omega t_i \approx 5.90, 4.11$) and mapping P_6 ($x_i = -1$, $y_i = 2$, $\Omega t_i \approx 2.76, 0.97$). The two local, grazing flows are of the first kind in domain Ω_2. It is observed that the grazing flows at the tangential points satisfy the conditions in Theorem 11. The detailed results can be referred to Luo (2005b).

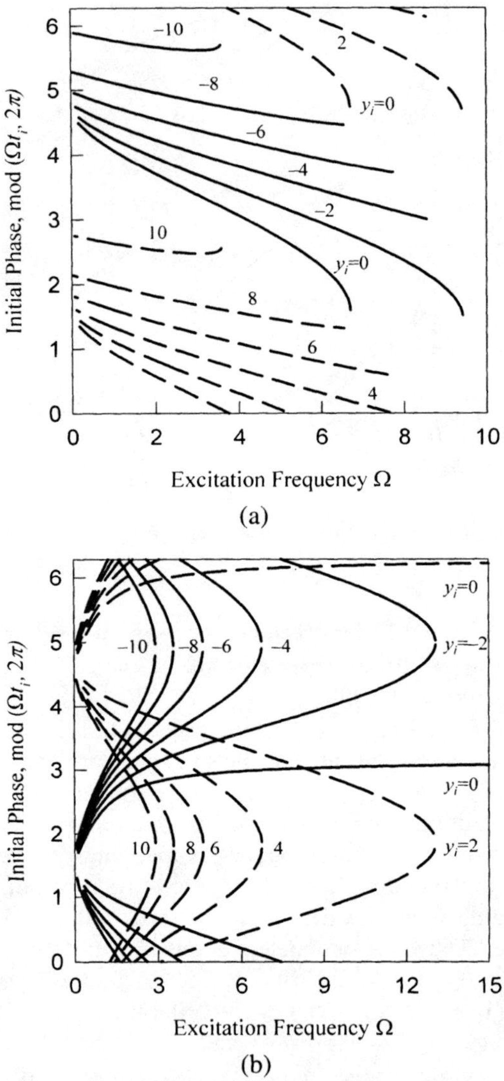

Figure 13. The grazing bifurcation conditions for (a) mappings P_2 ($y_i \leqslant 0$, solid) and P_4 ($y_i \geqslant 0$, dash) and (b) mappings P_5 ($y_i \leqslant 0$, solid) and P_6 ($y_i \geqslant 0$, dash). ($n = 1$, $a = 20$, $c = 100$, $E = 1$, $d = 0.5$.)

The fragmentation of strange attractors of chaotic motions in this system can be seen in Luo (2005c). The grazing in piecewise linear systems with impacting can be found in Luo and Chen (2006).

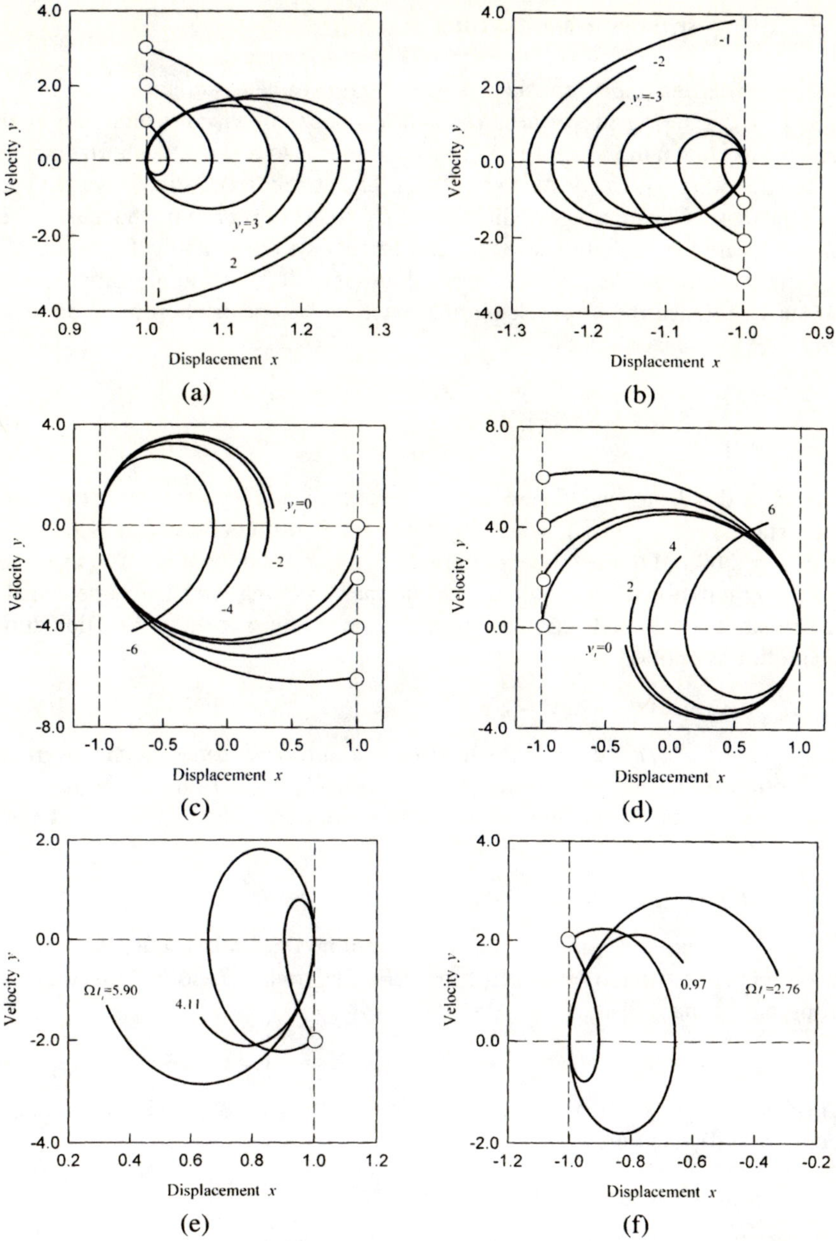

Figure 14. The grazing flows relative to (a) mapping P_1 ($x_i = 1$); (b) mapping P_3 ($x_i = -1$); (c) mapping P_2 ($x_i = -1$); (d) mapping P_4 ($x_i = 1$); (e) mapping P_5 ($x_i = 1$) and (f) mapping P_6 ($x_i = -1$). ($n = 1$, $a = 20$, $c = 100$, $E = 1$, $d = 0.5$.)

7. Friction-induced oscillators

For the second example, consider a periodically forced oscillator consisting of a mass (m), a spring of stiffness (k) and a damper of viscous damping coefficient (r), as shown in Figure 15(a). This oscillator also rests on the horizontal belt surface traveling with a constant speed (V). The absolute coordinate system (x, t) is for the mass. Consider a periodical force $Q_0 \cos \Omega t$ exerting on the mass, where Q_0 and Ω are the excitation strength and frequency, respectively. Since the mass contacts the moving belt with friction, the mass can move along or rest on the belt surface. Once the nonstick motion exists, a kinetic friction force shown in Figure 15(b) is described as

$$\overline{F}_{\mathrm{f}}(\dot{x}) \begin{cases} = \mu_k F_{\mathrm{N}}, & \dot{x} \in [V, \infty); \\ \in [-\mu_k F_{\mathrm{N}}, \mu_k F_{\mathrm{N}}], & \dot{x} = V; \\ = -\mu_k F_{\mathrm{N}}, & \dot{x} \in (-\infty, V]; \end{cases} \tag{62}$$

where $\dot{x} \triangleq \mathrm{d}x/\mathrm{d}t$, μ_k and F_{N} are friction coefficient and a normal force to the contact surface, respectively. The grazing flow for this problem was discussed in Luo and Gegg (2005). For the model in Figure 15, the friction force is $F_{\mathrm{N}} = mg$ where g is the gravitational acceleration. For the mass moving with the same speed of the belt surface, the nonfriction forces acting on the mass in the x-direction during this motion is defined as

$$F_{\mathrm{s}} = A_0 \cos \Omega t - 2dV - cx \quad \text{for } \dot{x} = V, \tag{63}$$

where $A_0 = Q_0/m$, $d = r/2m$ and $c = k/m$. This force cannot overcome the friction force during the stick motion, i.e., $|F_{\mathrm{s}}| \leqslant F_{\mathrm{f}}$ and $F_{\mathrm{f}} = \mu_k F_{\mathrm{N}}/m$. Therefore, the mass does not have any relative motion to the belt. No acceleration exists, i.e.,

$$\ddot{x} = 0 \quad \text{for } \dot{x} = V. \tag{64}$$

If $|F_{\mathrm{s}}| > F_{\mathrm{f}}$, the nonfriction force will overcome the static friction force on the mass and the nonstick motion will appear. For the nonstick motion, the total force acting on the mass is

$$F = A_0 \cos \Omega t - F_{\mathrm{f}} \operatorname{sgn}(\dot{x} - V) - 2d\dot{x} - cx \quad \text{for } \dot{x} \neq V; \tag{65}$$

$\operatorname{sgn}(\cdot)$ is the sign function. Therefore, the equation of the nonstick motion for this oscillator with friction is

$$\ddot{x} + 2d\dot{x} + cx = A_0 \cos \Omega t - F_{\mathrm{f}} \operatorname{sgn}(\dot{x} - V) \quad \text{for } \dot{x} \neq V. \tag{66}$$

Since the friction force is dependent on the direction of the relative velocity, the phase plane is partitioned into two regions in which the motion is described through the continuous dynamical systems, as shown in Figure 16. The two regions are expressed by Ω_α, $\alpha \in \{1, 2\}$. In phase plane, the following vectors are

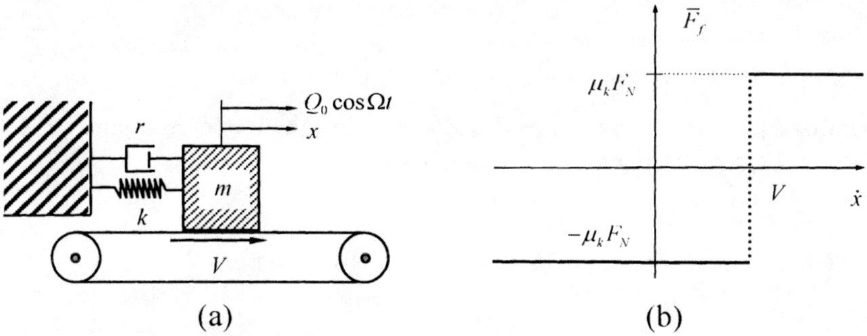

Figure 15. (a) Schematic of mechanical model and (b) friction force.

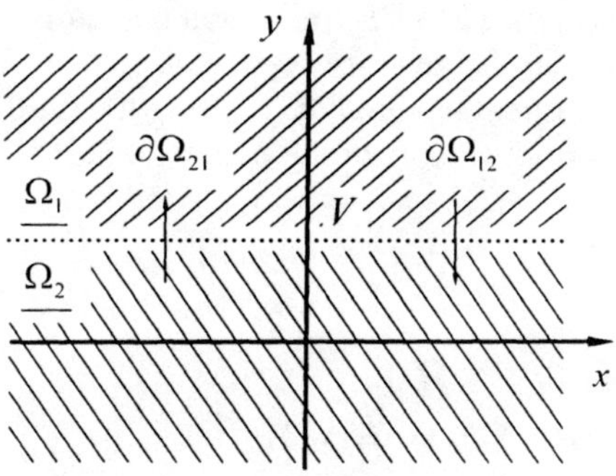

Figure 16. Domain partitions in phase plane.

introduced as

$$\mathbf{x} \triangleq (x, \dot{x})^{\mathrm{T}} \equiv (x, y)^{\mathrm{T}} \quad \text{and} \quad \mathbf{F} \triangleq (y, F)^{\mathrm{T}}. \tag{67}$$

The mathematical description of the regions and boundary

$$\begin{cases} \Omega_1 = \big\{(x, y) \mid y \in (V, \infty)\big\}, \\ \Omega_2 = \big\{(x, y) \mid y \in (-\infty, V)\big\}, \\ \partial\Omega_{12} = \big\{(x, y) \mid \varphi_{12}(x, y) \equiv y - V = 0\big\}, \\ \partial\Omega_{21} = \big\{(x, y) \mid \varphi_{21}(x, y) \equiv y - V = 0\big\}. \end{cases} \tag{68}$$

The subscript $(\cdot)_{\alpha\beta}$ denotes the boundary from Ω_α to Ω_β, $\alpha, \beta \in \{1, 2\}$ and $\alpha \neq \beta$. The equations of motion in equations (64) and (66) can be described as

$$\dot{\mathbf{x}} = \mathbf{F}^{(\alpha)}(\mathbf{x}, t) \quad \text{in } \Omega_\alpha, \alpha \in \{1, 2\}, \tag{69}$$

where

$$\mathbf{F}^{(\alpha)}(\mathbf{x}, t) = \big(y, F_\alpha(\mathbf{x}, \Omega t)\big)^{\mathrm{T}} \tag{70}$$

and

$$F_\alpha(\mathbf{x}, \Omega t) = A_0 \cos \Omega t - b_\alpha - 2d_\alpha y - c_\alpha x. \tag{71}$$

Note that $b_1 = -b_2 = \mu g$, $d_\alpha = d$ and $c_\alpha = c$ for the model in Figure 15. From Theorem 11, the grazing motion is guaranteed by

$$\begin{cases} \mathbf{n}^{\mathrm{T}}_{\partial\Omega_{\alpha\beta}} \cdot \mathbf{F}^{(\alpha)}(t_{m\pm}) = 0, \quad \alpha \in \{1, 2\}; \\ \mathbf{n}^{\mathrm{T}}_{\partial\Omega_{\alpha\beta}} \cdot D\mathbf{F}^{(1)}(t_{m\pm}) > 0; \qquad \mathbf{n}^{\mathrm{T}}_{\partial\Omega_{\alpha\beta}} \cdot D\mathbf{F}^{(2)}(t_{m\pm}) < 0; \end{cases} \tag{72}$$

where

$$D\mathbf{F}^{(\alpha)}(t) = \left(2F_\alpha(\mathbf{x}, t), \nabla F_\alpha(\mathbf{x}, t) \cdot \mathbf{F}^{(\alpha)}(t) + \frac{\partial F_\alpha(\mathbf{x}, t)}{\partial t}\right)^{\mathrm{T}}, \tag{73}$$

where $\nabla = (\partial/\partial x)\mathbf{i} + (\partial/\partial y)\mathbf{j}$ is the Hamiltonian operator. The time t_m represents the time for the motion on the velocity boundary. $t_{m\pm} = t_m \pm 0$ reflects the responses on the regions rather than boundary. Using the third and fourth equations of equation (68), the normal vectors of the boundary are

$$\mathbf{n}_{\partial\Omega_{12}} = \mathbf{n}_{\partial\Omega_{21}} = (0, 1)^{\mathrm{T}}. \tag{74}$$

Therefore, we have

$$\begin{cases} \mathbf{n}^{\mathrm{T}}_{\partial\Omega_{\alpha\beta}} \cdot \mathbf{F}^{(\alpha)}(t) = F_\alpha(\mathbf{x}, \Omega t), \\ \mathbf{n}^{\mathrm{T}}_{\partial\Omega_{\alpha\beta}} \cdot D\mathbf{F}^{(\alpha)}(t) = \nabla F_\alpha(\mathbf{x}, \Omega t) \cdot \mathbf{F}^{(\alpha)}(t) + \frac{\partial F_\alpha(\mathbf{x}, \Omega t)}{\partial t}. \end{cases} \tag{75}$$

From equations (74) and (75), the necessary and sufficient conditions for grazing motions are from Theorem 10:

$$F_\alpha(\mathbf{x}_m, \Omega t_{m\pm}) = 0, \qquad F_\alpha(\mathbf{x}_m, \Omega t_{m-\varepsilon}) \times F_\alpha(\mathbf{x}_m, \Omega t_{m+\varepsilon}) < 0, \tag{76}$$

or precisely,

$$\begin{cases} F_\alpha(\mathbf{x}_m, \Omega t_{m\pm}) = 0, \\ F_1(\mathbf{x}_m, \Omega t_{m-\varepsilon}) < 0, \qquad F_1(\mathbf{x}_m, \Omega t_{m+\varepsilon}) > 0; \\ F_2(\mathbf{x}_m, \Omega t_{m-\varepsilon}) > 0, \qquad F_2(\mathbf{x}_m, \Omega t_{m+\varepsilon}) < 0. \end{cases} \tag{77}$$

However, from Theorem 11, the necessary and sufficient conditions for grazing is given by

$$\begin{cases} F_\alpha(\mathbf{x}_m, \Omega t_{m\pm}) = 0, \\ \nabla F_\alpha(\mathbf{x}_m, \Omega t_{m\pm}) \cdot \mathbf{F}^{(\alpha)}(t_{m\pm}) + \frac{\partial F_\alpha(\mathbf{x}_m, \Omega t_{m\pm})}{\partial t} \begin{cases} > 0 & \text{for } \alpha = 1, \\ < 0 & \text{for } \alpha = 2. \end{cases} \end{cases} \tag{78}$$

A sketch of grazing motions in domain Ω_α ($\alpha = \{1, 2\}$) is illustrated in Figures 17(a) and 17(b). The grazing conditions are also presented, and the vector fields in Ω_1 and Ω_2 are expressed by the dashed and solid arrow-lines, respectively. The condition in equation (76) for the grazing motion in Ω_α is presented

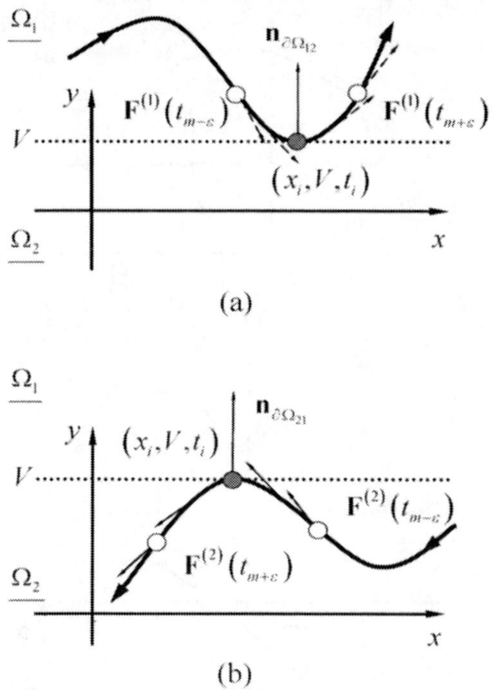

Figure 17. Vector fields for grazing motion in Ω_α, $\alpha = 1, 2$.

through the vector fields of $\mathbf{F}^{(\alpha)}(t)$. In addition to $F_\alpha(\mathbf{x}_m, t_{m\pm}) = 0$, the sufficient condition requires $F_1(\mathbf{x}_m, \Omega t_{m-\varepsilon}) < 0$ and $F_1(\mathbf{x}_m, \Omega t_{m+\varepsilon}) > 0$ in domain Ω_1; $F_2(\mathbf{x}_m, \Omega t_{m-\varepsilon}) > 0$ and $F_2(\mathbf{x}_m, \Omega t_{m+\varepsilon}) < 0$ in domain Ω_2. The detailed discussion can be refereed to Luo and Gegg (2005).

After grazing motion, the sliding motion will appear. Direct integration of equation (64) with initial condition (t_i, x_i, V) gives the sliding motion, i.e.,

$$x = V(t - t_i) + x_i. \tag{79}$$

Substitution of equations (79) into (71) gives the forces for the very small δ-neighborhood of the stick motion ($\delta \to 0$) in the two domains Ω_α, $\alpha \in \{1, 2\}$, i.e.,

$$F_\alpha(\mathbf{x}_m, \Omega t_{m-}) = -2d_\alpha V - c_\alpha\big[V(t_m - t_i) + x_i\big] + A_0 \cos \Omega t_m - b_\alpha. \tag{80}$$

For the nonstick motion, select the initial condition on the velocity boundary (i.e., $\dot{x}_i = V$), and then the coefficients of the solution in Appendix, $C_k^{(\alpha)}(x_i, \dot{x}_i, t_i) \triangleq$

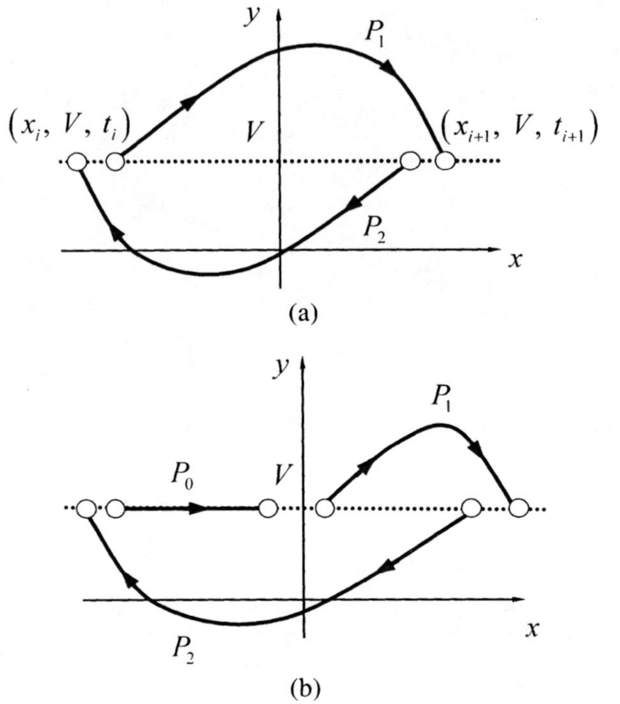

Figure 18. Regular mappings (P_1 and P_2) and stick mapping (P_0).

$C_k^{(\alpha)}(x_i, t_i)$ for $k = 1, 2$. The basic solutions in Appendix will be used for construction of mappings.

In phase plane, the trajectories in Ω_α starting and ending at the velocity boundary (i.e., from $\partial\Omega_{\beta\alpha}$ to $\partial\Omega_{\alpha\beta}$) are illustrated in Figure 18. The starting and ending points for mappings P_α in Ω_α are (x_i, V, t_i) and (x_{i+1}, V, t_{i+1}), respectively. The stick mapping is P_0. Define the switching planes as

$$\begin{cases} \Xi^0 = \{(x_i, \Omega t_i) \mid \dot{x}_i(t_i) = V\}, \\ \Xi^1 = \{(x_i, \Omega t_i) \mid \dot{x}_i(t_i) = V^+\}, \\ \Xi^2 = \{(x_i, \Omega t_i) \mid \dot{x}_i(t_i) = V^-\}, \end{cases} \tag{81}$$

where $V^- = \lim_{\delta\to 0}(V - \delta)$ and $V^+ = \lim_{\delta\to 0}(V + \delta)$ for arbitrarily small $\delta > 0$. Therefore,

$$P_1 : \Xi^1 \to \Xi^1, \qquad P_2 : \Xi^2 \to \Xi^2, \qquad P_0 : \Xi^0 \to \Xi^0. \tag{82}$$

From the foregoing two equations, we have

$$\begin{cases} P_0 : (x_i, V, t_i) \to (x_{i+1}, V, t_{i+1}), \\ P_1 : (x_i, V^+, t_i) \to (x_{i+1}, V^+, t_{i+1}), \\ P_2 : (x_i, V^-, t_i) \to (x_{i+1}, V^-, t_{i+1}). \end{cases} \tag{83}$$

The governing equations for P_0 and $\alpha \in \{1, 2\}$ are

$$-x_{i+1} + V(t_{i+1} - t_i) + x_i = 0, \tag{84a}$$

$$2d_\alpha V + c_\alpha\big[V(t_{i+1} - t_i) + x_i\big] - A_0 \cos \Omega t_{i+1} + b_\alpha = 0. \tag{84b}$$

The mapping P_0 described the starting and ending of the stick motion, the disappearance of stick motion requires $F_\alpha(\mathbf{x}_{i+1}, \Omega t_{i+1}) = 0$. This chapter will not use this mapping to discuss the grazing flow, which is presented herein as a generic mapping. For sliding motions, this mapping will be used, and such a discussion is arranged in Luo and Gegg (2006a, 2006b). From this problem, the two domains Ω_α, $\alpha = 1$ or 2, are unbounded. The flows of the dynamical systems on the corresponding domains should be bounded from assumptions (A1)–(A3) in nonsmooth dynamical systems. Therefore, for nonstick motion, there are three possible stable motions in the two domains Ω_α, $\alpha \in \{1, 2\}$, the governing equations of mapping P_α, $\alpha \in \{1, 2\}$, are obtained from the displacement and velocity responses for the three cases of motions in the Appendix. Therefore, the governing equations of mapping P_α, $\alpha \in \{0, 1, 2\}$, can be expressed by

$$f_1^{(\alpha)}(x_i, \Omega t_i, x_{i+1}, \Omega t_{i+1}) = 0,$$

$$f_2^{(\alpha)}(x_i, \Omega t_i, x_{i+1}, \Omega t_{i+1}) = 0. \tag{85}$$

If the grazing for the two nonstick mappings occurs at the final state (x_{i+1}, V, t_{i+1}), from equation (78), the grazing conditions based on mappings are obtained, i.e.,

$$\begin{cases} F_\alpha(x_{i+1}, V, \Omega t_{i+1}) = 0, \\ \nabla F_\alpha(\mathbf{x}_{i+1}, \Omega t_{i+1}) \cdot \mathbf{F}_\alpha^{(\alpha)}(t_{i+1}) + \frac{\partial F_\alpha(\mathbf{x}_{i+1}, \Omega t_{i+1})}{\partial t} \begin{cases} > 0 & \text{for } \alpha = 1, \\ < 0 & \text{for } \alpha = 2. \end{cases} \end{cases} \tag{86}$$

With equation (71), the grazing condition becomes

$$\begin{cases} A_0 \cos \Omega t_{i+1} - b_\alpha - 2d_\alpha V - c_\alpha x_{i+1} = 0, \\ -c_\alpha V - A_0 \Omega \sin \Omega t_{i+1} \begin{cases} > 0 & \text{for } \alpha = 1, \\ < 0 & \text{for } \alpha = 2. \end{cases} \end{cases} \tag{87}$$

The grazing conditions for the two nonstick mappings can be illustrated with parameters variation. The grazing conditions in equation (87) are given through the forces. Hence, both the initial and final switching sets of the two nonstick mappings will vary with system parameters. Because the grazing characteristics of the two nonstick mappings are different, illustrations of grazing conditions for the two mappings will be separated. The grazing conditions are computed through equations (85) and (87). Three equations plus an inequality with four unknowns require to fix one unknown. In illustrations, the initial displacement of mapping $P_\alpha, \alpha \in \{1, 2\}$, will be fixed to specific values. Therefore, three equations with three unknowns will give the grazing conditions. Namely, the initial switching phase, the final switching phase and displacement of mapping P_α, $\alpha = 1, 2$, will be determined by equations (85) and (87). To ensure the initial switching sets to be passable, from Luo (2005a, 2005b, 2006), the initial switching sets of mapping P_α, $\alpha \in \{1, 2\}$, should satisfy the following condition as in Luo and Gegg (2006a).

$$\begin{cases} F_1(x_i, V^+, \Omega t_i) < 0 \quad \text{and} \quad F_2(x_i, V^-, \Omega t_i) < 0 \\ \quad \text{for } \Omega_1 \to \Omega_2, \\ F_1(x_i, V^+, \Omega t_i) > 0 \quad \text{and} \quad F_2(x_i, V^-, \Omega t_i) > 0 \\ \quad \text{for } \Omega_2 \to \Omega_1. \end{cases} \tag{88}$$

To make sure motions relative to mappings P_α, $\alpha = 1, 2$, exist, the initial switching force product $F_1 \times F_2$ at the boundary should be nonnegative. The comprehensive discussion of the foregoing condition can be referred to Luo and Gegg (2006b). The condition of equation (88) guarantees the flow relative to the initial switching sets of mapping $P_\alpha, \alpha \in \{1, 2\}$, is passable on the discontinuous boundary (i.e., $y_i = V$). The force product for the initial switching sets is also illustrated to ensure the nonstick mapping exists. The force conditions for the final switching sets of mapping $P_\alpha, \alpha \in \{1, 2\}$, is presented in equation (76). However, the equivalent grazing conditions based on equation (78) give the inequality condition in equation (87), which is already embedded in the program for computation of the grazing. Therefore, such a force product of the final switching sets of the two mappings will not be presented. From the inequality of equation (87), the critical value for $\mathrm{mod}(\Omega t_{i+1}, 2\pi)$ is introduced through

$$\Theta_\alpha^{\mathrm{cr}} = \arcsin\left(-\frac{c_\alpha V}{A_0 \Omega}\right), \tag{89}$$

where the superscript "cr" represents a critical value relative to grazing and $\alpha \in \{1, 2\}$. From the second equation of (87), the final switching phase for mapping P_1 has the following six cases:

$$\begin{cases} \mathrm{mod}(\Omega t_{i+1}, 2\pi) \in \left(\pi + \left|\Theta_1^{\mathrm{cr}}\right|, 2\pi - \left|\Theta_1^{\mathrm{cr}}\right|\right) \subset (\pi, 2\pi) \\ \quad \text{for } V > 0 \text{ and } A_0\Omega > c_1 V, \\ \mathrm{mod}(\Omega t_{i+1}, 2\pi) \in \left(\pi - \Theta_1^{\mathrm{cr}}, 2\pi\right] \cup \left[0, \Theta_1^{\mathrm{cr}}\right) \\ \quad \text{for } V < 0 \text{ and } A_0\Omega > c_1|V|, \\ \mathrm{mod}(\Omega t_{i+1}, 2\pi) \in (\pi, 2\pi) \\ \quad \text{for } V = 0, \end{cases} \tag{90}$$

and

$$\begin{cases} \operatorname{mod}(\Omega t_{i+1}, 2\pi) \in \varnothing & \text{for } V > 0 \text{ and } A_0\Omega \leqslant c_1 V, \\ \operatorname{mod}(\Omega t_{i+1}, 2\pi) \in [0, 2\pi] & \text{for } V < 0 \text{ and } A_0\Omega < c_1|V|, \\ \operatorname{mod}(\Omega t_{i+1}, 2\pi) \in [0, 2\pi]/\{\pi/2\} & \text{for } V < 0 \text{ and } A_0\Omega = c_1|V|. \end{cases} \tag{91}$$

From the first equation of (87), the final displacement is bounded by

$$-A_0 \leqslant b_\alpha + 2d_\alpha V + c_\alpha x_{i+1} \leqslant A_0. \tag{92}$$

The spring and damper parameters ($d_1 = 1$, $d_2 = 0$, $c_1 = c_2 = 30$) are fixed as constants and the external parameters will vary. Consider the grazing variation of mapping P_1 with the belt speed V for the specified parameters $\Omega = 1$, $A_0 = 20$, $b_1 = -b_2 = 3$. When the initial displacements $x_i = \{-1, -2, \ldots, -9\}$ are specified, the initial switching phase and force products, and the final switch phase and displacement versus the belt speed are illustrated in Figure 19. The initial switching phase modulus (i.e., $\operatorname{mod}(\Omega t_i, 2\pi)$) is distributed in the interval $[0, 2\pi]$. With increasing both negative, initial switching displacements and negative belt speeds, the initial switching phase will drop in the interval $[\pi, 2\pi]$. The initial switching force product on the discontinuous boundary in Figure 19(b) is always positive, which is required in equation (88). For $x_i = -1$, the minimum value of the initial switching force product is close to zero. In an alike way, with increasing both negative, initial displacements and negative belt speeds, the initial force product becomes larger and larger. Such a result of the final switching phase in equations (90) and (91) is confirmed by the illustration in Figure 19(c). For $c_1 V > A_0\Omega$, the grazing will always occur as long as the solutions given by equations (90) and (91) exist in the domain Ω_1. The final switching displacement becomes positive and large with increasing the initial switching displacement, as shown in Figure 19(d). The final displacement is bounded through equation (92), which is depicted through the dark, dashed lines. Most of the grazing of mapping P_1 appears for negative belt speed. The final displacement is tangential to the upper bounded line.

Consider the grazing varying with the friction force b for mapping P_1 with $\Omega = 10$, $V = 1$, $A_0 = 20$, $b_1 = -b_2 = b$. The grazing condition is illustrated in Figure 20 for $x_i = \{-1, -2, \ldots, -5\}$ and $b \in [0, 100]$. The sufficient condition for grazing in the second equation of (90) is independent of the friction force. The boundary for the initial switching phase for the friction force must be two horizontal straight lines, which are observed in Figure 20(a). The initial switching force product is presented in Figure 20(b), and the appearance and disappearance boundaries of the grazing are two straight lines. Because the other parameters are given, the final switching phase should lie in $\operatorname{mod}(\Omega t_{i+1}, 2\pi) \in [3.29, 6.13]$. When the final switching phase for the grazing disappearance in Figure 20(c) reaches the upper critical boundary (i.e., $\operatorname{mod}(\Omega t_{i+1}, 2\pi) \approx 6.13$), at this time, the initial force product is not zero. In Figure 20(c), the lower boundary of the final

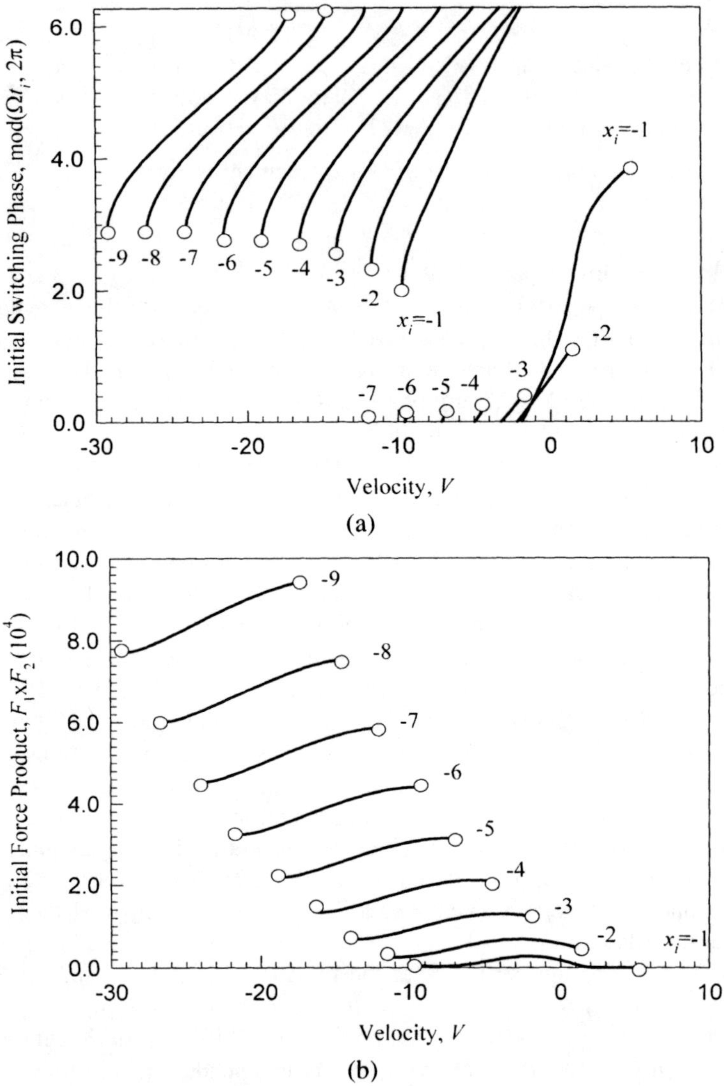

Figure 19. Grazing variation of mapping P_1 with belt speed V: (a) initial switching phase, (b) initial force product, (c) final switching phase and (d) final switching displacement for $x_i = -1, \ldots, -9$. ($\Omega = 1$, $A_0 = 20$, $d_1 = 1$, $d_2 = 0$, $b_1 = -b_2 = 3$, $c_1 = c_2 = 30$.)

switching phase does not reach the critical boundary because there is no solution for equations (90) and (91) in domain Ω_1. The final switching displacement is bounded in two parallel straight lines in Figure 20(d). The grazing appearance

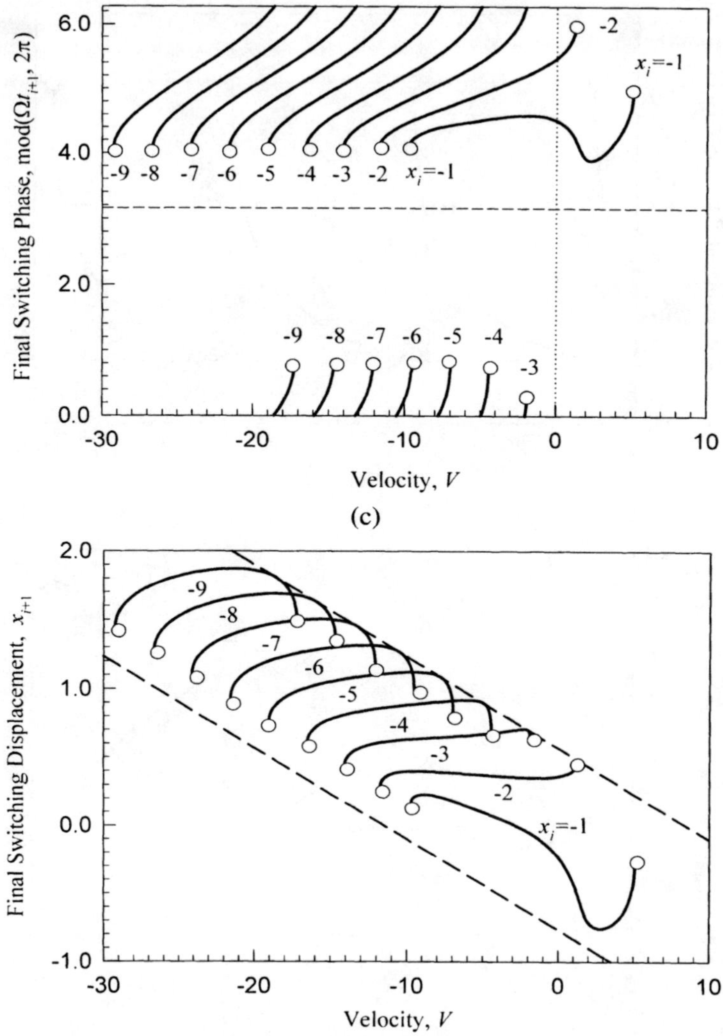

(c)

(d)

Figure 19. (*Continued.*)

and disappearance of mapping P_1 for the final switching displacement are linear to the friction force, which is given by equation (92). The dashed dark lines give the boundary. The final displacement arrives to the upper boundary before the force product becomes zero. The lower boundary is controlled by the sufficient conditions.

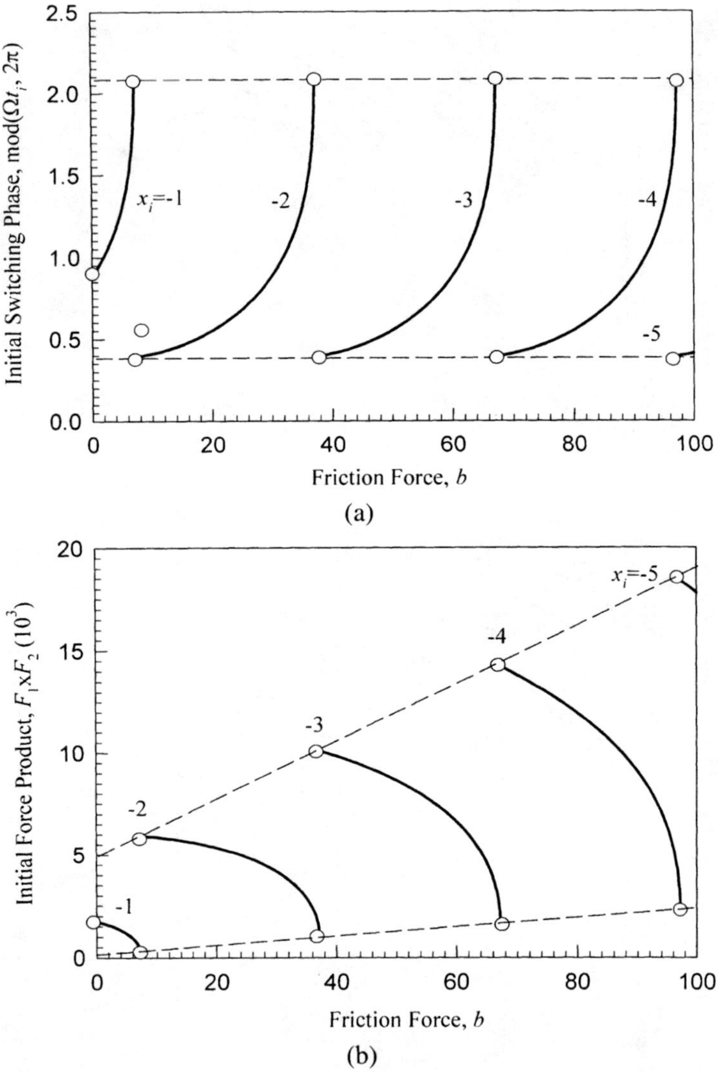

Figure 20. Grazing variation of mapping P_1 with friction force b: (a) initial switching phase, (b) initial force product, (c) final switching phase and (d) final switching displacement for $x_i = -1, -2, \ldots, -5$. ($\Omega = 10$, $A_0 = 20$, $V = 1$, $d_1 = 1$, $d_2 = 0$, $b_1 = -b_2 = b$, $c_1 = c_2 = 30$.)

Consider the grazing variation of mapping P_1 with excitation amplitude A_0 as the parameters $\Omega = 8$, $V = 1$, $b_1 = -b_2 = 3$ are used. Illustrations of the grazing condition is presented in Figure 21 for $x_i = -1, -2, \ldots, -6$. Owing to

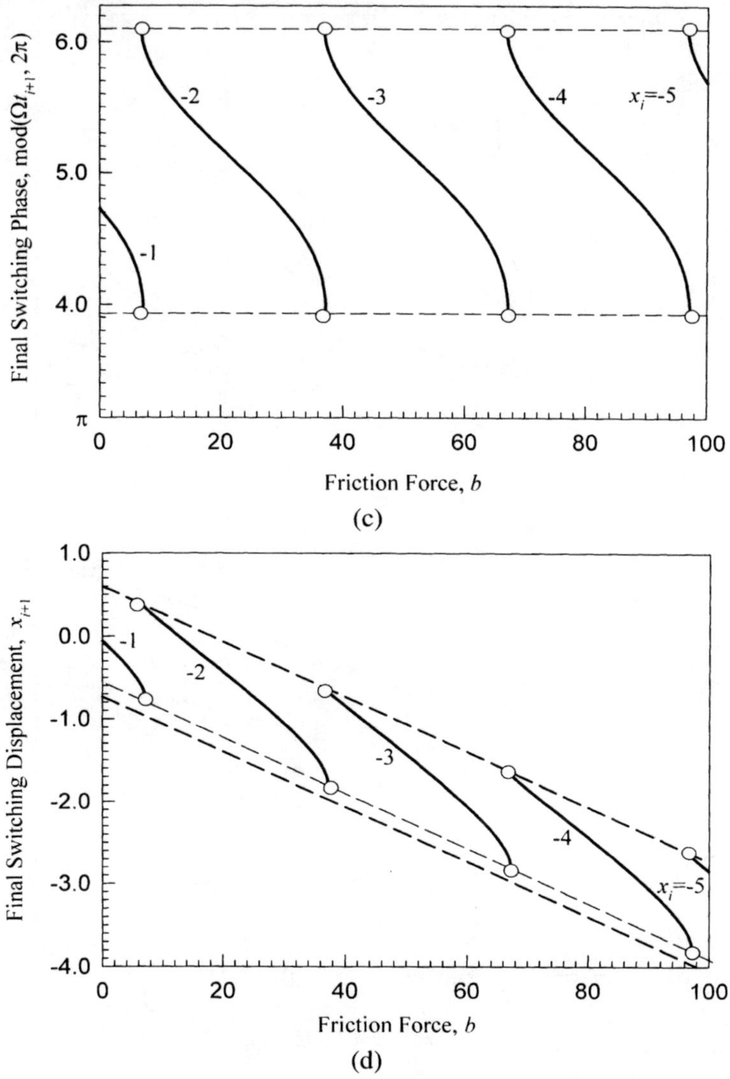

Figure 20. *(Continued.)*

$V = 1 > 0$, the sufficient condition for grazing requires $A_0 > 3.75$, and the final switching phase will be in the interval $(\pi, 2\pi)$ in equation (91). The corresponding initial force product is positive. The grazing band of the excitation amplitude becomes large with increasing initial switching displacements. The upper bounded line in the final displacement is to control the grazing existence. In

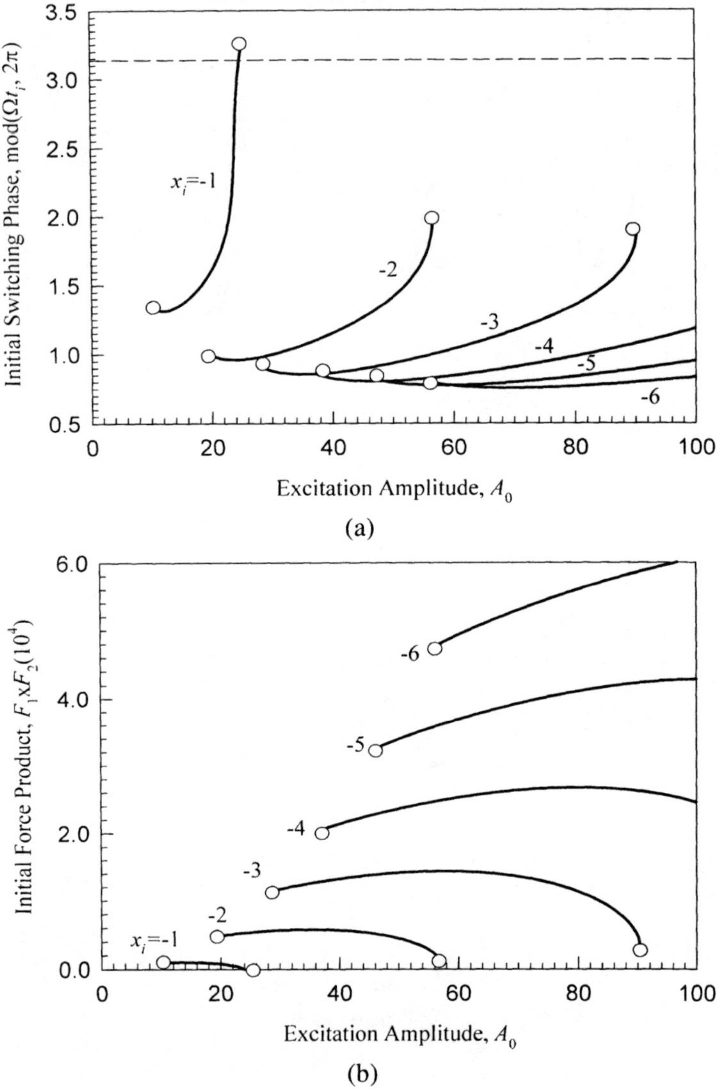

Figure 21. Grazing variation of mapping P_1 with excitation amplitude A_0: (a) initial switching phase, (b) initial force product, (c) final switching phase and (d) final switching displacement for $x_i = -1, -2, \ldots, -6$. ($\Omega = 8$, $V = 1$, $d_1 = 1$, $d_2 = 0$, $b_1 = -b_2 = 3$, $c_1 = c_2 = 30$.)

Figure 22, the grazing varying with excitation frequency Ω for mapping P_1 is shown for $A_0 = 20$, $V = 1$, $b_1 = -b_2 = 3$ with $x_i = -0.9, -1, \ldots, -2$. From Figure 22(b), the minimum of the initial force product approaches zero for

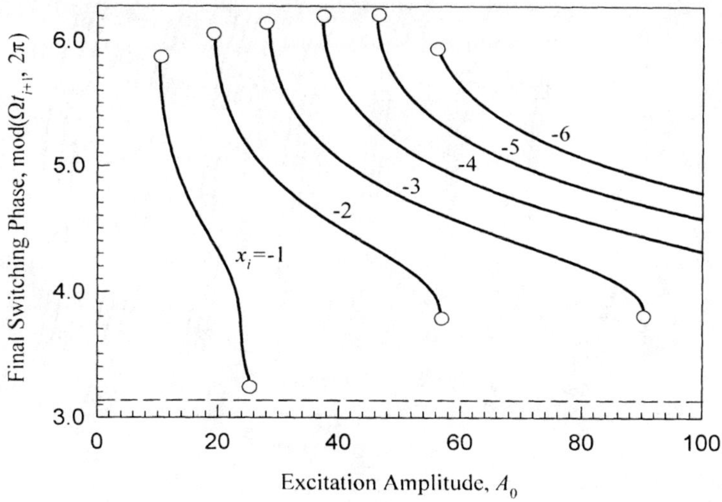

(c)

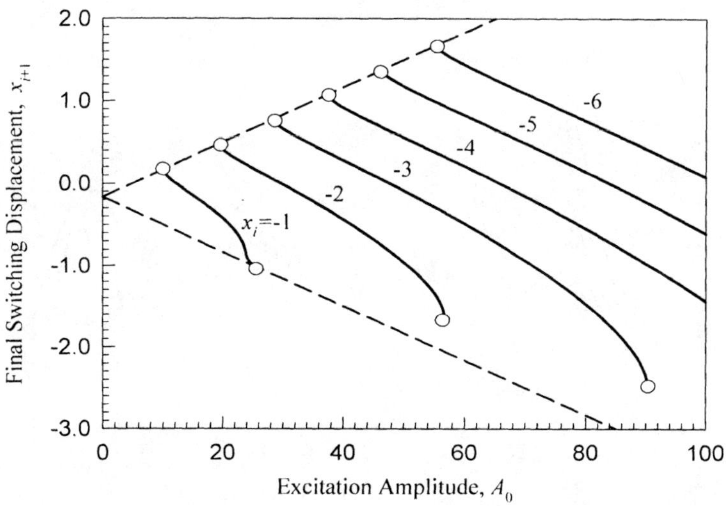

(d)

Figure 21. (*Continued.*)

$x_i = -0.9$. For $x_i > -0.9$, it is very difficult to find the wide spectrum of excitation frequency for grazing solutions obtained from equations (90) and (91). The initial switching phase for $x_i \in [-1.1, -0.9]$ takes the entire range of $[0, 2\pi]$. However, the final switching phase and displacement are in the narrow band for large excitation frequency. For $x_i < -1.25$, the grazing for mapping P_1 exists

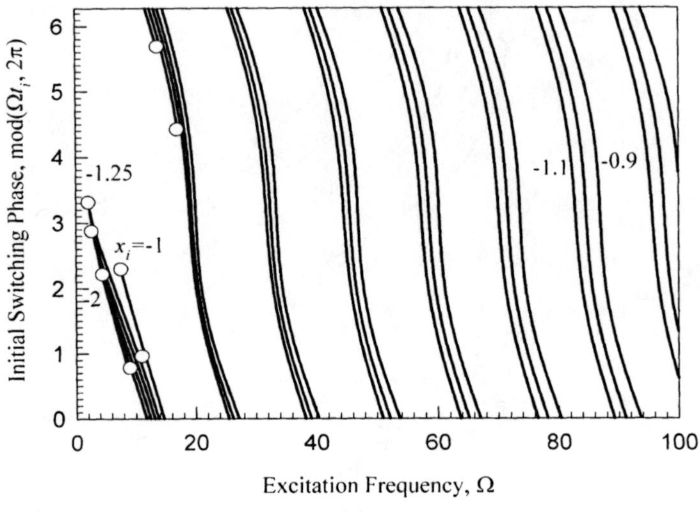

(a)

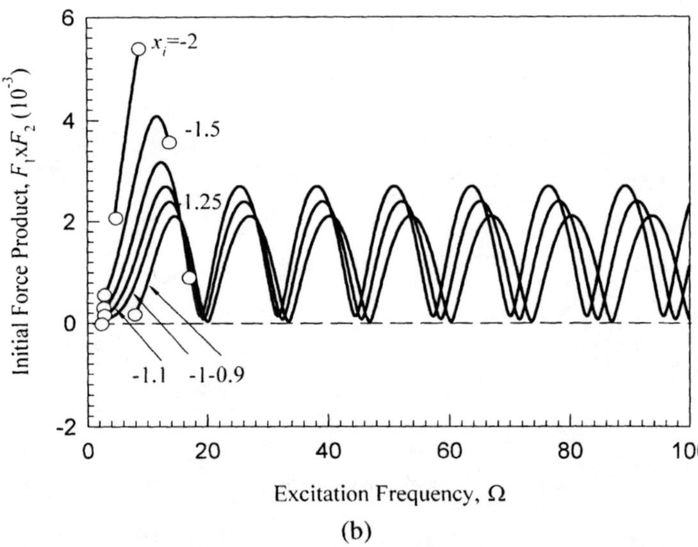

(b)

Figure 22. Grazing variation of mapping P_1 with excitation frequency Ω: (a) initial switching phase, (b) initial force product, (c) final switching phase and (d) final switching displacement for $x_i = -0.9, -1, \ldots, -2$. ($A_0 = 20$, $V = 1$, $d_1 = 1$, $d_2 = 0$, $b_1 = -b_2 = 3$, $c_1 = c_2 = 30$.)

only in the small range of excitation frequency because the grazing conditions catch the upper bounded lines. However, for $x_i \in [-1.1, -0.9]$, the grazing for mapping P_1 exists almost in the large range of excitation frequency. It implies that

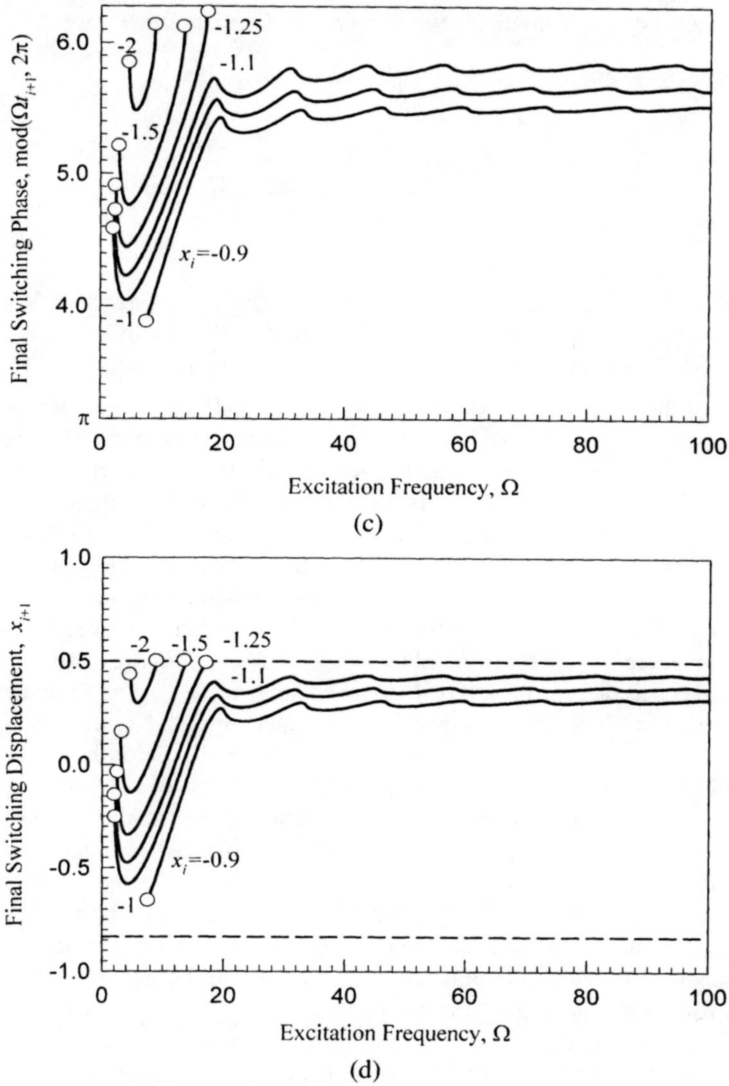

Figure 22. (*Continued.*)

the grazing exists in a wide spectrum of excitation for such initial displacements. From the parameter characteristics of grazing for mapping P_1, it is observed that the grazing is effected by many parameters. In the two domains, the system parameters are different. Therefore, the parameter characteristics of grazing for mapping P_2 will be presented as follows. Similarly, from the second equation of (87), the six cases of the final switching phase for mapping P_2 are

$$\mathrm{mod}(\Omega t_{i+1}, 2\pi) \in \left[0, \pi + \left|\Theta_2^{\mathrm{cr}}\right|\right) \cup \left(2\pi - \left|\Theta_2^{\mathrm{cr}}\right|, 2\pi\right] \\ \text{for } V > 0 \text{ and } A_0\Omega > c_2 V, \tag{93a}$$

$$\begin{cases} \mathrm{mod}(\Omega t_{i+1}, 2\pi) \in \left(\Theta_2^{\mathrm{cr}}, \pi - \Theta_2^{\mathrm{cr}}\right) \subset (0, \pi) \\ \qquad\qquad\qquad\qquad \text{for } V < 0 \text{ and } A_0\Omega > c_2|V|, \\ \mathrm{mod}(\Omega t_{i+1}, 2\pi) \in (0, \pi) \quad \text{for } V = 0, \end{cases} \tag{93b}$$

and

$$\begin{cases} \mathrm{mod}(\Omega t_{i+1}, 2\pi) \in [0, 2\pi] & \text{for } V > 0 \text{ and } A_0\Omega < c_2 V, \\ \mathrm{mod}(\Omega t_{i+1}, 2\pi) \in [0, 2\pi]/\{3\pi/2\} & \text{for } V > 0 \text{ and } A_0\Omega = c_2 V, \\ \mathrm{mod}(\Omega t_{i+1}, 2\pi) \in \varnothing & \text{for } V < 0 \text{ and } A_0\Omega < c_2|V|. \end{cases} \tag{94}$$

Consider the same system parameters $d_1 = 1, d_2 = 0, c_1 = c_2 = 30$. The grazing varying with the belt speed V for mapping P_2 are presented in Figure 23 for $x_i = 3, 4, \ldots, 8$ under the other parameters $A_0 = 90$, $\Omega = 1$, $b_1 = -b_2 = 30$. From the aforementioned parameters, the final switching phase for $V > 3$ will be in the entire interval of $[0, 2\pi]$ as long as the grazing solution of (85) and (87) exists for the domain Ω_2. In Figure 23(a), the initial switching phase lies in $(0, \pi/2) \cup (3\pi/2, 2\pi)$. With increasing the initial switching phase, the belt speeds for grazing of mapping P_2 will become large. For higher belt speeds, the positive, initial force product will be obtained, as shown in Figure 23(b). The final switching phase lies in the interval $(0, \pi)$ in Figure 23(c). The final switching displacement in Figure 23(d) is much smaller than the initial switching displacement (i.e., $x_{i+1} < x_i$). The final displacement is in the region bounded by the upper and lower boundaries by equation (92) (i.e., $x_{i+1} = -2$ and $x_{i+1} = 4$). Without damping, the upper and lower boundaries for the final displacement are independent of the velocity. The upper and lower bounded boundaries were not presented. For this case of mapping P_2, the boundaries of grazing are determined by the sufficient conditions. The grazing of mapping P_2 mostly occurs in the positive range of the belt speed. However, the grazing of mapping P_1 mostly exists in the negative range of the belt speed. In Figure 24, the grazing varying with friction force for mapping P_2 is illustrated for $x_i = 2, 4, \ldots, 6$ with $A_0 = 90$, $\Omega = 1.1$, $V = 1$, $b_1 = -b_2 = b$. The final switching phase should be in the interval $(0, 3.44946) \cup (5.96731, 2\pi)$. However, compared to the grazing of mapping P_1, the initial switching force product becomes zero before the final switching phase reaches the critical values, as illustrated in Figure 24(b). Again, the two switching phases for appearance and disappearance of grazing mappings are independent of the frictional force, as observed in Figures 24(a) and 24(c). Such a characteristic is determined by the grazing sufficient condition. The appearance and disappearance for the final switching displacement are linear to the frictional force, as shown in Figure 24(d). The final displacement lies in the region between the upper and lower bounded boundary of necessary conditions.

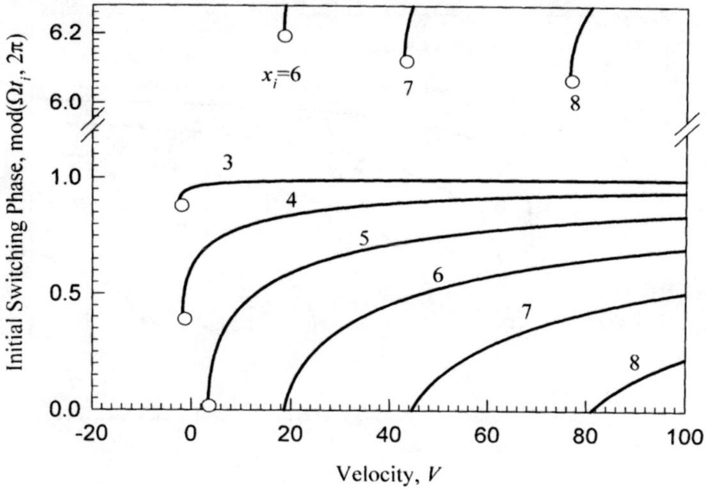

(a)

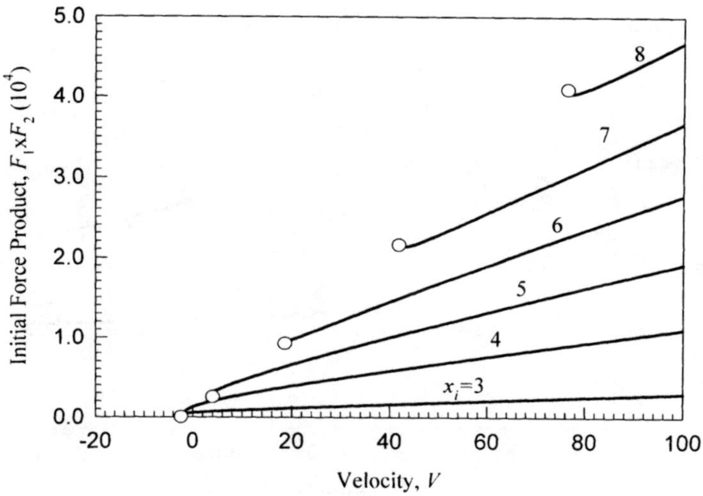

(b)

Figure 23. Grazing variation of mapping P_2 with belt speed V for $x_i = 3, 4, \ldots, 8$: (a) initial switching phase, (b) initial force product, (c) final switching phase and (d) final switching displacement. ($A_0 = 90$, $\Omega = 1$, $d_1 = 1$, $d_2 = 0$, $b_1 = -b_2 = 30$, $c_1 = c_2 = 30$.)

The grazing varying with excitation amplitude for mapping P_2 is also illustrated in Figure 25 for $x_i = 1.1, 1.5, \ldots, 8$ and $\Omega = 1.1$, $V = 1$, $b_1 = -b_2 = 30$. The upper boundary for grazing disappearance is generated by the zero of the initial force product. The final switching displacement is in the region between the up-

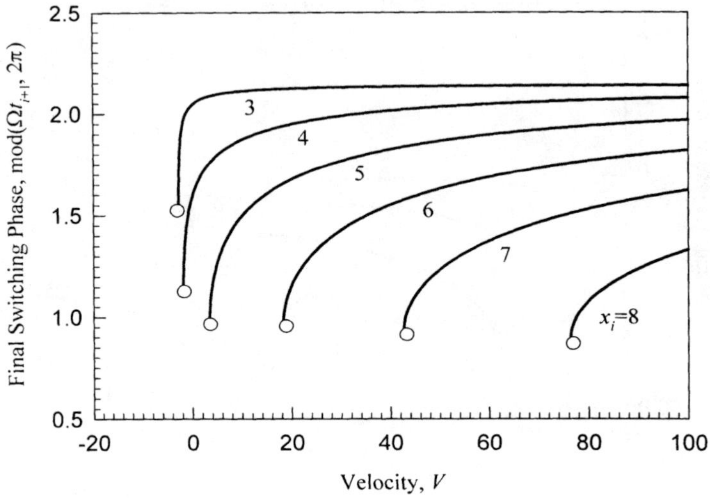

(c)

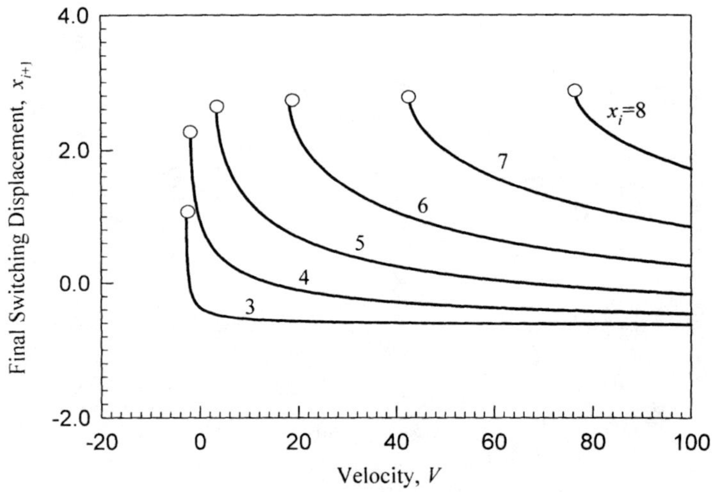

(d)

Figure 23. (*Continued.*)

per and lower boundaries of the necessary grazing conditions. It seems that the grazing is determined by its sufficient condition and solution existence. Finally, the grazing varying with excitation frequency Ω for mapping P_2 is presented in Figure 26 for $x_i = 1.5, 2, \ldots, 8.25$ and $A_0 = 90$, $V = 1$, $b_1 = -b_2 = 30$. The closed dashed line is the boundary for the grazing of mapping P_2. No more

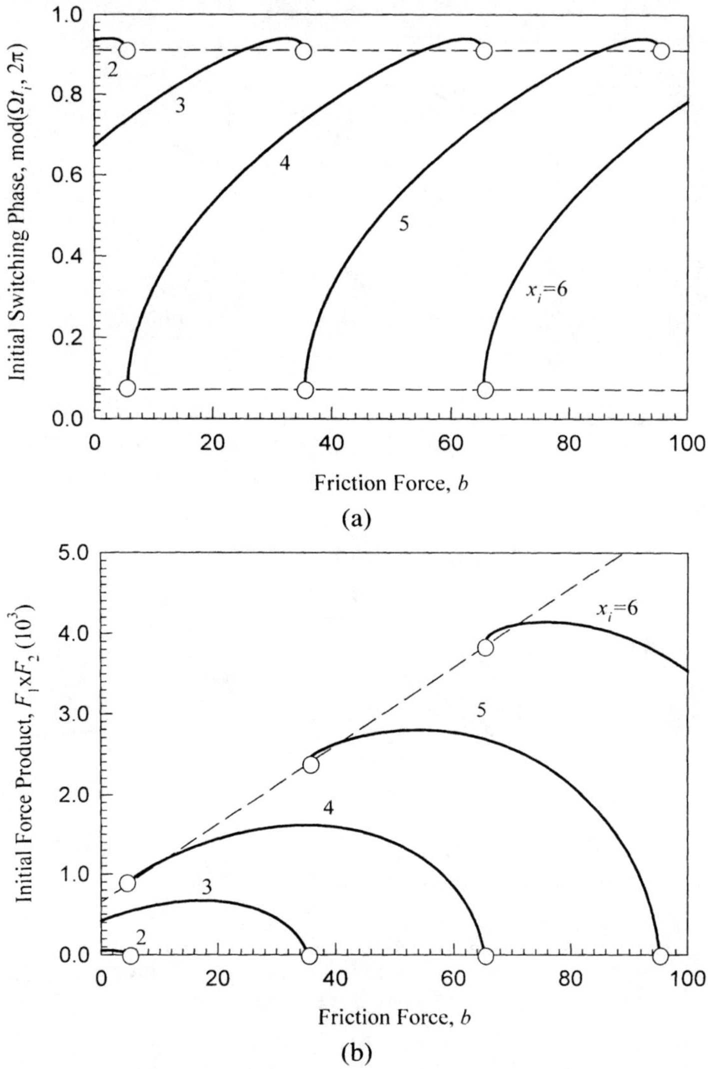

Figure 24. Grazing varying with friction force b for mapping P_2 with $x_i = 2, 4, \ldots, 6$: (a) initial switching phase, (b) initial force product, (c) final switching phase and (d) final switching displacement. ($A_0 = 90$, $\Omega = 1.1$, $V = 1$, $d_1 = 1$, $d_2 = 0$, $b_1 = -b_2 = 30$, $c_1 = c_2 = 30$.)

grazing of mapping P_2 will exist for $x_i > 8.25$. The initial switching phase and force product are presented in Figures 26(a) and 26(b), respectively. For small excitation frequencies, the disappearance for grazing mapping P_2 is caused by

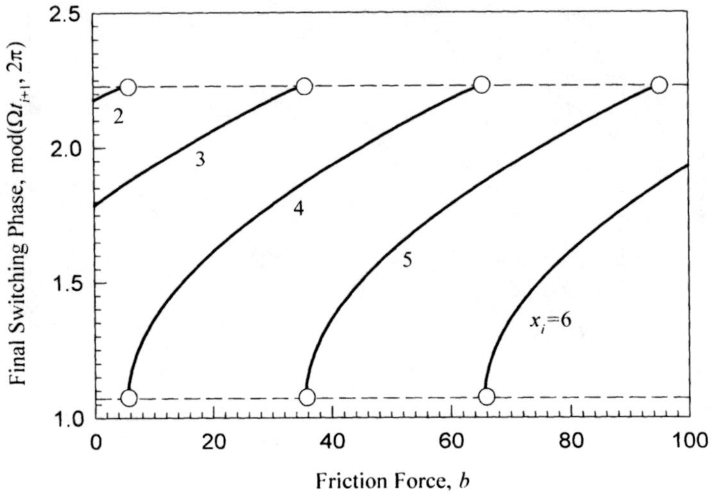

(c)

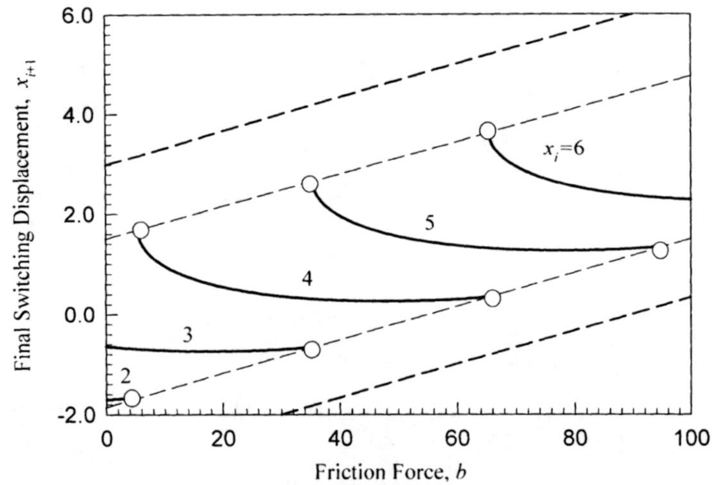

(d)

Figure 24. *(Continued.)*

the zero of the initial force product. For large excitation frequencies, the grazing disappearance is because the sufficient condition in (87) cannot be satisfied.

To verify the analytical prediction of the grazing motions, the motion responses of the oscillator will be demonstrated through time-history responses and trajectories in phase plane. The grazing strongly depends on the force responses in this discontinuous dynamical system. The force responses will be presented to illus-

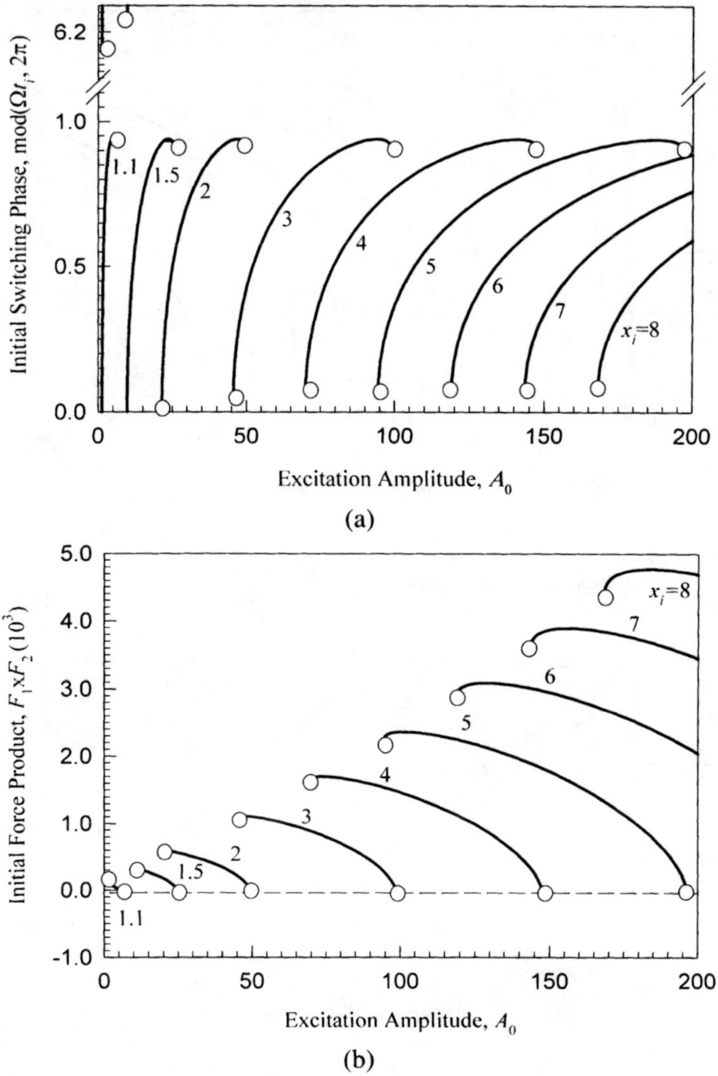

Figure 25. Grazing varying with excitation amplitude A_0 for mapping P_2 with $x_i = 1.1, 1.5, \ldots, 8$: (a) initial switching phase, (b) initial force product, (c) final switching phase and (d) final switching displacement. ($\Omega = 1.1$, $V = 1$, $d_1 = 1$, $d_2 = 0$, $b_1 = -b_2 = 30$, $c_1 = c_2 = 30$.)

trate the force criteria for the grazing motions in such a friction-induced oscillator. The starting and grazing points of mapping P_α, $\alpha \in \{1, 2\}$, are represented by the large, hollow and dark-solid circular symbols, respectively. The switching points

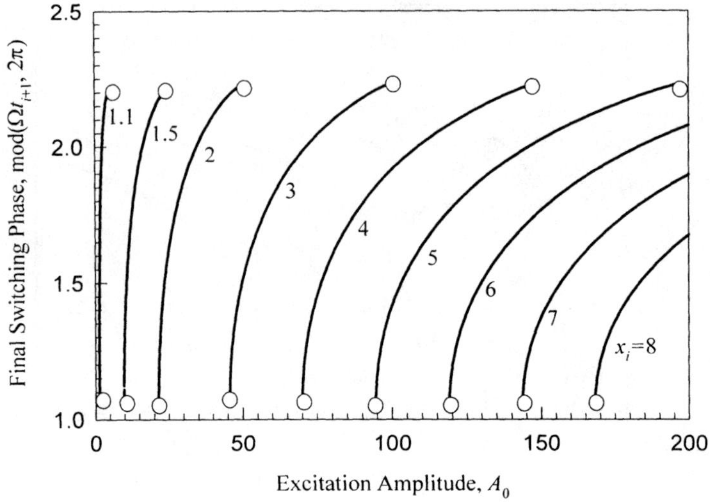

(c)

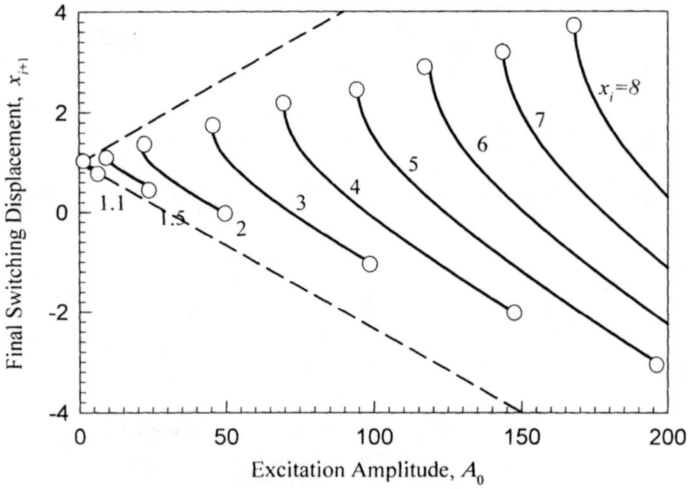

(d)

Figure 25. (*Continued.*)

from domain α to β, $\alpha, \beta \in \{1, 2\}$ but $\alpha \neq \beta$, are depicted by smaller circular symbols.

In Figure 27, phase trajectories, forces distribution along displacement, velocity time-history and forces distribution on velocity are presented for the grazing motion of mapping P_1. The parameters $\Omega = 8$, $V = 1$, $b_1 = -b_2 = 3$

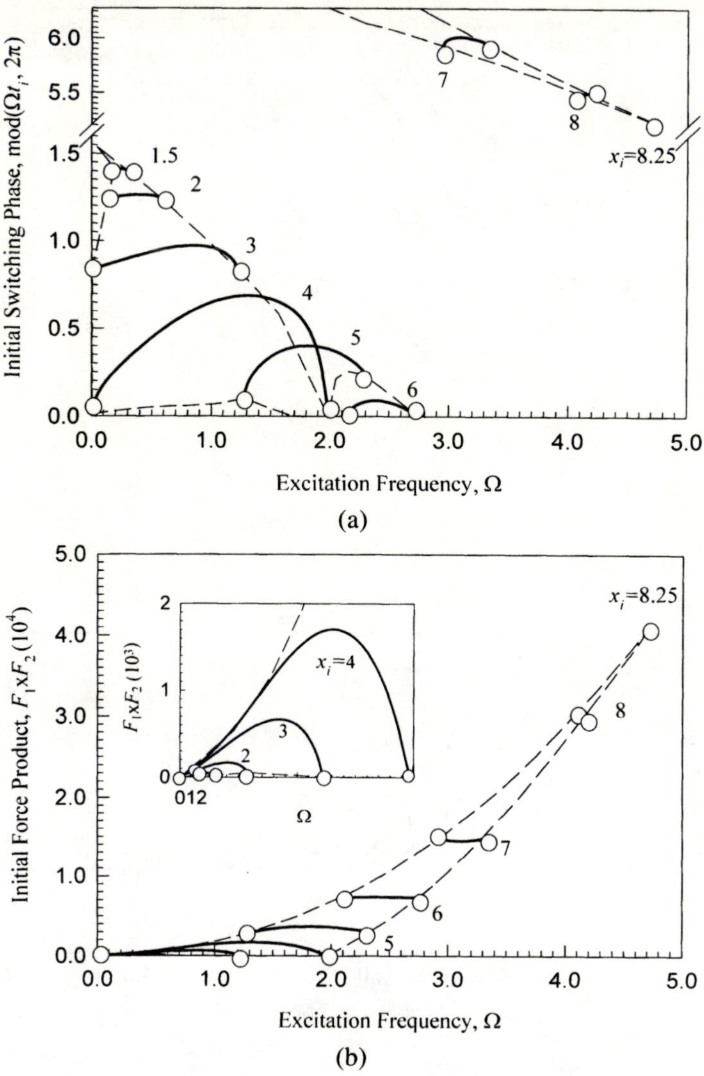

Figure 26. Grazing varying with excitation frequency Ω for mapping P_2 with $x_i = 1.5, 2, \ldots, 8.25$: (a) initial switching phase, (b) initial force product, (c) final switching phase and (d) final switching displacement. ($A_0 = 90$, $V = 1$, $d_1 = 1$, $d_2 = 0$, $b_1 = -b_2 = 30$, $c_1 = c_2 = 30$.)

plus the initial conditions $(x_i, y_i) = (-1, 1)$ and $\Omega t_i \approx 1.3617, 1.6958, 1.4830$ corresponding to $A_0 = 15, 18, 21$ are used. In phase plane, the three grazing trajectories are tangential to the discontinuous boundary (i.e., $y = V$), which are seen in Figure 27(a). In Figure 27(b), the thick and thin solid curves repre-

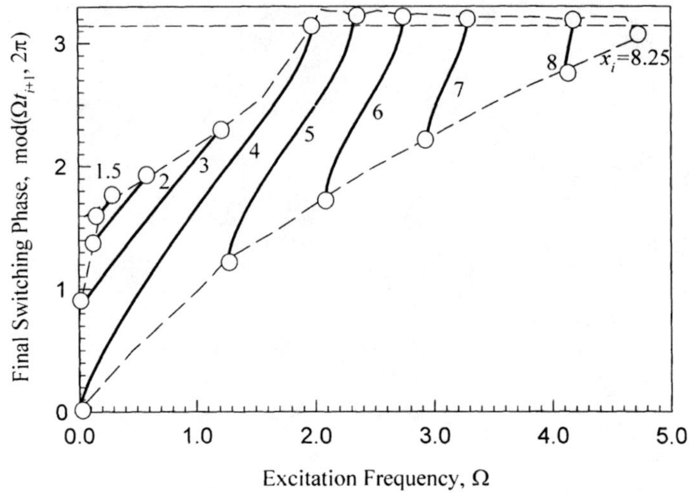

(c)

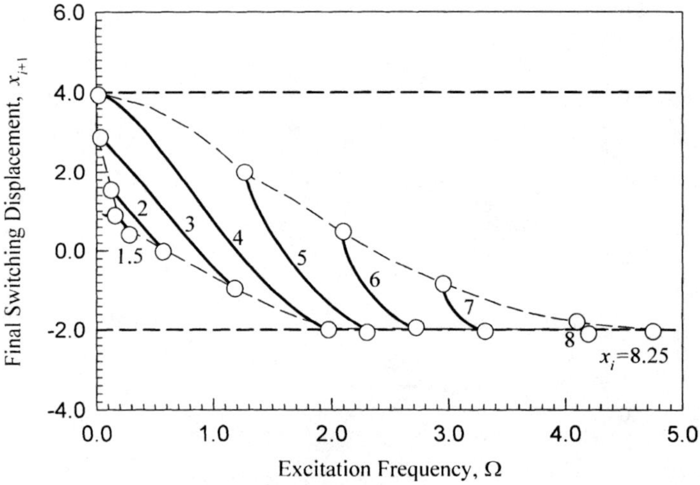

(d)

Figure 26. *(Continued.)*

sent the forces $F_1(t)$ and $F_2(t)$, respectively. From the force distribution along displacement, the force $F_1(t)$ has a sign change from negative to positive. This indicates that the grazing conditions in equation (76) are satisfied. The forces $F_1(t)$ and $F_2(t)$ at the switching points from domain Ω_1 to Ω_2 have a jump with the same sign, which satisfies equation (77). In the velocity time-history plot, the velocity curves are tangential to the discontinuous boundary (see Fig-

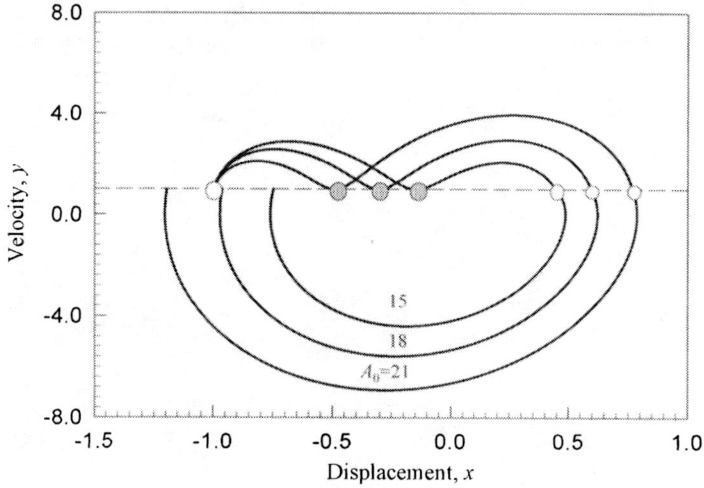

(a)

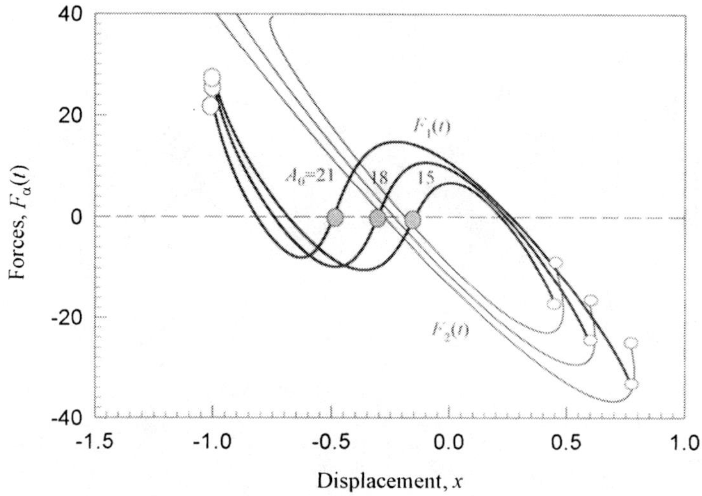

(b)

Figure 27. Grazing motion of mapping P_1 for $A_0 = 15, 18, 21$: (a) phase trajectory, (b) forces distribution along displacement, (c) velocity-time history and (d) forces distribution on velocity. ($\Omega = 8$, $V = 1$, $d_1 = 1$, $d_2 = 0$, $b_1 = -b_2 = 3$, $c_1 = c_2 = 30$.) The initial conditions are $(x_i, y_i) = (-1, 1)$ and $\Omega t_i \approx 1.3617, 1.6958, 1.4830$; accordingly.

ure 27(c)). Finally, the forces distributions along velocity are presented in Figure 27(d). The force $F_1(t)$ at the grazing points is zero. The force jump from domain Ω_1 to Ω_2 is observed as well. The phase trajectories for mapping P_1

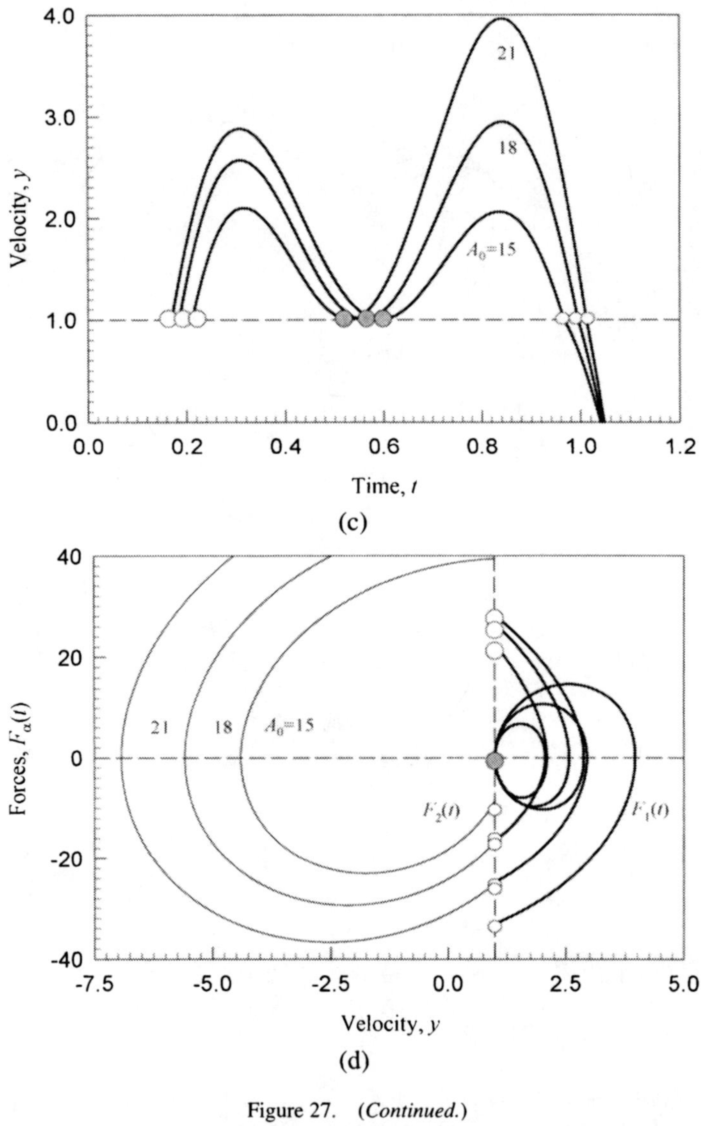

Figure 27. *(Continued.)*

are presented in Figure 28. The parameter $A_0 = 20$, $\Omega = 8$, $b_1 = -b_2 = b$ are used. The initial conditions $x_i = -1$, $y_i = -1, 0, 1$ are adopted in Figure 28(a) for $\Omega t_i \approx 0.3970, 0.9060, 1.6092$. The initial conditions $(x_i, y_i) = (-3, 1)$ with $\Omega t_i \approx 0.3285, 0.4405, 0.7065$ are used for $b = 40, 50, 60$ and $V = 1$, respectively. Similarly, the phase trajectories of grazing motions for

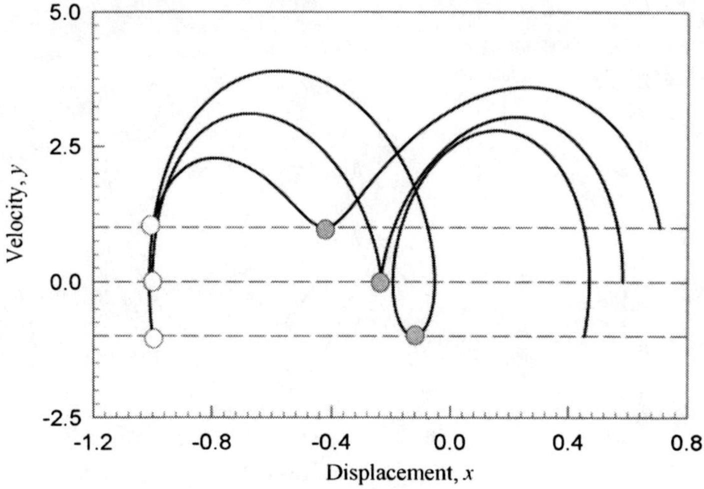

(a)

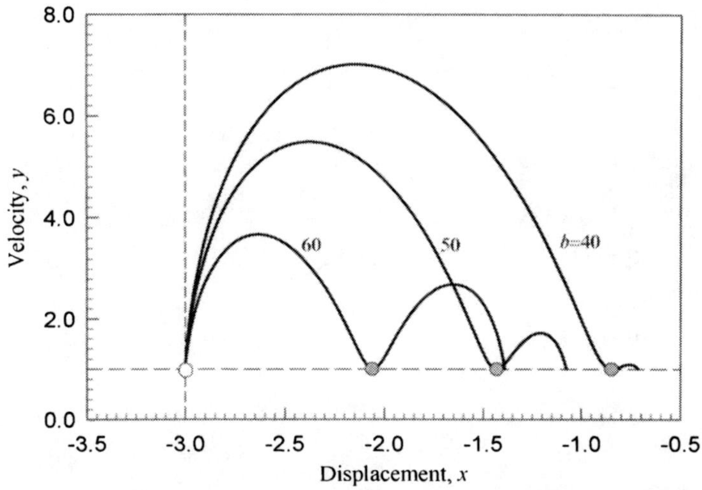

(b)

Figure 28. Grazing phase trajectories of mapping P_1 ($A_0 = 20$, $\Omega = 8$, $d_1 = 1$, $d_2 = 0$, $b_1 = -b_2 = b$, $c_1 = c_2 = 30$): (a) $x_i = -1$, $\Omega t_i \approx 0.3970, 0.9060, 1.6092$ for $y_i \equiv V = -1, 0, 1$ with $b = 3$, accordingly. (b) $(x_i, y_i) = (-3, 1)$ with $\Omega t_i \approx 0.3285, 0.4405, 0.7065$ for $b = 40, 50, 60$ and $V = 1$, respectively.

mapping P_2 are illustrated in Figure 29. The two sets of initial conditions and parameters are $\{x_i = 3$, $\Omega t_i \approx 0.9297, 0.9551, 0.9652$ for $y_i = -2, 0, 2$ and $\Omega = 1$, respectively$\}$ and $\{(x_i, y_i) = (4, 1)$, $\Omega t_i \approx 0.4459, 0.6495, 0.6700$ for

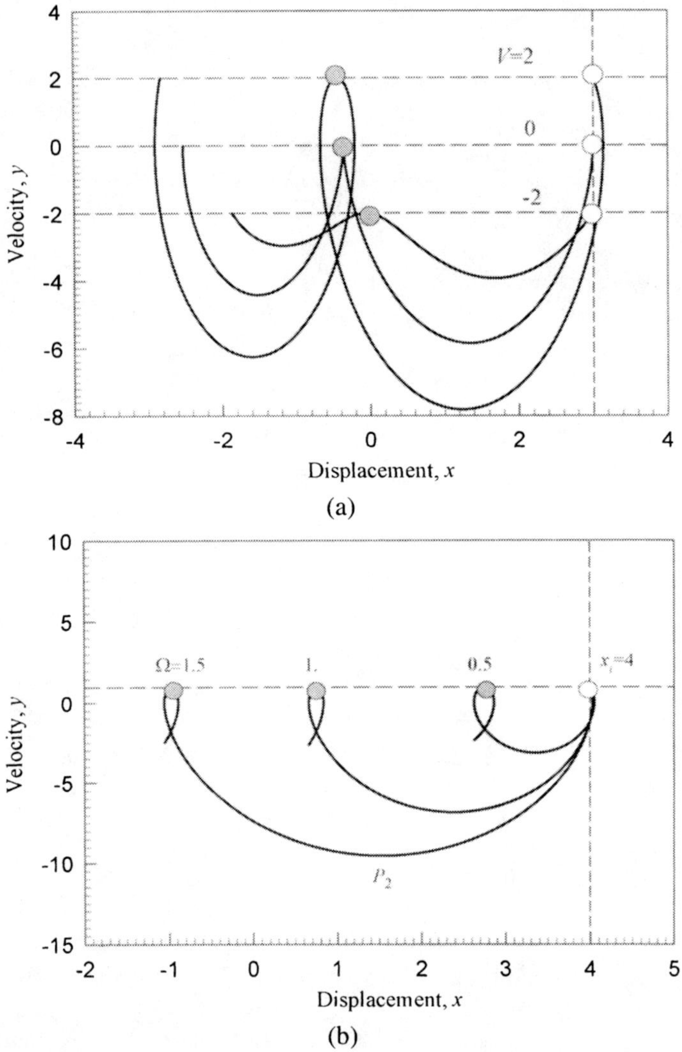

Figure 29. Grazing phase trajectories of mapping P_2 ($A_0 = 90, d_1 = 1, d_2 = 0, b_1 = -b_2 = 30$): (a) $x_i = 3$, $\Omega t_i \approx 0.9297, 0.9551, 0.9653$ for $y_i \equiv V = -2, 0, 2$ and $\Omega = 1$, respectively, (b) $(x_i, y_i) = (4, 1)$, $\Omega t_i \approx 0.4459, 0.6495, 0.6700$ for $\Omega = 0.5, 1, 1.5$ and $V = 1$, respectively.

$\Omega = 0.5, 1, 1.5$ and $V = 1$, accordingly}, and the other parameters $A_0 = 90$, $b_1 = -b_2 = 30$ are employed as well. The periodic motion with stick and non-stick of this oscillator can be found in Luo and Gegg (2006a, 2006b). This methodology can be applied other discontinuous dynamical systems to determine the grazing flows.

8. Conclusions

In this chapter, the singularity in the vicinity of the discontinuous boundary was discussed. The accessible and inaccessible subdomains were introduced. On the accessible domain, the corresponding dynamic systems are defined. The oriented boundary sets and singular sets caused by the separation boundary were presented. The local singularity and tangency of a flow on the separation boundary were discussed. The necessary and sufficient conditions for such a local singularity and tangency were given. The two examples were investigated to demonstrate the tangential flows in discontinuous dynamical systems.

Appendix

Consider a second-order linear dynamic system in domain Ω_j as

$$\ddot{x} + 2d_j\dot{x} + c_jx = b_j + a\cos\Omega t \tag{A.1}$$

with an initial condition $(t, x, \dot{x}) = (t_i, x_i, y_i)$. The total solutions for such a system are given as follows.

CASE I. $d_j^2 - c_j > 0$.

$$\left\{\begin{aligned} x &= \mathrm{e}^{-d_j(t-t_i)}\big(C_1^{(j)}\mathrm{e}^{\omega_d(t-t_i)} + C_2^{(j)}\mathrm{e}^{-\omega_d(t-t_i)}\big) \\ &\quad + D_1^{(j)}\cos\Omega t + D_2^{(j)}\sin\Omega t + D_0^{(j)}, \\ \dot{x} &= \mathrm{e}^{-d_j(t-t_i)}\big[(\omega_d - d_j)C_1^{(j)}\mathrm{e}^{\omega_d(t-t_i)} - (\omega_d + d_j)C_2^{(j)}\mathrm{e}^{-\omega_d(t-t_i)}\big] \\ &\quad - D_1^{(j)}\Omega\sin\Omega t + D_2^{(j)}\Omega\cos\Omega t, \end{aligned}\right. \tag{A.2}$$

where

$$\left\{\begin{aligned} C_1^{(j)} &= \frac{1}{2\omega_d^{(j)}}\big\{\dot{x}_i + \big(x_i - D_0^{(j)}\big)\big(d_j + \omega_d^{(j)}\big) \\ &\qquad - \big[D_1^{(j)}\big(d_j + \omega_d^{(j)}\big) + D_2^{(j)}\Omega\big]\cos\Omega t_i \\ &\qquad + \big[D_1^{(j)}\Omega - D_2^{(j)}(d_j + \omega_d)\big]\sin\Omega t_i\big\}, \\ C_2^{(j)} &= \frac{1}{2\omega_d^{(j)}}\big\{-\dot{x}_i + \big(x_i - D_0^{(j)}\big)\big(-d_j + \omega_d^{(j)}\big) \\ &\qquad - \big[D_1^{(j)}\Omega + D_2^{(j)}\big(-d_j + \omega_d^{(j)}\big)\big]\sin\Omega t_i \\ &\qquad - \big[D_1^{(j)}\big(-d_j + \omega_d^{(j)}\big) - D_2^{(j)}\Omega\big]\cos\Omega t_i\big\} \end{aligned}\right. \tag{A.3}$$

and

$$\begin{cases} D_0^{(j)} = -\frac{b_j}{c_j}, \qquad D_1^{(j)} = \frac{a(c_j-\Omega^2)}{(c_j-\Omega^2)^2+(2d_j\Omega)^2}, \\ D_2^{(j)} = \frac{a(2d_j\Omega)}{(c_j-\Omega^2)^2+(2d_j\Omega)^2}, \qquad \omega_d^{(j)} = \sqrt{c_j - d_j^2}. \end{cases} \tag{A.4}$$

CASE II. $d_j^2 - c_j < 0$.

$$\begin{cases} x = \mathrm{e}^{-d_j(t-t_i)}\big[C_1^{(j)}\cos\omega_d^{(j)}(t-t_i) + C_2^{(j)}\sin\omega_d^{(j)}(t-t_i)\big] \\ \qquad + D_1^{(j)}\cos\Omega t + D_2^{(j)}\sin\Omega t + D_0^{(j)}, \\ \dot{x} = \mathrm{e}^{-d_j(t-t_i)}\big[\big(C_2^{(j)}\omega_d^{(j)} - d_j C_1^{(j)}\big)\cos\omega_d^{(j)}(t-t_i) \\ \qquad - \big(C_1^{(j)}\omega_d^{(j)} + d_j C_2^{(j)}\big)\sin\omega_d^{(j)}(t-t_i)\big] \\ \qquad - D_1^{(j)}\Omega\sin\Omega t + D_2^{(j)}\Omega\cos\Omega t, \end{cases} \tag{A.5}$$

where

$$\begin{cases} C_1^{(j)} = x_i - D_1^{(j)}\cos\Omega t_i - D_2^{(j)}\sin\Omega t_i - D_0^{(j)}, \\ C_2^{(j)} = \frac{1}{\omega_d^{(j)}}\big[d_j\big(x_i - D_1^{(j)}\cos\Omega t_i - D_2^{(j)}\sin\Omega t_i - D_0^{(j)}\big) \\ \qquad + \dot{x}_i + D_1^{(j)}\Omega\sin\Omega t_i - D_2^{(j)}\Omega\cos\Omega t_i\big]. \end{cases} \tag{A.6}$$

CASE III. $d_j^2 - c_j = 0$.

$$\begin{cases} x = \mathrm{e}^{-d_j(t-t_i)}\big[C_1^{(j)}(t-t_i) + C_2^{(j)}\big] \\ \qquad + D_1^{(j)}\cos\Omega t + D_2^{(j)}\sin\Omega t + D_1^{(j)}, \\ \dot{x} = \mathrm{e}^{-d_j(t-t_i)}\big[C_1^{(j)} - C_1^{(j)}d_j(t-t_i) - d_j C_2^{(j)}\big] \\ \qquad - D_1^{(j)}\Omega\sin\Omega t + D_2^{(j)}\Omega\cos\Omega t, \end{cases} \tag{A.7}$$

where

$$\begin{cases} C_2^{(j)} = x_i - D_1^{(j)}\cos\Omega t_i - D_2^{(j)}\sin\Omega t_i - D_0^{(j)}, \\ C_1^{(j)} = x_i + \cos\Omega t_i\big(D_2^{(j)}\Omega - d_j D_1^{(j)}\big) \\ \qquad - \sin\Omega t_i\big(D_1^{(j)}\Omega + d_j D_2^{(j)}\big) - d_j D_1^{(j)}. \end{cases} \tag{A.8}$$

CASE IV. $d_j \neq 0, c_j = 0$.

$$\begin{cases} x = C_1^{(j)}\mathrm{e}^{-2d_j(t-t_i)} + D_1\cos\Omega t + D_2^{(j)}\sin\Omega t + D_0^{(j)}t + C_2^{(j)}, \\ \dot{x} = -2d_j C_1^{(j)}\mathrm{e}^{-2d_j(t-t_i)} - D_1^{(j)}\Omega\sin\Omega t \\ \qquad + D_2^{(j)}\Omega\cos\Omega t + D_0^{(j)}, \end{cases} \tag{A.9}$$

where

$$\begin{cases} C_1^{(j)} = -\frac{1}{2d_j}\left(\dot{x}_i + D_1^{(j)}\Omega \sin \Omega t_i - D_2^{(j)}\Omega \cos \Omega t_i - D_0^{(j)}\right), \\ C_2^{(j)} = \frac{1}{2d_j}\left[2d_j x_i + \dot{x}_i + \left(D_1^{(j)}\Omega - 2d_j D_2^{(j)}\right) \sin \Omega t_i \right. \\ \qquad \left. - \left(2d_j D_1^{(j)} + D_2^{(j)}\Omega\right) \cos \Omega t_i - 2d_j D_0^{(j)} t_i - D_0^{(j)}\right]. \end{cases} \tag{A.10}$$

CASE V. $d_j = 0, c_j = 0$.

$$\begin{cases} x = -\frac{a}{\Omega^2} \cos \Omega t - \frac{1}{2} b_j t^2 + C_1^{(j)} t + C_2^{(j)}, \\ \dot{x} = \frac{a}{\Omega} \sin \Omega t - b_j t + C_1^{(j)}, \end{cases} \tag{A.11}$$

where

$$\begin{cases} C_1^{(j)} = \dot{x}_i - \frac{a}{\Omega} \sin \Omega t_i + b_j t_i, \\ C_2^{(j)} = x_i - \dot{x} t_i + \frac{a}{\Omega^2} \cos \Omega t_i + \frac{a}{\Omega} t_i \sin \Omega t_i - \frac{1}{2} b_j t_i^2. \end{cases} \tag{A.12}$$

References

Bapat, C.N., Popplewell, N., Mclachlan, K., 1983. Stable periodic motion of an impact pair. J. Sound Vib. 87, 19–40.

Birkhoff, G.D., 1927. On the periodic motions of dynamical systems. Acta Math. 50, 359–379.

Broucke, M., Pugh, C., Simic, S.N., 2001. Structural stability of piecewise smooth systems. Comput. Appl. Math. 20 (1–2), 51–89.

Den Hartog, J.P., 1930. Forced vibration with combined viscous and Coulomb damping. Philos. Mag. VII (9), 801–817.

Den Hartog, J.P., Mikina, S.J., 1932. Forced vibrations with non-linear spring constants. ASME J. Appl. Mech. 58, 157–164.

di Bernardo, M., Budd, C.J., Champney, A.R., 2001. Normal formal maps for grazing bifurcation in n-dimensional piecewise-smooth dynamical systems. Physica D 160, 222–254.

di Bernardo, M., Kowalczyk, K., Nordmark, A., 2002. Bifurcations of dynamical systems with sliding: derivation of normal formal mappings. Physica D 170, 175–205.

Fillippov, A.F., 1988. Differential Equations with Discontinuous Right Hand Sides. Kluwer Academic, Dordrecht.

Han, R.P.S., Luo, A.C.J., Deng, W., 1995. Chaotic motion of a horizontal impact pair. J. Sound Vib. 181, 231–250.

Kleczka, M., Kreuzer, E., Schiehlen, W., 1992. Local and global stability of a piecewise linear oscillator. Philos. Trans. R. Soc. London Ser. A 338 (1651), 533–546.

Kunze, M., 2000. Non-Smooth Dynamical Systems. Lecture Notes in Math., vol. 1744. Springer-Verlag, Berlin.

Leine, R.I., Van Campen, D.H., 2002. Discontinuous bifurcations of periodic solutions. Math. Comput. Modelling 36, 250–273.

Luo, A.C.J., 1995. Analytical modeling of bifurcations, chaos and fractals in nonlinear dynamics. Doctoral dissertation, University of Manitoba, Winnipeg, Canada.

Luo, A.C.J., 2002. An unsymmetrical motion in a horizontal impact oscillator. ASME J. Vib. Acoust. 124, 420–426.

Luo, A.C.J., 2005a. A theory for nonsmooth dynamic systems on the connectable domains. Commun. Nonlinear Sci. Numer. Simul. 10, 1–55.

Luo, A.C.J., 2005b. A periodically forced, piecewise, linear system, Part I: Local singularity and grazing bifurcation. Commun. Nonlinear Sci. Numer. Simul., in press.

Luo, A.C.J., 2005c. A periodically forced, piecewise, linear system, Part II: The fragmentation mechanism of strange attractor. Commun. Nonlinear Sci. Numer. Simul., in press.

Luo, A.C.J., 2006. Grazing and chaos in a periodically forced, piecewise linear system. ASME J. Vib. Acoust. 128, 28–34.

Luo, A.C.J., Chen, L.D., 2005. Periodic motions and grazing in a periodically forced, piecewise linear oscillator with impacts. Chaos Solitons Fractals 24, 567–578.

Luo, A.C.J., Chen, L.D., 2006. Grazing phenomena and fragmented strange attractors in a harmonically forced, piecewise, linear system with impacts. IMeChE Part K: Journal of Multi-body Dynamics 220, 35–51.

Luo, A.C.J., Gegg, B.C., 2005. Grazing phenomena in a periodically forced, friction-induced, linear oscillator. Commun. Nonlinear Sci. Numer. Simul., in press.

Luo, A.C.J., Gegg, B.C., 2006a. On the mechanism of stick and non-stick, periodic motions in a forced linear oscillator including dry friction. ASME J. Vib. Acoust. 128, 97–105.

Luo, A.C.J., Gegg, B.C., 2006b. Stick and non-stick, periodic motions of a periodically forced, linear oscillator with dry friction. J. Sound Vib. 291, 132–168.

Luo, A.C.J., Han, R.P.S., 1996. The dynamics of a bouncing ball with a sinusoidally vibrating table revisited. Nonlinear Dynam. 10, 1–18.

Luo, A.C.J., Menon, S., 2004. Global chaos in a periodically forced, linear system with a dead-zone restoring force. Chaos Solitons Fractals 19, 1189–1199.

Masri, S.F., 1970. General motion of impact dampers. J. Acoust. Soc. Am. 47, 229–237.

Masri, S.F., Caughey, T.K., 1966. On the stability of the impact damper. ASME J. Appl. Mech. 33, 586–592.

Menon, S., Luo, A.C.J., 2005. An analytical prediction of the global period-1 motion in a periodically forced, piecewise linear system. Internat. J. Bifur. Chaos 15 (6), 1945–1957.

Natsiavas, S., 1989. Periodic response and stability of oscillators with symmetric trilinear restoring force. J. Sound Vib. 134 (2), 315–331.

Nordmark, A.B., 1991. Non-periodic motion caused by grazing incidence in an impact oscillator. J. Sound Vib. 145, 279–297.

Poincare, H., 1892. Les Methods Nouvelles de la Mecanique Celeste, vol. 1. Gauthier-Villars, Paris.

Popp, K., 2000. Non-smooth mechanical systems. J. Appl. Math. Mech. 64, 765–772.

Senator, M., 1970. Existence and stability of periodic motions of a harmonically forced impacting system. J. Acoust. Soc. Am. 47, 1390–1397.

Shaw, S.W., Holmes, P.J., 1983a. A periodically forced impact oscillator with large dissipation. ASME J. Appl. Mech. 50, 849–857.

Shaw, S.W., Holmes, P.J., 1983b. A periodically forced piecewise linear oscillator. J. Sound Vib. 90 (1), 121–155.

Utkin, V.I., 1978. Sliding Modes and Their Application in Variable Structure Systems. Mir, Moscow.

Utkin, V.I., 1981. Sliding Regimes in Optimization and Control Problem. Nauka, Moscow.

Ye, H., Michel, A., Hou, L., 1998. Stability theory for hybrid systems. IEEE Trans. Automat. Control 43 (4), 461–474.

Chapter 3

Global Bifurcations of Complex Nonlinear Dynamical Systems with Cell Mapping Methods

Ling Hong and Jian-Qiao Sun

Department of Mechanical Engineering, University of Delaware, Newark, DE 19716, USA

E-mail: *jqsun@udel.edu*

Contents

1. Introduction 192
2. Cell mapping methods 193
    2.1. Simple cell mapping 194
        Properties of SCM 194
    2.2. Generalized cell mapping 195
        Properties of GCM 197
3. Crises in deterministic systems 198
    3.1. A chaotic boundary crisis 198
    3.2. Chaotic boundary and interior crises 200
    3.3. Wada fractal boundary and indeterminate crisis 202
    3.4. Double crises 210
4. Bifurcations of nonlinear systems with small random disturbances 212
    4.1. Logistic map with random coefficients 215
    4.2. A two-dimensional random map 217
    4.3. Duffing oscillator with small random excitations 220
    4.4. Noisy crisis in a twin-well Duffing system 221
5. Fuzzy bifurcations 230
    5.1. Fuzzy generalized cell mapping 231
        Topological property of FGCM 233
        A backward algorithm 234
        Finite convergence of membership 235
    5.2. Bifurcation of one-dimensional fuzzy systems 235

Edited Series on Advances in Nonlinear Science and Complexity
Volume 1 ISSN: 1574-6909 DOI: 10.1016/S1574-6909(06)01003-3

5.3. Bifurcation of fuzzy nonlinear oscillators 240
A forced nonlinear Mathieu oscillator 240
A Duffing–Van der Pol oscillator 245
The DVP equation with $(\mu_1, \mu_2) = (1, -1)$ 245
The DVP equation with $(\mu_1, \mu_2) = (0.64, 0.1)$ 253
5.4. Conjectures 255
6. Effect of bifurcation on semiactive optimal controls 258
6.1. Optimal control problem 259
General control database 260
Search algorithm for fixed final state problems 261
6.2. Saddle-node bifurcation 262
6.3. Supercritical Pitchfork bifurcation 265
6.4. Subcritical Hopf bifurcation 268
A summary 272
References 274

1. Introduction

Bifurcation has been an active research topic for several decades now. Fundamental concepts and mathematical theory on bifurcation are well documented in the textbooks (Guckenheimer and Holmes, 1983; Hale and Kocak, 1991). Bifurcation analysis is usually focused on the existence, stability and interactions of solutions of nonlinear dynamical systems governed by ordinary differential equations. These solutions can include equilibrium states, periodic motion, chaotic motion and quasiperiodic motion. However, with a few exceptions, it is often difficult to obtain these solutions in analytical form for nonlinear dynamical systems. Numerical methods are frequently used to obtain the solutions. Consequently, bifurcation analysis has to be done numerically as well.

Furthermore, when nonlinear dynamical systems are subject to stochastic excitations, or contain random or fuzzy uncertainties, the system response becomes far more difficult to obtain because it must be handled by a method which can describe the global properties of nonlinear dynamical systems. The cell mapping methods are created just for this purpose.

This chapter deals with global bifurcations of nonlinear dynamical systems that can be deterministic, stochastic and fuzzy by using the cell mapping methods. Before the global bifurcation analysis can be conducted, one needs to be able to obtain the global solutions of the system including the topology of the stable attractors, boundaries of the domains of attraction, and unstable solutions. The global bifurcation analysis amounts to studying the changes of these global solution characteristics as the control parameter varies.

In this chapter, we first review the cell mapping methods in Section 2. In Section 3, we study several scenarios of boundary and interior crises of deterministic nonlinear dynamical systems. In Section 4, we present a study of bifurcations when the system is either uncertain with random parameter or is subject to small random excitations. The effect of random noise on bifurcation is investigated. In Section 5, we extend the generalized cell mapping (GCM) method to the fuzzy dynamical systems, and present the fuzzy GCM method. We then apply the fuzzy GCM method to study bifurcation of nonlinear dynamical systems with fuzzy parameters or fuzzy excitations. Finally, in Section 6, we present a study of the effect of bifurcation on the semiactive optimal controls.

2. Cell mapping methods

Cell mapping methods for global analysis of nonlinear systems were introduced by Hsu in the early 1980s. He first introduced the simple cell mapping (SCM) (Hsu, 1980) and subsequently the generalized cell mapping (GCM) (Hsu, 1981). Early developments of the cell mapping methods are well documented in his book (Hsu, 1987). There have been many studies and applications of cell mapping methods since then (Foale and Thompson, 1991; Guder and Kreuzer, 1999a, 1999b; Hsu and Chiu, 1986; Kim and Hsu, 1986; Sun and Hsu, 1987; Ushio and Hsu, 1987). A variety of random vibration and stochastic optimal problems have been studied by Hsu, Sun and associates (Crespo and Sun, 2003; Sun and Hsu, 1990a). A theory of the subdomain-to-subdomain for global transient analysis was developed in Hsu (1992). Hsu (1995) further extended the generalized cell mapping method by introducing the graph theory to the analysis of dynamical systems.

An interpolated cell mapping was developed by Tongue (1987), which uses the cell mappings as a database to interpolate the trajectories not passing the grid points of the cell states, leading to continuous time trajectories of the response of nonlinear systems. The point mapping using the cell state as a reference was developed by Jiang and Xu (1994) to reduce the error due to the coarseness of the cell discretization.

The cell mapping method was first applied to optimal control problems by Hsu (1985). It has since become a very popular tool for fuzzy control designs. Chen and Tsao (1989) pioneered the work of fuzzy control design with the simple cell mapping method. Smith and his associates further extended the simple cell mapping method to fuzzy optimal control problems in higher-dimensional state space (Smith and Comer, 1990; Song and Smith, 2000; Song, Smith and Rizk, 1999a, 1999b). The generalized cell mapping was also extended to fuzzy dynamical systems. Edwards and Choi (1997) proposed to impose the initial probability distribution in each cell that had the same form as a typical fuzzy membership function. The resulting generalized cell mapping, however, was still a Markov chain. Sun and Hsu (1990b) extended the generalized cell mapping to fuzzy systems by applying Zadeh's extension principle.

As the computers grow faster with bigger memory space than ever, the cell mapping methods become a feasible tool for bifurcation analysis of complex nonlinear dynamical systems. Crises in deterministic dynamical systems have been investigated by Hong and Xu (1999, 2001a, 2001b) using the GCM with digraphs. Bifurcations of nonlinear oscillators subject to stochastic disturbances were investigated extensively in Xu, Fang and Rong (2003, 2004, 2005). The GCM method was applied to study bifurcations of fuzzy nonlinear dynamical systems (Hong and Sun, 2006a, 2006b). This chapter summarizes the recent results of these bifurcation analyses.

2.1. Simple cell mapping

Consider a nonlinear dynamical system,

$$\dot{\mathbf{x}} = \mathbf{F}(\mathbf{x}, t, \mu), \quad \mathbf{x} \in \mathbb{R}^n, t \in \mathbb{R}, \mu \in \mathbb{R}^m, \tag{1}$$

where $\mathbf{x}$ is an n-dimensional state vector, t is the time variable, μ is an m-dimensional parameter vector and $\mathbf{F}$ is a vector-valued function of its arguments, and is periodic in t. A point mapping known as the Poincaré map can be constructed to describe the dynamic evolution of the system response over one period,

$$\mathbf{x}(k+1) = \mathbf{G}\big(\mathbf{x}(k), \mu\big), \tag{2}$$

where $\mathbf{x}(k)$ denotes the state of the system at the beginning of a period and $\mathbf{G}$ is the mapping, which maps an n-tuple of real numbers to another n-tuple of real numbers.

The cell mapping methods propose to discretize the continuous state space into a collection of small cells, known as the cell state space (Hsu, 1987). A straightforward way to construct the cell state space is to divide a state variable x_i into equal intervals. Let the total number of cells be N. We can label each cell with an integer. The simple cell mapping is described by

$$j(k+1) = C\big(j(k), \mu\big), \tag{3}$$

where C maps an integer of the cell state $j(k)$ to another one $j(k+1)$ in one step. The simple cell mapping is illustrated in Figure 1 for a two-dimensional state space.

Properties of SCM

1. *Equilibrium cells and periodic motions.* A cell j^* satisfying $j^* = C(j^*, \mu)$ is said to be an equilibrium cell. Let C^k denote the cell mapping C applied

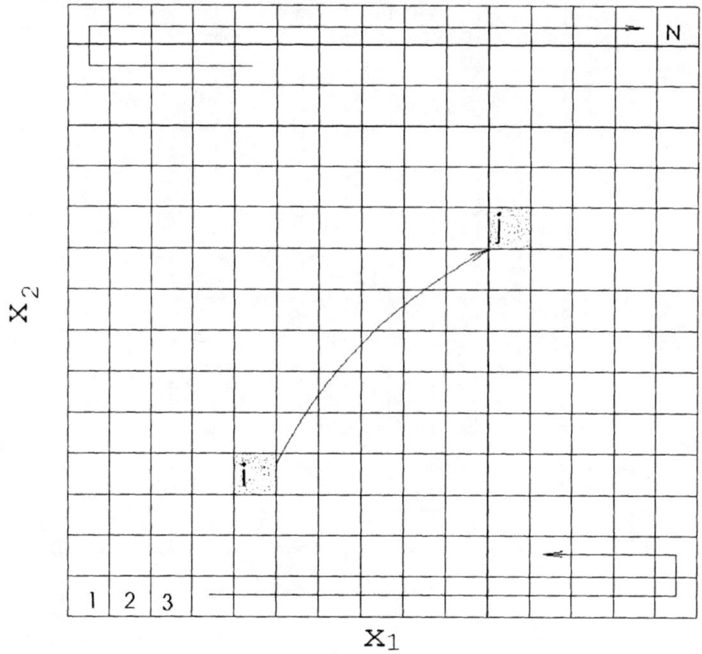

Figure 1. An illustration of the simple cell mapping in a two-dimensional state space.

k times with $C^0 = 1$. A sequence of distinct cells $j(k)$ $(1 \leqslant k \leqslant K)$ is said to constitute a periodic motion (or solution) of period K and each of the element $j(k)$ is a periodic cell if

$$j(k+1) = C^k\big(j(1), \mu\big), \quad 1 \leqslant k \leqslant K-1,$$
$$j(1) = C^K\big(j(1)\big). \tag{4}$$

In particular, the equilibrium cell is period one.

2. *Transient cells and domains of attraction*. Cells which are not periodic or equilibrium cells are eventually mapped to periodic or equilibrium cells in a finite number of steps. These are called transient cells. The set of all the transient cells which are mapped to the kth periodic group of cells is called the domain of attraction of that periodic solution.

2.2. Generalized cell mapping

In the generalized cell mapping, a cell is allowed to have several image cells with each image cell having a fraction of the probability, as illustrated in Figure 2.

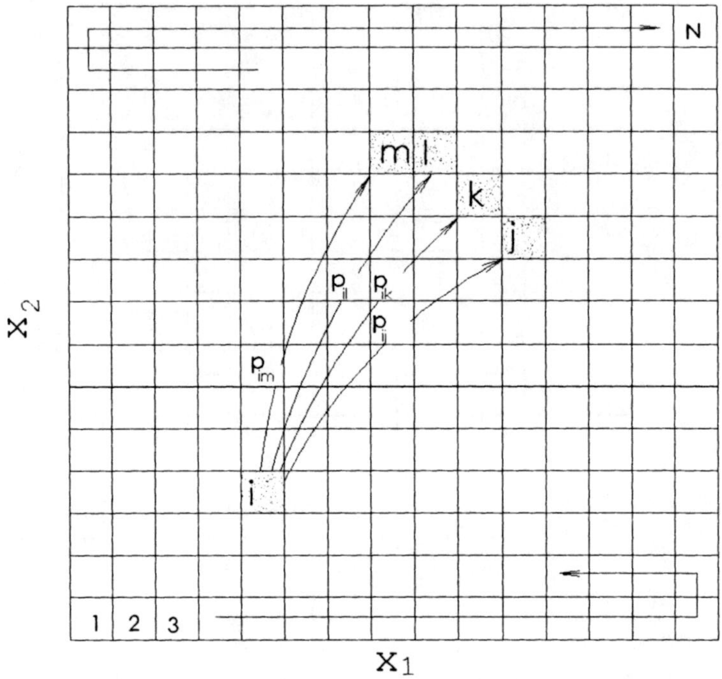

Figure 2. An illustration of GCM concept in a two-dimensional state space.

Consider a cell, say cell i and M uniformly distributed initial conditions in the cell. We integrate the equation of motion for one mapping step starting from these initial conditions. Let M_j denote the number of trajectories falling in cell j at the end of the mapping step. The ratio $p_{ij} = M_j/M$ is known as the one-step transition probability from cell i to cell j. This approach of computing the transition probability matrix is called the sampling point method (Hsu, 1987). Let $\mathbf{P} = [p_{ij}]^{\mathrm{T}}$ denote a matrix with elements p_{ij}. It is known as the one-step transition probability matrix. Let $p_i(k)$ denote the probability of the system residing in cell i at time k, and $\mathbf{p}(k) = \{p_i(k)\}$ be the vector consisting of the probabilities of all the cells. Then, the generalized cell mapping system is described by a Markov chain as

$$\mathbf{p}(k+1) = \mathbf{P}\mathbf{p}(k) \quad \text{or} \quad \mathbf{p}(k) = \mathbf{P}^k\mathbf{p}(0), \tag{5}$$

where $\mathbf{p}(0)$ is the initial probability distribution vector.

Properties of GCM

The classical theory of Markov chain can be used to describe the GCM method (Chung, 1967). Here, we list several relevant concepts and properties of the GCM method.

1. *Lead-to and communicate-with.* If cell i is mapped to cell j in finite number of steps, we say cell i leads to cell j. If cell i leads to cell j and cell j leads to cell i, it is said that cell i communicates with cell j and vice versa.
2. *Self-cycling set.* A set of cells that communicate with all the other members in the set is a self-cycling set.
3. *Closed set.* Let A be a set of cells. A is called a closed set if for every $i \in A$ and $j \notin A$, we have $p_{ij}^k = 0$ for all $k \geqslant 1$.
4. *Persistent and transient self-cycling sets.* If a self-cycling set is a closed set, it is called a persistent self-cycling set, or simply a persistent set. The cell belonging to a persistent set is called a persistent cell. Persistent sets represent possible long-term stable motions of the system including attractors. The self-cycling set that is not closed is called a transient self-cycling set. Transient self-cycling sets are usually associated with unstable fixed points and periodic solutions of the system.
5. *Transient cell.* All the cells which are not persistent cells are called transient cells.
6. *Single-domicile and multiple-domicile transient sets.* If a transient cell j leads to a persistent set denoted as $P(l)$, $P(l)$ is said to be a domicile of cell j. A transient cell can have several domiciles. Transient cells may be classified according to the number of domiciles they have. In particular, the transient cells that have only one domicile are called single-domicile transient cells, and those that have more than one domiciles are called multiple-domicile transient cells. The single-domicile transient cells form basins of attraction, and the multiple-domicile transient cells form the boundaries separating the basins of attraction.

   The domiciles of the transient cells provide a very natural way of partitioning the transient cells into transient sets. Let there be N_p number of persistent self-cycling sets denoted by $P(l)$, $l = 1, 2, \dots, N_p$. The set consisting of all the single-domicile transient cells having the lth persistent self-cycling set as its domicile is denoted by $B(l)$, $l = 1, 2, \dots, N_p$. It represents the domain of attraction of the lth persistent self-cycling set.

   The set consisting of all the two-domicile transient cells having lth and mth persistent self-cycling sets as their domiciles is denoted by $B(l, m)$ where $l, m = 1, 2, \dots, N_p$ and $l \neq m$. The region represents the boundary between $B(l)$ and $B(m)$, i.e., the lth and mth domains of attraction. There are altogether $C_{N_p}^2$ number of two-domicile transient sets where $C_{N_p}^2 = N_p(N_p - 1)/2$.

The set consisting of all the three-domicile transient cells having lth, mth and nth persistent self-cycling sets as their only domiciles is denoted by $B(l, m, n)$ where $l, m, n = 1, 2, \ldots, N_{\mathrm{p}}$ and $l \neq m \neq n$. The region represents the boundary between $B(l)$, $B(m)$ and $B(n)$, i.e., the lth, mth and nth domains of attraction. There are altogether $C^3_{N_{\mathrm{p}}}$ number of three-domicile transient sets where $C^3_{N_{\mathrm{p}}} = N_{\mathrm{p}}(N_{\mathrm{p}} - 1)(N_{\mathrm{p}} - 2)/6$.

The transient sets with four or more domiciles can be defined in a similar way. Finally, there is one set of transient cells that have all the persistent self-cycling sets as their domiciles. Hence, the transient cells can be classified into $2^{N_{\mathrm{p}}} - 1$ subsets according to the number of domiciles.

3. Crises in deterministic systems

Almost all the sudden discontinuous changes of chaotic attractors, as a control parameter of the system varies, are due to crises (Grebogi, Ott and Yorke, 1983, 1986a). According to the nature of the discontinuous change of chaotic attractors, crises can be divided into two categories: a boundary crisis and an interior crisis, or a catastrophe and an explosion (Thompson and Stewart, 2000; Thompson, Stewart and Ueda, 1994). In the case of a boundary crisis, a chaotic attractor, together with its basin of attraction, is eradicated from the phase space. Such events have been extensively studied in discrete mappings (Grebogi, Ott and Yorke, 1983, 1986a). It is showed that a boundary crisis results from the collision of a chaotic attractor with an unstable periodic orbit at its basin boundary. In differential equations, the abrupt disappearance of a chaotic attractor has been investigated in Abraham and Stewart (1986) and Stewart (1988). While in the case of an interior crisis, there is an abrupt increase (or decrease) in the size of a chaotic attractor, as first observed by Ueda (1980) in differential equations. In discrete mappings, such a sudden change in the size of a chaotic attractor has also been studied (Grebogi, Ott and Yorke, 1983; Lai, Grebogi and Yorke, 1992). It is showed that an interior crisis results from the collision of a chaotic attractor with an unstable periodic orbit (Grebogi, Ott and Yorke, 1983) or a chaotic saddle (Lai, Grebogi and Yorke, 1992) in its basin interior.

Crises in deterministic systems are studied by means of the GCM method in this section. A crisis happens when a chaotic attractor collides with a regular saddle or a chaotic saddle and is called a regular crisis or a chaotic crisis, respectively.

3.1. A chaotic boundary crisis

Consider a sinusoidally driven twin-well Duffing oscillator,

$$\frac{\mathrm{d}^2x}{\mathrm{d}t^2} + 0.25\frac{\mathrm{d}x}{\mathrm{d}t} - x + x^3 = A \sin t. \tag{6}$$

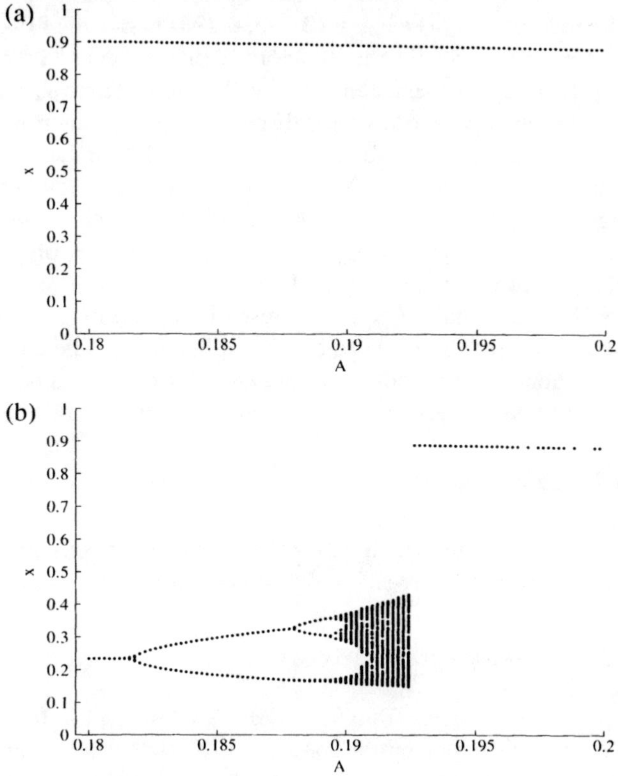

Figure 3. Bifurcation diagrams of different initial conditions for the Duffing equation (6). (a) $x(0) = 0.8$, $\dot{x}(0) = 0.1$; (b) $x(0) = 0.3$, $\dot{x}(0) = -0.2$.

Bifurcation diagrams of the system are generated for the parameter A between 0.18 and 0.2, and are shown in Figure 3. The one with $x(0) = 0.8$, $\dot{x}(0) = 0.1$ is shown in Figure 3(a), the other with $x(0) = 0.3$, $\dot{x}(0) = -0.2$ is shown in Figure 3(b). The integration time step of the fourth-order Runge–Kutta algorithm is 1.0×10^{-4}. The first 5×10^6 transient steps are ignored. In Figure 3(b), as the parameter A increases, the chaotic attractor disappears at $A \approx A_{cr} = 0.192$. At $A \approx A_{cr}$, there is a measurable distance between the chaotic attractor and the period-1 attractor, which suggests the existence of a chaotic saddle. We shall use the GCM method to validate this finding.

A domain of interest $\mathbf{D} = \{-0.1 \leqslant x \leqslant 1.0; -0.5 \leqslant \dot{x} \leqslant 0.4\}$ is discretized into 105×105 cells. The 25×25 interior points of each cell are sampled to compute transition probabilities. The region outside of $\mathbf{D}$ is referred to as the sink cell.

A chaotic boundary crisis when $A \in (0.192, 0.193)$ is shown in Figure 4. When $A = 0.192$, there are two coexistent attractors, a chaotic one and a period-1 one, represented by two acyclic persistent self-cycling sets. There is a chaotic saddle represented by an acyclic transient self-cycling set as shown in Figure 4(a). It should be noted that the basin boundary is a fractal boundary, which can be verified by numerical simulations. The chaotic saddle is embedded in the fractal boundary. Figure 4(a) shows that the chaotic attractor touches the chaotic saddle in the fractal boundary and disappears. When $A = 0.193$, only the period-1 attractor is observed, as shown in Figure 4(b).

Hence, this chaotic boundary crisis is caused by the collision of a chaotic attractor with a chaotic saddle in the fractal basin boundary. The chaotic attractor together with its domain of attraction is suddenly destroyed and becomes part of the new chaotic saddle as the system parameter passes through the critical value.

3.2. *Chaotic boundary and interior crises*

We present another example which exhibits both a chaotic boundary crisis and a chaotic interior crisis in the cosinusoidally forced Duffing oscillator,

$$\frac{d^2x}{dt^2} + 0.25\frac{dx}{dt} + \alpha x + x^3 = 8.5 \cos t. \tag{7}$$

When the parameter α varies from 0.15 to 0.3, we find the bifurcation diagrams in Figure 5 for two different initial conditions. Note that bifurcation diagrams are obtained by using a fourth-order Runge–Kutta algorithm with a time step 1.0×10^{-4}. The first 5×10^6 transient steps are ignored. Figure 5(a) indicates that a discontinuity of the response occurs at $\alpha \approx \alpha_{cr1} = 0.181$. This discontinuity also exists in Figure 5(b) for a different initial conditions. When $\alpha_{cr1} < \alpha < 0.22$, two chaotic attractors coexist. At $\alpha \approx \alpha_{cr1}$, there is a measurable distance between the two chaotic attractors, which suggests the existence of a chaotic saddle. As α decreases through $\alpha \approx \alpha_{cr1}$, the two chaotic attractors merge into a larger chaotic attractor. By means of the GCM method, we can describe this bifurcation event. A domain of interest $\mathbf{D} = \{2.3 \leqslant x \leqslant 3.3; -1.5 \leqslant \dot{x} \leqslant 3.5\}$ is discretized into 105×105 cells. The 25×25 interior points of each cell are sampled to compute transition probabilities.

We first find a chaotic boundary crisis when $\alpha \in (0.206, 0.207)$ as shown in Figure 6. The chaotic boundary crisis results from a collision between the chaotic attractor in the upper-right part of Figure 6(a) and a chaotic saddle in the fractal basin boundary. As α further decreases toward α_{cr1}, the chaotic saddle eventually collides with the chaotic attractor in the lower-left part of Figure 6(a). After the collision at $\alpha = 0.206$, only the chaotic attractor in the lower-left survives in Figure 6(b).

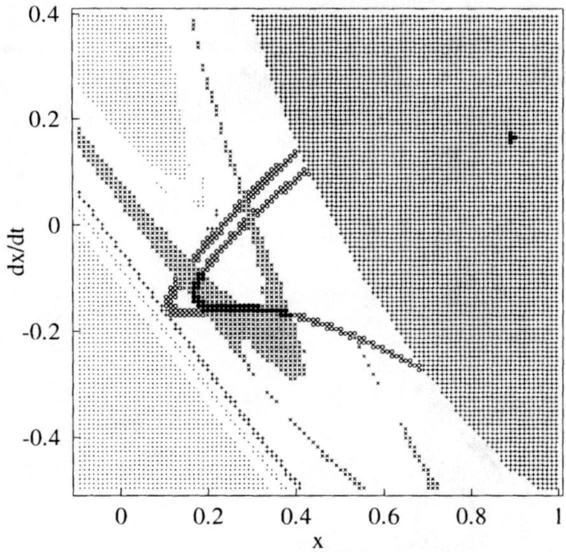

(a)

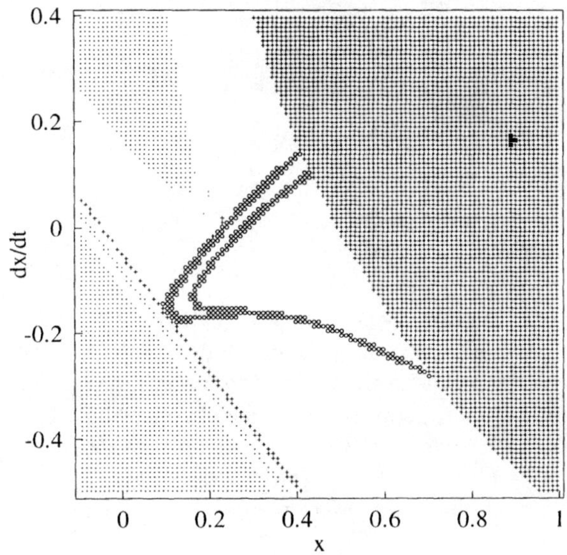

(b)

Figure 4. A chaotic boundary crisis of the Duffing equation (6) when the parameter $A \in (0.192, 0.193)$. (a) $A = 0.192$; (b) $A = 0.193$. The symbol • denotes the chaotic attractor; ×, the domain of the chaotic attractor; ■, the period-1 attractor; +, the domain of the period-1 attractor; ◯, the chaotic saddle; ·, the domain of attraction of the sink cell. The blank space stands for the basin boundary.

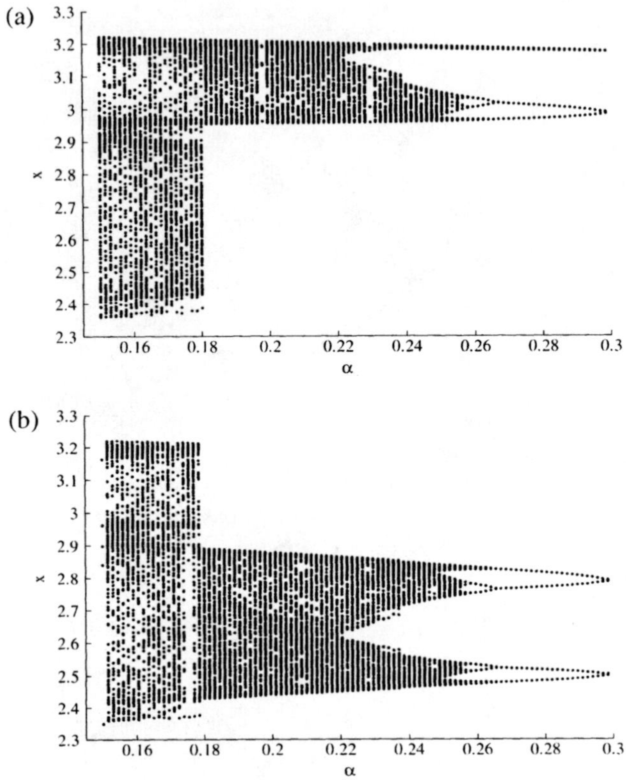

Figure 5. Bifurcation diagrams of the Duffing equation (7) starting from different initial conditions. (a) $x(0) = 3.0$, $\dot{x}(0) = 2.5$; (b) $x(0) = 2.6$, $\dot{x}(0) = -1.0$.

We find a chaotic interior crisis when $\alpha \in (0.18, 0.181)$ as shown in Figure 7. At $\alpha = 0.181$, we find a chaotic attractor shown in the lower-left part of Figure 7(a) and a chaotic saddle in the upper-right part of the figure. As α decreases to 0.18, the chaotic attractor connects with the chaotic saddle in its basin interior, forming a larger chaotic attractor. This is the chaotic interior crisis.

3.3. Wada fractal boundary and indeterminate crisis

We have discussed two types of dynamical chaotic invariant sets, namely, chaotic attractors and chaotic saddles. Chaotic saddles are nonattracting dynamical invariant sets that can lead to a variety of physical phenomena such as chaotic scattering and fractal basin boundaries (Ott and Tél, 1993; McDonald, Grebogi, Ott and Yorke, 1985). Just as chaotic attractors can undergo discontinuous bifurcations or crises (Grebogi, Ott and Yorke, 1982, 1983), so can chaotic saddles. The

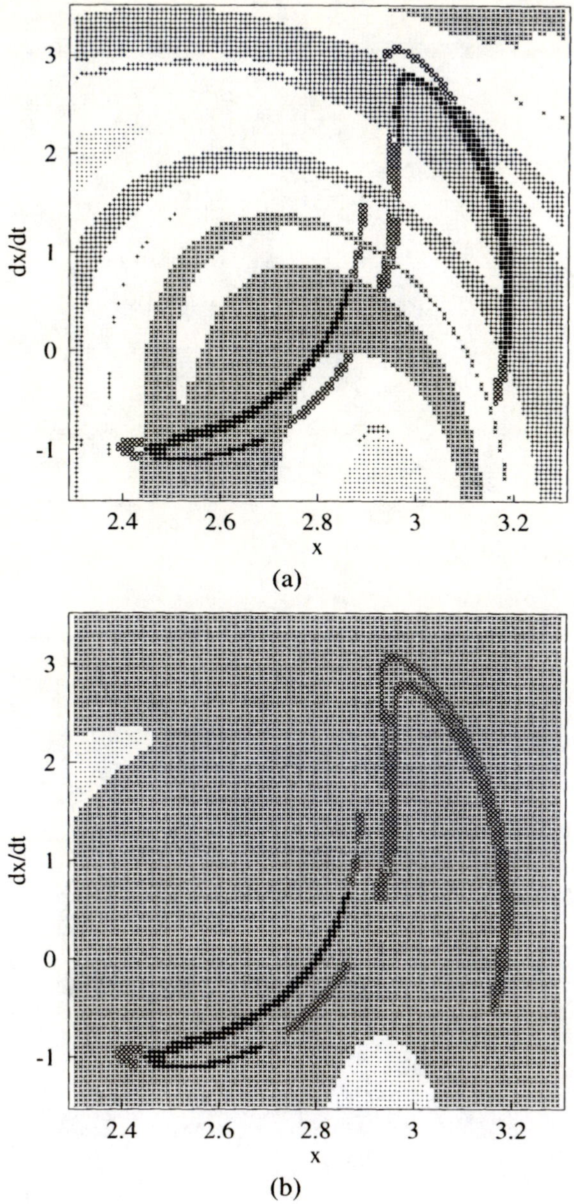

Figure 6. A chaotic boundary crisis of the Duffing equation (7) when the parameter $\alpha \in (0.207, 0.206)$. (a) $\alpha = 0.207$; (b) $\alpha = 0.206$. The symbol • denotes the chaotic attractor in the lower-left part of the figure; ×, its domain; ■, the chaotic attractor in the upper-right part of the figure; +, its domain; ○, the chaotic saddle; ·, the domain of attraction of the sink cell. The blank space stands for the basin boundary.

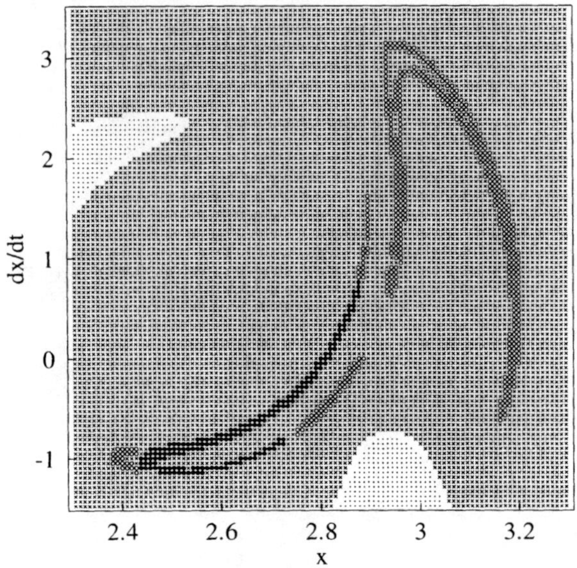

(a)

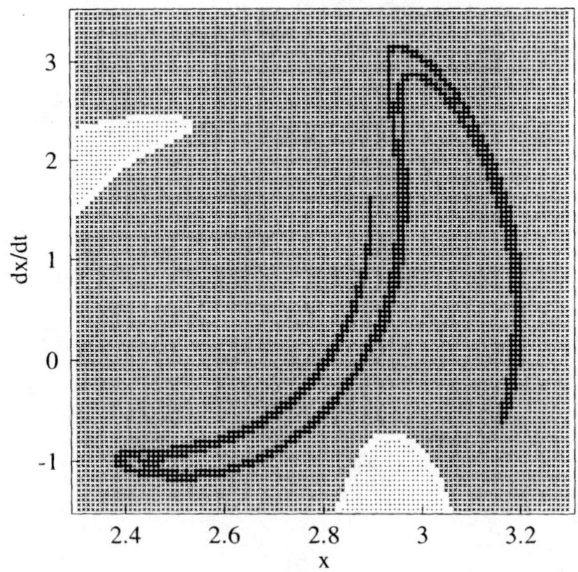

(b)

Figure 7. A chaotic interior crisis of the Duffing equation (7) when parameter $\alpha \in (0.18, 0.181)$. (a) $\alpha = 0.181$; (b) $\alpha = 0.18$. The symbol • denotes the chaotic attractor; ×, its domain; ○, the chaotic saddle; ·, the domain of attraction of the sink cell. The blank space stands for the basin boundary.

discontinuous bifurcations or metamorphoses of chaotic saddles result in physically observable phenomena such as enhancement of chaotic scattering (Lai, Grebogi, Blüel and Kan, 1993) and basin boundary metamorphoses (Grebogi, Ott and Yorke, 1986b, 1987). Fractal basin boundary may have the Wada property that every point on the common boundary of the basins of two attractors is also on the boundary of the basin of the third attractor (Kennedy and Yorke, 1991; Nusse, Ott and Yorke, 1995; Nusse and Yorke, 1996).

Here, we use Thompson's escape equation as an example to study Wada fractal boundary and indeterminate crisis. The canonical escape equation was introduced by Thompson and associates to model the typical escape from a potential well (Soliman and Thompson, 1992; Thompson, 1992; Thompson and Soliman, 1991). The canonical escape equation describing the sinusoidally forced motions of a particle in a potential $V(x) = x^2/2 - x^3/3$ with linear viscous damping is given by Thompson and Soliman (1991),

$$\frac{\mathrm{d}^2x}{\mathrm{d}t^2} + \beta\frac{\mathrm{d}x}{\mathrm{d}t} + x - x^2 = F\sin(\omega t + \phi), \tag{8}$$

where damping coefficient $\beta > 0$, forcing magnitude F and frequency ω are system parameters. Thompson has studied the forced oscillator (8) in considerable detail for the case of $\beta = 0.1$, and has plotted a bifurcation diagram in the (ω, F) parameter space (Thompson and Soliman, 1991). Here we are primarily interested in the region around the optimal escape at Q, where the system can escape with a minimum forcing magnitude F. The bifurcation diagram near the optimal escape is reproduced in Figure 8.

Figure 8 shows six bifurcation curves, namely, the two saddle-node fold curves A and B, the dashed flip curve C for the period doubling cascade, the boundary crisis curve E, the homoclinic tangency curve M and the heteroclinic tangency curve H. Two coexisting attractors within the well are found in the triangular region bounded by curves A, B and E. The shaded region in Figure 8 defines parameter values for which all motions must escape to infinity. We consider the vertical escape route $\omega = 0.827$ that is very close to the optimal escape Q.

By using the cell mapping method, we can confirm the previously reported results. Furthermore, we find a chaotic saddle, and show that the chaotic saddle is embedded in a Wada basin boundary, and that the chaotic saddle in the Wada basin boundary plays an extremely important role in the bifurcations governing escape. We also investigate the origin and evolution of the chaotic saddle in the Wada basin boundary. A domain of interest $\mathbf{D} = \{-0.8 \leqslant x \leqslant 1.2; -0.9 \leqslant \dot{x} \leqslant 0.8\}$ is discretized into 105×105 cells. The 25×25 interior points of each cell are sampled to compute transition probabilities.

Along the escape route $\omega = 0.827$, there are three bifurcations besides the saddle-node bifurcation crossing the fold curve B. The first one, when the heteroclinic tangency curve H is crossed, occurs in the interval $F \in (0.0705, 0.071)$

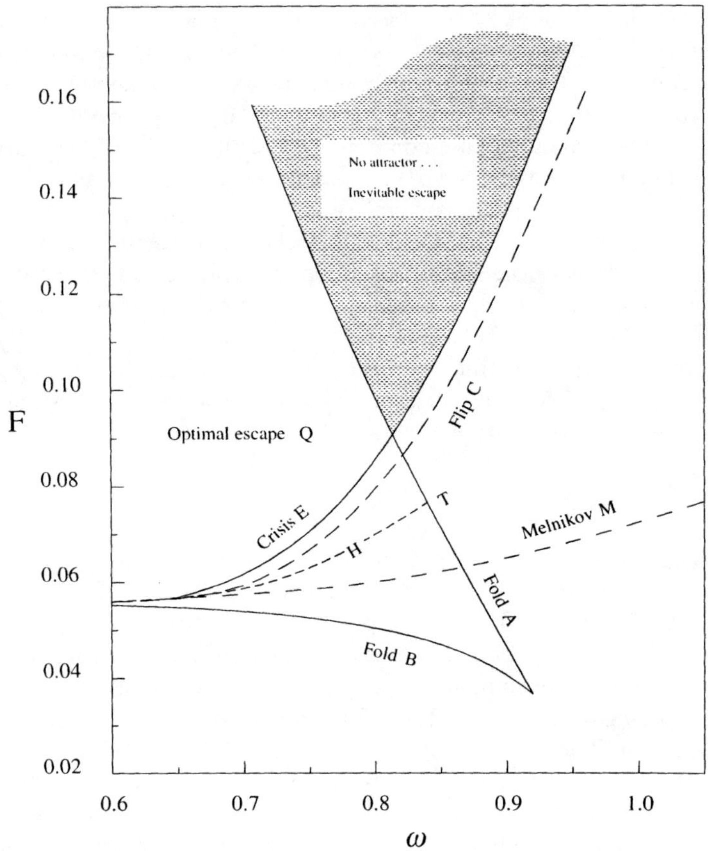

Figure 8. A bifurcation diagram of the escape equation (8) in the (ω, F) parameter space with $\beta = 0.1, \phi = 180°$. Escape from the potential well is inevitable in the shaded region.

that creates a chaotic saddle in a Wada basin boundary. The second one, when the crisis curve E is crossed, is a chaotic boundary crisis with indeterminate outcome in the interval $F \in (0.081, 0.0815)$. The third one, when the fold curve A is crossed, is a saddle-node fold bifurcation in the interval $F \in (0.0815, 0.082)$.

Let $P(0)$ denote the union of the attractor at infinite and the sink cell, $P(1)$ and $P(2)$ denote the two attractors within the potential well, $B(0)$, $B(1)$ and $B(2)$ denote the domains of attraction of $P(0)$, $P(1)$ and $P(2)$, $B(0, 1)$, $B(0, 2)$, $B(1, 2)$ and $B(0, 1, 2)$ denote the boundaries separating the domains.

The first bifurcation that creates a chaotic saddle in a Wada basin boundary occurs when $F \in (0.0705, 0.071)$. When $F = 0.0705$, there exist the two period-1 attractors $P(1)$ and $P(2)$, two chaotic saddles and a period-1 saddle D_r consisting of only four cells. The boundary $B(0, 1)$ is an empty set. The boundary $B(0, 2)$ has only one cell. The boundaries $B(1, 2)$ and $B(0, 1, 2)$ are fractal, and can be further confirmed by numerical simulations. A chaotic saddle is embedded in the fractal boundary $B(0, 1, 2)$, and the other one is in the fractal boundary $B(1, 2)$. The saddle D_r is on the edge of the basin interior $B(1)$. Figure 9(a) shows that the two chaotic saddles are about to touch each other. In Figure 9(a), the stable manifold of the chaotic saddle forms the boundary $B(0, 1, 2)$, and its unstable manifold has three branches extending toward three attractors $P(0)$, $P(1)$ and $P(2)$. Similarly, the stable manifold of the chaotic saddle forms the boundary $B(1, 2)$, and its unstable manifold has two branches extending toward two attractors $P(1)$ and $P(2)$. Due to the collision between the two chaotic saddles, the boundary $B(1, 2)$ is attracted to the chaotic saddle, and then to the attractor $P(0)$ along the unstable manifold of the chaotic saddle.

When $F = 0.071$, there exist two period-1 attractors $P(1)$ and $P(2)$, a chaotic saddle and a period-1 saddle D_r having only four cells. In this case, all the two-domicile transient sets $B(0, 1)$, $B(0, 2)$ and $B(1, 2)$ are empty. The boundary $B(0, 1, 2)$ is fractal and has the Wada property that every point on the common boundary of the basins of two attractors is also on the boundary of the basin of the third attractor, namely, $\overline{B}(0) = \overline{B}(1) = \overline{B}(2) = B(0, 1, 2)$ where $\overline{B}(i)$ is the boundary of the basin $B(i)$. Furthermore, a chaotic saddle is embedded in the Wada basin boundary $B(0, 1, 2)$. Figure 9(b) shows that the chaotic saddle in the Wada basin boundary is created by the collision between the two chaotic saddles in different fractal basin boundaries.

Figure 10 shows the chaotic boundary crisis with indeterminate outcome when $F \in (0.081, 0.0815)$. When $F = 0.081$, the period-1 attractor $P(2)$ becomes a chaotic attractor. The period-1 attractor $P(1)$, the period-1 saddle D_r and the chaotic saddle in the Wada basin boundary are all present. In Figure 10(a), the stable manifold of the chaotic saddle forms the Wada boundary $B(0, 1, 2)$, three branches of its unstable manifold are directed to three attractors $P(0)$, $P(1)$ and $P(2)$. Figure 10(a) shows that the part of the chaotic attractor $P(2)$ touches the stable manifold of the saddle, the basin boundary $B(0, 1, 2)$ in several places at the same time, suggesting that homoclinic tangency occurs. Hence, the points of the regions $B(2)$ and $P(2)$ will eventually touch the chaotic saddle, and then be sent to two remaining attractors $P(0)$ and $P(1)$ along the unstable manifold of the chaotic saddle.

When $F = 0.0815$, there exist the period-1 attractor $P(1)$, the period-1 saddle D_r and a chaotic saddle. The boundary $B(0, 1)$ is fractal. The chaotic boundary crisis with indeterminate outcome is caused by the collision between the chaotic attractor and the chaotic saddle in the Wada basin boundary. The Wada basin boundary $B(0, 1, 2)$ disappears after the collision, and is converted into the fractal basin boundary $B(0, 1)$ resulting in the Wada-fractal boundary bifurcation. After the crisis, the chaotic saddle lies in the fractal basin boundary $B(0, 1)$ of

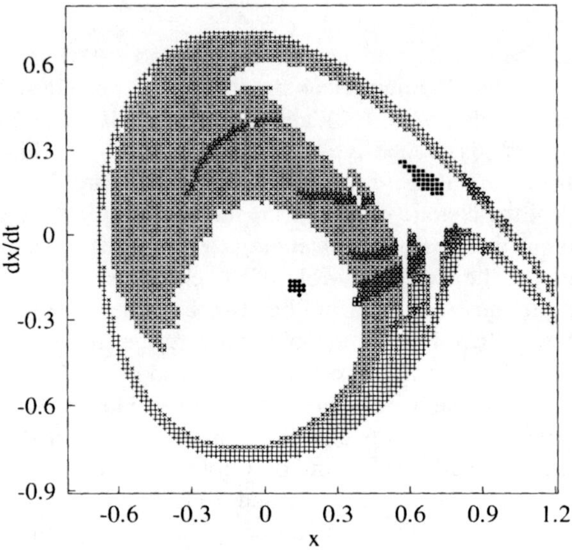

(a)

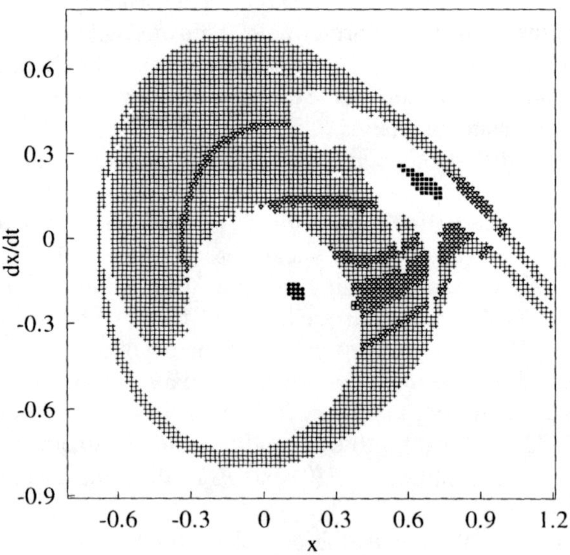

(b)

Figure 9. A bifurcation of the escape equation (8) that creates a chaotic saddle in the Wada basin boundary with $\omega = 0.827$, $\beta = 0.1$, $\phi = 180°$. (a) $F = 0.0705$; (b) $F = 0.071$. The symbol • denotes the attractor $P(1)$; ■, the attractor $P(2)$; □, the period-1 saddle D_{r}; ▽, the chaotic saddle on the boundary $B(0, 1, 2)$; △, the chaotic saddle on the boundary $B(1, 2)$; ·, the boundaries $B(0, 1)$ and $B(0, 2)$; ×, the boundary $B(1, 2)$; +, the boundary $B(0, 1, 2)$.

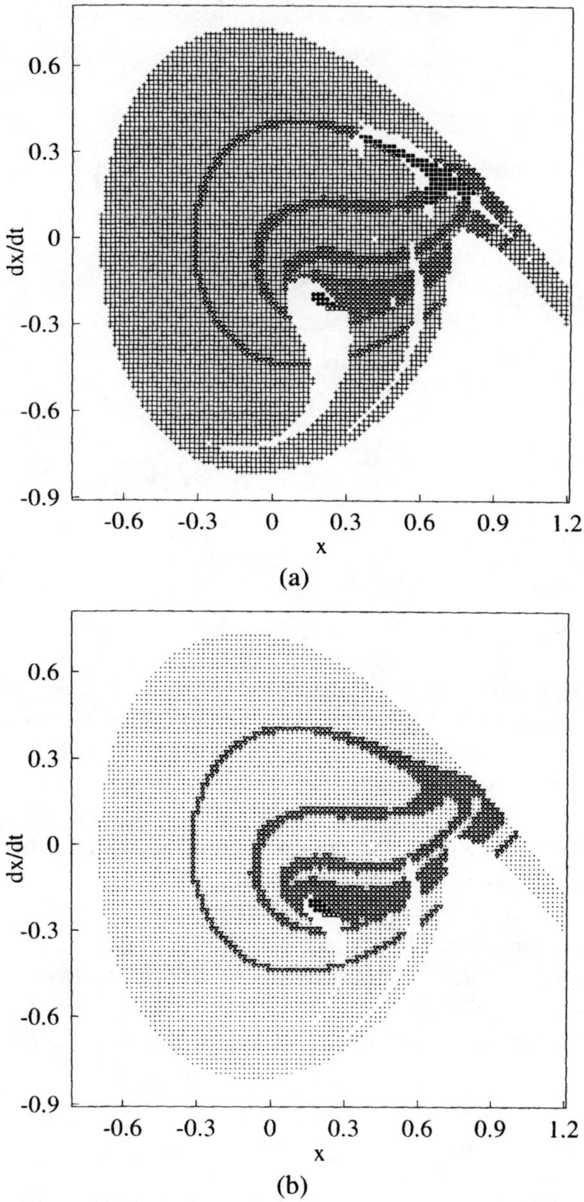

Figure 10. A chaotic boundary crisis with indeterminate outcome for the escape equation (8) with $\omega = 0.827$, $\beta = 0.1$, $\phi = 180°$. Legends are the same as in Figure 9. (a) $F = 0.081$; (b) $F = 0.0815$.

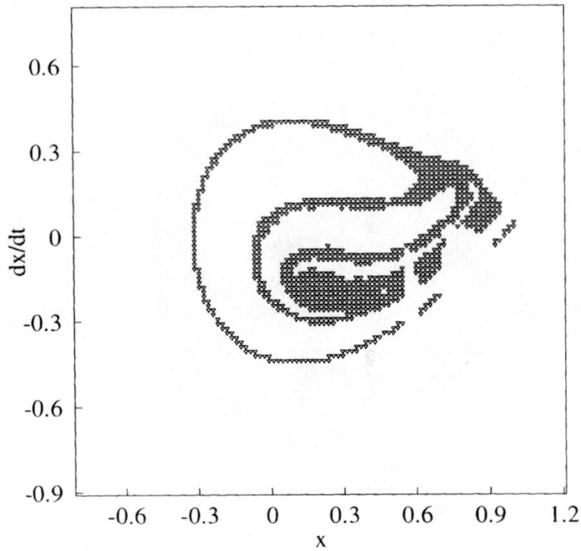

Figure 11. A chaotic saddle in the basin interior of the attractor at infinity after a final saddle-node bifurcation for the escape equation (8) with $\omega = 0.827$, $\beta = 0.1$, $\phi = 180°$ when $F = 0.082$. Legends are the same as in Figure 9.

the two remaining attractors $P(0)$ and $P(1)$, which implies that the outcome of the chaotic boundary crisis is indeterminate, namely, after the crisis, which of the two remaining attractors the orbit goes to is indeterminate. Figure 10(b) shows the phase portrait of the system after the chaotic boundary crisis.

Figures 10(b) and 11 show the saddle-node fold bifurcation when $F \in (0.0815, 0.082)$. When $F = 0.082$, we find only one attractor $P(0)$ and a chaotic saddle. The chaotic saddle is in the basin interior $B(0)$.

In summary, when the system has three or more attractors, there may exist the Wada basin boundary. The chaotic saddle residing in the Wada basin boundary can collide with a chaotic attractor, leading to a chaotic boundary crisis with indeterminate outcome. The chaotic saddle in the Wada basin boundary is created by the collision between two chaotic saddles in different fractal basin boundaries.

3.4. Double crises

Physical systems usually contain several control parameters. When more than one parameters are changed simultaneously, a great variety of new bifurcation phenomena may appear. In two-parameter dissipative systems, for example, one of the most interesting phenomena is the coincidence and interaction of two distinct crises. Such an event has been called double crises and was first highlighted by Gallas, Grebogi and Yorke (1993) and Stewart, Ueda, Grebogi and Yorke (1995), where a pattern involving a boundary crisis, an interior crisis, and a basin metamorphosis was examined in the laser ring cavity map and the Hénon map. Another early example was due to Rössler, Stewart and Wiesenfeld (1990) who studied a codimension two bifurcation of chaotic attractors using the forced damped pendulum with a DC bias.

Here, we study double crises or codimension two bifurcations of a sinusoidally forced damped pendulum,

$$\frac{d^2\theta}{dt^2} + \kappa\frac{d\theta}{dt} + \sin\theta = A\sin\omega t + C, \tag{9}$$

where θ is the angle from the vertical position of the pendulum, ω is the forcing frequency, A is the amplitude and C is a DC bias. Rössler, Stewart and Wiesenfeld (1990) have studied bifurcations of the pendulum equation (9) in the (C, A) parameter space where $-0.002 < C < 0.002$ and $0.87 < A < 0.88$, while $\kappa = 0.5$ and $\omega = 0.55$. A codimension two bifurcation of chaotic attractors was found by numerical simulations, and was presented in Figure 4 of Rössler, Stewart and Wiesenfeld (1990).

With the GCM method, we can confirm the existence of codimension two bifurcations reported in Rössler, Stewart and Wiesenfeld (1990). Furthermore, we provide a global view of crises, namely the collision of a chaotic attractor with a regular saddle in the smooth basin boundary (a regular boundary crisis) or with a chaotic saddle in the basin interior (a chaotic interior crisis), and further demonstrate that Rössler's codimension two point is the vertex where two regular boundary crises and two chaotic interior crises coincide. When applying the GCM method, we discretize the domain $\mathbf{D} = \{-3.14 \leqslant \theta \leqslant -1.57; -0.58 < \dot{\theta} \leqslant 0\}$ into 105×105 cells. The 25×25 interior points are sampled within each cell.

Figure 12 shows two curves of a regular boundary crisis $RBC1$ and $RBC2$ and two curves of a chaotic interior crisis $CIC1$ and $CIC2$, corresponding to two different chaotic attractors. The four curves meet at a point in the (C, A) parameter space, approximately located at $(C, A) \approx (0, 0.875)$. This point is called the vertex of double crises.

In the quadrant bounded by $RBC1$ and $RBC2$, there are two coexisting chaotic attractors a_1 and a_2. In the quadrant bounded by $RBC1$ and $CIC2$, there is only one chaotic attractor a_2. In the quadrant bounded by $RBC2$ and $CIC1$, there is only one chaotic attractor a_1. In the quadrant bounded by $CIC1$ and $CIC2$, there is a single large chaotic attractor a_{large}. The disappearance–appearance of the attractor a_1 or a_2 occurs when $RBC1$ or $RBC2$ is crossed, respectively. The explosion–implosion of the attractor a_1 or a_2 occurs when $CIC1$ or $CIC2$ is crossed, respectively. When $C \neq 0$, as we vary A to cross these four curves, we find boundary crises and interior crises (Hong and Xu, 2003).

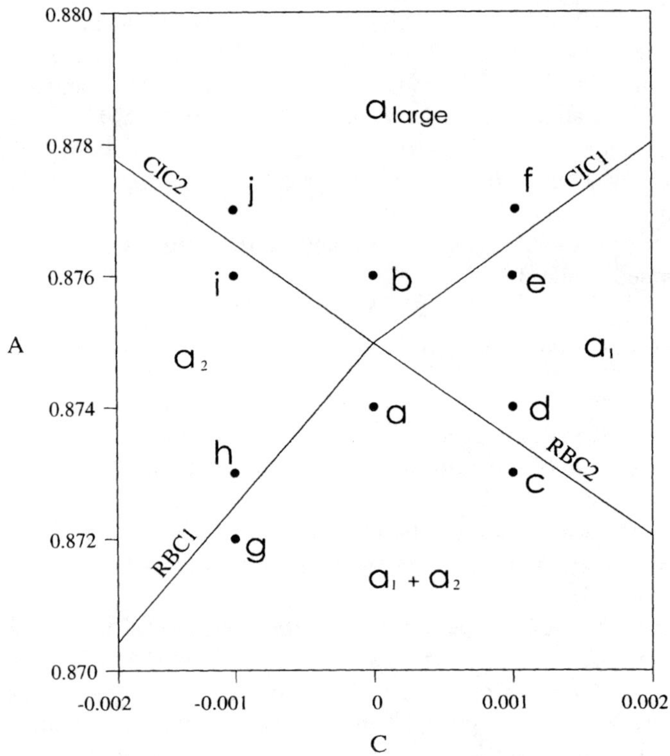

Figure 12. The vertex of double crises of the forced pendulum equation (9) with $\kappa = 0.5$, $\omega = 0.55$ in the (C, A) parameter space.

We find double crises when $C = 0$ and $A \in (0.874, 0.876)$. The result is shown in Figure 13 where two chaotic attractors a_1 and a_2 simultaneously collide with a period-1 saddle in their smooth basin boundary, and merge to form one large chaotic attractor a_{large}.

4. Bifurcations of nonlinear systems with small random disturbances

In designing mechanical and structural systems it is important to take into account system uncertainties that may come from system parameters and external loadings. These system uncertainties will undoubtedly affect the dynamic response of the system, its bifurcations, and ultimately the reliability of the system. Much attention has been paid to studying systems with uncertainties in various areas (Brambilla, Lugiato, Strini and Narducci, 1986; Chiang, Dong and Wong, 1987;

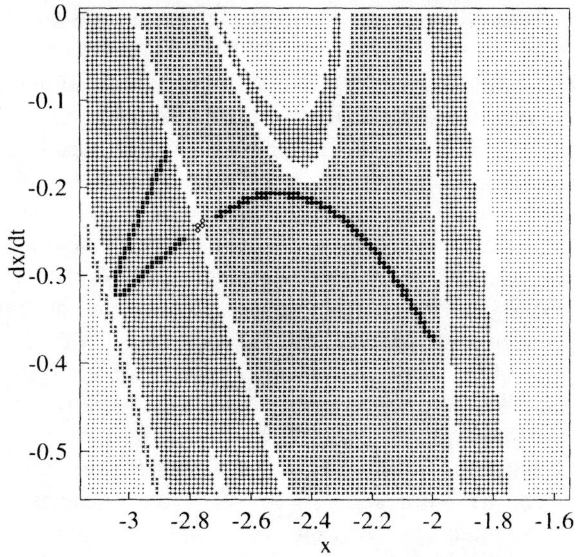

(a)

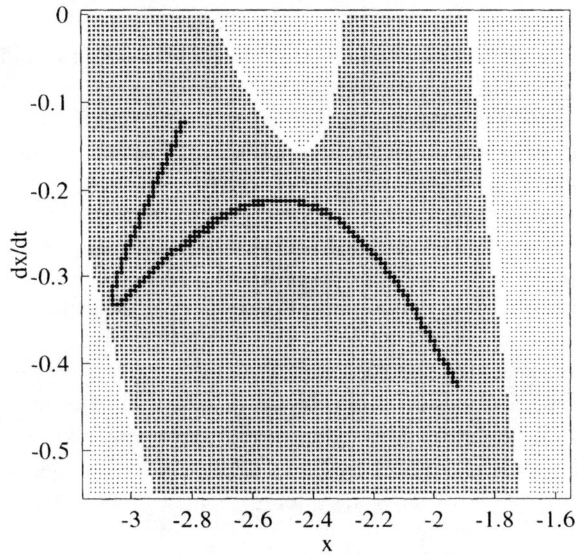

(b)

Figure 13. Double crises of the forced pendulum equation (9) with $\kappa = 0.5$, $\omega = 0.55$ when $A \in (0.874, 0.876)$ and $C = 0$. The symbol ■ denotes the chaotic attractor a_1; +, its domain of attraction; •, the chaotic attractor a_2 and the large chaotic attractor a_{large}; ×, their domains of attraction; ·, the domain of attraction of the sink cell; ◊, the period-1 saddle on the basin boundary. The blank space stands for the boundary.

Crutchfield and Huberman, 1980; Crutchfield, Farmer and Huberman, 1982; Dong, Chiang and Wong, 1987; Ibrahim, 1987; Kapitaniak, 1988; Kotulski and Sobczyk, 1987; Udwadia, 1987a, 1987b). When the uncertainty is random in nature, the theory of random vibration may be applied to the system (Caughey, 1971; Lin, 1976; Nigam, 1983; Soong, 1973). The presence of noise in a nonlinear system greatly enhances the difficulty of the bifurcation problem. There are mainly two kinds of definitions and related analyses for stochastic bifurcation available. One is focused upon the abrupt change of shape for stationary probability density function, such as the change of shape of a stationary probability density function, for example, a mono-peak function is changed into a double-peak one at a critical bifurcation parameter value (Arnold, 1998). The other is focused on the sudden change of the sign of the largest Lyapunov exponent of the system (Baxendale, 1986). Sometimes these two analyses may lead to different results. For instance, Baxendale (1986) provided an example in which the largest Lyapunov exponent does change its sign, while the shape of a stationary probability density does not depend on the bifurcation parameter. On the other hand, Crauel and Flandoli (1998) presented a system in which the largest Lyapunov exponent does not change its sign, while the shape of a stationary probability density does change from a mono-peak one into a double-peak one at a critical parameter value. According to Meunier and Verga (1988), it is difficult to describe the true change of topological property of a stochastic system simply based on the shape change of the stationary probability density function.

When the system is random, the global analysis of the stochastic attractors must be used to describe stochastic bifurcations and crises. Within the context of the GCM method, a stochastic attractor is an invariant set of persistent cells. Stochastic bifurcations are described by topological changes of the persistent groups. Note that the global analysis is quite often the ultimate task of a complete study of a nonlinear system. Global analysis of a system having multiple stable steady-state solutions may provide important information for the reliability design of the system.

In this section we study some examples of nonlinear systems with either parametric or external random excitations by using the GCM method. We have chosen the systems whose deterministic dynamics is reasonably well known. By comparing the deterministic solutions with the solutions of the system with random excitations, we are able to show the effects of random excitations on the system response and its bifurcation.

In computing the one step transition probability matrix, the Monte Carlo simulation method is used to generate random uncertainties. We have determined the number of samples of random uncertainties in the computation so that by adding more samples no appreciable change in the numerical results of the one step transition probability matrix can be found. As can be expected, when the intensity of uncertainties is low, the number of samples for computing the one step transition probability matrix need not be excessively large. This point needs, of course, careful interpretation and understanding. Even when the intensity of uncertainties is low, large deviations from deterministic responses are possible. To capture these happenings of large deviations, a large number of samples is necessary. However, when the uncertainties are low in intensity, these large deviations are very rare events with extremely low probabilities. The details of computational procedures of the GCM method, and the Monte Carlo simulation of uncertainties in connection with the GCM method can be found in Chiu and Hsu (1986) and Sun and Hsu (1988).

4.1. Logistic map with random coefficients

Consider the one-dimensional logistic map

$$x(n+1) = sx(n)\big[1 - x(n)\big], \quad x(n) \in [0, 1]. \tag{10}$$

This map has been thoroughly and frequently studied in the literature when the system is deterministic (Feigenbaum, 1978; May, 1976). One of the reasons for the popularity of this map is perhaps that it has many interesting dynamic features universal to one-dimensional nonlinear maps. In Crutchfield, Farmer and Huberman (1982), the influence of fluctuations modeled as additive noise excitation of the map in equation (10) on the period doubling bifurcations to chaos was studied. Here we assume that the parameter s is random with a given probability distribution.

We present two bifurcation diagrams of the map for the mean value $\bar{s}$ ranging from 2.5 to 4.0, and the standard deviation σ_s of s being 0.001 and 0.01, respectively. It can be shown that when the standard deviation σ_s is small, the results to be presented are insensitive to the distribution of s. For this reason, we shall only report the results for uniform distribution. However, it is of interest to note that any distribution of s can be easily incorporated in the GCM method.

In the computation of bifurcation diagrams, we divide the interval [0, 1] into 2000 uniform cells. We uniformly sample five points from each cell, and generate twenty uniform random numbers of s with the mean value $\bar{s}$ and the standard deviation σ_s by using a linear congruential uniform random number generator (Knuth, 1969). Thus, we have a total of a hundred sample image points of each cell. These image points are used to compute the transition probabilities (Chiu and Hsu, 1986; Hsu, 1981, 1987; Hsu and Chiu, 1986; Sun and Hsu, 1988). The persistent groups representing stable steady-state motions of the system are plotted in Figures 14 and 15 as a function of the mean value $\bar{s}$. These diagrams should be compared with the bifurcation diagram of the deterministic system (Crutchfield, Farmer and Huberman, 1982; Hsu, 1987). In this example the period of a persistent group is literally the number of bands of the group. It is observed that the uncertainty in s blurs the bifurcation diagram, reduces the observability of periodicity of

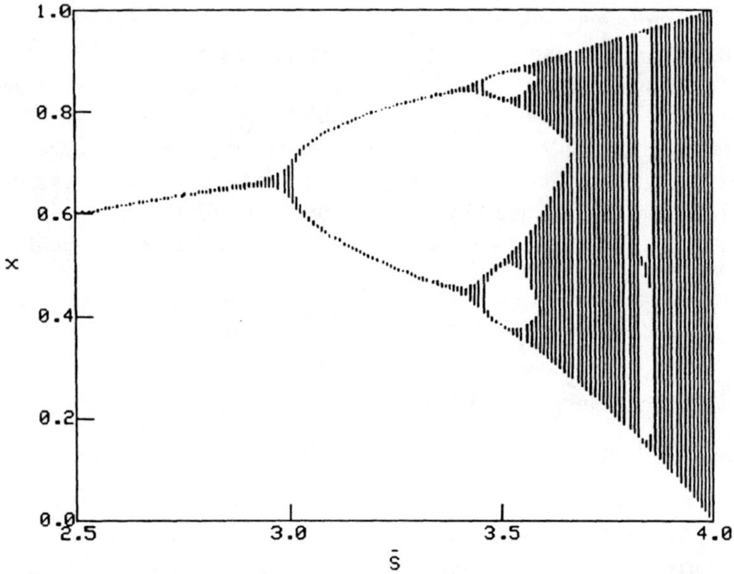

Figure 14. The persistent groups representing the steady-state solutions of the logistic map (10) with the mean value $\bar{s}$ being from 2.5 to 4.0, and the standard deviation $\sigma_s = 0.001$.

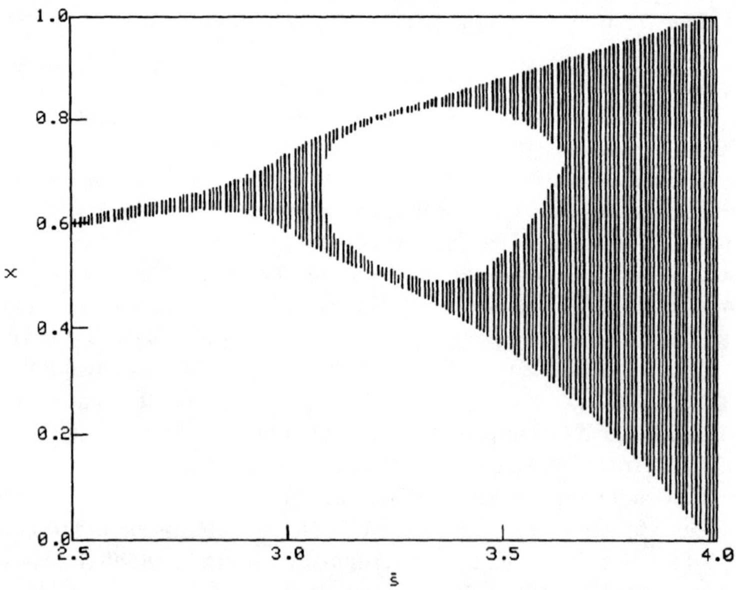

Figure 15. The persistent groups representing the steady-state solutions of the logistic map (10) with the mean value $\bar{s}$ being from 2.5 to 4.0, and the standard deviation $\sigma_s = 0.01$.

steady-state solutions, and induces early transition to "chaos" in the bifurcation sequences. These observations are qualitatively the same as those in Crutchfield, Farmer and Huberman (1982).

It is of special interest to note that when $\bar{s}$ is equal to a bifurcation value of the deterministic system, the blurring effect of the uncertainty in s is more prominent. This suggests that the system is very sensitive to uncertainties when $\bar{s}$ is equal to a bifurcation value of the deterministic system.

4.2. A two-dimensional random map

Next, we study nonlinear oscillations of a rigid, damped bar hinged at one end and subjected to a random impulsive load at the other free end. In general, the random impulsive load can have random amplitude and random arriving time with certain correlation between the two. Here we assume that the impulsive load is periodic with independent random amplitude. By integrating the equation of motion over a period, we obtain a two-dimensional nonlinear map (Hsu, 1987)

$$\begin{aligned} x_1(n+1) &= x_1(n) - C_1\alpha(n)\sin\big[x_1(n)\big] + C_1x_2(n), \\ x_2(n+1) &= -D_1\alpha(n)\sin\big[x_1(n)\big] + D_1x_2(n), \end{aligned} \tag{11}$$

where

$$C_1 = \frac{1-\mathrm{e}^{-2\mu}}{2\mu}, \qquad D_1 = \mathrm{e}^{-2\mu}, \tag{12}$$

where μ is a damping factor. x_1 and x_2 are respectively the dimensionless angular displacement and angular velocity of the bar. $\alpha(n)$ represents the random amplitude of the impulsive load at the nth period. The deterministic map (11) with $\alpha(n) = \text{const}$ has been studied in Hsu (1987), and can be shown to be related to the Zaslavskii map in physics, which is used to model the motion of charged particles or the Fermi acceleration problem. It is known that for a given damping factor μ, the deterministic system undergoes the period doubling bifurcations to chaos as the amplitude of the impulsive load varies. In this example we set $\mu = 0.1\pi$, and assume that $\alpha(n)$ are independent random variables of uniform distribution with constant mean value and constant standard deviation.

We use 16,215 (115 × 151) uniform cells to cover the domain $\mathbf{D} = \{-\pi < x_1 < \pi; -7.0 < x_2 < 7.0\}$. Nine points are uniformly sampled from each cell, and twenty random samples of $\alpha(n)$ are generated, resulting in 180 sample image points of each cell. We present a case with the mean value $\bar{\alpha}$ of $\alpha(n)$ equal to 5.7. It is known that for $\bar{\alpha} = 5.7$, the deterministic system has a stable P-2 solution, and two stable advancing P-1 solutions (Hsu, 1987). Some of the numerical results are presented in Figures 16–18. It is found that the general observations of the effects of uncertainties on the dynamic response and the bifurcation property of the system made in the earlier example still hold in this example.

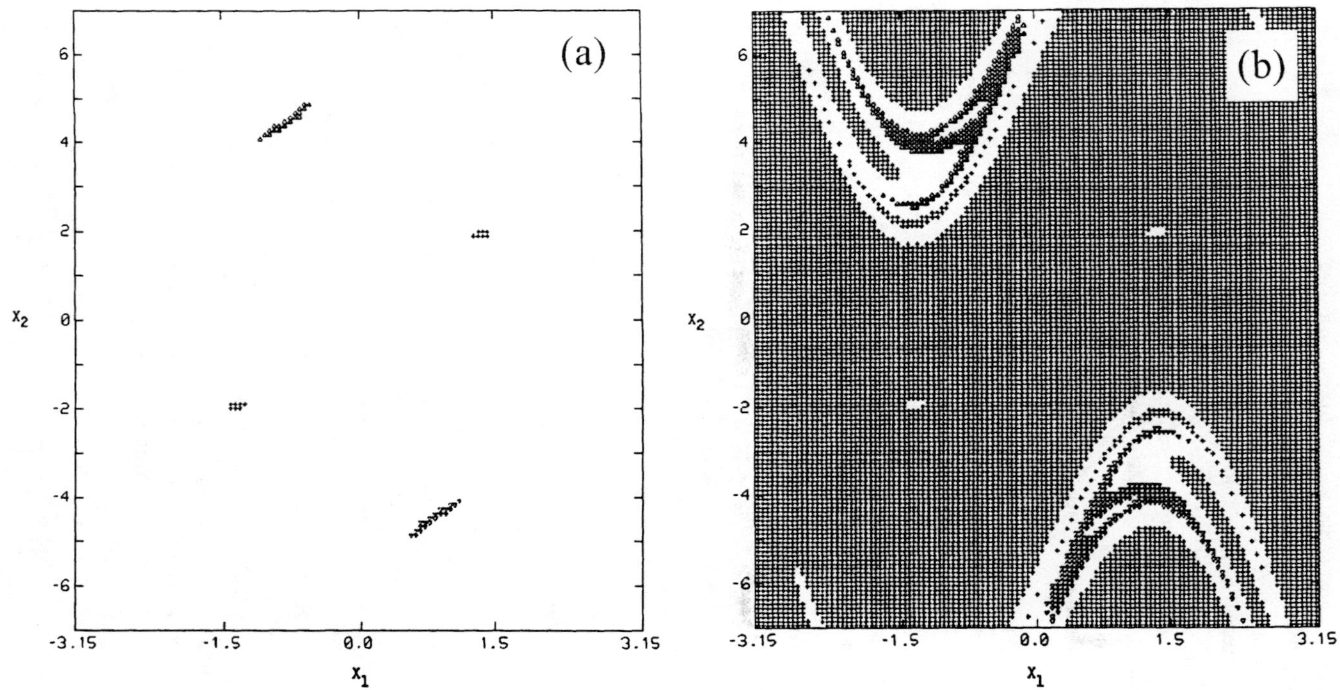

Figure 16. Deterministic case of the system (11) with $\bar{\alpha} = 5.7$ and $\sigma_\alpha = 0.0$. (a) The persistent groups; (b) the 100% domains of attraction. The symbol + denotes the persistent group of period two representing the P-2 solution in (a) or its domain in (b); Δ, the persistent group of period one representing the upper advancing P-1 solution in (a) or its domain in (b); ∇, the persistent group of period one representing the lower advancing P-1 solution in (a) or its domain in (b).

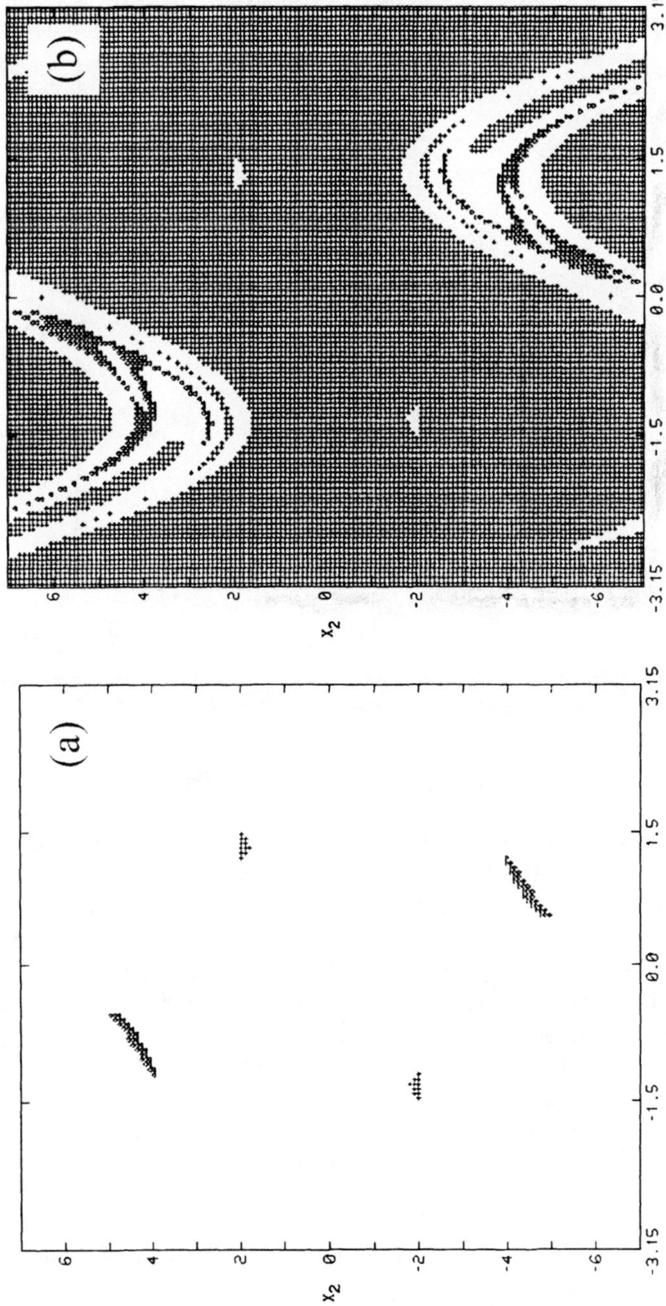

Figure 17. Solutions of the system (11) with $\bar{\alpha} = 5.7$ and $\sigma_\alpha = 0.01$. The legends are the same as in Figure 16.

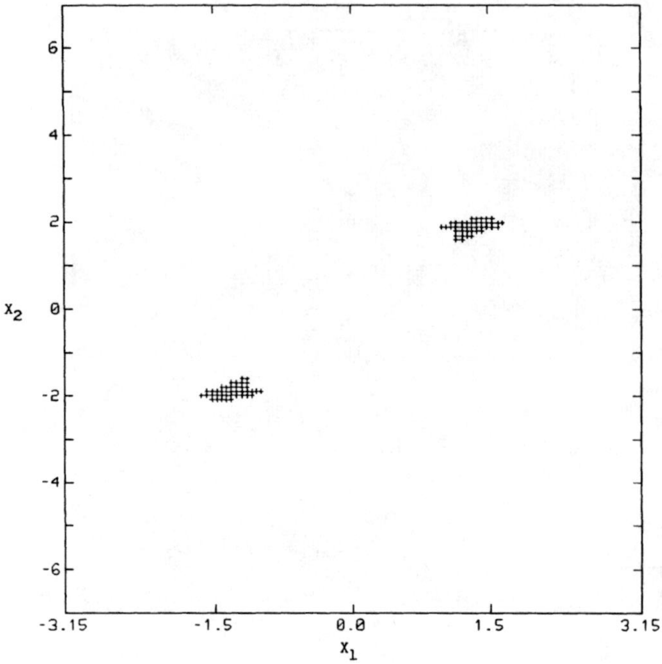

Figure 18. The persistent group of period two representing P-2 solution of the system (11) with $\bar{\alpha} = 5.7$ and $\sigma_\alpha = 0.05$. The legends are the same as in Figure 16.

From extensive numerical studies of the case $\bar{\alpha} = 5.7$ with various noise levels of σ_α, we find that the stable steady-state solutions, such as two advancing P-1 solutions in this example, "protected" by a thinner layer of the 100% domain of attraction are more "vulnerable" to noise fluctuations of the system. At $\sigma_\alpha = 0.05$, the two advancing P-1 solutions go unstable and the P-2 solution remains stable. A stochastic bifurcation takes place. This observation is of importance to reliability design of mechanical and structural systems. It suggests that not only should the local stability of steady-state solutions, but also the global domains of attraction be taken into account in the reliability design of the system if it has multiple steady-state solutions and is expected to endure parametric and external random excitations and uncertainties.

4.3. *Duffing oscillator with small random excitations*

We now study a Duffing oscillator excited by a sinusoidal forcing and random fluctuations. The system is governed by an ordinary differential equation

$$\ddot{x} + k\dot{x} + \alpha x + x^3 = B\cos t + w(t), \tag{13}$$

where $w(t)$ is a Gaussian white noise such that

$$E\left[w(t)\right] = 0, \qquad E\left[w(t)w(t+\tau)\right] = \sigma_w^2 \delta(\tau). \tag{14}$$

This system, in the absence of $w(t)$, was studied in Hsu and Chiu (1987). It is found that for $k = 0.25$ and $B = 8.5$, the system may have a third-order subharmonic response, or a strange attractor, or both, depending on the value of α. In the case of coexistence of the third-order subharmonic response and the strange attractor, the domains of attraction and the basin boundaries are obtained by Hsu and Chiu (1987) using the compatible cell mapping method. Here we use the GCM method to study the system with random fluctuation $w(t)$. We also pick $k = 0.25$ and $B = 8.5$, and set $\alpha = -0.05$ and 0.02. These two α values correspond to the case of coexistence of the third-order subharmonic response and the strange attractor. We study the influence of random fluctuation on these two solutions, their domains of attraction and bifurcations.

The domain $\mathbf{D} = \{-1.5 < x < 4.2; -3.0 < \dot{x} < 6.0\}$ is divided into 151×151 cells. Nine points are uniformly sampled from each cell, and ten random sample functions of $w(t)$ are simulated by using the method due to Shinozuka (1972) in the same way as described in Chiu and Hsu (1986). A total of ninety sample trajectories of duration $T = 2\pi$ out of each cell are integrated from equation (13) by using the fourth-order Runge–Kutta method. Some of the results are plotted in Figures 19–22.

It is found that as the noise level increases, the transient group of multiple-domicile cells and the persistent groups become larger. As the noise level increases further to a critical value, the 100% domain of attraction of one of these two solutions is pierced somewhere by the solution, and that solution becomes unstable so that the persistent group representing that solution disappears. For $\alpha = -0.05$ the strange attractor becomes unstable first, while for $\alpha = 0.02$ the third-order subharmonic response becomes unstable first. It is seen that the noise fluctuation of the system does not always promote chaotic motion of the system. As has been discussed in the previous examples, when the system experiences a sequence of period doubling bifurcations to chaos, the noise fluctuation of the system promotes early transition to chaos. In the case of coexistence of multiple steady-state solutions, including chaotic motions or not, the noise fluctuation first destabilizes the solution with "weaker protection", meaning that the solution is closer to the boundary of the domain of attraction than other coexisting solutions. The above discussion is physically and intuitively quite reasonable, once the picture is cast in its global form.

4.4. Noisy crisis in a twin-well Duffing system

We now reconsider crisis of nonlinear dynamical systems subject to small random noise excitations. In the following example, the GCM method with digraph is

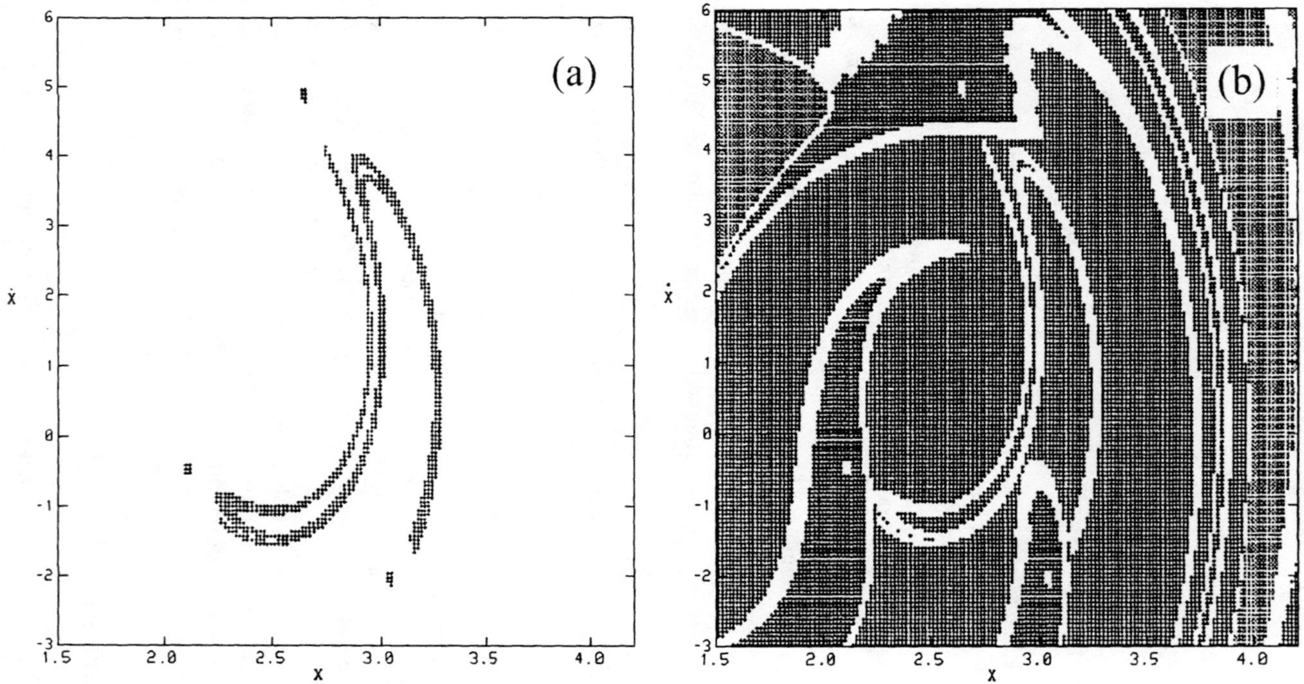

Figure 19. The Duffing oscillator (13) excited by a sinusoidal forcing and random fluctuations with $\alpha = -0.05$ and $\sigma_w = 0.01$. (a) The persistent groups; (b) the domains of attraction. The symbol + denotes the persistent group of period one representing the strange attractor or its domain; Δ, the persistent group of period three representing the third-order subharmonic response or its domain; $\times$, the domain of sink cell.

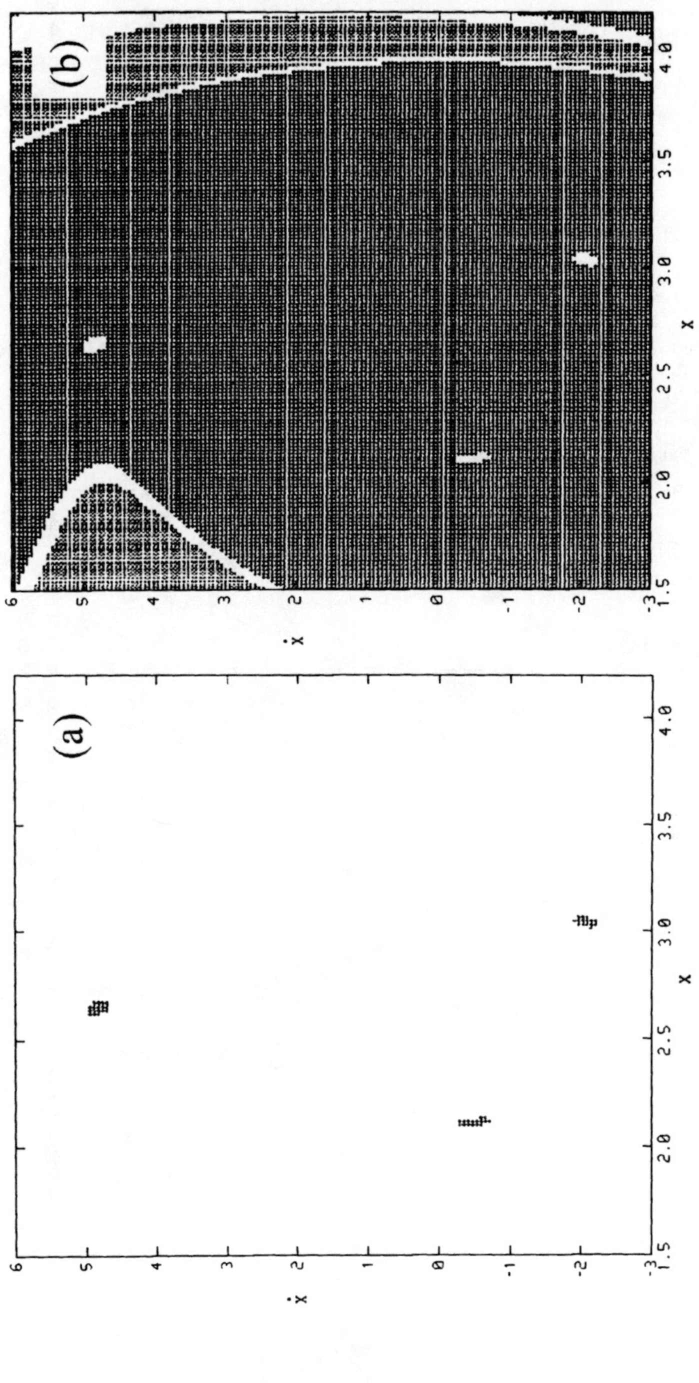

Figure 20. The Duffing oscillator (13) excited by a sinusoidal forcing and random fluctuations with $\alpha = -0.05$ and $\sigma_w = 0.05$. The legends are the same as in Figure 19.

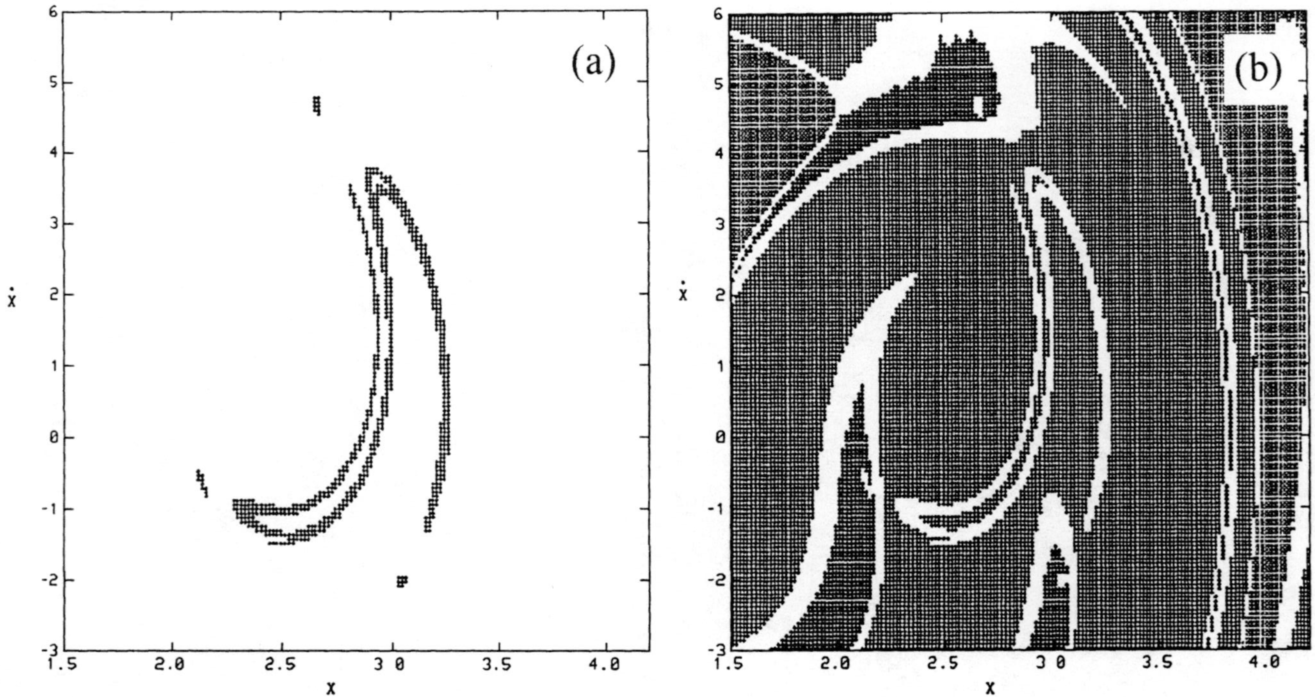

Figure 21. The Duffing oscillator (13) excited by a sinusoidal forcing and random fluctuations with $\alpha = 0.02$ and $\sigma_w = 0.01$. The legends are the same as in Figure 19.

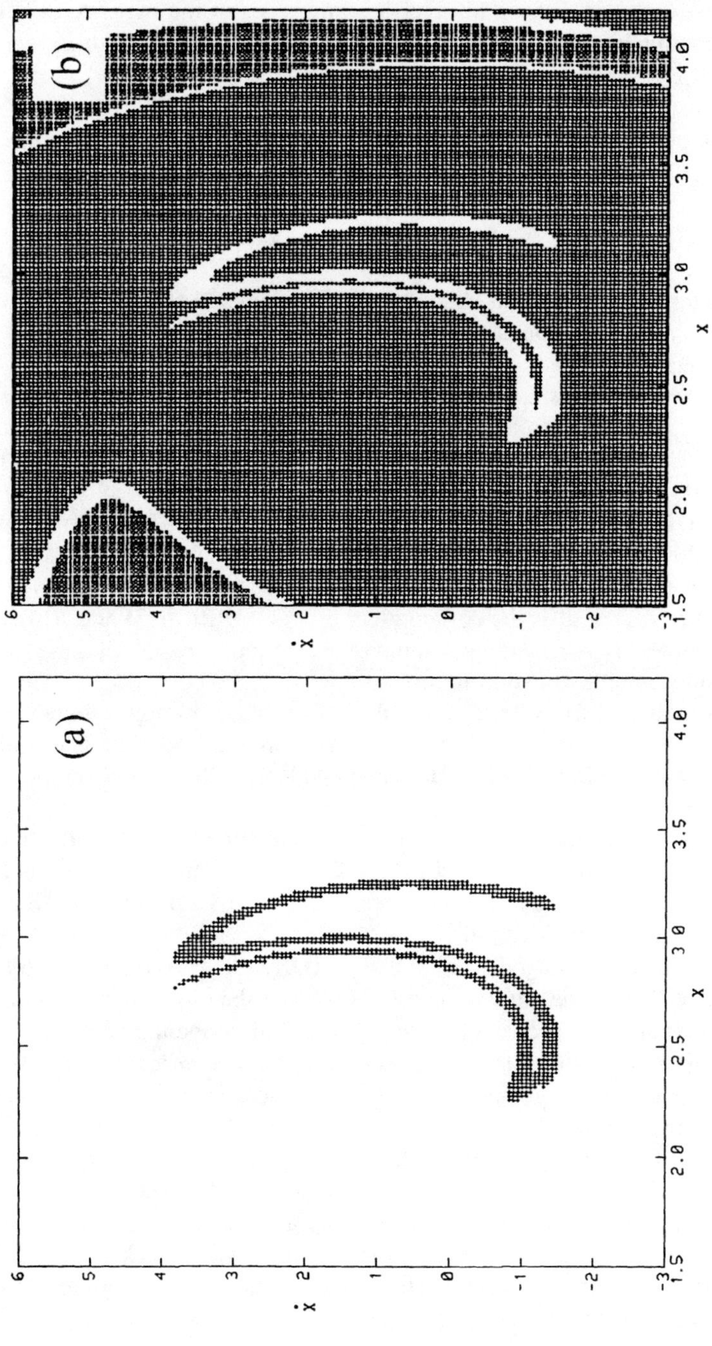

Figure 22. The Duffing oscillator (13) excited by a sinusoidal forcing and random fluctuations with $\alpha = 0.02$ and $\sigma_w = 0.05$. The legends are the same as in Figure 19.

applied to investigate stochastic bifurcations and crisis of the twin-well Duffing system subject to a harmonic excitation in the presence of noise (Hsu, 1995),

$$\ddot{x} + \alpha\dot{x} - \beta x + \delta x^3 = F \cos \omega t + \sigma\xi(t), \tag{15}$$

where $\xi(t)$ is the zero-mean univariant Gaussian process and $\sigma > 0$. In the numerical example, we have taken $\alpha = 0.11$, $\beta = 0.5$, $\delta = 0.5$, $F = 0.1405$ and $\omega = 0.68$.

A 400×400 cell space is used to partition the domain $\mathbf{D} = \{-2 \leqslant x \leqslant 2; -2 \leqslant \dot{x} \leqslant 2\}$. The 15×15 points are sampled from each cell. The 10 random sample trajectories are generated starting from each sampled point leading to $15 \times 15 \times 10 = 2250$ image points. Every trajectory is integrated by means of the 6th-order Runge–Kutta method for one period.

For the deterministic system when $\sigma = 0.0$, we find that there are three persistent self-cycling sets (attractors) denoted by $A(1)$, $A(2)$ and $A(3)$, five transient self-cycling sets (saddles) denoted by $S(1)$, $S(2)$, $S(3)$, $S(1,2)$ and $S(1,2,3)$, four single-domicile transient cell sets (domains of attraction) denoted by $B(1)$, $B(2)$, $B(3)$ and $B(4)$ and four multiple-domicile transient cell sets (the basin boundaries) denoted by $B(1,2)$, $B(1,2,3)$, $B(1,2,4)$ and $B(1,2,3,4)$. $B(1,2)$ is the boundary between $B(1)$ and $B(2)$, $B(1,2,3)$ is the co-boundary of $B(1)$, $B(2)$ and $B(3)$ and $B(1,2,3,4)$ is the co-boundary of $B(1)$, $B(2)$, $B(3)$ and $B(4)$. The sink cell is denoted by $A(4)$ and its domain of attraction by $B(4)$. All these global solution features are illustrated in Figure 23.

When σ is changed from 0.0 to 0.01, the system goes through a sequence of bifurcations. These include boundary and interior crises. More details of the bifurcation study can be found in Xu, He, Fang and Rong (2005). As examples, we present two cases.

First, when σ is changed from 0.003 to 0.004, the attractor $A(2)$ collides with the saddle $S(2,3)$ on the boundary $B(2)$. $A(2)$ and $B(2)$ are destroyed, and the attraction basin $B(3)$ becomes a bigger one. A boundary crisis occurs. Figures 24 and 25 describe this bifurcation.

Second, when σ is changed from 0.009 to 0.01, the attractor $A(3)$ collides with the saddle $S(3)$ in the attraction basin absorbing the saddle to form a bigger attractor. This is an interior crisis. Figures 26 and 27 describe this result.

In summary, we have found that system uncertainties decrease the predictability of the system and blur the geometrical structure of solutions of the deterministic system by making more multiple-domicile cells near the boundary of domains of attraction and producing larger persistent groups representing stable steady-state solutions of the system. The noise fluctuations of the system induce early transition to "chaos" for the system undergoing a sequence of period doubling bifurcations to chaos, and destabilize first the solution with "weakest protection" for the system having multiple stable steady-state solutions. Random excitations also cause collisions of attractors with saddles leading to boundary and interior

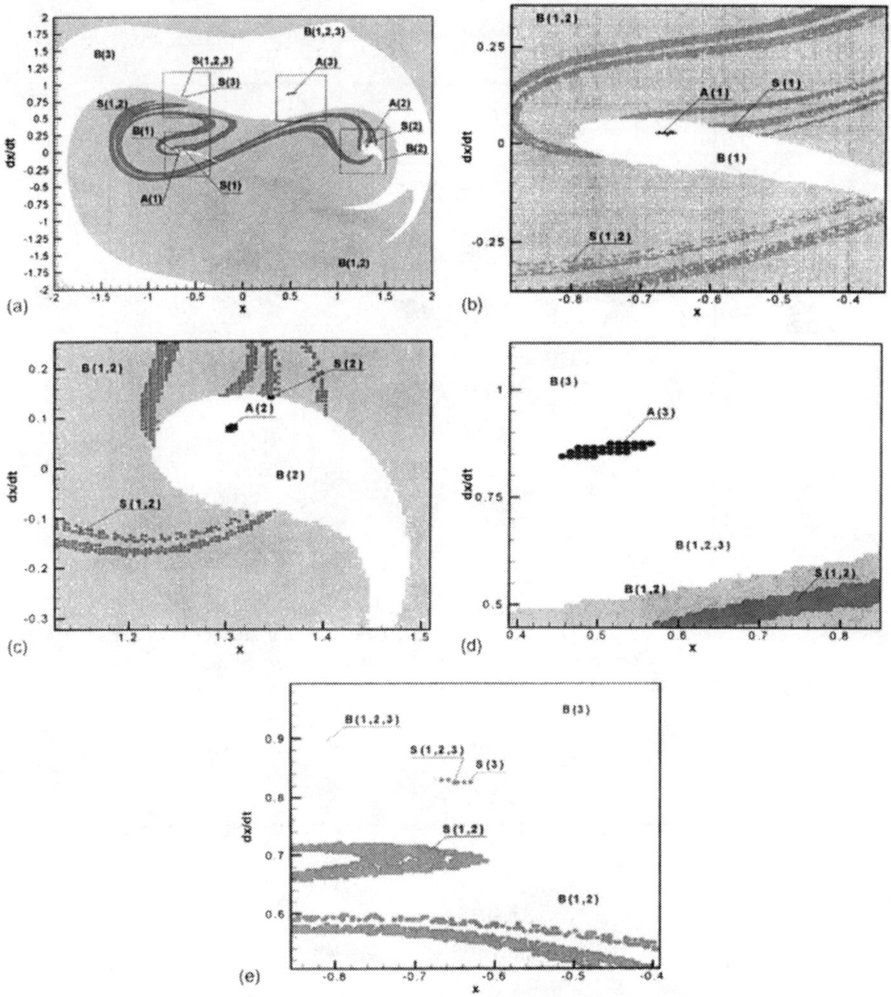

Figure 23. (a) The attractors, the attraction basins and the saddles of the system (15) with $\sigma = 0.0$. (b) The attractor $A(1)$, attraction basin $B(1)$, the saddle $S(1)$, the chaotic saddle $S(1, 2)$ and the boundary $B(1, 2)$. (c) The attractor $A(2)$, attraction basins $B(2)$ and $B(3)$, the saddle $S(2)$, the chaotic saddle $S(1, 2)$ and the boundaries $B(1, 2)$ and $B(1, 2, 3)$. (d) The attractor $A(3)$, attraction basin $B(3)$ and the boundaries $B(1, 2)$ and $B(1, 2, 3)$. (e) The saddles $S(3)$ and $S(1, 2, 3)$, the chaotic saddle $S(1, 2)$ and the boundaries $B(1, 2)$ and $B(1, 2, 3)$. Courtesy of Xu, He, Fang and Rong (2005).

crises. We have demonstrated that the GCM method can effectively describe the crises or stochastic bifurcations of nonlinear dynamical systems under random excitations.

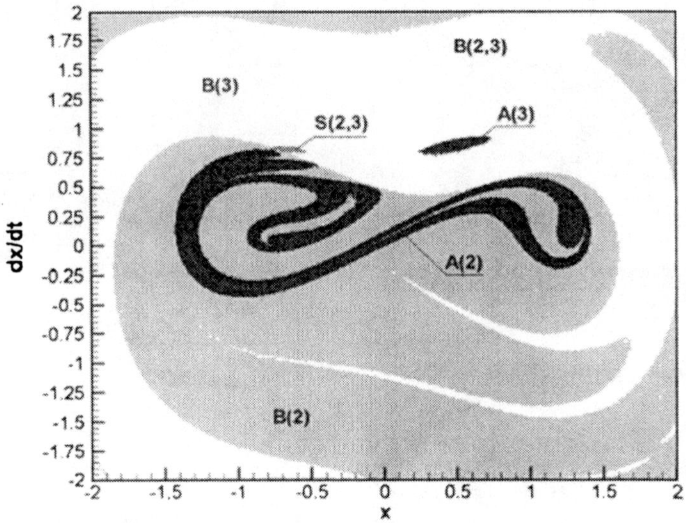

Figure 24. The attractors, the attraction basins and the saddles of the system (15) with $\sigma = 0.003$. Courtesy of Xu, He, Fang and Rong (2005).

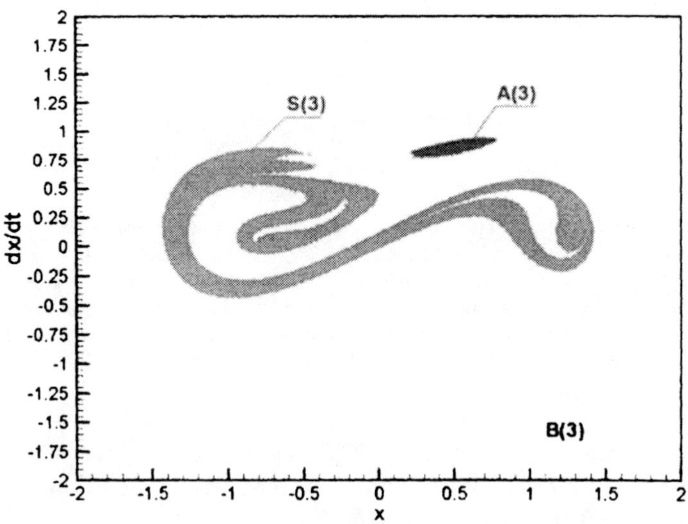

Figure 25. The attractors, the attraction basins and the saddles of the system (15) with $\sigma = 0.004$. Courtesy of Xu, He, Fang and Rong (2005).

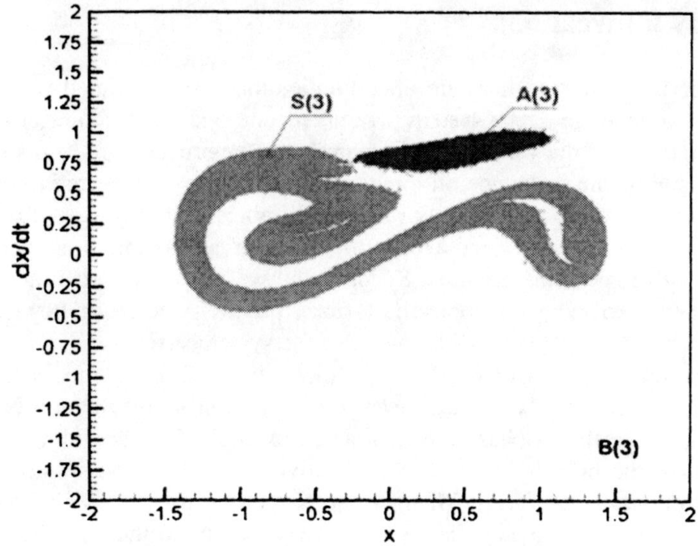

Figure 26. The attractors, the attraction basins and the saddles of the system (15) with $\sigma = 0.009$. Courtesy of Xu, He, Fang and Rong (2005).

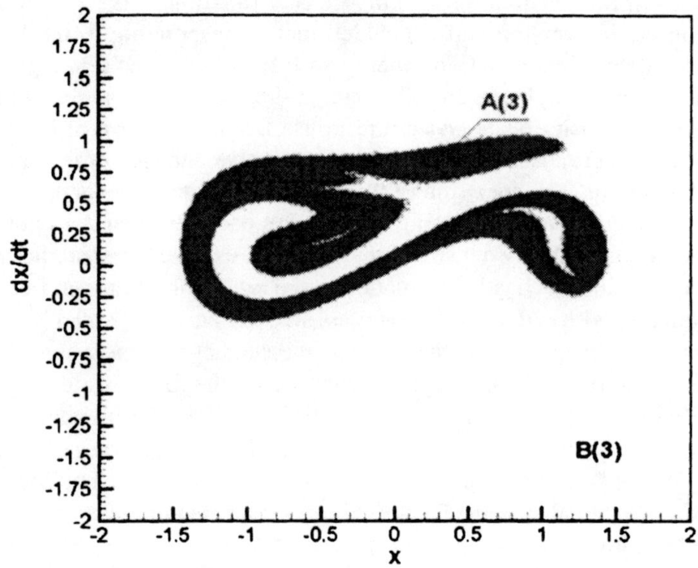

Figure 27. The attractors, the attraction basins and the saddles of the system (15) with $\sigma = 0.01$. Courtesy of Xu, He, Fang and Rong (2005).

5. Fuzzy bifurcations

In engineering systems, there are uncertainties that are associated with the lack of precise knowledge of the system parameters and operating conditions and that are originated from the variabilities in manufacturing processes. The uncertainties can have significant influence on the dynamic response and the reliability of the system, and are often modeled as random variables or fuzzy sets. This section proposes a method to analyze the response and bifurcation of nonlinear dynamical systems with fuzzy uncertainties. Specifically, we are interested in a nonlinear dynamical system whose response is a fuzzy process, and study how the fuzzy response changes as the fuzzy parameter of the system varies.

Bifurcation analysis of uncertain nonlinear dynamical systems is in general a difficult subject, partly because even the definition of bifurcation is open to discussion. Take the stochastic system as an example. The commonly accepted definition of the bifurcation is the "qualitative change" of the system response as a control parameter varies. Meunier and Verga (1988) studied pitchfork bifurcation of a stochastic dynamical system. They examined the quantities such as invariant measures, Lyapunov exponents, correlation functions and exit times. It turns out that the behavior of all these quantities near the deterministic bifurcation point changes for different values of the bifurcation parameter, making them a poor indicator of bifurcation in some cases. They proposed an effective potential function of the invariant probability density function of the system response to describe the bifurcation and concluded that corresponding to the bifurcation point of the deterministic system, there is a bifurcation transition region for the stochastic system. Doi, Inoue and Kumagai (1998) have also found that the invariant probability density of the system response is not indicative of bifurcation in some cases, and proposed to examine the qualitative changes of the spectrum of a Markov operator. The spectrum of the Markov operator reveals the bifurcation of the stochastic phase lockings of a Van der Pol oscillator that does not show up in the invariant probability density of the system response. Other studies phenomenologically define stochastic bifurcation based on the observation of collision of stable attractors with saddle nodes, as presented in Sections 3 and 4.

For fuzzy nonlinear dynamical systems, the subject is even more difficult because the evolution of the membership function of the fuzzy response process is not readily obtained analytically. There is little study in the literature on the bifurcation of fuzzy nonlinear dynamical systems. There are studies of bifurcations of fuzzy control systems where the fuzzy control law leads to a nonlinear and deterministic dynamical system. The bifurcation studies are practically the same as that of deterministic systems. See the articles by Cuesta, Ponce and Aracil (2001) and Tomonaga and Takatsuka (1998), for example. The work by Satpathy, Das and Gupta (2004) is the only one that deals with bifurcation of fuzzy dynamical systems having a fuzzy response. Numerical simulations are used to simulate the system response with a given parameter and fuzzy membership grade. The eigenvalues and the membership distribution are both used to describe the bifurcation. For a given membership grade, the bifurcation of the system is defined in the same manner as for the deterministic system.

This section studies bifurcations of fuzzy dynamical systems by the fuzzy generalized cell mapping (FGCM) method, which has been developed for nonlinear dynamical systems with fuzzy uncertainties (Sun and Hsu, 1990b). We first present some recent advances in the development of the GCM method for fuzzy dynamical systems. We study the topological equivalence of the FGCM with that of the traditional GCM described by Markov chains, and a finite convergence property of the FGCM to the invariant distribution. Furthermore, we describe a backward algorithm for locating the unstable solutions. Subsequently, we study bifurcations of fuzzy dynamical systems within the context of generalized cell mapping.

5.1. *Fuzzy generalized cell mapping*

Here, we first review the FGCM method for nonlinear dynamical systems with fuzzy uncertainties (Sun and Hsu, 1990b). Consider a dynamical system with a fuzzy parameter

$$\dot{\mathbf{x}} = \mathbf{f}(\mathbf{x}, t, S), \quad \mathbf{x} \in \mathbf{D}, \tag{16}$$

where $\mathbf{x}$ is the state vector, t the time variable, S a fuzzy set with a membership function $\mu_S(s) \in (0, 1]$ where $s \in S$ and $\mathbf{f}$ is a vector-valued nonlinear function of its arguments. Function $\mathbf{f}$ is assumed to be periodic in t with period T for all $s \in S$ and to satisfy the Lipschitz condition for all $s \in S$, and $\mathbf{D}$ is a bounded domain of interest in the state space. A fuzzy Poincaré map can be obtained from equation (16) as

$$\mathbf{x}(n+1) = \mathbf{G}\big(\mathbf{x}(n), S\big), \quad n = 0, 1, 2, \ldots. \tag{17}$$

The cell mapping method proposes to further discretize the state space in searching for the global solution of the system (Hsu, 1987). In order to apply the cell mapping method, we also need to discretize the fuzzy set S. Suppose that S is a finite interval in $\mathbb{R}$. We divide S into M segments of appropriate length and sample a value $s_k \in S$ ($k = 1, \ldots, M$) in the middle of each segment. The division of S is such that there is at least one s_k with membership grade equal to one. The domain $\mathbf{D}$ is then discretized into N small cells. Each cell is identified by an integer ranging from 1 to N. For a cell, say cell j, N_{p} points are uniformly sampled from cell j, $M \times N_{\mathrm{p}}$ fuzzy sample trajectories are computed for one period T, or one mapping step. Each trajectory carries a membership grade determined by that of s_k's. We then find the cells in which the end points of the trajectories fall. Assume that cell i is one of the image cells of cell j, and that there are m $(0 < m \leqslant MN_{\mathrm{p}})$ trajectories falling in cell i. Define a quantity

$$p_{ij} = \max_{i_k}\left[\mu_S(s_{i_k})\right], \quad 0 < p_{ij} \leqslant 1, \tag{18}$$

where i_k $(k = 1, 2, \ldots, m)$ are referred to the trajectories falling in cell i, and $\mu_S(s_{i_k})$ are the membership grades of the corresponding trajectories. This procedure for computing p_{ij} is known as the sampling point method in the context of generalized cell mapping (Hsu, 1987).

Now, assume that the membership grade of the system being in cell j at the nth mapping step is $p_j(n)$ $(0 < p_j(n) \leqslant 1)$. Cell j is mapped in one step to cell i with the membership grade given by

$$\begin{aligned}
&\max\left\{\min\left[\mu_S(s_{i_1}), p_j(n)\right], \min\left[\mu_S(s_{i_2}), p_j(n)\right], \ldots, \min\left[\mu_S(s_{i_m}), p_j(n)\right]\right\}\\
&\quad = \min\left[\max_{i_k}\left(\mu_S(s_{i_k})\right), p_j(n)\right]\\
&\quad = \min\left[p_{ij}, p_j(n)\right].
\end{aligned} \tag{19}$$

Considering all possible pre-images of cell i, we have the membership grade of the system being in cell i at the $(n+1)$th step as

$$p_i(n+1) = \max_j \min\left[p_{ij}, p_j(n)\right]. \tag{20}$$

Let $\mathbf{p}(n)$ be a vector with components $p_i(n)$, and let $\mathbf{P}$ be a matrix with components p_{ij}. Equation (20) can be written in a compact matrix notation

$$\mathbf{p}(n+1) = \mathbf{P} \circ \mathbf{p}(n), \qquad \mathbf{p}(n) = \mathbf{P}^n \circ \mathbf{p}(0), \tag{21}$$

where $\mathbf{P}^{n+1} = \mathbf{P} \circ \mathbf{P}^n$ and $\mathbf{P}^0 = \mathbf{I}$. The matrix $\mathbf{P}$ is called the one-step transition membership matrix. The vector $\mathbf{p}(n)$ is called the n-step membership distribution vector, and $\mathbf{p}(0)$ is called the initial membership distribution vector. The (i, j)th element p_{ij} of the matrix $\mathbf{P}$ is called the one-step transition membership from cell j to cell i.

Equation (21) is called a fuzzy generalized cell mapping system, which describes the evolution of the fuzzy solution process $\mathbf{x}(n)$ and its membership function, and is a finite approximation to the mapping (17) in $\mathbf{D}$.

Consider the master equation for the possibility transition of continuous fuzzy processes (Friedman and Sandler, 1996, 1999; Yoshida, 2000),

$$p(\mathbf{x}, t) = \sup_{\mathbf{x}_0 \in \mathbf{D}}\left[\min\left\{p(\mathbf{x}, t|\mathbf{x}_0, t_0), p(\mathbf{x}_0, t_0)\right\}\right], \quad \mathbf{x} \in \mathbf{D}, \tag{22}$$

where $\mathbf{x}$ is a fuzzy process, $p(\mathbf{x}, t)$ is the membership function of $\mathbf{x}$ and $p(\mathbf{x}, t|\mathbf{x}_0, t_0)$ is the transition possibility function, also known as a fuzzy relation (Yoshida, 2000). Equation (20) of the FGCM can be viewed as a discrete representation of equation (22). Friedman and Sandler (1996, 1999) have derived a partial differential equation from equation (22) for continuous time processes. This equation is analogous to the Fokker–Planck–Kolmogorov equation for the probability density function of stochastic processes (Risken, 1996). The solution to this equation is in general very difficult to obtain analytically. Numerically, the FGCM offers a very effective method for solutions to this equation, particularly, for *fuzzy nonlinear* dynamical systems.

Topological property of FGCM

In the GCM method for deterministic or stochastic systems, the solution of the system is completely characterized by the one-step transition probability matrix of the Markov chain. In the FGCM method, the one-step transition membership matrix **P** plays the same role as that of the one-step transition probability matrix. Hsu (1995) has described in great detail that the topological structure of the one-step transition probability matrix represents a digraph and determines the stable equilibrium solutions. In the following, we shall establish the topological equivalence of the one-step transition membership matrix **P** to that of the one-step transition probability matrix of the GCM method. In other words, the topological structure of the one-step transition membership matrix **P** also forms a digraph and determines the stable equilibrium solutions.

Recall that the min–max operation in equation (20) really represents the intersection (product) and union (summation) of fuzzy sets in the form of cells in **D**. Hence, $\mathbf{P} \circ \mathbf{p}$ is an inner product of fuzzy sets, as is the case of the Markov chains with the matrix inner product $\mathbf{P} \cdot \mathbf{p}$.

The topological matrix of the one-step transition membership matrix **P** is denoted by $\overline{\mathbf{P}} = [\bar{p}_{ij}]$, where

$$\bar{p}_{ij} = \begin{cases} 1, & p_{ij} > 0; \\ 0, & p_{ij} = 0. \end{cases} \tag{23}$$

Furthermore, we denote the topological vector of the fuzzy membership distribution as

$$\bar{p}_i(n) = \begin{cases} 1, & p_i(n) > 0; \\ 0, & p_i(n) = 0; \end{cases} \qquad n \geqslant 0. \tag{24}$$

Equation (21) becomes

$$\bar{\mathbf{p}}(n+1) = \overline{\mathbf{P}} \circ \bar{\mathbf{p}}(n), \qquad \bar{\mathbf{p}}(n) = \overline{\mathbf{P}}^n \circ \bar{\mathbf{p}}(0), \tag{25}$$

where $\overline{\mathbf{P}}^{n+1} = \overline{\mathbf{P}} \circ \overline{\mathbf{P}}^n$ and $\overline{\mathbf{P}}^0 = \mathbf{I}$. Note that the min–max operation is equivalent to the logic operations of multiplication $\wedge$ and addition $\vee$ of binary numbers. For example, we have $0 \wedge 1 = 0$, $1 \wedge 0 = 0$, $0 \wedge 0 = 0$, $1 \wedge 1 = 1$, $0 \vee 1 = 1$, $1 \vee 0 = 1$, $0 \vee 0 = 0$ and $1 \vee 1 = 1$. Hence, the min–max operation in equation (25) leads to the identical result to that of the topological matrix of the Markov chains.

We can partition $\overline{\mathbf{P}}$ into the standard normal form as

$$\overline{\mathbf{P}} = \begin{bmatrix} \overline{\mathbf{P}}_1 & & & \overline{\mathbf{T}}_1 \\ & \ddots & & \vdots \\ & & \overline{\mathbf{P}}_n & \overline{\mathbf{T}}_n \\ & & & \overline{\mathbf{Q}} \end{bmatrix}, \tag{26}$$

where $\overline{\mathbf{P}}_k$ $(1 \leqslant k \leqslant n)$ represent the stable steady-state solutions in the form of persistent groups of cells, $\overline{\mathbf{Q}}$ represents all the transient cells including the unstable steady-state solutions and $\overline{\mathbf{T}}_k$ $(1 \leqslant k \leqslant n)$ determine how the transient cells are absorbed to the persistent groups.

The matrix $\overline{\mathbf{Q}}$ can be further partitioned into the same form as that of $\overline{\mathbf{P}}$ with cyclic groups of cells. These transient cyclic groups contain unstable solutions. When the system is lightly damped or when the system is close to bifurcation, the transient cyclic groups contain cells that do not belong to the unstable solutions. This makes it difficult to identify the unstable solutions.

A backward algorithm

To overcome the difficulty of the cell mapping methods in finding the unstable solutions, we propose a backward search algorithm for the FGCM method that can identify the unstable solutions just as easily as looking for the stable solutions. This backward algorithm works equally well for the GCM method with the Markov chains. It should be pointed out that Hsu (1995) has examined the backward search scheme for simple cell mapping.

Recall the meaning of the topological one-step transition membership matrix. When $\bar{p}_{ij} = 1$, this signifies that cell j is mapped to cell i in one step with a certain possibility. In other words, if the system is found to be in cell i with a possibility, it must be in cell j one step before with another possibility. We introduce a notation $\bar{b}_{ji}$ to represent this backward mapping relationship. When $\bar{b}_{ji} = 1$, cell j is a pre-image of cell i. When $\bar{b}_{ji} = 0$, it is not a pre-image. The matrix defined by these elements such that $\overline{\mathbf{B}} = [\bar{b}_{ij}]$ is called the backward one-step transition topological matrix of the FGCM system. By definition, we have

$$\overline{\mathbf{B}} = [\bar{b}_{ij}] = [\bar{p}_{ji}] = \overline{\mathbf{P}}^{\mathrm{T}}. \tag{27}$$

Hence,

$$\bar{\mathbf{p}}(n-1) = \overline{\mathbf{B}} \circ \bar{\mathbf{p}}(n). \tag{28}$$

Since the matrix $\overline{\mathbf{B}}$ describes the dynamics of the original system backward in time topologically, the stable solutions of the original system appear to be unstable and the unstable solutions appear to be stable in equation (28). We can partition the matrix $\overline{\mathbf{B}}$ into the normal form in the same manner as for the matrix $\overline{\mathbf{P}}$,

$$\overline{\mathbf{B}} = \begin{bmatrix} \overline{\mathbf{Q}}_1 & & & \overline{\mathbf{R}}_1 \\ & \ddots & & \vdots \\ & & \overline{\mathbf{Q}}_m & \overline{\mathbf{R}}_m \\ & & & \overline{\mathbf{P}}_\mathrm{b} \end{bmatrix}. \tag{29}$$

Here, $\overline{\mathbf{Q}}_k$ $(1 \leqslant k \leqslant m)$ denote the persistent groups in the backward dynamics and represent unstable steady-state solutions and $\overline{\mathbf{P}}_\mathrm{b}$ includes all the remaining transient cells and the persistent groups of the original system. Hence, the backward algorithm allows us to identify the unstable solutions in the state space with an exception of saddle nodes that require additional processing steps to identify. This will be studied in a future paper.

Finite convergence of membership

It is interesting to note that under the min–max operation in equation (20), the membership function of the cells is nonincreasing in the sense that

$$\|\mathbf{p}(n+1)\| \leqslant \|\mathbf{p}(n)\|, \quad \forall n \geqslant 0, \tag{30}$$

where the norm of the fuzzy membership vector is defined as $\|\mathbf{p}(n)\| = \max_i\{p_i(n)\}$. Furthermore, the min–max operator does not introduce any new numbers that are not the entries of the matrix $\mathbf{P}$ or the vector $\mathbf{p}(0)$. This implies that $\mathbf{p}(n)$ will assume only finite number of possible values as $n \to \infty$ (Sun and Hsu, 1990b). In the steady state, $\mathbf{p}(n)$ will either converge to a constant vector, or to a set of vectors which form a periodic group and repeat themselves as the iteration goes on. In either case, we consider the system to have converged to the steady state. Because there are only finite number of possible values for $\mathbf{p}(n)$, equation (20) will converge in finite number of iterations. This is a sharp contrast to the Markov chains, which theoretically converge to steady state in infinite iterations.

5.2. *Bifurcation of one-dimensional fuzzy systems*

Next, we investigate fuzzy bifurcations of one-dimensional dynamical systems, that will illustrate the unique characteristics of fuzzy nonlinear dynamical systems and provide a foundation for further discussions of fuzzy bifurcations.

Consider a one-dimensional nonlinear system given by

$$\dot{x} = Sx + \beta x^3, \tag{31}$$

where S is a fuzzy parameter with a triangular membership function,

$$\mu_S(s) = \begin{cases} \frac{s-(s_0-\varepsilon)}{\varepsilon}, & s_0 - \varepsilon \leqslant s < s_0; \\ -\frac{s-(s_0+\varepsilon)}{\varepsilon}, & s_0 \leqslant s < s_0 + \varepsilon; \\ 0, & \text{otherwise}; \end{cases} \tag{32}$$

where $\varepsilon > 0$ is a parameter characterizing the degree of fuzziness of S, and s_0 is the nominal value of S with membership grade $\mu_S(s_0) = 1$. When S is a deterministic parameter, the system has a supercritical pitchfork bifurcation at $S = 0$ with $\beta = -1$ and a subcritical pitchfork bifurcation with $\beta = 1$. The bifurcation diagram of the deterministic system is shown in Figure 28 for both cases.

When S is a fuzzy parameter, the system response $x(t)$ is a fuzzy process. To define the bifurcation for a fuzzy process is no longer an easy matter. First of all, we need to be able to describe the evolution of the membership function of the fuzzy process $x(t)$. At present, there is a lack of mathematical theory that describes the evolution of the membership function of the response of nonlinear dynamical systems. The GCM method can obtain the membership function numerically as demonstrated in Sun and Hsu (1990b). In the following, we apply

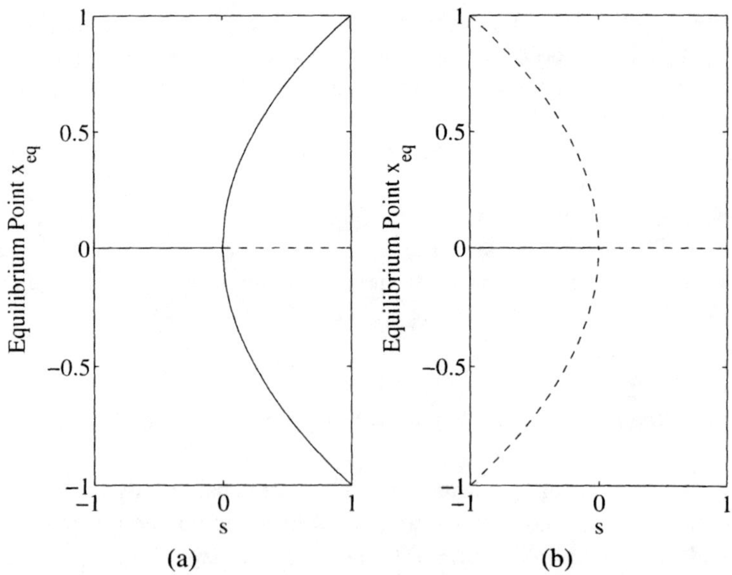

Figure 28. Supercritical (with $\beta = -1$ (a)) and subcritical (with $\beta = 1$ (b)) bifurcation of the one-dimensional nonlinear system with a deterministic parameter S ($\varepsilon = 0$) (31). Solid line: stable equilibrium; dashed line: unstable equilibrium.

the FGCM method to compute the solutions of the fuzzy process $x(t)$ and its membership function for various fuzzy parameters.

In the computation, we have fixed $\varepsilon = 0.2$, and allow s_0 to vary such that $-1 < s_0 < 1$. We have chosen the domain $\mathbf{D} = \{-2 \leqslant x \leqslant 2\}$. The mapping time step is $\pi/4$. This domain is divided into 1501 cells of equal size. From each cell, five points ($N_{\text{sample}} = 5$) are sampled. The membership function is discretized into 201 segments ($M = 201$). Hence, out of each cell, there are 1005 trajectories with varying membership grades. These trajectories are then used to compute the transition membership matrix.

Figure 29(a) shows the fuzzy sets of the stable equilibrium of the system which is identified as the persistent groups of the forward dynamics of the FGCM system. The unstable equilibrium has been found as the persistent groups of the backward dynamics of the FGCM system. It is seen that the fuzzy set representing the stable equilibrium can have a broad support near the deterministic bifurcation point $S = 0$.

Another useful information of the fuzzy response is the α-cut of the steady-state membership function distribution of the persistent groups representing the stable solutions. The α-cut of a fuzzy variable, say S, is defined as a set $S_\alpha = \{s\colon \mu_S(s) \geqslant \alpha\}$. Figure 29(b) shows the α-cut of the steady-state membership

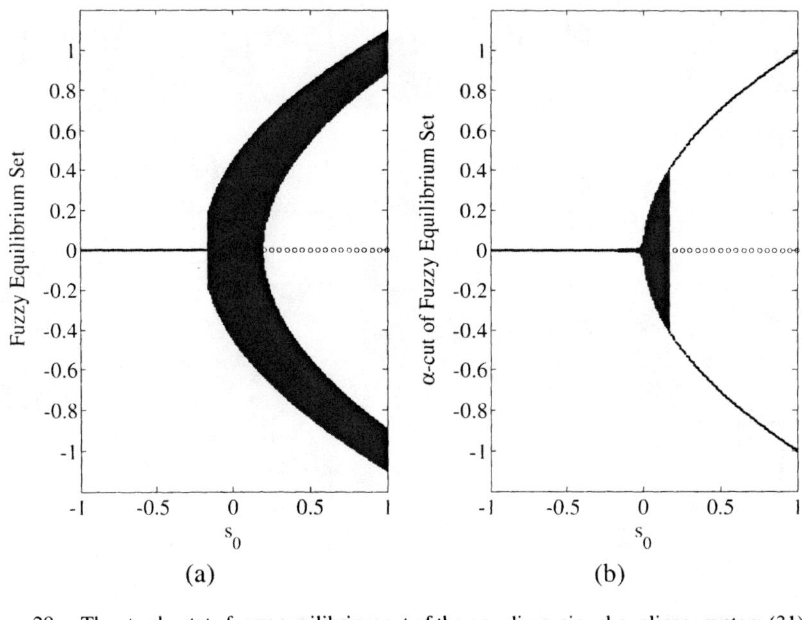

Figure 29. The steady-state fuzzy equilibrium set of the one-dimensional nonlinear system (31) with a fuzzy parameter S ($\varepsilon = 0.2$) and $\beta = -1$ (a), and the α-cut of its membership function with $\alpha = 1$ (b). The symbol o denotes the unstable equilibrium of the system.

function distribution of the persistent groups with $\alpha = 1$. It should be pointed out that the α-cut with $\alpha = 1$ would be corresponding to the deterministic system response with a crispy parameter $S = s_0$. The figure shows that when S is away from the bifurcation point $S = 0$, the α-cut with $\alpha = 1$ indeed matches the response of the corresponding deterministic system. Furthermore, the steady-state membership function is invariant in the sense that the distribution is independent of initial conditions as long as the system starts in the domain of attraction of the corresponding stable solution, excluding the boundary.

On the other hand, the steady-state membership function for S values near the bifurcation point $S = 0$ depends on the initial condition. For this reason, we have chosen an uniformly distributed membership function over the cell in the persistent group as the initial condition for all the results presented in this chapter. This result is thus difficult to duplicate with pointwise simulations. The α-cut with $\alpha = 1$ of the steady-state membership function starting from the uniform initial condition is substantially different from that of the deterministic system when S is near the bifurcation point $S = 0$. The system dynamics is going through an interesting and complex transition in this region as illustrated by the steady-state membership functions in Figure 30. The membership changes from a single

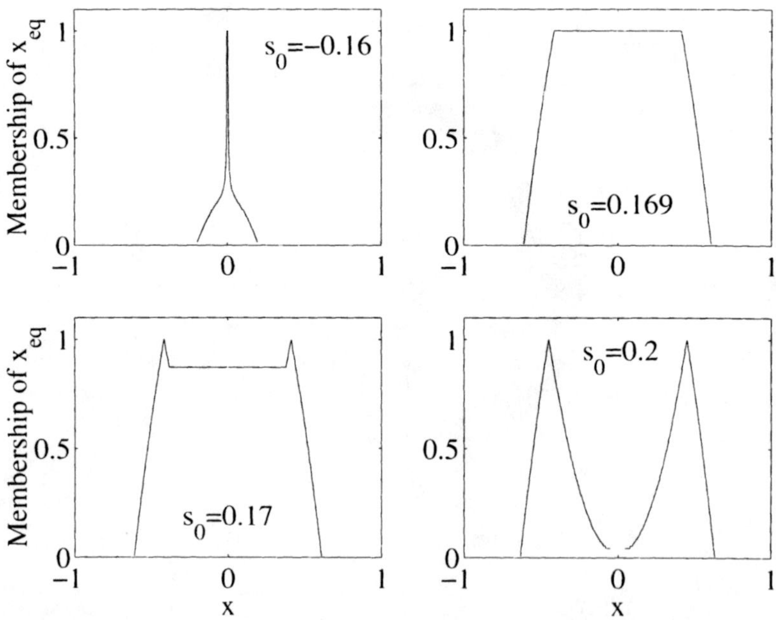

Figure 30. The steady-state membership distributions of the one-dimensional nonlinear system (31) with $\beta = -1$ for the fuzzy parameter S ($\varepsilon = 0.2$) in the transition region of bifurcation. The initial condition is the uniform membership distribution in the persistent group.

peak *invariant* distribution corresponding to a stable equilibrium with $S \ll 0$ to a two-peak disjoint *invariant* distribution representing the two stable equilibria with $S \gg 0$. In between, the membership function assumes a flat platform for some parameters of S because of the complex behavior of the equilibrium at the origin $x = 0$. For example, for $s_0 = 0.169$, out of the cell centered at $x = 0$, there are both stable trajectories when $S < 0$ with a certain possibility and unstable trajectories when $S > 0$. The cell also maps to itself with possibility one. This is a unique property of the fuzzy dynamical system within the context of generalized cell mapping.

The complex transition of the fuzzy response near the bifurcation point $S = 0$ suggests that the bifurcation of fuzzy dynamical system is characterized by this transition region, as is the case for stochastic dynamical systems discussed in Meunier and Verga (1988). The beginning of this transition region is marked by the change of the measure of the α-cut of the steady-state membership function distribution of the persistent groups with $\alpha = 1$ from zero to a finite measure at $s_0 = -\varepsilon$, and the ending of the transition region is marked by another sudden change of the measure of the α-cut from a finite measure to zero at $s_0 = \varepsilon$, accompanied by a topological change of the response from having one persistent group to two groups. Furthermore, in this transition region, the steady-state membership distribution is dependent on the initial condition. For this example, the fuzzy bifurcation region is given by $-2\varepsilon \leqslant S \leqslant 2\varepsilon$.

It should be noted that the above discussion of the bifurcation region is not intended to provide a universal definition of fuzzy bifurcation since there may be various different bifurcation scenarios that are not described by the current concept.

Consider now the subcritical bifurcation with $\beta = 1$. Figure 31(a) shows the α-cut of the steady-state membership function distribution of the persistent group with $\alpha = 1$, representing the stable equilibrium of the system. At $s_0 = -\varepsilon$, the stable solution disappears. This is a topological change, marking the beginning of the bifurcation region. If one stays with the forward dynamics of the system, this would be the end of the analysis with the cell mapping method. However, with the backward algorithm, we can further study the bifurcation of unstable equilibria that cannot be found with the forward dynamics.

Figure 31(b) shows the fuzzy sets representing the unstable equilibria obtained from the backward dynamics. At $s_0 = \varepsilon$, the two separate unstable equilibria are merged into one. The measure of the fuzzy sets changes from a finite number to zero at this point. Hence, $s_0 = \varepsilon$ marks the ending of the bifurcation region. The fuzzy bifurcation region is again given by $-2\varepsilon \leqslant S \leqslant 2\varepsilon$, although the content of the bifurcation is drastically different from that of the supercritical bifurcation.

It appears that near the deterministic bifurcation point, the fuzzy system undergoes a complex transition as the control parameter varies. In this transition region, the steady-state membership distribution is dependent on the initial condition. If

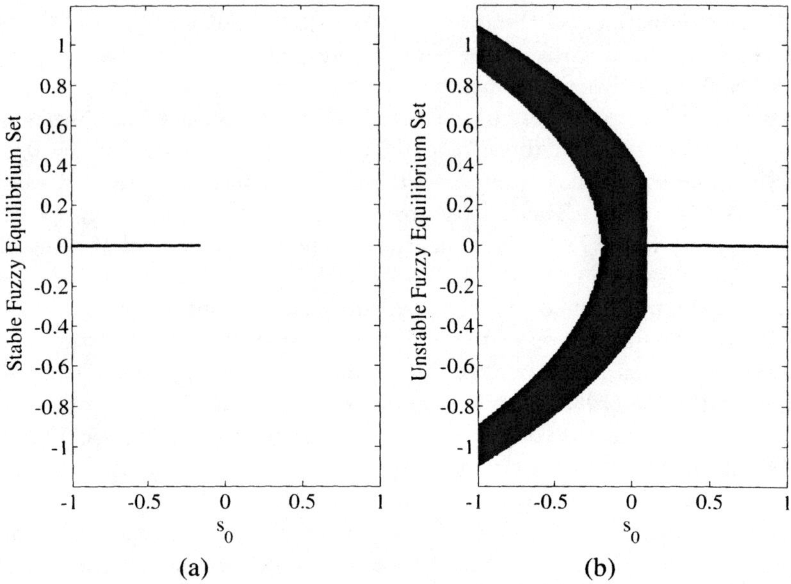

Figure 31. The steady-state stable fuzzy equilibrium (a) and unstable equilibrium (b) of the nonlinear system (31) with a fuzzy parameter S ($\varepsilon = 0.2$) and $\beta = 1$.

we use the measure and topology of the α-cut ($\alpha = 1$) of the steady-state membership function of the persistent group representing the stable fuzzy equilibrium solution to characterize the fuzzy bifurcation, assuming the uniform initial condition within the persistent group, the bifurcation of the fuzzy dynamical system is then completed within an interval of the control parameter, rather than at a point as is the case of deterministic systems. The FGCM method can be applied to study the bifurcation characterized by the qualitative changes of unstable equilibrium solutions by considering the backward topological mapping of the system.

5.3. Bifurcation of fuzzy nonlinear oscillators

A forced nonlinear Mathieu oscillator

Consider a forced nonlinear Mathieu oscillator with fuzzy uncertainties,

$$\begin{aligned}\dot{x}_1 &= x_2,\\ \dot{x}_2 &= -25x_1^3 - 0.173x_2 - \big[2.62 - 0.456S(1 - \cos 2t)\big]x_1\\ &\quad + 0.92S(1 - \cos 2t),\end{aligned} \tag{33}$$

where S is a fuzzy set with a triangular membership function defined in equation (32). When the system is deterministic, it has a boundary crisis and a saddle-

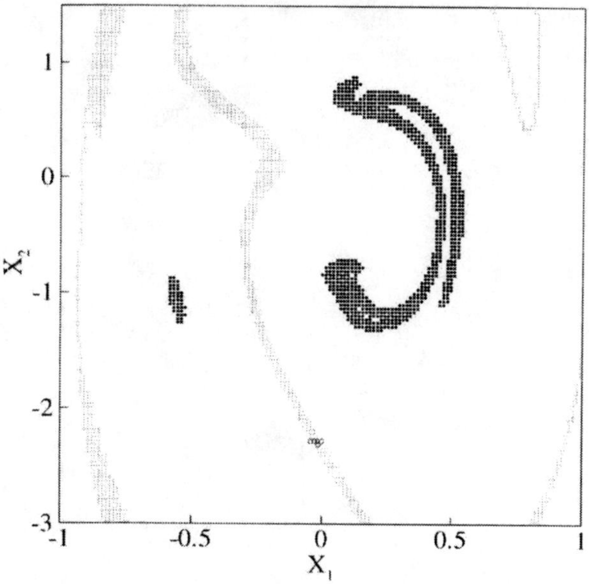

Figure 32. Global phase portrait of the forced Mathieu equation (33) with a deterministic parameter $S = 3.8$.

node bifurcation for the parameter in the range (3.5, 4.2) (Gong and Xu, 1998; Hong and Xu, 2001a).

We discretize the domain $\mathbf{D} = \{-1.0 \leqslant x_1 \leqslant 1.0; -3.0 \leqslant x_2 \leqslant 1.5\}$ into 141×141 cells. The 5×5 points are sampled in each cell. The membership function is discretized into 21 segments ($M = 21$). Hence, out of each cell, there are 525 trajectories with varying membership grades.

When S is a deterministic number 3.8, the system has two stable solutions: a chaotic attractor and a period-1 attractor as shown in Figure 32.

Consider now S as a fuzzy set with $\varepsilon = 0.2$. When s_0 varies between 3.765 and 3.962, the two deterministic stable solutions, i.e., the chaotic attractor and the period-1 attractor, become fuzzy sets. Furthermore, a saddle on the basin boundary also becomes a fuzzy set. These results for $s_0 = 3.8$ are shown in Figure 33.

When s_0 increases, the fuzzy chaotic attractor moves towards the fuzzy saddle. The boundary crisis occurs when the fuzzy chaotic attractor touches the fuzzy saddle at $s_0 = 3.962$. After the collision, say, at $s_0 = 3.963$, the fuzzy chaotic attractor together with its basin of attraction suddenly disappears, leaving behind a fuzzy chaotic saddle in the place of the original chaotic attractor in the phase space. This boundary crisis is illustrated in Figure 34. In Figure 34(a), the stable manifolds of the fuzzy saddle form the basin boundary, while two unstable manifolds are directed, respectively, to the fuzzy attractors. The fuzzy chaotic at-

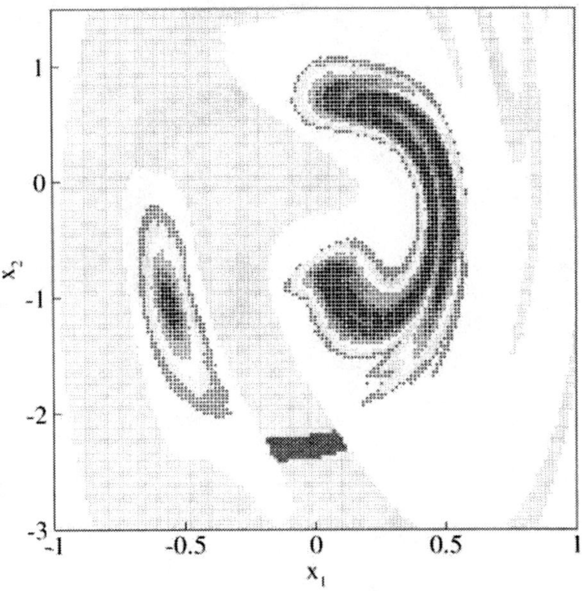

Figure 33. Global phase portrait of the forced Mathieu equation (33) with a fuzzy parameter $\varepsilon = 0.2$ and $s_0 = 3.8$. In the figure the fuzzy attractors are marked by the symbol ■. The membership distribution of the fuzzy attractors is coded with grayscale. Black represents membership one. Each adjacent level steps down by 0.2. The saddle is marked by the symbol ○. The basin boundary is marked by the symbol +.

tractor, when viewed as an extension of the unstable manifold of the saddle, is about to touch the stable manifold of the saddle, suggesting that the homoclinic tangency is about to happen when the boundary crisis occurs (Grebogi, Ott and Yorke, 1986a). In Figure 34(b), after the boundary crisis, the fuzzy chaotic attractor becomes a fuzzy chaotic saddle, and together with its basin of attraction, is absorbed to the remaining fuzzy period-1 attractor. Hence, the attraction basin of the remaining fuzzy period-1 attractor is substantially expanded.

When s_0 decreases, the fuzzy period-1 attractor moves toward the fuzzy saddle and collides with it at $s_0 = 3.765$, leading to the saddle-node bifurcation. After the collision, at $s_0 = 3.764$, the fuzzy period-1 attractor together with its basin of attraction is suddenly destroyed and becomes a fuzzy period-1 saddle in the interior of the attraction basin of the remaining fuzzy chaotic attractor. This scenario is illustrated in Figure 35. Figure 35(a) also shows a homoclinic tangency when the saddle-node bifurcation occurs.

The fuzzy bifurcations are characterized by the topological change of the persistent groups of the generalized cell mapping, and the stability change of one

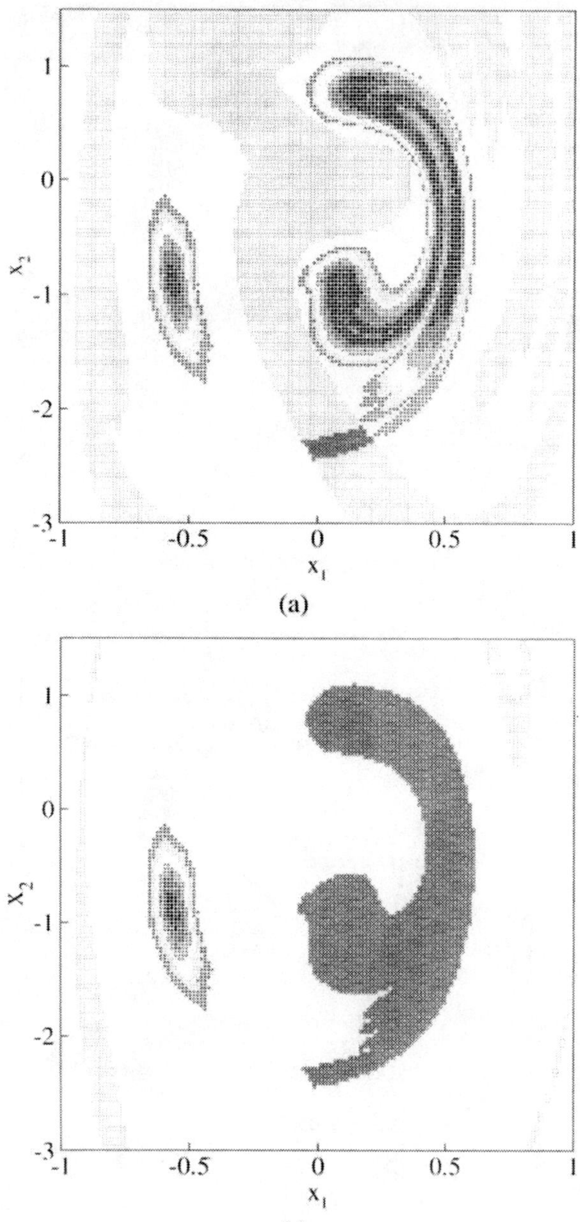

Figure 34. The fuzzy boundary crisis of the Mathieu system (33) with $\varepsilon = 0.2$ when $s_0 \in (3.962, 3.963)$. (a) $s_0 = 3.962$; (b) $s_0 = 3.963$. Legends are the same as in Figure 33.

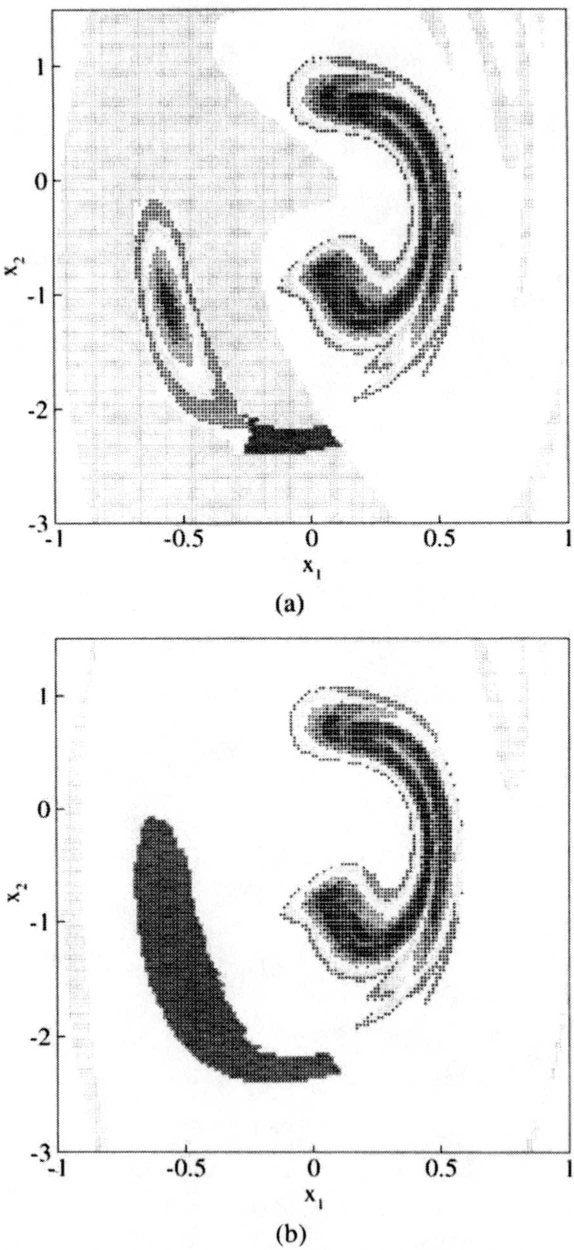

Figure 35. The fuzzy saddle-node bifurcation of the Mathieu system (33) with $\varepsilon = 0.2$ when $s_0 \in (3.765, 3.764)$. (a) $s_0 = 3.765$; (b) $s_0 = 3.764$. Legends are the same as in Figure 33.

of the attractors, caused by its collision with the saddle. In this case, there is no transition region of the fuzzy parameter near the deterministic bifurcation points.

In summary, we have demonstrated that a boundary crisis results from the collision of a fuzzy chaotic attractor with a fuzzy saddle on the basin boundary. In this case, the fuzzy chaotic attractor together with its basin of attraction is eradicated as the fuzzy control parameter reaches a critical point. We have also showed that a saddle-node bifurcation is caused by the collision of a fuzzy period-1 attractor with a fuzzy saddle on the basin boundary, in which the fuzzy attractor together with its basin of attraction suddenly disappears as the fuzzy control parameter passes through a critical value.

A Duffing–Van der Pol oscillator

Consider a Duffing–Van der Pol (DVP) oscillator,

$$\ddot{x} + (x^2 - \mu_2)\dot{x} + (x^2 - \mu_1)x = 0. \tag{34}$$

This system is used in physics, engineering, biology and many other disciplines and is one of the most intensively studied systems in nonlinear dynamics. Its importance in modeling physical phenomena and its seemingly "simple" structure have given rise to many thorough studies of its local and global bifurcations (Guckenheimer and Holmes, 1983; Holmes and Rand, 1980). The influence of stochastic noise on the DVP oscillator exhibiting codimension one and two bifurcations has also been studied (Namachchivaya, 1990, 1991; Schenk-Hoppé, 1996).

The DVP system has very rich dynamics exhibited in the parameter space spanned by (μ_1, μ_2), see, for example, the paper by Schenk-Hoppé (1996) who has presented a comprehensive bifurcation diagram of the system. In this section, *we are only interested in codimension two bifurcations* and consider two sets of parameters: $(\mu_1, \mu_2) = (1, -1)$ and $(\mu_1, \mu_2) = (0.64, 0.1)$, when the DVP equation has two stable fixed points at $(\pm\sqrt{\mu_1}, 0)$ and an unstable saddle at the origin $(0, 0)$ for $\mu_1 > 0$ as shown in Figure 36 (Guckenheimer and Holmes, 1983). Note that $(\mu_1, \mu_2) = (0.64, 0.1)$ lies in the upper-right quadrant of the parameter space (μ_1, μ_2), where there exist more complicated global bifurcations involving the coalescence of closed orbits as well as saddle connections.

The DVP equation with $(\mu_1, \mu_2) = (1, -1)$

Additive fuzzy noise

We will consider the DVP equation driven by additive fuzzy noise,

$$\begin{aligned} \dot{x}_1 &= x_2, \\ \dot{x}_2 &= x_1 - x_2 - x_1^3 - x_1^2 x_2 + S, \end{aligned} \tag{35}$$

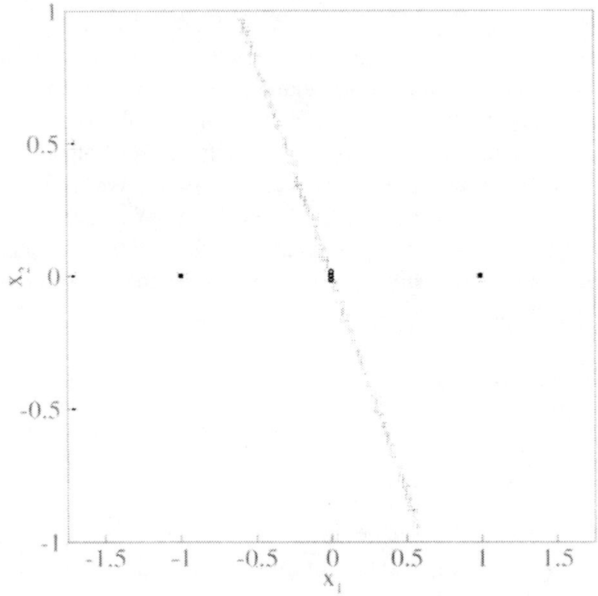

Figure 36. Global phase portrait of the deterministic Duffing–Van der Pol equation (34) with $(\mu_1, \mu_2) = (1, -1)$. Two stable fixed points denoted by ■ are on the left and right side of the origin. At the origin lies the saddle ○. The line passing through the saddle is the stable manifold of the saddle delineating the boundary + of the domains of attraction of the stable fixed points.

where S is a fuzzy parameter with a triangular membership function given by equation (32).

The domain $\mathbf{D} = \{-1.75 \leqslant x_1 \leqslant 1.75; -1.0 \leqslant x_2 \leqslant 1.0\}$ is discretized into 141×141 cells when applying the FGCM method, 5×5 sampling points are used within each cell. The membership function is discretized into 201 segments ($M = 201$). Hence, out of each cell, there are 5025 trajectories with varying membership grades. These trajectories of π second long are then used to compute the transition membership matrix.

When we fix $s_0 = 0$ and allow the fuzzy noise intensity ε to vary, the rotational symmetry holds in the noisy DVP equation. As ε increases, a merging explosion bifurcation of two fuzzy period-1 attractors is discovered in the interval (0.385, 0.386). The phase portraits of the attractors with membership distribution are shown in Figure 37. In the explosion, two fuzzy period-1 attractors touch simultaneously the fuzzy saddle on their basin boundary at $\varepsilon = 0.385$, the two fuzzy period-1 attractors merge into a bigger fuzzy attractor at $\varepsilon = 0.386$ after the collision. At the same time, the membership distribution changes from a two-peak disjoint function representing the two fuzzy period-1 attractors to a single peak function with a flat platform representing the bigger fuzzy attractor.

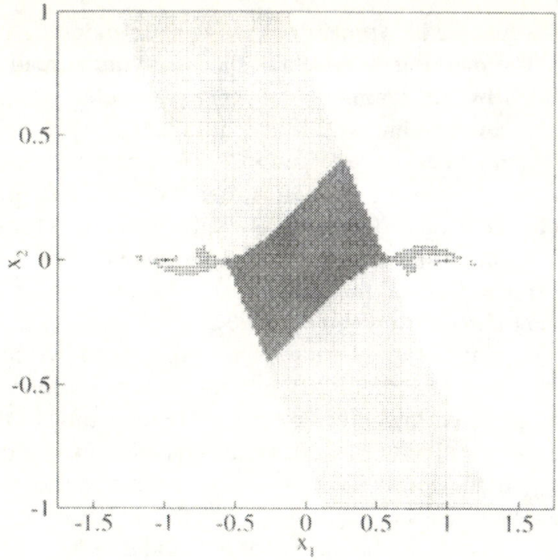

(a)

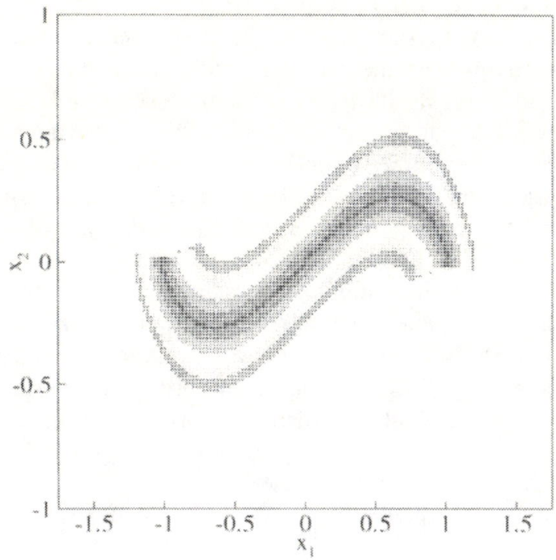

(b)

Figure 37. A merging explosion bifurcation of the Duffing–Van der Pol equation (35) with additive fuzzy noise. (a) $s_0 = 0$, $\varepsilon = 0.385$; (b) $s_0 = 0$, $\varepsilon = 0.386$. In both figures the fuzzy period-1 attractors are marked by the symbol ■. The membership distribution of the fuzzy attractors is coded with grayscale. Black represents membership one. Each adjacent level steps down by 0.2. The saddle is marked by the symbol ○. The basin boundary is marked by the symbol +.

We now consider the case of symmetry breaking by introducing a small nonzero parameter $s_0 \neq 0$. The merging explosion of the two symmetrically related fuzzy attractors is unfolded by the symmetry-breaking parameter, which prevents the two fuzzy attractors from touching the saddle simultaneously, and thus eliminates the merging explosion. When the two period-1 fuzzy attractors touch the saddle at a different ε value, this results in a catastrophic bifurcation and an explosive bifurcation in sequence. Hence, the merging explosion of two fuzzy attractors is not generic under one control parameter (codimension one). For the example with $s_0 = \pm 0.0005$ and $\varepsilon = 0.3855$, the global phase portraits of the attractors with membership distribution are shown in Figure 38.

A bifurcation diagram in the s_0–ε parameter plane is shown in Figure 39. In Figure 39, there are two coexistent fuzzy period-1 attractors denoted by a_{left} and a_{right} in the south quadrant (a), only one fuzzy attractor a_{left} in the west quadrant (c), only one fuzzy attractor a_{right} in the east quadrant (d) and a single large fuzzy attractor a_{large} in the north quadrant (b). a_{left} denotes the attractor on the left of the origin of the phase space, a_{right} represents the attractor on the right of the origin, while a_{large} results from the merger of a_{left} and a_{right}.

The disappearance–appearance of the fuzzy period-1 attractor a_{right} or a_{left} occurs crossing the lower left or right thick curve, respectively, indicating a catastrophic bifurcation. In such a case, the fuzzy period-1 attractor a_{right} or a_{left} collides respectively with the period-1 saddle on the basin boundary before the bifurcation, and suddenly disappears, leaving behind a fuzzy saddle in the place of the original fuzzy attractor in the phase space after the bifurcation. The explosion–implosion of the fuzzy period-1 attractor a_{left} or a_{right} occurs crossing the upper left or right thin curve, respectively, representing an explosive bifurcation. In such a case, the fuzzy period-1 attractor a_{left} or a_{right} collides respectively with a fuzzy saddle in its basin interior before the bifurcation, and suddenly increases its size after the bifurcation. At the same time, the membership distribution changes from a one-peak function representing the fuzzy period-1 attractor a_{left} or a_{right} to a single peak function with a flat platform corresponding to the large fuzzy attractor a_{large}. The codimension two bifurcation vertex is found to be at $(s_0, \varepsilon) = (0, 0.3855)$ at which four distinct bifurcations coincide and interact.

Multiplicative fuzzy noise

Consider now the DVP equation driven by multiplicative fuzzy noise,

$$\begin{aligned}\dot{x}_1 &= x_2,\\ \dot{x}_2 &= x_1 - x_2 - x_1^3 - x_1^2 x_2 + S x_1 + C,\end{aligned} \tag{36}$$

where C is a small symmetry-breaking parameter, S is a fuzzy parameter with the membership function given by equation (32).

The domain $\mathbf{D} = \{-1.75 \leqslant x_1 \leqslant 1.75; -1.0 \leqslant x_2 \leqslant 1.0\}$ is discretized into 141×141 cells. The 5×5 sampling points are used within each cell. The

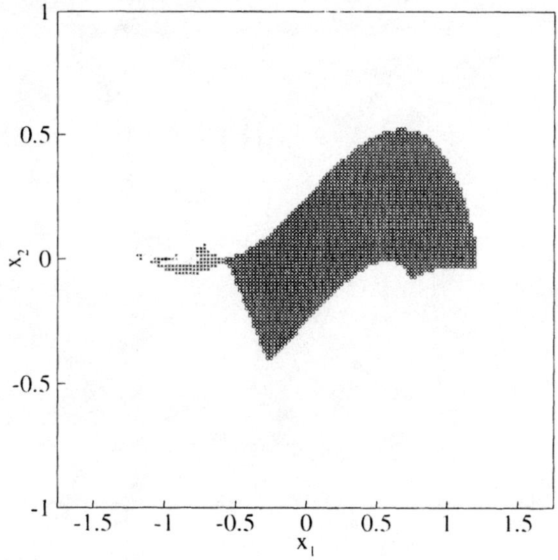

(a)

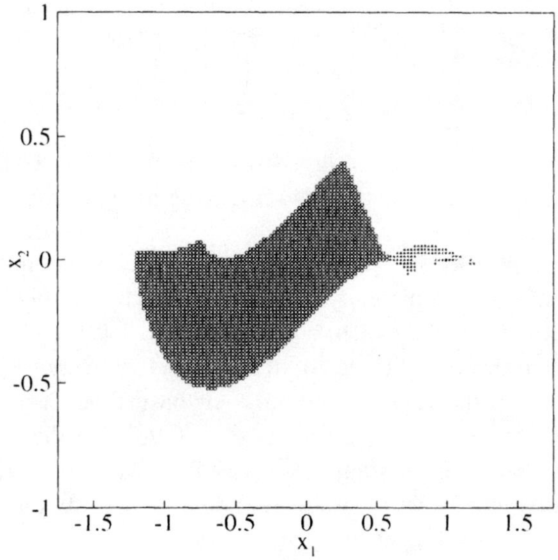

(b)

Figure 38. A symmetry breaking bifurcation of the Duffing–Van der Pol equation (35) with additive fuzzy noise. (a) $s_0 = -0.0005$, $\varepsilon = 0.3855$; (b) $s_0 = 0.0005$, $\varepsilon = 0.3855$. Legends are the same as those in Figure 37.

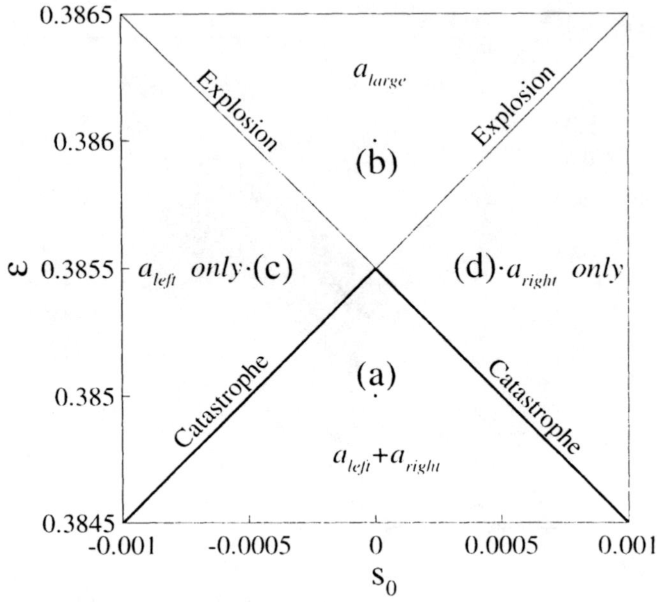

Figure 39. The codimension two bifurcation diagram in the s_0–ε parameter plane for the Duffing–Van der Pol equation (35) with additive fuzzy noise.

membership function is discretized into 401 segments ($M = 401$). Hence, out of each cell, there are 10,025 trajectories with varying membership grades.

When we fix $C = 0$, $s_0 = 0$, and allow the fuzzy intensity ε to vary, the symmetry with respect to the origin of the phase space holds in the noisy DVP equation. As ε increases, a merging explosion bifurcation of two fuzzy period-1 attractors is discovered in the ε interval (0.979, 0.98). The membership functions of the attractors are shown in Figure 40. In this case, two fuzzy period-1 attractors simultaneously touch the fuzzy saddle on their basin boundary at $\varepsilon = 0.979$, and merge into a bigger fuzzy attractor at $\varepsilon = 0.98$ after the collision. At the same time, the membership distribution changes from a two-peak disjoint function representing the two fuzzy period-1 attractors to a single peak function with a flat platform representing the bigger fuzzy attractor.

When $C = \pm 0.0005$ and $\varepsilon = 0.9795$, the global phase portraits of the attractors with membership distribution are shown in Figure 41. A bifurcation diagram in the C–ε parameter plane is shown in Figure 42. The codimension two bifurcation occurs at $(C, \varepsilon) = (0, 0.9795)$ where four distinct bifurcations coincide and interact.

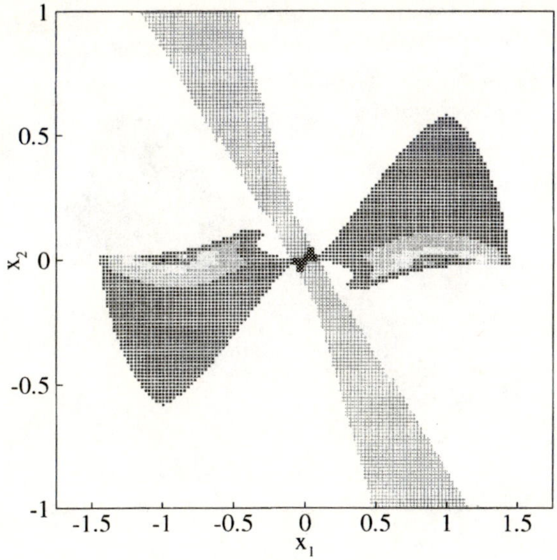

(a)

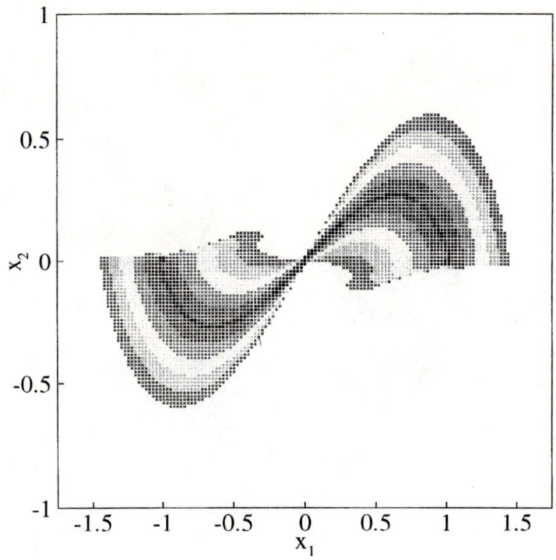

(b)

Figure 40. A merging explosion bifurcation of the Duffing–Van der Pol equation (36) with multiplicative fuzzy noise. (a) $C = 0$, $\varepsilon = 0.979$; (b) $C = 0$, $\varepsilon = 0.98$. Legends are the same as those in Figure 37.

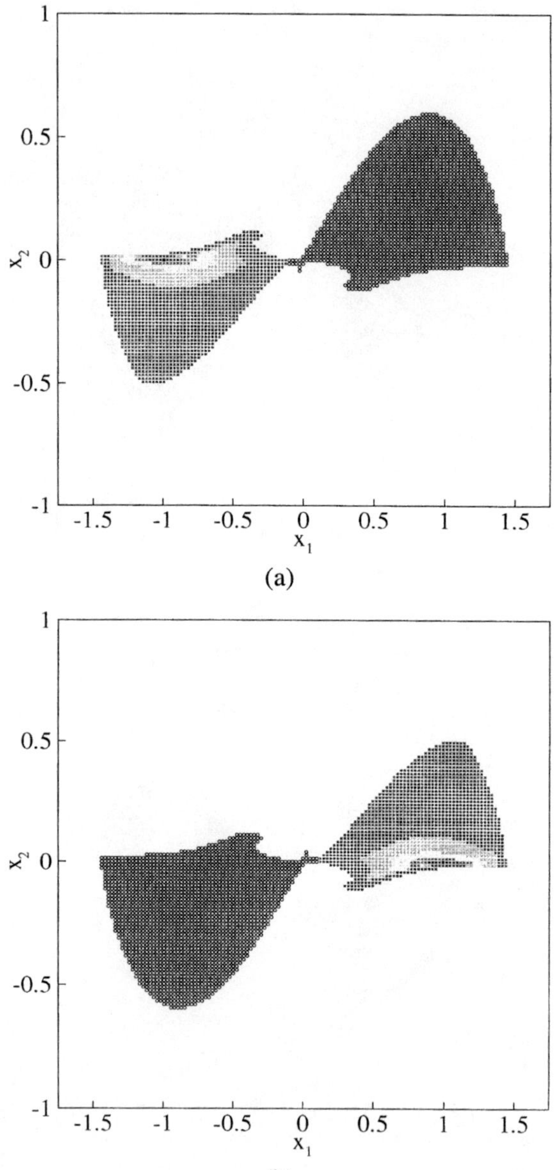

Figure 41. A symmetry breaking bifurcation of the Duffing–Van der Pol equation (36) with multiplicative fuzzy noise. (a) $C = -0.0005$, $\varepsilon = 0.9795$; (b) $C = 0.0005$, $\varepsilon = 0.9795$. Legends are the same as those in Figure 37.

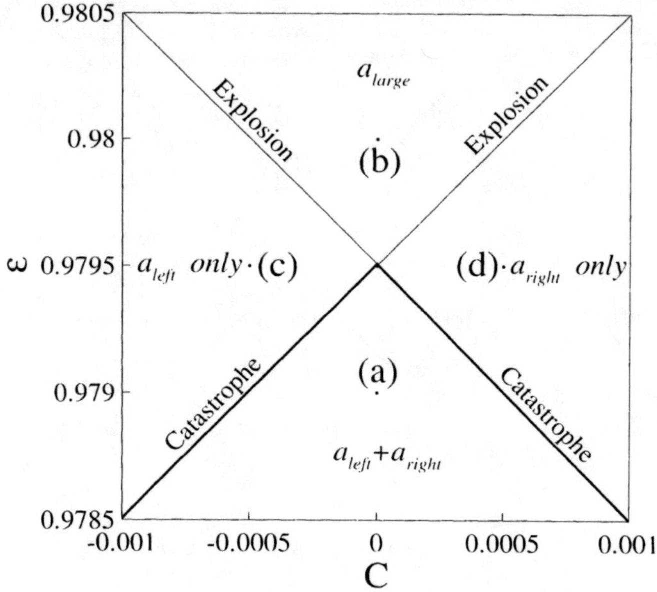

Figure 42. The codimension two bifurcation diagram in the C–ε parameter plane for the Duffing–Van der Pol equation (36) with multiplicative fuzzy noise.

The DVP equation with $(\mu_1, \mu_2) = (0.64, 0.1)$

We have considered the case when $(\mu_1, \mu_2) = (1, -1)$. The case $(\mu_1, \mu_2) = (1, -1)$ is in the lower right quadrant of the parameter space (μ_1, μ_2) where there exist only three fixed points and no closed orbits. Now, we consider a case when $(\mu_1, \mu_2) = (0.64, 0.1)$. This case is in the upper-right quadrant which contains more complicated global bifurcations, involving the coalescence of closed orbits encircling all three fixed points and saddle connections. We consider the DVP equation with additive fuzzy noise,

$$\begin{aligned} \dot{x}_1 &= x_2, \\ \dot{x}_2 &= 0.64x_1 + 0.1x_2 - x_1^3 - x_1^2 x_2 + S, \end{aligned} \tag{37}$$

and with multiplicative fuzzy noise,

$$\begin{aligned} \dot{x}_1 &= x_2, \\ \dot{x}_2 &= 0.64x_1 + 0.1x_2 - x_1^3 - x_1^2 x_2 + Sx_1 + C, \end{aligned} \tag{38}$$

where S has the membership function given by equation (32) and C is a symmetry-breaking parameter. In applying the FGCM method, the domain $\mathbf{D} = \{-1.75 \leqslant x_1 \leqslant 1.75; -1.0 \leqslant x_2 \leqslant 1.0\}$ is discretized into 141×141 cells, 5×5

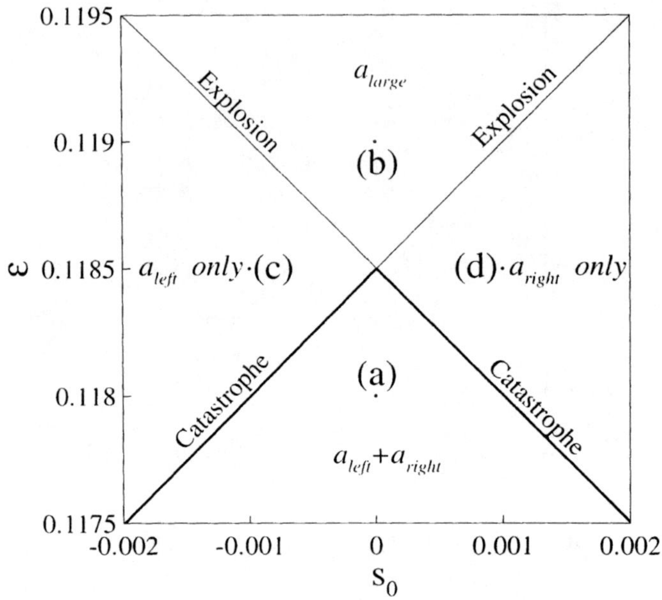

Figure 43. The codimension two bifurcation diagram in the s_0–ε parameter plane for the Duffing–Van der Pol equation (37) with additive fuzzy noise.

sampling points are used within each cell. The membership function is discretized into 201 segments.

The codimension two bifurcation diagrams are presented in Figures 43 and 44. The vertex in Figure 43 is found to be at $(s_0, \varepsilon) = (0, 0.1185)$ and in Figure 44 is at $(C, \varepsilon) = (0, 0.1985)$. Figures 43 and 44 contain four curves of codimension one bifurcations. On each curve, two distinct catastrophes occur on the lower left and right thick curves and two distinct explosions on the upper left and right thin curves.

The global phase portraits for the case with additive fuzzy noise are shown in Figures 45 and 46 to illustrate the codimension two bifurcation process. For the case $(\mu_1, \mu_2) = (1, -1)$, in Figures 37(a) and 40(a), the saddle on the boundary changes from a fixed point at the origin into a domain under the influence of fuzzy noise. For the case $(\mu_1, \mu_2) = (0.64, 0.1)$, however, the saddle is broken into several disjoint segments by the fuzzy noise, as illustrated in Figure 45(a), since there exist a homoclinic saddle connection and a closed orbit outside the connection when $\mu_2 > 0$.

In the above examples, a merging explosion is observed as the fuzzy noise intensity is varied, when two symmetrically related fuzzy attractors collide simultaneously with a fuzzy saddle on the basin boundary. By increasing a small

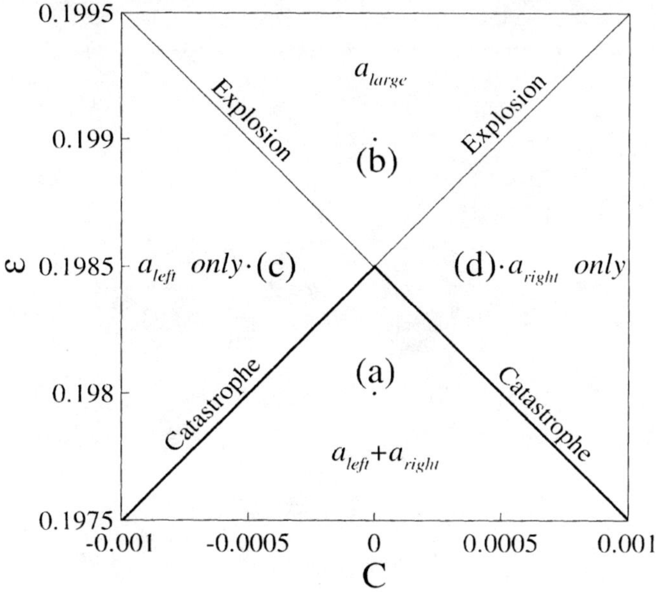

Figure 44. The codimension two bifurcation diagram in the C–ε parameter plane for the Duffing–Van der Pol equation (38) with multiplicative fuzzy noise.

symmetry-breaking parameter and considering both the fuzzy noise intensity and the symmetry-breaking parameter together as control parameters, a codimension two bifurcation is determined. The codimension two bifurcation corresponds to a vertex in a two-parameter plane at which catastrophic and explosive bifurcations coincide. The dynamics of fuzzy systems is extremely rich at the vertex. The codimension two bifurcation is fuzzy noise-induced, which is not seen in deterministic systems.

5.4. Conjectures

In the computation of the examples studied in this section, we have observed that the steady-state membership distribution of fuzzy attractors sometimes is dependent on the initial condition and can assume a platform under certain initial conditions. In other words, the steady-state membership distribution is not always invariant. Then, *under what conditions is the steady-state membership distribution of an attractor invariant?*

It is, in general, difficult to answer this question either analytically or numerically. Based on the observations of the fuzzy response characteristics in the context of cell mapping, we propose the following conjectures.

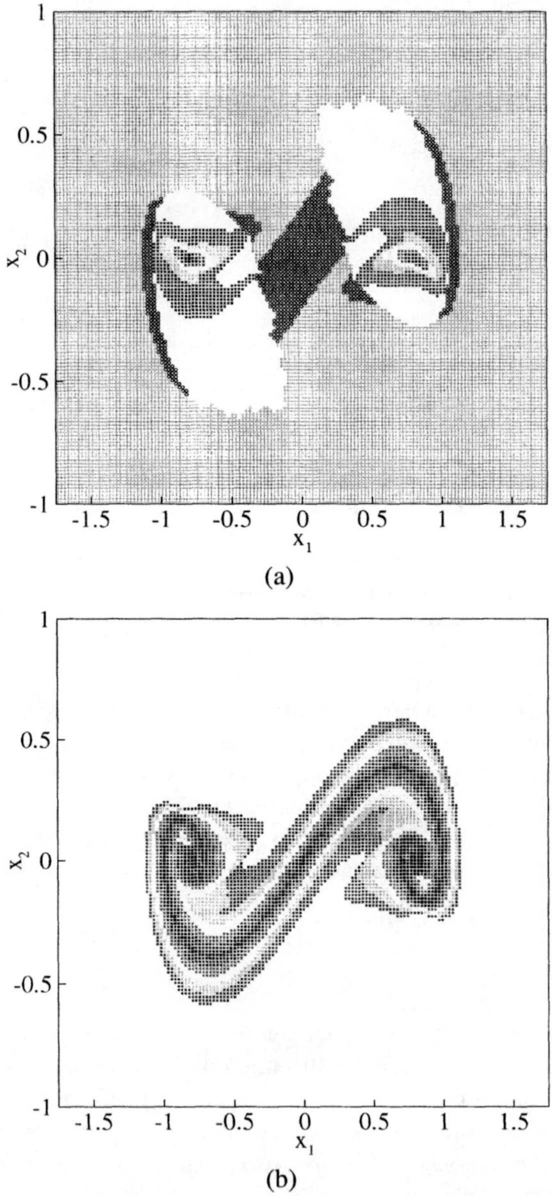

Figure 45. A merging explosion bifurcation of the Duffing–Van der Pol equation (37) with additive fuzzy noise. (a) $s_0 = 0$, $\varepsilon = 0.118$; (b) $s_0 = 0$, $\varepsilon = 0.119$. Legends are the same as those in Figure 37.

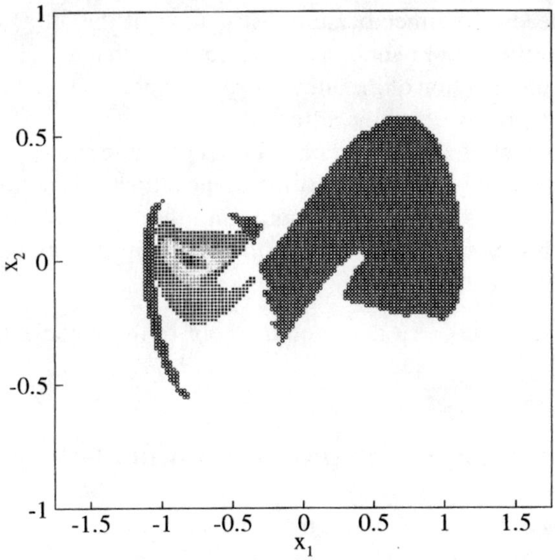

(a)

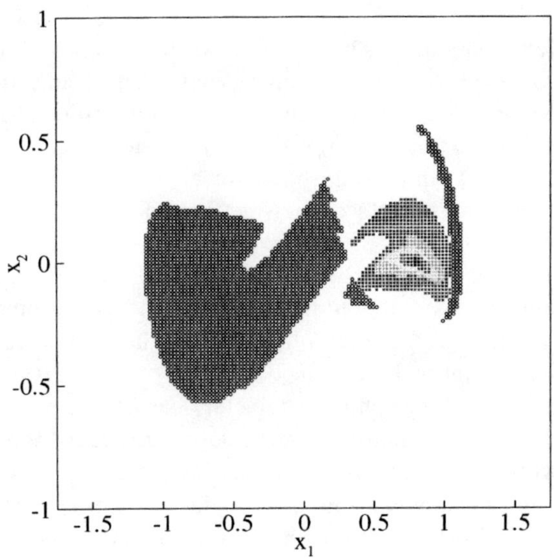

(b)

Figure 46. A symmetry breaking bifurcation of the Duffing–Van der Pol equation (37) with additive fuzzy noise. (a) $s_0 = -0.001$, $\varepsilon = 0.1185$; (b) $s_0 = 0.001$, $\varepsilon = 0.1185$. Legends are the same as those in Figure 37.

1. When the fuzzy set of an attractor does not contain an unstable solution of the system, the steady-state membership distribution of the attractor is invariant with respect to the initial membership distributions that are either completely in the domain of attraction of the attractor and do not contain unstable solutions or are within the fuzzy set of the attractor.
2. When the fuzzy set of an attractor contains an unstable solution of the system, the steady-state membership distribution of the attractor is not invariant.
3. When the attractor is chaotic, it contains an infinite number of unstable trajectories. In this case, the steady-state membership distribution can still be invariant.

It should be pointed out that the rigorous proof of these conjectures is elusive at this time.

6. Effect of bifurcation on semiactive optimal controls

There is one class of control problems that involve adjusting system parameters to achieve control objectives. This class of controls is known as semiactive control (Sun, 1995). For example, vibrations of mechanical systems can be suppressed by varying the stiffness elements of the system (Crespo and Sun, 2000b; Douay and Hagood, 1993; Kobs and Sun, 1997; Onoda, Endo, Tamoaki and Watanabe, 1991). The populations of competing species can be controlled by adjusting parameters that influence the reproduction rate of the species, or the competition among them (Crespo and Sun, 2002; Hrinca, 1997). When the system under control is linear with one globally stable equilibrium point for all the values considered in the control range, the domain of attraction of the fixed point does not change when the control is tuned. However, when the system is nonlinear and has multiple stable equilibrium states, the topological behavior of the system is much more complex. The domains of attraction of the stable fixed points dictate the ability to regulate the dynamics of the system by adjusting the parameter. For nonlinear systems, two cases are normally examined separately.

First, when the range of the control parameter does not include any bifurcation point of the system, the domains of attraction of the stable equilibrium states change slightly with the control. The controllability region from where the control can drive the system towards a stable node is determined by the separatrices of the system corresponding to various control parameters (Crespo and Sun, 2000b, 2002). Furthermore, this controllability region changes with the control parameter in a continuous manner.

Second, when bifurcations of the system occur within the range of the control parameter, the controllability region and the optimal solution may change discontinuously. Because these changes occur through differential changes in the control effort, discrete control strategies about the bifurcation point capture the essence of the problem.

This section examines how bifurcations affect the performance of nonlinear semiactive optimal control systems. There have been many studies about the effect of bifurcations on the control of nonlinear dynamic systems. The equilibrium sets, controllability and stabilizability of dynamical systems near bifurcation have been studied in Wei (1998a, 1998b) using a state feedback strategy. Chen, Yap and Lu (1998) have proposed a state-feedback strategy to control vibrations near Hopf bifurcations for both continuous time and discrete time systems. In this work, the control strategy is either to shift an existing bifurcation or to create a new one. Parametric state feedback control and the harmonic balance method are used in Genesio, Tesi, Wang and Abed (1993) to develop a technique which delays and stabilizes period doubling bifurcations in continuous time nonlinear systems. In Nayfeh and Balachandran (1995), different bifurcation control methods and the pole placement technique for controlling chaotic systems are presented. In Cheng, Yang, Hackl and Chajes (1993), the stability, bifurcation and chaos of autonomous and nonautonomous nonlinear structures are studied by using linear feedback control. Chaos control via small perturbations is applied to the shipboard cranes problem in Ott, Kostelich, Yuan, Hunt, Grebogi and Yorke (1996).

In the following, we first review the optimal control problem formation, and the solution strategy by using the SCM method. Three examples are presented subsequently.

6.1. Optimal control problem

Consider a dynamical system governed by the nonlinear time-varying differential equation,

$$\dot{\mathbf{x}}(t) = \mathbf{f}\big(\mathbf{x}(t), \mathbf{u}(t), t\big), \tag{39}$$

where $\mathbf{x}(t) \in \mathbb{R}^n$ is the state vector and $\mathbf{u}(t)$ is a control vector of the system. When $\mathbf{u}(t)$ consists of a set of system parameters, rather than external excitations, it is called a semiactive control. Define a performance index as,

$$J(\mathbf{u}) = \phi\big(\mathbf{x}(T), T\big) + \int_{t_0}^{T} L\big(\mathbf{x}(t), \mathbf{u}(t), t\big)\, \mathrm{d}t, \tag{40}$$

where $[t_0, T]$ is the time interval of interest, $\phi(\mathbf{x}(T), T)$ is the terminal cost and $L(\mathbf{x}(t), \mathbf{u}(t), t)$ is the positive Lagrangian function. The optimal semiactive control is to find a sequence of parametric controls $\mathbf{u}(t)$ within a given admissible set $\mathbf{U} \subset \mathbb{R}^m$ on the time interval $[t_0, T]$ that drives the system from a given initial condition $\mathbf{x}(t_0) = \mathbf{x}_0$ to the target set defined by $\mathbf{\Psi}(\mathbf{x}(T), T) = 0$ such that the cost function $J(\mathbf{u})$ is minimized. The target set $\mathbf{\Psi}$ describes any simply or multiply connected region in $\mathbb{R}^n$ where the controller must drive the system. The optimal control solution is determined by a set of coupled equations for the state and co-state vectors subject to initial and terminal conditions. A detailed discussion of these equations can be found in Lewis and Syrmos (1995).

Assume that the state equation (39) can be discretized in time domain leading to a point-to-point mapping given by

$$\mathbf{x}(k) = \mathbf{F}\big(\mathbf{x}(k-1), \mathbf{u}(k), \Delta t(k)\big), \tag{41}$$

where $\mathbf{x}(k) \in \mathbb{R}^n$ is the state vector at the kth mapping step and $\mathbf{u}(k) \in \mathbb{U}$ is the control in the kth time interval. $\mathbb{U}$ is a discrete representation of the set $\mathbf{U}$. The simple cell mapping transforms the point-to-point mapping into a cell-to-cell mapping by discretizing the phase space. By allowing only a finite set of time intervals and a finite set of admissible control inputs, the performance index $J(t_0)$ is discretized accordingly. We denote the set of all admissible control vectors as $\mathbb{U} = \{\mathbf{u}_1, \mathbf{u}_2, \ldots, \mathbf{u}_{N_c}\}$, and the set of all mapping time steps as $\mathbb{T} = \{\Delta t_1, \Delta t_2, \ldots, \Delta t_{N_t}\}$.

After discretizing the state space, the point-to-point mapping (41) is replaced with the simple cell mapping,

$$j(k) = C\big(j(k-1), \mathbf{u}(k), \Delta t(k)\big). \tag{42}$$

General control database

The approximation of equation (41) with (42) is a source of error in the long term behavior of the system response. Such a problem affects the accuracy and convergence of the control solution considerably (Bursal and Hsu, 1989; Hsu, 1985, 1987; Wang and Lever, 1994). An iterative method that uses nonuniform time steps has been proposed by Crespo and Sun (2000a, 2000b) to overcome this difficulty. A mapping time step is chosen for each cell such that the end point of the trajectory starting from the center of this cell within a specified duration of time is closest to the center of its image cell.

When applying the SCM method to the optimal control problem, we first construct a database of cell mappings under all allowable controls in $\mathbb{U}$ by integrating the state equations. Recall that N_c is the number of admissible control values in the set $\mathbb{U}$. Let $\mathbf{N}_m$ denote the set of mappings from a pre-image cell to its first N_m consecutive image cells along the trajectory under a given control. These mappings have nonuniform mapping time steps. The general control database contains the following elements: for each pre-image cell i, there will be $N_c \times N_m$ image cells j, the corresponding controls $\mathbf{u}_l$, the associated mapping time steps Δt_{ijl} and the incremental control costs

$$\Delta J_{ijl} = \int_{t_0}^{t_0+\Delta t_{ijl}} L\big(\mathbf{x}(t), \mathbf{u}(t), t\big)\, dt. \tag{43}$$

We denote this database of the complete set of mappings by $\mathbf{M}$. A special subset of all the mappings $\mathbf{M}$ denoted by $\mathbf{N}$ contains the image cells in the closest neighborhood of every pre-image cell i under all allowable controls. The following is a description of the search algorithm based on the SCM method with nonuniform time steps.

Search algorithm for fixed final state problems

Let $\mathbf{\Omega} \in \mathbb{R}^n$ denote the set of cells representing the target set defined by $\mathbf{\Psi}(\mathbf{x}(T), T) = 0$. Let N_b be the number of backward search iterations that we would like to carry out. Initially, the search is over the mapping set $\mathbf{N}$ according to the following steps:

1. Search through all the mappings to find all the cells in the set that are mapped into $\mathbf{\Omega}$ in one mapping step.
2. Assign a cumulative cost to the cells found in step 1. The cumulative cost is the smallest cost for the system to move from the current cell to the original target set. It is calculated by adding the cumulative cost of the image cell and the incremental cost of the current cell. If more than one image cell reached the target, the mapping with smallest cumulative cost is taken. Note that the cells in the original target set have a cumulative cost given by $\phi(\mathbf{x}(T), T)$.
3. Expand the target set $\mathbf{\Omega}$ by including the cells found in step 1 with less cumulative cost.
4. Repeat the search from step 1 until all the cells in the state space are processed.
5. Examine the cumulative costs of all N_m consecutive image cells j for every pre-image cell i and for every control $\mathbf{u}_l$ in $\mathbb{U}$. We retain the image cell that has the smallest cumulative cost. This cell is stored in a set $\mathbf{M}^*$.
6. Repeat from step 1 over the set $\mathbf{M}^*$ for $N_\mathrm{b} - 1$ times.

It has been shown in Crespo and Sun (2000b) that the average cost over the entire phase space is reduced with the number of backward searches. In other words, the above backward search is a converging process. Note that the final set $\mathbf{M}^*$ contains the information on the location of the switching curves and the optimal controls for each cell. A discriminating function which considers smoothness and local continuity of the trajectories may be used to break the cost ties in the backward search. The main objective of the section is to apply the above methodology to study the effect of bifurcation on the performance of parametric controls of nonlinear systems. It should be noted that this algorithm can also deal with fixed final time optimal control problems (Crespo and Sun, 2000a).

In order to isolate the effect of bifurcations on the control solution, bang–bang control sets with values that encompass the bifurcation point are used. In the following examples, a scalar parametric control u with two admissible values is considered. Let the bifurcation point of the system with respect to the parameter u be denoted by u_bf. The set $\mathbb{U}_1 = \{u_1, u_2\}$ consists of the parametric controls such that $u_1 < u_2 < u_{\rm bf}$. The set $\mathbb{U}_2 = \{u_1, u_3\}$ involves two parametric controls that encompass the bifurcating value, i.e. $u_2 < u_{\rm bf} < u_3$. The effect of bifurcation on semiactive controls is studied by comparing the solutions of the control problem with these two admissible sets.

6.2. Saddle-node bifurcation

Consider a nonlinear dynamical system given by

$$\begin{aligned} \dot{x}_1 &= -ux_1 + x_2, \\ \dot{x}_2 &= \frac{x_1^2}{1 + x_1^2} - bx_2, \end{aligned} \tag{44}$$

where u is the parametric control. This system has been discussed in Griffith (1971) as a model for a genetic control system. When the control satisfies $-\infty < u < u_{\rm bf} = 1/(2b)$, there are three fixed points: a stable node at the origin $\mathbf{x}_0 = (0, 0)$, a saddle node at

$$\mathbf{x}_u = \left(\frac{1 - \sqrt{1 - 4u^2b^2}}{2ub}, \frac{1 - \sqrt{1 - 4u^2b^2}}{2b} \right), \tag{45}$$

and a stable node at

$$\mathbf{x}_s = \left(\frac{1 + \sqrt{1 - 4u^2b^2}}{2ub}, \frac{1 + \sqrt{1 - 4u^2b^2}}{2b} \right). \tag{46}$$

When $u > u_{\rm bf}$, there is only one fixed point, which is the stable node at the origin. As u approaches $u_{\rm bf}$ from below, the fixed points $\mathbf{x}_u$ and $\mathbf{x}_s$ collide and disappear. When $u < u_{\rm bf}$, the unstable manifold of the saddle $\mathbf{x}_u$ is trapped between the nullclines of the system and the stable manifold defines a separatrix. The stable manifold of $\mathbf{x}_s$ outlines the basin of attraction of $\mathbf{x}_0$.

We have taken $b = 1$, $\mathbb{U}_1 = \{0.4, 0.495\}$ and $\mathbb{U}_2 = \{0.4, 0.505\}$ in the numerical simulations reported next. The region $\mathbf{D} = \{-0.5 < x_1 < 1.5; -0.5 < x_2 < 1.5\}$ in the state space is discretized with 7922 square cells. First, let us take the cell that contains the point $\mathbf{x}_0$ as the target set. The objective is thus to drive the system from any initial condition to $\mathbf{x}_0$ in minimum time by adjusting the parameter u. The optimal control solution for $u \in \mathbb{U}_1$ is shown in Figure 47. Figure 48 presents the solution for the same optimal control problem with $u \in \mathbb{U}_2$. It can be observed that the controllable region of the optimal control problem with $u \in \mathbb{U}_2$ is significantly enlarged due to the change of the basin of attraction of $\mathbf{x}_0$ after bifurcation. Due to the annihilation of the saddle and the fixed point, the boundary of the original controllability area disappears, making the entire phase space the new domain of attraction. This result indicates that a significant change in the controllability area can be made through a small change in the control. In this

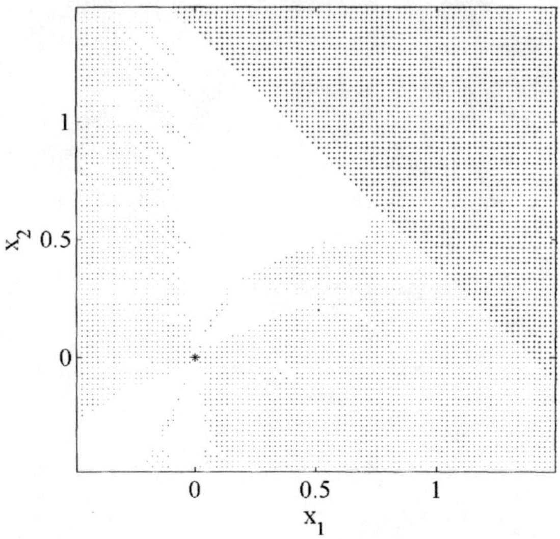

Figure 47. Global distribution of the optimal control solution with $\mathbb{U}_1$ for the system (44). The target (0, 0) is tagged with an asterisk. Cells with u_2 as the starting optimal control are marked with dots. Cells with u_1 as the starting optimal control are not marked. Cells marked with circles form the uncontrollable region.

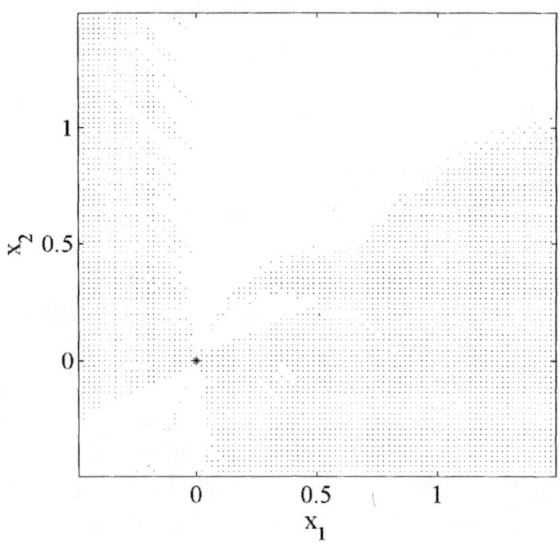

Figure 48. Global distribution of the optimal control solution with $\mathbb{U}_2$ for the system (44). The target (0, 0) is tagged with an asterisk. Cells with u_3 as the starting optimal control are marked with dots. Cells with u_1 as the starting optimal control are not marked. Cells marked with circles form the uncontrollable region.

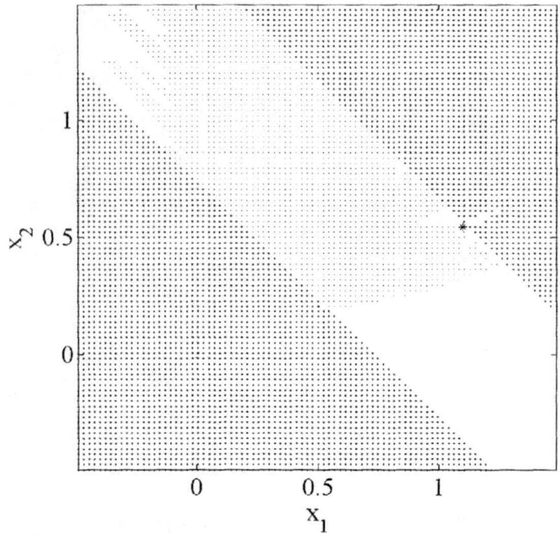

Figure 49. Global distribution of the optimal control solution with $\mathbb{U}_1$ for the system (44). The target $(1.1, 0.546)$ is tagged with an asterisk. Cells with u_2 as the starting optimal control are marked with dots. Cells with u_1 as the starting optimal control are not marked. Cells marked with circles form the uncontrollable region.

way, a much higher control authority can be achieved by a small increase in the control effort. Note also that away from the separatrix created by the stable manifold of the saddle, the optimal control solution seems to be locally insensitive to the change in the parameter. The optimal control solution in the enlarged area is a smooth continuation of the control distribution found for $\mathbb{U}_1$.

Figure 49 shows the optimal solution for a target cell containing the point $(1.1, 0.56)$ with the control set $\mathbb{U}_1$. In this case, the controllability area is bounded on the left by the stable manifold of the system when $u = u_1$. The other boundary is determined by the vector field. Different targets will have different right boundaries. For the control set $\mathbb{U}_2$, the left boundary remains the same, but the right one disappears as seen in Figure 50. The flow of this dynamical system has a volume contraction in the phase space, and the trajectories converge to the unstable manifold of the saddle node. The bifurcation introduces substantial changes to the vector field along the unstable manifold affecting the global distribution of optimal controls. Far from these locations, only minor changes in the control solution occur. In this case, the distribution of the optimal control solution has the same geometry for both control sets $\mathbb{U}_1$ and $\mathbb{U}_2$.

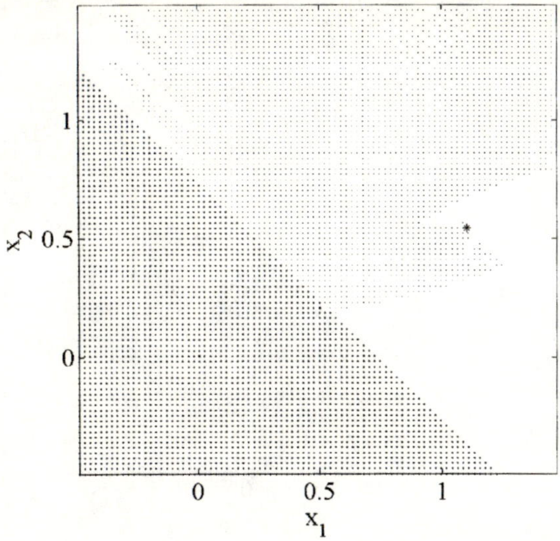

Figure 50. Global distribution of the optimal control solution with $\mathbb{U}_2$ for the system (44). The target (1.1, 0.546) is tagged with an asterisk. Cells with u_3 as the starting optimal control are marked with dots. Cells with u_1 as the starting optimal control are not marked. Cells marked with circles form the uncontrollable region.

6.3. *Supercritical Pitchfork bifurcation*

Consider the nonlinear dynamical system studied in Strogatz (1997),

$$\begin{aligned} \dot{x}_1 &= -ux_1 + x_2 + \sin(x_1), \\ \dot{x}_2 &= x_1 - x_2, \end{aligned} \tag{47}$$

where u is a parametric control. The system is symmetric about the origin. When $-\infty < u < u_{bf} = -2$, there is only one fixed point. This point is a stable node $\mathbf{x}_0 = (0, 0)$. In this range, the eigendirection associated with the minimum eigenvalue at $\mathbf{x}_0$ is given by $[u/2 - (1/2)(u^2 + 4u + 8)^{1/2} + 1, 1]$. When $u > u_{bf}$, the system has three fixed points: a saddle node at $\mathbf{x}_{sn} = (0, 0)$, and two stable nodes $\mathbf{x}_{s_1}$ and $\mathbf{x}_{s_2}$. For values of u slightly bigger than u_{bf} these points are located approximately at $\mathbf{x}_{s_1} \approx (\sqrt{6(u+2)}, \sqrt{6(u+2)})$ and $\mathbf{x}_{s_2} \approx (-\sqrt{6(u+2)}, -\sqrt{6(u+2)})$. As u approaches u_{bf} from above, the fixed points $\mathbf{x}_{s_1}$ and $\mathbf{x}_{s_2}$ merge at the origin and then disappear. At bifurcation, the eigenvectors of the fixed point at the origin are $\mathbf{v}_1 = [1, 1]$ and $\mathbf{v}_2 = [1, -1]$. Because the sign of the eigenvalue associated with the vector $\mathbf{v}_1$ changes at this instance, the bifurcation occurs in this direction. It is important to remark that once the stable node $\mathbf{x}_0$ becomes a saddle $\mathbf{x}_{sn}$, its stable manifold divides the phase space into two basins of attraction.

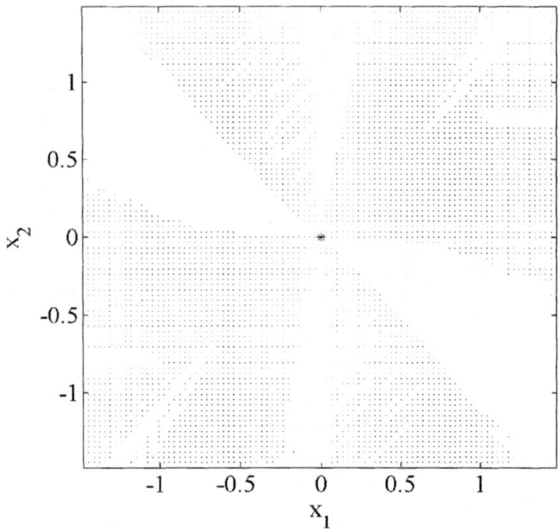

Figure 51. Global distribution of the optimal control solution with $\mathbb{U}_1$ for the system (47). The target (0, 0) is tagged with an asterisk. Cells with u_2 as the starting optimal control are marked with dots. Cells with u_1 as the starting optimal control are not marked. Cells marked with circles form the uncontrollable region.

In the cell mapping analysis, the region $\mathbf{D} = \{-1.5 < x_1 < 1.5; -1.5 < x_2 < 1.5\}$ is discretized into 7922 square cells. Let us take the origin $\mathbf{x}_0 = (0, 0)$ as the control target. Hence, the objective is to drive the system from any initial condition to the origin in minimum time by adjusting the parameter u. The bang–bang controls are considered with $\mathbb{U}_1 = \{-4, -2.05\}$ and $\mathbb{U}_2 = \{-4, -1.95\}$. Figure 51 shows the optimal control solution for $\mathbb{U}_1$. Figure 52 presents the solution for the same optimal control problem with $u \in \mathbb{U}_2$. By comparing Figures 51 and 52, it can be seen that changes in the optimal control solution propagate mostly in the directions of $\mathbf{v}_1$ and $\mathbf{v}_2$. Away from the manifolds of the saddle, the distribution of optimal control solution is the same for both control sets. The separatrix in the direction of $\mathbf{v}_2$ leads to a fuzzy distribution of the optimal control solution as can be seen in Figure 52. This is due to multiple control actions with relatively close cumulative costs. Far away from these locations, the optimal control solution remains insensitive to bifurcation. If the target cell is placed on the unstable manifold of the saddle just before bifurcation, i.e., $u = u_{bf} - \varepsilon$, $0 < \varepsilon \ll 1$, the optimal control solution and the controllability area change substantially. Away from it, the solution seems to be insensitive to bifurcation.

As a second example, we take a target cell located on the manifold associated with the slow eigendirection of $\mathbf{x}_0$ for $u = u_{bf}$. Figures 53 and 54 present the control solution for the target cell containing the point $\mathbf{x}_t = (-0.3, -0.34)$.

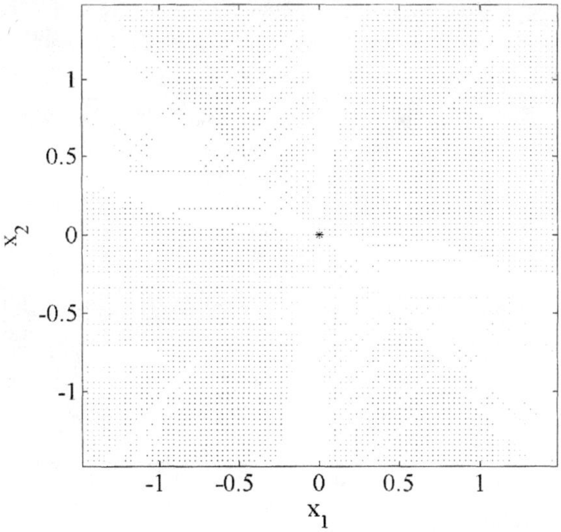

Figure 52. Global distribution of the optimal control solution with $\mathbb{U}_2$ for the system (47). The target (0, 0) is tagged with an asterisk. Cells with u_3 as the starting optimal control are marked with dots. Cells with u_1 as the starting optimal control are not marked. Cells marked with circles form the uncontrollable region.

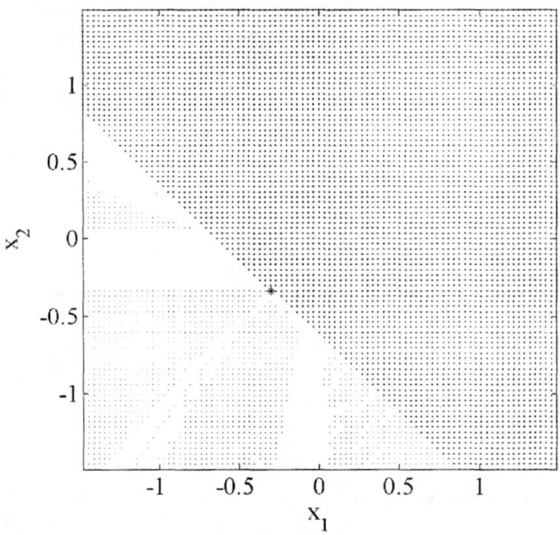

Figure 53. Global distribution of the optimal control solution with $\mathbb{U}_1$ for the system (47). The target $(-0.3, -0.34)$ is tagged with an asterisk. Cells with u_2 as the starting optimal control are marked with dots. Cells with u_1 as the starting optimal control are not marked. Cells marked with circles form the uncontrollable region.

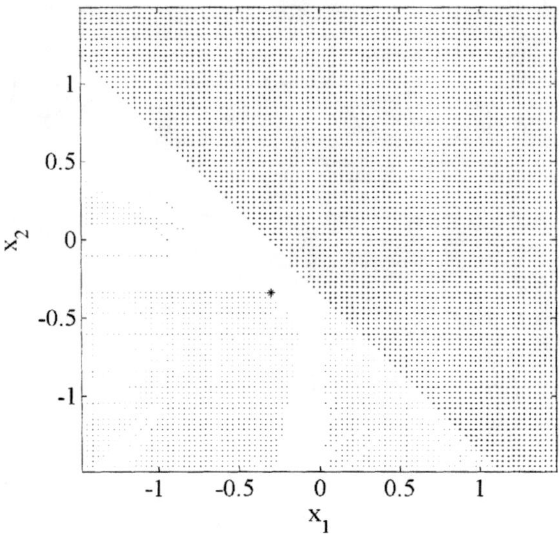

Figure 54. Global distribution of the optimal control solution with $\mathbb{U}_2$ for the system (47). The target $(-0.3, -0.34)$ is tagged with an asterisk. Cells with u_3 as the starting optimal control are marked with dots. Cells with u_1 as the starting optimal control are not marked. Cells marked with circles form the uncontrollable region.

A considerable enlargement of the controllability region takes place. In addition, a smooth continuation of the control distribution into the added region can be observed. In Figure 53, the boundary of the controllable area is determined by the eigenvector associated with the fast eigenvalue of the linearized system about the point $\mathbf{x}_t$.

6.4. Subcritical Hopf bifurcation

Consider the following nonlinear dynamical system (Strogatz, 1997),

$$\dot{x}_1 = ux_1 - x_2 + x_1x_2^2,$$
$$\dot{x}_2 = x_1 + ux_2 + x_2^3, \tag{48}$$

where u is the control. When $-\infty < u < u_{\mathrm{bf}} = 0$, there is only one fixed point. This point is a stable spiral located at the origin $\mathbf{x}_0 = (0, 0)$. The basin of attraction of this point is bounded by an elliptical limit cycle. For $u > u_{\mathrm{bf}}$, $\mathbf{x}_0$ becomes an unstable spiral. When u reaches u_{bf} from below, the stable spiral collides with the limit cycle becoming unstable, hence the elliptical domain of attraction disappears.

In applying the cell mapping method, the region $\mathbf{D} = \{-0.9 < x_1 < 0.9; -0.9 < x_2 < 0.9\}$ is discretized into 7922 square cells. In order to avoid the inci-

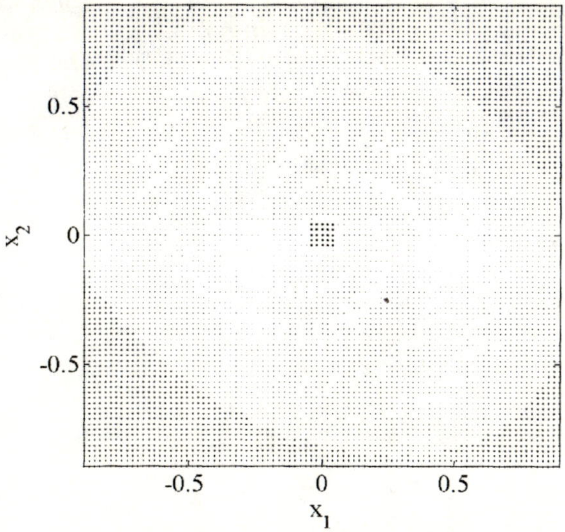

Figure 55. Global distribution of the optimal control solution with $\mathbb{U}_1$ of the system (48). The target set around the point $(0, 0)$ is tagged with asterisks. Cells with u_2 as the starting optimal control are marked with dots. Cells with u_1 as the starting optimal control are not marked. Cells marked with circles form the uncontrollable region.

dence of the error due to the local similarities of the vector fields associated with different values of u, a uniform and relatively big mapping time step $\Delta t = 0.175$.

First, we assume that the target set $\Psi(x(T), T)$ consists of a group of cells in the vicinity of the origin. Let us denote this set of cells as $\mathbb{S}$. The goal is to drive the system from any initial condition to $\mathbb{S}$ in minimum time by adjusting the parameter u. For the bang–bang control problems considered here, we use $\mathbb{U}_1 = \{-0.4, -0.1\}$ and $\mathbb{U}_2 = \{-0.4, 0.1\}$. The optimal control solution for the $\mathbb{U}_1$ problem is shown in Figure 55. It is found that for this target set, the distributions of the optimal control solutions for $\mathbb{U}_1$ and $\mathbb{U}_2$ are identical. As u approaches and exceeds u_{bf} from below, the stable spiral inside the limit cycle becomes unstable. In this transition, the sign of the trace of the Jacobian changes from negative to positive. Because near bifurcation the rates of convergence and divergence of the system with respect to the limit cycle are very small, the orbits of the system when $u = u_2$ and $u = u_3$ become noticeably different only over a relatively longer time.

Figures 56 and 57 show the optimal solutions for the control sets $\mathbb{U}_1$ and $\mathbb{U}_2$ when a single target cell containing the point $(0.25, 0)$ is used. This target cell is located inside both limit cycles. Both controls in $\mathbb{U}_1$ drive the system to the origin for initial conditions inside them. The controllability area is externally bounded by the limit cycle associated with u_1. The shape of the inner boundary depends on the vector field of the system when $u = u_1$ and $u = u_2$. The control solution

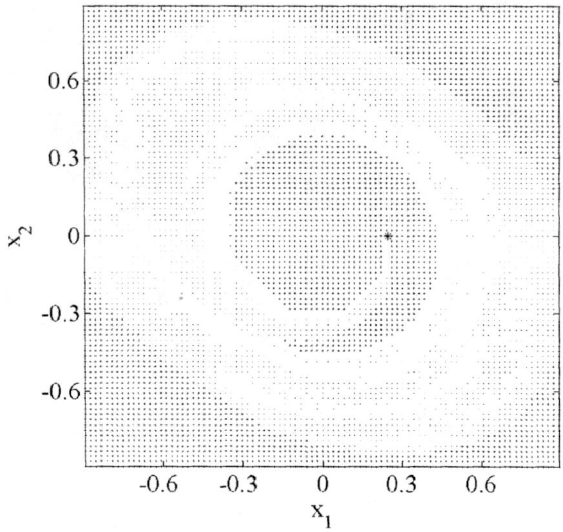

Figure 56. Global distribution of the optimal control solution with $\mathbb{U}_1$ for the system (48). The target (0.25, 0) is tagged with an asterisk. Cells with u_2 as the starting optimal control are marked with dots. Cells with u_1 as the starting optimal control are not marked. Cells marked with circles form the uncontrollable region.

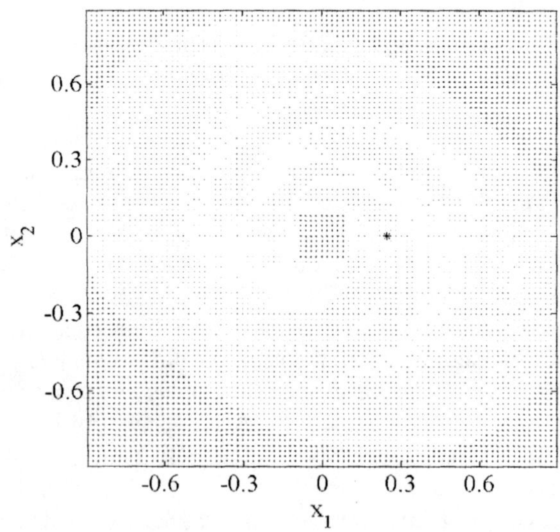

Figure 57. Global distribution of the optimal control solution with $\mathbb{U}_2$ for the system (48). The target (0.25, 0) is tagged with an asterisk. Cells with u_3 as the starting optimal control are marked with dots. Cells with u_1 as the starting optimal control are not marked. Cells marked with circles form the uncontrollable region.

for all targets inside of both limit cycles will have a controllability area containing the entire region between them. With the control set $\mathbb{U}_2$, the controllable area and the optimal control solution change substantially. At bifurcation, the interior limit cycle disappears, and a new diverging spiral at the origin enlarges the controllable area with $\mathbb{U}_1$ by driving points within the limit cycle with $u = u_2$ to the regions where u_1 can move the system to the target. By comparing both solutions, one can see that the distributions of controls for the cases in which $\mathbb{U}_1$ and $\mathbb{U}_2$ are used are not closely related as before.

It should be noted that in Figure 57 and later in Figures 59 and 61, the set $\mathbb{S}$ is avoided in the backward search algorithm. Within the current cell mapping approximation, $\mathbb{S}$ is a closed set meaning that all the cells map to themselves even when the origin is unstable. In the absence of $\mathbb{S}$, the controllability area in these figures would not have a hole around the origin. The set $\mathbb{S}$ can be made smaller by using smaller cells and much shorter mapping time steps.

Figures 58 and 59 show the optimal solutions for the control sets $\mathbb{U}_1$ and $\mathbb{U}_2$ when the target is a single cell containing the point $(-0.6, 0.4)$. This target cell is located between the limit cycles associated with u_1 and u_2. The optimal control solution for $\mathbb{U}_1$ is bounded by the limit cycles. In the controllability area, u_1 drives the system towards the origin, while u_2 does the opposite. The optimal control solution is a nontrivial combination of both controls. In contrast, the

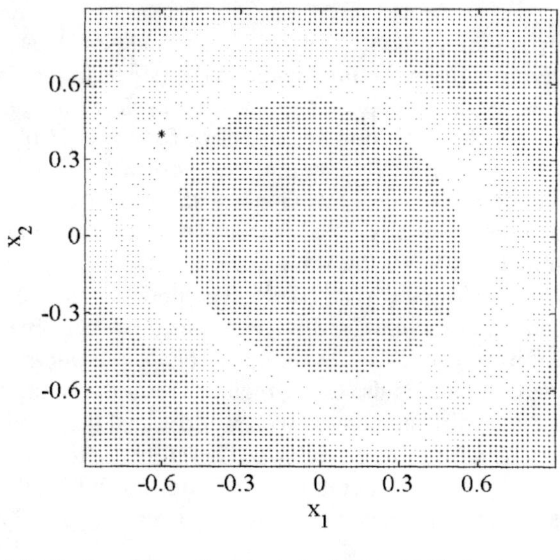

Figure 58. Global distribution of the optimal control solution with $\mathbb{U}_1$ for the system (48). The target $(-0.6, 0.4)$ is tagged with an asterisk. Cells with u_2 as the starting optimal control are marked with dots. Cells with u_1 as the starting optimal control are not marked. Cells marked with circles form the uncontrollable region.

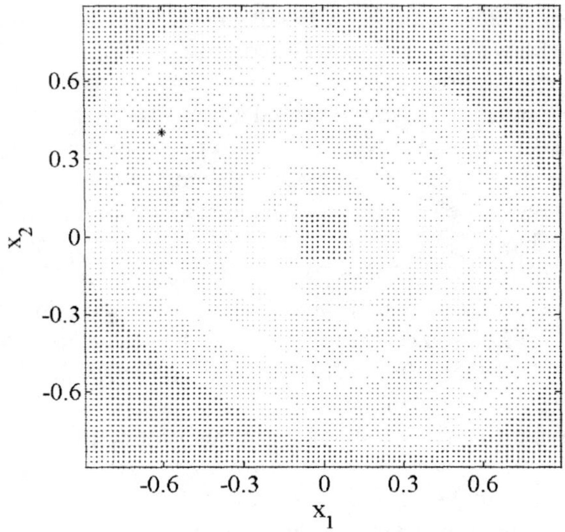

Figure 59. Global distribution of the optimal control solution with $\mathbb{U}_2$ for the system (48). The target $(-0.6, 0.4)$ is tagged with an asterisk. Cells with u_3 as the starting optimal control are marked with dots. Cells with u_1 as the starting optimal control are not marked. Cells marked with circles form the uncontrollable region.

problem with $\mathbb{U}_2$ exhibits a completely different behavior. Now, the control set drives all the interior point of the limit cycle to the target. In the absence of $\mathbb{S}$, the controllability area would be bounded by the limit cycle with u_1 only.

Figures 60 and 61 show the optimal solutions for the control sets $\mathbb{U}_1$ and $\mathbb{U}_2$ when the target is a single cell containing the point $(0.4, 0.8)$. This target cell is located outside the limit cycles associated with both u_1 and u_2. In this region all trajectories diverge to infinity. The distribution of the optimal control solution with $\mathbb{U}_1$ is bounded inside by the limit cycle associated with u_2. Outside, the boundary is determined by the vector field. The optimal control problem for all targets outside both limit cycles will have a controllability area containing the region between the limit cycles. On the other hand, the controllability area of the problem with $\mathbb{U}_2$ contains the entire region within the limit cycle associated with u_1. The unstable spiral associated with u_3 drives all the interior points of the limit cycle associated with u_1 to the outside. Therefore, both controls together can drive the system to the target. By comparing Figures 60 and 61, it can be seen that bifurcation causes a substantial change in the controls in a global and local basis.

A summary

In order to capture the effect of bifurcation on the global distribution of optimal controls, bang–bang control sets with values that closely encompass the bifurca-

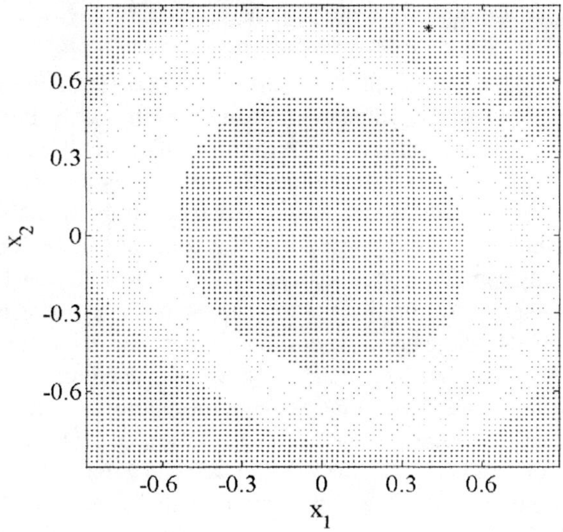

Figure 60. Global distribution of the optimal control solution with $\mathbb{U}_1$ for the system (48). The target (0.4, 0.8) is tagged with an asterisk. Cells with u_2 as the starting optimal control are marked with dots. Cells with u_1 as the starting optimal control are not marked. Cells marked with circles form the uncontrollable region.

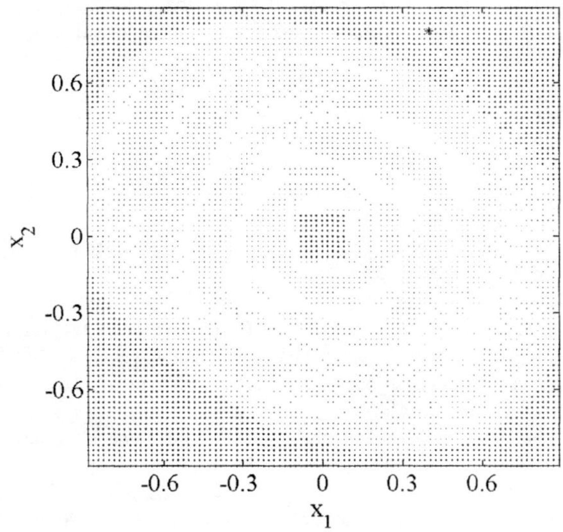

Figure 61. Global distribution of the optimal control solution with $\mathbb{U}_2$ for the system (48). The target (0.4, 0.8) is tagged with an asterisk. Cells with u_3 as the starting optimal control are marked with dots. Cells with u_1 as the starting optimal control are not marked. Cells marked with circles form the uncontrollable region.

tion point are used. Mathematical models that exhibit saddle node, supercritical pitchfork Hopf and subcritical Hopf bifurcations are considered in the examples.

For the saddle-node bifurcation, the distribution of optimal controls is insensitive to the bifurcation locally. Substantial changes in the control distribution take place only when the target state is close to the unstable manifold of the saddle. For the supercritical pitchfork bifurcation, the solution is dependent on the state location of the target as well. If the target is placed on the unstable manifold of the saddle node, but not on the fixed point, the control distribution and the controllability of the system change considerably. Away from the unstable manifold, the control is insensitive to the bifurcation. The subcritical Hopf bifurcation has more dramatic effects on the control solution. In this case, the control is quite sensitive to the location of the target set. If the target is placed on the fixed point, the control distribution remains unchanged during bifurcation. For a target set placed elsewhere, substantial changes in the optimal solution occur locally and globally.

Since bifurcations and target states strongly affect the control solution, the control parameter and target locations can be manipulated to achieve the best control performance and the largest controllability region of the system with the smallest effort.

References

Abraham, R.H., Stewart, H.B., 1986. A chaotic blue sky catastrophe in forced relaxation oscillations. Physica D 19, 394–400.

Arnold, L., 1998. Random Dynamical Systems. Springer-Verlag, New York.

Baxendale, P., 1986. Asymptotic behavior of stochastic flows of diffeomorphisms. In: Itô, K., Hida, T. (Eds.), Proceedings of Stochastic Processes and Their Applications, Lecture Notes in Math., vol. 1203. Springer-Verlag, Berlin–Heidelberg, pp. 1–19.

Brambilla, M., Lugiato, L.A., Strini, G., Narducci, L.M., 1986. Influence of phase diffusion on spontaneous oscillations in driven optical systems. Phys. Rev. A 34, 1237–1241.

Bursal, F.H., Hsu, C.S., 1989. Application of a cell-mapping method to optimal control problems. Internat. J. Control 49 (5), 1505–1522.

Caughey, T.K., 1971. Nonlinear theory of random vibration. Adv. Appl. Mech. 11, 209–253.

Chen, G., Yap, K.C., Lu, J., 1998. Feedback control of Hopf bifurcations. In: Proceedings of the IEEE International Symposium on Circuits and Systems, Monterrey, CA.

Chen, Y.Y., Tsao, T.C., 1989. Description of the dynamical behavior of fuzzy systems. IEEE Trans. Systems Man Cybernet. 19 (4), 745–755.

Cheng, A., Yang, C., Hackl, K., Chajes, M., 1993. Stability bifurcation and chaos of nonlinear structures with control. Internat. J. Non-Linear Mech. 28 (5), 549–565.

Chiang, W.L., Dong, W.M., Wong, F.S., 1987. Dynamic response of structures with uncertain parameters: A comparative study of probabilistic and fuzzy sets models. Probab. Eng. Mech. 2 (2), 82–91.

Chiu, H.M., Hsu, C.S., 1986. A cell mapping method for nonlinear deterministic and stochastic systems – part II: Examples of application. J. Appl. Mech. 53, 702–710.

Chung, K.L., 1967. Markov Chains with Stationary Transition Probabilities. Springer-Verlag, New York.

Crauel, H., Flandoli, F., 1998. Additive noise destroys a pitchfork bifurcation. J. Dynam. Differential Equations 10, 259–274.

Crespo, L.G., Sun, J.Q., 2000a. An improved strategy to solve fixed final time optimal control problems via simple cell mapping. In: Proceedings of the Conference on Nonlinear Vibrations Stability and Dynamics of Structures, Blacksburg, VA.

Crespo, L.G., Sun, J.Q., 2000b. Solution of fixed final state optimal control problems via simple cell mapping. Nonlinear Dynam. 23, 391–403.

Crespo, L.G., Sun, J.Q., 2002. Optimal control of populations of competing species. Nonlinear Dynam. 27, 197–210.

Crespo, L.G., Sun, J.Q., 2003. Stochastic optimal control of nonlinear dynamic systems via Bellman's principle and cell mapping. Automatica 39 (12), 2109–2114.

Crutchfield, J.P., Farmer, J.D., Huberman, B.A., 1982. Fluctuations and simple chaotic dynamics. Phys. Rep. 92 (2), 45–82.

Crutchfield, J.P., Huberman, B.A., 1980. Fluctuations and the onset of chaos. Phys. Lett. A 77 (6), 407–410.

Cuesta, F., Ponce, E., Aracil, J., 2001. Local and global bifurcations in simple Takagi–Sugeno fuzzy systems. IEEE Trans. Fuzzy Syst. 9 (2), 355–368.

Doi, S., Inoue, J., Kumagai, S., 1998. Spectral analysis of stochastic phase lockings and stochastic bifurcations in the sinusoidally forced Van der Pol oscillator with additive noise. J. Stat. Phys. 90 (5–6), 1107–1127.

Dong, W.M., Chiang, W.L., Wong, F.S., 1987. Propagation of uncertainties in deterministic systems. Comput. & Structures 26 (3), 415–423.

Douay, A.C., Hagood, N.W., 1993. Evaluation of optimal variable stiffness feedback control authority, stability, feasibility and implementability. In: Proceedings of Fourth International Conference on Adaptive Structures, Cologne, Germany.

Edwards, D., Choi, H.T., 1997. Use of fuzzy logic to calculate the statistical properties of strange attractors in chaotic systems. Fuzzy Sets and Systems 88 (2), 205–217.

Feigenbaum, M.J., 1978. Qualitative universality for a class of nonlinear transformations. J. Stat. Phys. 19 (1), 25–52.

Foale, S., Thompson, J.M.T., 1991. Geometrical concepts and computational techniques of nonlinear dynamics. Comput. Methods Appl. Mech. Engrg. 89, 381–394.

Friedman, Y., Sandler, U., 1996. Evolution of systems under fuzzy dynamic laws. Fuzzy Sets and Systems 84, 61–74.

Friedman, Y., Sandler, U., 1999. Fuzzy dynamics as an alternative to statistical mechanics. Fuzzy Sets and Systems 106, 61–74.

Gallas, J.A.C., Grebogi, C., Yorke, J.A., 1993. Vertices in parameter space: Double crises which destroy chaotic attractors. Phys. Rev. Lett. 71 (9), 1359–1362.

Genesio, R., Tesi, A., Wang, H., Abed, E., 1993. Control of period doubling bifurcations using harmonic balance. In: Proceedings of the IEEE Conference on Decision and Control, San Antonio, TX.

Gong, P.L., Xu, J.X., 1998. On the multiple-attractor coexisting system with parameter uncertainties using generalized cell mapping method. Appl. Math. Mech. 19 (12), 1179–1187.

Grebogi, C., Ott, E., Yorke, J.A., 1982. Chaotic attractors in crisis. Phys. Rev. Lett. 48, 1507–1510.

Grebogi, C., Ott, E., Yorke, J.A., 1983. Crises, sudden changes in chaotic attractors, and transient chaos. Physica D 7, 181–200.

Grebogi, C., Ott, E., Yorke, J.A., 1986a. Critical exponents of chaotic transients in nonlinear dynamical systems. Phys. Rev. Lett. 57, 1284–1287.

Grebogi, C., Ott, E., Yorke, J.A., 1986b. Metamorphoses of basin boundaries in nonlinear dynamical systems. Phys. Rev. Lett. 56, 1011–1014.

Grebogi, C., Ott, E., Yorke, J.A., 1987. Basin boundary metamorphoses: Changes in accessible boundary orbits. Physica D 24, 243–262.

Griffith, J.S., 1971. Mathematical Neurobiology. Academic Press, New York.
Guckenheimer, J., Holmes, P.J., 1983. Nonlinear Oscillations, Dynamical Systems and Bifurcations of Vector Fields. Springer-Verlag, New York.
Guder, R., Kreuzer, E., 1999a. Control of an adaptive refinement technique of generalized cell mapping by system dynamics. Nonlinear Dynam. 20 (1), 21–32.
Guder, R., Kreuzer, E., 1999b. Using generalized cell mapping to approximate invariant measures on compact manifolds. Internat. J. Bifur. Chaos 7, 2487–2499.
Hale, J.K., Kocak, H., 1991. Dynamics and Bifurcations. Springer-Verlag, New York.
Holmes, P., Rand, D., 1980. Phase portraits and bifurcations of the non-linear oscillator. Internat. J. Non-Linear Mech. 15, 449–458.
Hong, L., Sun, J.Q., 2006a. Bifurcations of forced oscillators with fuzzy uncertainties by the generalized cell mapping method. Chaos Solitons Fractals 27 (4), 895–904.
Hong, L., Sun, J.Q., 2006b. Bifurcations of fuzzy nonlinear dynamical systems. Commun. Nonlinear Sci. Numer. Simul. 11 (1), 1–12.
Hong, L., Xu, J.X., 1999. Crises and chaotic transients studied by the generalized cell mapping digraph method. Phys. Lett. A 262, 361–375.
Hong, L., Xu, J.X., 2001a. Crises and transient chaos in a forced nonlinear Mathieu oscillator. Acta Mech. Sinica 33 (3), 423–429.
Hong, L., Xu, J.X., 2001b. Discontinuous bifurcations of chaotic attractors in forced oscillators by generalized cell mapping digraph (GCMD) method. Internat. J. Bifur. Chaos 11, 723–736.
Hong, L., Xu, J.X., 2003. Double crises in two-parameter forced oscillators by generalized cell mapping digraph method. Chaos Solitons Fractals 15, 871–882.
Hrinca, I., 1997. An optimal control problem for the Lotka–Volterra system with delay. Nonlinear Anal. 28 (2), 247–262.
Hsu, C.S., 1980. A theory of cell-to-cell mapping dynamical systems. J. Appl. Mech. 47, 931–939.
Hsu, C.S., 1981. A generalized theory of cell-to-cell mapping for nonlinear dynamical systems. J. Appl. Mech. 48, 634–642.
Hsu, C.S., 1985. A discrete method of optimal control based upon the cell state space concept. J. Optim. Theory Appl. 46 (4), 547–569.
Hsu, C.S., 1987. Cell-to-Cell Mapping: A Method of Global Analysis for Non-Linear Systems. Springer-Verlag, New York.
Hsu, C.S., 1992. Global analysis by cell mapping. Internat. J. Bifur. Chaos 2 (4), 727–771.
Hsu, C.S., 1995. Global analysis of dynamical systems using posets and digraphs. Internat. J. Bifur. Chaos 5 (4), 1085–1118.
Hsu, C.S., Chiu, H.M., 1986. A cell mapping method for nonlinear deterministic and stochastic systems – part I: The method of analysis. J. Appl. Mech. 53, 695–701.
Hsu, C.S., Chiu, H.M., 1987. Global analysis of a system with multiple responses including a strange attractor. J. Sound Vib. 114 (2), 203–218.
Ibrahim, R.A., 1987. Structural dynamics with parameter uncertainties. Appl. Mech. Rev. 40 (3), 309–328.
Jiang, J., Xu, J.X., 1994. A method of point mapping under cell reference for global analysis of nonlinear dynamical systems. Phys. Lett. A 188 (2), 137–145.
Kapitaniak, T., 1988. Chaos in Systems with Noise. World Scientific, New Jersey.
Kennedy, J., Yorke, J.A., 1991. Basins of Wada. Physica D 51, 213–225.
Kim, M.C., Hsu, C.S., 1986. Computation of the largest Liapunov exponent by generalized cell mapping. J. Stat. Phys. 45, 49–61.
Knuth, D.E., 1969. The Art of Computer Programming. Addison–Wesley, Reading, MA.
Kobs, T., Sun, J.Q., 1997. A non-linear variable stiffness feedback control with tuning range and rate saturation. J. Sound Vib. 205 (2), 243–249.
Kotulski, Z., Sobczyk, K., 1987. Effects of parameter uncertainty on the response of vibratory systems to random excitation. J. Sound Vib. 119 (1), 159–171.

Lai, Y.C., Grebogi, C., Blüel, R., Kan, I., 1993. Crisis in chaotic scattering. Phys. Rev. Lett. 71, 2212–2215.

Lai, Y.C., Grebogi, C., Yorke, J.A., 1992. Sudden change in the size of chaotic attractors: How does it occur? In: Kim, J.H., Stringer, J. (Eds.), Applied Chaos. Wiley, New York, pp. 441–455.

Lewis, F.L., Syrmos, V.L., 1995. Optimal Control. Wiley, New York.

Lin, Y.K., 1976. Probabilistic Theory of Structural Dynamics. Kreiger, Huntington, NY.

May, R.M., 1976. Simple mathematical models with very complicated dynamics. Nature 261, 459–467.

McDonald, S.W., Grebogi, C., Ott, E., Yorke, J.A., 1985. Fractal basin boundaries. Physica D 17, 125–153.

Meunier, C., Verga, A.D., 1988. Noise and bifurcations. J. Stat. Phys. 50 (1–2), 345–375.

Namachchivaya, N.S., 1990. Stochastic bifurcation. J. Appl. Math. Comput. 38, 101–159.

Namachchivaya, N.S., 1991. Co-dimension two bifurcations in the presence of noise. J. Appl. Mech. 58, 259–265.

Nayfeh, A.H., Balachandran, B., 1995. Applied Nonlinear Dynamics: Analytical, Computational, and Experimental Methods. Wiley, New York.

Nigam, N.C., 1983. Introduction to Random Vibrations. MIT Press, Cambridge, MA.

Nusse, H.E., Ott, E., Yorke, J.A., 1995. Saddle-node bifurcations on fractal basin boundaries. Phys. Rev. Lett. 75, 2482–2485.

Nusse, H.E., Yorke, J.A., 1996. Basins of attraction. Science 271, 1376–1380.

Onoda, J., Endo, T., Tamoaki, H., Watanabe, N., 1991. Vibration suppression by variable-stiffness members. AIAA Journal 29, 977–983.

Ott, E., Kostelich, E., Yuan, G., Hunt, B., Grebogi, C., Yorke, J.A., 1996. Control of shipboard cranes. In: Proceedings of the National Conference on Noise Control Engineering, Bellevue, WA.

Ott, E., Tél, T., 1993. Chaotic scattering: An introduction. Chaos 3, 417–423.

Risken, H., 1996. The Fokker–Planck Equation. Springer-Verlag, New York.

Rössler, O.E., Stewart, H.B., Wiesenfeld, K., 1990. Unfolding a chaotic bifurcation. Proc. Roy. Soc. London Ser. A 431, 371–383.

Satpathy, P.K., Das, D., Gupta, P.B.D., 2004. A fuzzy approach to handle parameter uncertainties in Hopf bifurcation analysis of electric power systems. Int. J. Elec. Power 26 (7), 531–538.

Schenk-Hoppe, K.R., 1996. Bifurcation scenarios of the noisy Duffing–Van der Pol oscillator. Nonlinear Dynam. 11, 255–274.

Shinozuka, M., 1972. Digital simulation of random processes and its applications. J. Sound Vib. 25 (1), 111–128.

Smith, S.M., Comer, D.J., 1990. Self-tuning of a fuzzy logic controller using a cell state space algorithm. In: Proceedings of the IEEE International Conference on Systems, Man and Cybernetics, vol. 6, pp. 445–450.

Soliman, M.S., Thompson, J.M.T., 1992. Global dynamics underlying sharp basin erosion in nonlinear driven oscillators. Phys. Rev. A 45, 3425–3431.

Song, F., Smith, S.M., 2000. Cell state space based incremental best estimate directed search algorithm for Takagi–Sugeno type fuzzy logic controller automatic optimization. In: Proceedings of the 9th IEEE International Conference on Fuzzy Systems, San Antonio, TX, vol. 1, pp. 19–24.

Song, F., Smith, S.M., Rizk, C.G., 1999a. Fuzzy logic controller design methodology for 4D systems with optimal global performance using enhanced cell state space based best estimate directed search method. In: Proceedings of the IEEE International Conference on Systems, Man and Cybernetics, Tokyo, Japan, vol. 6, pp. 138–143.

Song, F., Smith, S.M., Rizk, C.G., 1999b. Optimized fuzzy logic controller design for 4D systems using cell state space technique with reduced mapping error. In: Proceedings of the IEEE International Fuzzy Systems Conference, Seoul, South Korea, vol. 2, pp. 691–696.

Soong, T.T., 1973. Random Differential Equations in Science and Engineering. Academic Press, New York.

Stewart, H.B., 1988. A chaotic saddle catastrophe in forced oscillators. In: Proceedings of Dynamical Systems Approaches to Nonlinear Problems in Systems and Circuits, Philadelphia, pp. 138–149.

Stewart, H.B., Ueda, Y., Grebogi, C., Yorke, J.A., 1995. Double crises in two-parameter dynamical systems. Phys. Rev. Lett. 75 (13), 2478–2481.

Strogatz, S.H., 1997. Nonlinear Dynamics and Chaos. Addison–Wesley, Reading, MA.

Sun, J.Q., 1995. Random vibration analysis of a non-linear system with dry friction damping by the short-time Gaussian cell mapping method. J. Sound Vib. 180 (5), 785–795.

Sun, J.Q., Hsu, C.S., 1987. Cumulant-neglect closure method for nonlinear systems under random excitations. J. Appl. Mech. 54, 649–655.

Sun, J.Q., Hsu, C.S., 1988. First-passage time probability of nonlinear stochastic systems by generalized cell mapping method. J. Sound Vib. 124 (2), 233–248.

Sun, J.Q., Hsu, C.S., 1990a. The generalized cell mapping method in nonlinear random vibration based upon short-time Gaussian approximation. J. Appl. Mech. 57, 1018–1025.

Sun, J.Q., Hsu, C.S., 1990b. Global analysis of nonlinear dynamical systems with fuzzy uncertainties by the cell mapping method. Comput. Methods Appl. Mech. Engrg. 83 (2), 109–120.

Thompson, J.M.T., 1992. Global unpredictability in nonlinear dynamics: Capture, dispersal and the indeterminate bifurcations. Physica D 58, 260–272.

Thompson, J.M.T., Soliman, M.S., 1991. Indeterminate jumps to resonance from a tangled saddle-node bifurcation. Proc. Roy. Soc. London Ser. A 432, 101–111.

Thompson, J.M.T., Stewart, H.B., 2000. Nonlinear Dynamics and Chaos. Wiley, Chichester, NY.

Thompson, J.M.T., Stewart, H.B., Ueda, Y., 1994. Safe, explosive, and dangerous bifurcations in dissipative dynamical systems. Phys. Rev. E 49 (2), 1019–1027.

Tomonaga, Y., Takatsuka, K., 1998. Strange attractors of infinitesimal widths in the bifurcation diagram with an unusual mechanism of onset. Nonlinear dynamics in coupled fuzzy control systems. II. Physica D 111 (1–4), 51–80.

Tongue, B.H., 1987. On obtaining global nonlinear system characteristics through interpolated cell mapping. Physica D 28, 401–408.

Udwadia, F.E., 1987a. Response of uncertain dynamic systems I. Appl. Math. Comput. 22, 115–150.

Udwadia, F.E., 1987b. Response of uncertain dynamic systems II. Appl. Math. Comput. 22, 151–187.

Ueda, Y., 1980. Explosion of strange attractors exhibited by Duffing's equation. In: Helleman, R.H.G. (Ed.), Nonlinear Dynamics. Annals of the New York Academy of Sciences, vol. 357. New York Academy of Sciences, pp. 422–434.

Ushio, T., Hsu, C.S., 1987. Chaotic rounding error in digital control systems. IEEE Trans. Circuits Syst. 34, 133–139.

Wang, F.Y., Lever, P.J.A., 1994. A cell mapping method for general optimum trajectory planning of multiple robotic arms. Robot. Auton. Syst. 12, 15–27.

Wei, K., 1998a. Bifurcation and normal form of nonlinear control systems, Part I. SIAM J. Control Optim. 36 (1), 193–212.

Wei, K., 1998b. Bifurcation and normal form of nonlinear control systems, Part II. SIAM J. Control Optim. 36 (1), 213–232.

Xu, W., He, Q., Fang, T., Rong, H., 2003. Global analysis of stochastic bifurcation in Duffing system. Internat. J. Bifur. Chaos 13 (10), 3115–3123.

Xu, W., He, Q., Fang, T., Rong, H., 2004. Stochastic bifurcation in Duffing system subject to harmonic excitation and in presence of random noise. Internat. J. Non-Linear Mech. 39, 1473–1479.

Xu, W., He, Q., Fang, T., Rong, H., 2005. Global analysis of crisis in twin-well Duffing system under harmonic excitation in presence of noise. Chaos Solitons Fractals 23, 141–150.

Yoshida, Y., 2000. A continuous-time dynamic fuzzy system. (I) A limit theorem. Fuzzy Sets and Systems 113, 453–460.

Chapter 4

Bifurcation Analysis of Nonlinear Dynamic Systems with Time-Periodic Coefficients

Alexandra Dávid and S.C. Sinha

Nonlinear Systems Research Laboratory,
Department of Mechanical Engineering, Auburn University,
202 Ross Hall, Auburn, AL 36849, USA
E-mail: szandra@comcast.net; ssinha@eng.auburn.edu

Contents

1. Introduction 280
2. Formulation of the problem 286
3. Local stability and conditions for bifurcations: Floquét theory 287
4. Lyapunov–Floquét transformation 289
5. Nonlinear analysis 291
   5.1. Time-periodic center manifold reduction 292
   5.2. Time-dependent normal form theory 294
   5.3. Versal deformation of the normal form 296
      5.3.1. Finding an α–μ relationship by a sensitivity analysis of the Floquét transition matrix 297
      5.3.2. Finding an approximate α–μ relationship by a curve fitting technique 299
   5.4. Solution in the original (physical) variables 300
6. The codimension one bifurcations 301
   6.1. Flip bifurcation 301
   6.2. Transcritical and symmetry breaking bifurcations 304
   6.3. Cyclic fold bifurcation 307
   6.4. Secondary Hopf bifurcation 309
7. Applications 313
   7.1. A system with an exact solution: an example of the flip bifurcation 313
   7.2. A system with a small parameter: a comparison with averaging method 317
   7.3. A simple pendulum with periodic base excitation: an example of the symmetry breaking bifurcation 319

Edited Series on Advances in Nonlinear Science and Complexity
Volume 1 ISSN: 1574-6909 DOI: 10.1016/S1574-6909(06)01004-5

7.4. An example of the secondary Hopf bifurcation: a double inverted pendulum with a periodic follower load 327
8. Summary and conclusions 332
References 335

1. Introduction

In this chapter a general methodology for the local bifurcation analysis of nonlinear ordinary differential equations with periodic coefficients is presented.

In order to predict the outcome or the future course of a process we need to solve a differential equation that describes our system. A *general solution* of a differential equation is a solution that depends on the system parameters and initial conditions of the given problem. Therefore, it can tell us the evolution as time progresses and the outcome for any given set of parameters starting from any particular initial state. Such solutions, however, cannot always be found. In fact, the few cases when we can find general solutions are rather the exception than the rule. Even so, we still wish to be able to say something about the systems we study. There are some aspects of dynamic behavior that are very important and might be predictable without obtaining a general solution. Among the most important aspects are such as long term behavior, that tells us whether the system is going to some equilibrium state, will remain within certain boundaries, or will grow without bounds. Parameter dependence is also a very important feature. A change in system parameters may lead to a change in the system's long term behavior. Such a change is what we call a bifurcation. The aim of bifurcation analysis is to provide exact or (more often) approximate information about how the possible outcomes of a dynamical process depend on the system parameters, both qualitatively and quantitatively.

Dynamic systems governed by ordinary differential equations with periodically varying coefficients have been studied for the past one and a half centuries with increasing interest. The study of such problems is of great theoretical and practical importance, because these systems play significant roles in several branches of engineering as well as in various other branches of science. Some important engineering applications include:

- dynamic stability of structures subjected to periodic loads (Bolotin, 1964; Hsu, 1974; Evan-Ivanowski, 1976; Nayfeh and Mook, 1979);
- asymmetric rotor-bearing systems (Roseau, 1987; Lalanne and Ferraris, 1990), with magnetic bearings (Szasz and Flowers, 1999);
- helicopter rotor blade vibration in forward flight (Johnson, 1980; McKillip, 1985; Friedmann, 1986, 1987);
- dynamics of fluid flows in microgravity environment (gravity modulation) or with temperature modulation on the boundary (Farooq and Homsy, 1996;

Grassia and Homsy, 1998; Widmann, Gorman and Robbins, 1989; Roppo, Davis and Rosenblat, 1984);
- ship dynamics, capsizing (Sanchez and Nayfeh, 1990);
- attitude stability of satellites (Mingori, 1969; Lindh and Likins, 1970);
- robots performing repetitive tasks (Streit, Krousgrill and Bajaj, 1989; Lindtner, Steindl and Troger, 1990);
- electrical circuits (Richards, 1983; Hawthorne, 1951).

Also, applications can be found in other areas of science such as Physics, Chemistry, Biology and Medical Sciences. Here we only mention a few examples:

- quantum mechanics (Powell and Crasemann, 1961);
- laser molecule interactions (Thuraisingham and Meath, 1985; Hirschfelder, Wyatt and Coalson, 1989);
- elliptical restricted three-body problem (Szebehely, 1967; Danby, 1964);
- cardiac rhythms (Guevara, Shrier and Glass, 1990; Glass, 1991; Rigney and Goldberger, 1989).

These problems are naturally very complex due to their nonlinear and time-varying nature. Except for a very few special cases, general solutions of these equations cannot be found. Therefore, it is very important to develop tools of analysis that enable us to understand at least certain aspects of the dynamic behavior of periodic systems. From an engineering point of view, some of the most significant aspects are the long term behavior (stability, steady-state solutions) and parameter dependence (bifurcations and chaos). Stability and bifurcation analysis is an essential step before design and control of these systems can be attempted. It is no surprise then that most of the methods that have been developed are focused on the study of these two particular facets of system dynamics.

It was Mathieu in 1868 (Mathieu, 1868) who first encountered a differential equation with periodic coefficients, while studying wave motions on a lake. He modeled the problem as vibrations of an elliptical membrane to derive his famous equation, the so-called Mathieu equation. Hill in 1886 (Hill, 1886) was also faced with periodic differential equation when he investigated the Moon's perigee. It was Hill who devised the first ever solution technique of linear periodic equations although he did not prove the convergence of the method. The proof of convergence was given by Poincaré in 1892 (Poincaré, 1892). This approach, called the Hill's method of infinite determinants is suitable for obtaining stability boundaries of periodic systems. It is in active use even these days, although it is not very practical for digital computation, especially for systems with large degrees of freedom. It was only a couple of years after Hill's contribution that Floquét (1883) developed the complete stability theory of linear periodic differential equations. According to Floquét theory, the stability of a linear periodic system can be determined from the eigenvalues of the state transition (fundamental solution) matrix evaluated at the principal period of the system. These eigenvalues are often called the Floquét or characteristic multipliers. A linear periodic system is asymptotically stable if and only if all its Floquét multipliers have magnitudes less than one. As an extension to Floquét theory, Lyapunov (1896) showed that there exists an invertible, linear and periodic transformation which converts a linear periodic system into a dynamically equivalent time-invariant form. It is now known as the Lyapunov–Floquét transformation and it is of extreme importance since it allows a large number of techniques developed for autonomous systems to be applied to periodic systems.

The roots of nonlinear analysis lie in the 18th century, when physicists were trying to implement Newton's theory of gravity to explain the observed motions of planets and satellites. Describing the dynamics of two celestial bodies considering only the forces between them and neglecting any effect of others is a simple solvable problem. However, it became clear very early that a two-body approximation does not provide accurate description of the dynamics of the Solar system. On the other hand, three- (and multi-)body problems are very complex and were, at that time, unsolvable. It seemed a reasonable compromise to assume that the effects of other planets or satellites can be treated as small perturbations of the two-body problem. This idea led to the formulation of perturbation equations. Since those equations are still unsolvable, in general, averaging and perturbation methods were developed to provide approximate solutions. Perturbation equations and some elements of averaging appeared first in the work of Clairaut in 1754 (Clairaut, 1754), but it was Lagrange in 1788 (Lagrange, 1788) who presented the idea of averaging first in a clear fashion. One of the classical perturbation methods, so-called the method of strained coordinates originates in the works of Stokes (1847) and Lindstedt (1882). In 1892 Poincaré presented an extended version of Lagrange's and Lindstedt's ideas and described perturbation and averaging methods in a clear and comprehensive way. It was also he who first justified the use of asymptotic expansions. Poincaré introduced other concepts of nonlinear analysis also, such as point mapping and the theory of normal forms. All of these have proved to be very useful tools for nonlinear systems, and most of the currently used techniques are based on these ideas.

In the analysis of linear systems with periodic coefficients two of the most important tools are the Floquét theory and the Lyapunov–Floquét (L–F) transformation. Floquét theory can be used to determine the stability of linear periodic systems. The L–F transformation, which converts a linear periodic system into a dynamically equivalent time-invariant one, can be used to find solutions, forced responses, or even to design linear feedback control systems. However, application of the Floquét theory as well as computation of the L–F transformation requires computation of the state transition (or fundamental solution) matrix (STM) associated with the linear system. Obtaining the STM essentially means finding n independent exact solutions of a periodic system, which, in general, cannot be done. Except for the special case of commutative systems (Wu, 1978; Lukes, 1982), approximations are needed. Until recently even obtaining approximate solutions has been a challenge (Bellman, 1970; Verhulst, 1990). Sinha and his associates (Sinha and Wu, 1991; Sinha and Juneja, 1991; Wu and Sinha, 1994) have devised a very efficient computational method for finding the STM of linear periodic systems. The technique is based on an expansion of the solution vector into shifted Chebyshev polynomials and provides the STM with any desired accuracy, in a form suitable for further algebraic manipulations.

In nonlinear systems a number of phenomena can occur which are impossible to predict from linear analysis alone. Linear theory cannot predict what kind of motion follows after a bifurcation and whether the system is stable or not in the critical situations when a multiplier's magnitude is exactly equal to one. Post-bifurcation behavior and features such as limit cycles, quasiperiodic attractors and chaos have great significance in both theory and practical applications and can be treated by nonlinear analysis techniques only. From a design or control point of view it is essential, in addition to qualitative descriptions of these phenomena, to be able to provide quantitative analyses.

The currently used tools of nonlinear analysis include the averaging method, perturbation methods, point mapping techniques and the theory of normal forms. Averaging and perturbation methods are based on the basic idea that if the problem at hand differs from a simple solvable problem only by small additional terms, then we can assume that the solutions of the two problems will be close as well. The periodic and nonlinear terms are assumed to be small compared to the time-invariant linear part and treated as slowly varying perturbation of the time-invariant linear part of the equation. Solution of the time-invariant linear problem can, of course, be found easily. It is considered a crude first approximation to the whole problem. This approximation can be further refined. In the averaging method the system is first transformed into a standard form; then the periodic and nonlinear terms are averaged out over the principal period to obtain a time-invariant nonlinear equation. The averaged system may sometimes become simple enough to be solved in a closed form or it can be further simplified by time-invariant methods such as the normal form theory. Milestone works on modern averaging method with detailed description of the theory and numerous practical applications include those by Bogoliubov and Mitropolsky (1961) and Sanders and Verhulst (1985). In perturbation methods, the solution is assumed to be expanded into asymptotic series in terms of the small parameter, where the zero-order term of the approximation is the solution of the unperturbed problem. A set of differential equations is obtained by substituting the expansion into the original problem and collecting the coefficients of the like powers of the small parameter. These equations are generally simpler than the original problem and can be solved successively. However, straightforward asymptotic expansions are often not valid uniformly for the entire domain of the variables. The principal aim of perturbation methods is to render the asymptotic expansion uniformly valid. A detailed description of the different techniques as well as many examples and practical applications can be found in the works of Nayfeh and his associates (Nayfeh, 1973, 1985; Nayfeh and Mook, 1979; Nayfeh and Balachandran, 1995). Both averaging and perturbation methods are, however, restricted to systems with a nonzero constant linear part (for a generating solution) and very small periodic and nonlinear terms.

For the analysis of general periodic systems the so-called point mapping method may be used. For periodic systems, we can formulate the motion of the system as events in discrete time instants. The goal here is to eliminate the explicit time-dependence from the equations by replacing the dynamic system with a set of difference equations sampled at the principal period T. This process links the state of the system at the beginning of a period to the state of the system at the end of the period by a point mapping transformation. This formulation allows an easier determination of the various solutions and their stability properties. Another advantage is that the difference equations are easy to simulate on digital computers. One of the main obstacles in applying the point mapping technique to real problems is in obtaining the corresponding difference equations. In order to derive the complete difference equation representation one has to find a solution of a nonautonomous nonlinear system of differential equations over one period. This is not possible except in few particular cases. Thus, in general, one is forced to accept an approximation to the point mapping transformation. Obtaining such approximations usually involves rather elaborate numerical integrations, which makes the method impractical for larger systems. Also, the results obtained in the discrete time domain cannot be transformed back to the original, physically meaningful coordinates. Point mapping techniques, based on Poincaré's idea, have been developed by Birkhoff (1966), Bernussou (1977) and Arnold (1988). Some of the most recent studies include those by Flashner and Hsu (1983), Hsu (1987) and Guttalu and Flashner (1989, 1990).

Another alternative for the analysis of general nonlinear periodic systems is the theory of normal forms. This theory provides a powerful tool for simplification. By means of a near-identity transformation most of the nonlinear terms of the equation can be eliminated, and the resulting normal form is often simple enough to solve. In fact, for hyperbolic systems (systems that are either asymptotically stable or unstable in the linear sense) the normal forms reduce to completely linear equations. The original time-invariant normal form theory (first introduced by Poincaré) applies to time-invariant systems only. It has been used recently together with averaging and perturbation methods (Sanders and Verhulst, 1985; Nayfeh, 1993), to further simplify the problem in case the averaged equations or the perturbation formulation were still unsolvable. The time-dependent normal form theory (Arnold, 1988; Arrowsmith and Place, 1990) is applicable to periodic systems with a time-invariant linear part; therefore, it becomes useful only after an application of the Lyapunov–Floquét transformation. Although the theory can greatly simplify even the nonhyperbolic systems, in those cases the normal form remains nonlinear due to resonance conditions, and especially for systems with relatively large dimensions (systems with multiple degrees of freedom), it may still be unsolvable. In such cases another powerful simplification technique comes to the rescue: the theory of center manifold reduction. The generalization of the time-invariant center manifold reduction to periodic systems was introduced by Malkin (1962) who presented the mathematical theory and but it was Pandiyan and Sinha (1995) who developed it into a practical analytic tool by devising a computational method to obtain the necessary transformations. The aim of center manifold theory is to reduce the dimension of the system to the dimension of its critical part while preserving the stability and bifurcation characteristics. The time-periodic center manifold reduction, similar to the time-dependent normal form theory, is only applicable to periodic systems with a constant linear term, that is, only after an application of the L–F transformation. This makes the Lyapunov–Floquét transformation extremely important not only for linear periodic systems but also in the analysis of nonlinear time-periodic systems. In theory, a sequential application of the L–F transformation, the center manifold reduction and the normal form theory together with versal deformation theory has proved to be a superb tool of bifurcation analysis. It has been shown by several authors (Arrowsmith and Place, 1990; Chow, Li and Wang, 1994) that the normal forms so obtained are completely time-invariant for codimension one cases and simple enough to be solved in closed form. In the same works even normal forms of higher codimensional bifurcations have been derived (see, for example, Chow and Wang, 1986). However, all these studies have been purely theoretical, and only provide qualitative information, which (unnecessary to say) are of very little use for practical engineering and scientific problems. Practical problems, especially the design and control of dynamic systems always require quantitative information. Actual calculation of these normal forms at the critical points have been made possible only very recently by constructing efficient techniques for the computation of the L–F transformation (Sinha, Pandiyan and Bibb, 1996) and of the center manifold relations and normal form transformations (Pandiyan and Sinha, 1995). These advances now allow us to determine the stability of critical systems and even to obtain solutions at the bifurcation points. However, the main goal of bifurcation analysis, namely, to provide quantitative information about the post-bifurcation dynamics, involves the application of versal deformation theory. By constructing the versal deformation of the normal form we obtain a description of the dynamics valid not only at the critical point but also in a small neighborhood of it. Construction of such versal deformations involves finding the relationship between the original system parameters (bifurcation parameters) and the eigenvalues of the time-invariant linear part after the application of the L–F transformation. Due to the time-varying nature of the problem, such relationships cannot be found in exact forms, in general. Dávid and Sinha (2000) presented two techniques for computing such approximate relationships and several examples to illustrate that by using these techniques very accurate quantitative predictions of post-bifurcation dynamics can be made.

In this chapter we present a detailed description of bifurcation analysis of periodic systems based on a construction of dynamically equivalent normal forms employing normal form reduction, center manifold theory and versal deformations. The method is illustrated by several examples and also compared with other existing methods.

2. Formulation of the problem

A parameter dependent nonlinear dynamic system with time-periodic coefficients in its most general form is represented as

$$\dot{x} = F(x, \alpha, t), \tag{1}$$

where $x,\ F \in \mathbb{R}^n$, $F(x, \alpha, t) = F(x, \alpha, t + T)$ and T is the principal period. The vector $\alpha \in \mathbb{R}^m$ contains the parameters of the system and we have a given range of interest $\alpha \in [\alpha_l, \alpha_h]$. We assume that either there exists $x_0 = 0$ such that $F(x_0, \alpha, t) = 0$, $\forall \alpha \in [\alpha_l, \alpha_h]$ and $\forall t$, that is to say that $x_0 = 0$ is an equilibrium point; or there exists a known periodic solution $x_0(t) = x_0(t+pT)$, $p = 1, 2, \ldots$, such that $x_0 = F(x_0, \alpha, t)$, $\forall \alpha \in [\alpha_l, \alpha_h]$ and $\forall t$. The zero equilibrium point can be assumed without any loss of generality, since any nonzero equilibrium solution can be transformed to the origin by a simple coordinate change.

In order to study the stability and bifurcations of either the equilibrium or the periodic solution we need to construct the so-called variational equation (or, in other words, we need to localize) about the given equilibrium or periodic solution by expanding equation (1) into Taylor series up to the kth order. The local equation has the form

$$\dot{x} = A(\alpha, t)x + f_2(x, \alpha, t) + \cdots + f_k(x, \alpha, t) + \mathrm{O}\big(|f_{k+1}|\big), \tag{2}$$

where $A = \frac{\partial F}{\partial x}|_{x=x_0}$ is an $n \times n$ matrix and is T- (or pT-)periodic in time. The higher-order terms $f_i = \frac{\partial^i F}{\partial x^i}|_{x=x_0}$ are ith-order monomials of the states x and are also T-periodic in time. Note, that because of the time-varying nature of the original problem (1), equation (2) has the same structure, in general, whether it is expanded about an equilibrium or a periodic solution. In periodic systems, we can study the stability and bifurcations of limit cycles as if they were equilibria. The entire method of analysis described in the following sections is valid for both cases.

Further, we assume that the stability of the equilibrium point or periodic orbit depends on the parameter α and there exists a critical value, $\alpha_c \in [\alpha_l, \alpha_h]$, for which the trivial solution, $x = 0$, of equation (2) undergoes a bifurcation. This critical value may be given in the actual physical problem at hand, or as in most cases, we need to find it. One approach, of course, is to perform a numerical search and find α_c by a trial and error method. An alternate approach employs an analytical technique to compute the state transition matrix associated with the linear part of equation (2) symbolically in terms of the system parameters. Such a technique has been developed by Sinha and Butcher (1997) and Butcher and Sinha (1998). The computation involves *Picard iteration* and *Chebyshev polynomial expansion*. The bifurcation surfaces of equation (1) can be computed in the m-dimensional space of the parameter α and sets of critical values α_c can be determined for various bifurcations using stability criteria either for maps, such as the Shur–Cohn criterion, or after a suitable transformation applied to the multipliers, criteria for autonomous systems such as the Routh–Hurwitz criterion.

Our goal in this section is to determine the stability of the system at the critical point $\alpha = \alpha_c$ using a nonlinear analysis, and study the bifurcation phenomena in the neighborhood of the critical point, that is, for $\alpha \neq \alpha_c$, $\alpha - \alpha_c =$ small. We would like to obtain not only qualitative information about the post-bifurcation behavior but quantitative results as well.

3. Local stability and conditions for bifurcations: Floquét theory

Consider the linear part of equation (2),

$$\dot{x} = A(\alpha, t)x. \tag{3}$$

Let us study this equation at any given fixed $\alpha \in [\alpha_l, \alpha_h]$. If we fix the parameter value, it becomes a parameter-independent system

$$\dot{x} = A(t)x. \tag{4}$$

The definition of stability we use in this chapter is given as follows.

DEFINITION (Asymptotic stability in the Lyapunov sense). The $x_0 = 0$ trivial solution of equation (3) is stable in the Lyapunov sense if and only if for all $\varepsilon > 0$ there exists $\delta > 0$ such that if $|x_i(0) - x_{0i}(0)| < \delta$, $i = 1, 2, \ldots, n$, then $|x_i(t) - x_{0i}(t)| < \varepsilon$ for all $t > 0$ and if $t \to \infty$ then $|x_i(t) - x_{0i}(t)| \to 0$.

The general linear stability theory of systems with periodic coefficients is due to Floquét (1883). We define the fundamental solution matrix or state transition matrix (STM) $\Phi(t)$ as

$$\dot{\Phi}(t) = A(t)\Phi(t), \quad \Phi(0) = I, \tag{5}$$

where I is the n-dimensional identity matrix. The STM computed at the end of the principal period T, $\Phi(T)$, is called the Floquét transition matrix (FTM). The eigenvalues of the FTM, also called the Floquét multipliers, determine the stability of the trivial solution in equation (4) as follows. If all the eigenvalues lie inside the unit circle of the complex plane the system is asymptotically stable. If at least one of them has a magnitude greater than one, the zero equilibrium is unstable. A more detailed description of Floquét theory, including proofs, can be found in Yakubovich and Starzhinskii (1975) and Iooss and Joseph (1990).

The cases when some of the Floquét multipliers of the linear system (4) lie on the unit circle and all the others have magnitudes less than one are called critical. If our system was linear to begin with, this would be the case of marginal stability. But if equation (4) is the linearization of a nonlinear system then in a critical case the linear stability theory is not valid anymore as we know from the Hartman–Grobman theorem. To determine whether the equilibrium point or periodic solution is stable or not we must take the nonlinear effects into account and perform a nonlinear analysis.

As the parameter α varies in equation (2), the stability of the equilibrium point and the behavior of solutions in a small neighborhood around the equilibrium point might change. When such a change occurs it is called a bifurcation and the parameter value where it occurs is called a critical value, α_c. The loss of stability in a periodic system can occur through several different routes. The most frequent cases are when only one real or one pair of complex multipliers leaves the unit circle. Such bifurcations can be caused by a change in only one of the system parameters and called codimension one bifurcations. (A definition of codimension one bifurcations can be found in Hale and Koçak (1991) on p. 45.) There are three basic codimension one routes in periodic systems through which stability can be lost (Arnold, 1988). When a real multiplier crosses the unit circle at -1, it is called a flip (also called a period doubling) bifurcation. A real multiplier crossing at $+1$ can give rise to three different bifurcations, the symmetry breaking, the transcritical and the cyclic fold bifurcations, depending on the nonlinear structure of the problem. If a pair of complex multipliers reaches the unit circle it is the case of secondary Hopf (also called the Neimark–Sacker) bifurcation. A description of these phenomena can be found in Nayfeh and Balachandran (1995). All five of these bifurcations will be discussed in detail in following sections. Both an equilibrium point and a periodic orbit can undergo all of these bifurcations except the cyclic fold bifurcation which is restricted to limit cycles only. In the following we describe the general methodology, the mathematical tools needed to perform such analyses, provide detailed results and illustrative examples for all of them.

4. Lyapunov–Floquét transformation

There exists a corollary of the Floquét theory called the Lyapunov–Floquét theorem (Yakubovich and Starzhinskii, 1975) which states that the state transition matrix can be factored into a product of a periodic matrix and an exponential one as

$$\begin{aligned} &\Phi(t) = L(t)\mathrm{e}^{Ct}, \\ &L(t) = L(t+T), \qquad L(t) \in \mathbb{C}^{n\times n} \quad \forall t, \qquad C \in \mathbb{C}^{n\times n}, \end{aligned} \tag{6}$$

or as

$$\begin{aligned} &\Phi(t) = Q(t)\mathrm{e}^{Rt}, \\ &Q(t) = Q(t+2T), \qquad Q(t) \in \mathbb{R}^{n\times n} \quad \forall t, \qquad R \in \mathbb{R}^{n\times n}. \end{aligned} \tag{7}$$

The T-periodic matrix $L(t)$ and the $2T$-periodic $Q(t)$ matrix are called, respectively, the complex and the real Lyapunov–Floquét (L–F) transformations. From now in this chapter we will use the real transformation, $Q(t)$. This matrix has a couple of very special properties. If we apply the change of coordinates

$$x = Q(t)y \tag{8}$$

to the localized system (2), we obtain

$$\dot{y} = Ry + Q^{-1}(t)\big(f_2(y,t) + \cdots + f_k(y,t)\big), \tag{9}$$

where $R = Q^{-1}(t)A(t)Q(t) - Q^{-1}(t)\dot{Q}(t)$ is a time-invariant real matrix. As we can see, the linear part of the time-periodic equation (2) has been converted into a completely autonomous linear part while the nonlinear terms have become $2T$-periodic. Since this linear transformation fully preserves all the stability and bifurcation characteristics, equation (9) is dynamically equivalent to the original equation (2).

Another special feature of the real L–F transformation is an important symmetry property. If all the Floquét multipliers lie in the left half of the complex plane, then $Q(t)$ is $2T$-periodic and has the symmetry $Q(t+T) = -Q(t)$. This means that it has a zero time-average over the period $2T$ and its Fourier series expansion contains only odd integer multiples of the principal frequency. On the other hand, if at least one of the Floquét multipliers lie in the right half plane, then the real and complex L–F transformations coincide, both are T-periodic and real.

It is evident that the L–F transformation is a very valuable tool to handle periodic systems. The only problem is that except for a set of very special systems, it cannot be computed in a closed form. Since the transformation is obtained from a factorization of the state transition matrix, computing it essentially involves finding n independent solutions of a time-varying linear differential equation, and that, in general, cannot be done. So we need to resort to approximations. Besides numerical integration there are several ways in the literature to obtain approximate solutions. Currently the most efficient approach is a technique based on Chebyshev polynomial expansion. This method, first proposed by Sinha and Wu (1991), reduces the problem from integrating differential equations to solving a set of linear algebraic equations. It yields the L–F transformation matrix in the form of a superconvergent power series, up to any desired accuracy and in a form that is very convenient for further algebraic manipulations (also see Sinha, Pandiyan and Bibb, 1996). We will outline the idea only very briefly here.

From equations (6) and (7) the complex and real L–F transformation can be expressed as

$$L(t) = \Phi(t)\mathrm{e}^{-Ct}, \qquad Q(t) = \Phi(t)\mathrm{e}^{-Rt}. \tag{10}$$

Also, we have

$$\Phi(T) = \mathrm{e}^{CT}, \qquad \Phi(2T) = \Phi^2(T) = \mathrm{e}^{CT}\mathrm{e}^{C^*t} = \mathrm{e}^{2RT}. \tag{11}$$

Therefore, once the STM is obtained the matrices C and R can be computed by performing an eigenvalues analysis on the FTM, and then the Lyapunov–Floquét transformation can be calculated according to equation (10).

Since, in general, exact solutions cannot be found, let us seek a power series approximation of the state transition matrix. Following Sinha and Wu (1991), we expand the unknown solution $x(t)$ and the periodic matrix $A(t)$ from equation (4) into shifted Chebyshev polynomials:

$$\begin{aligned} x_i(t) &\approx \sum_{r=0}^{s-1} b_r^i S_r^*(t) \equiv S^{*\mathrm{T}}(t)b^i, \quad i = 1, 2, \ldots, n, \\ a_{ij}(t) &\approx \sum_{r=0}^{s-1} d_r^{ij} S_r^*(t) \equiv S^{*\mathrm{T}}(t)d^{ij}, \quad i, j = 1, 2, \ldots, n, \end{aligned} \tag{12}$$

where b_r^i are the unknown Chebyshev coefficients of $x_i(t)$, d_r^{ij} are the known Chebyshev coefficients of $A_{ij}(t)$. The shifted Chebyshev polynomials of the first kind, $S_r^*(t)$ on the interval [0, 1] can be generated by the recursive rule:

$$\begin{aligned} &S_0^* = 1, \qquad S_1^* = 1 - 2t, \\ &S_{r+1}^* = 2(1 - 2t)S_r^* - S_{r-1}^*, \quad r = 1, \ldots, m - 1. \end{aligned} \tag{13}$$

For convenience in algebraic manipulation we define the $n \times mn$ Chebyshev polynomial matrix as

$$\widehat{S}(t) = I \otimes S^{*\mathrm{T}}(t), \tag{14}$$

where $\otimes$ represents the Kronecker product and I is the identity matrix. Two special features of polynomial expansions make the application very easy. The product of two functions expanded into shifted Chebyshev polynomials can be expressed as

$$f(t)g(t) \approx \widehat{S}(t)\bar{f}\widehat{S}(t)\bar{g} = \widehat{S}(t)\overline{Q}_f\bar{g}, \tag{15}$$

where $\overline{Q}_f$ is the product operation matrix associated with $f(t)$. The integral of the Chebyshev polynomial matrix can be written in an algebraic form

$$\int_0^t \widehat{S}(\tau)\,\mathrm{d}\tau = \widehat{S}(t)\overline{G}^{\mathrm{T}}, \tag{16}$$

where $\overline{G}$ is the integral operational matrix. The definitions of these operational matrices can be found in Sinha and Wu (1991). Now, using the definitions in equation (12), $x(t)$ and $A(t)$ can be rewritten as

$$x(t) = \widehat{S}(t)\bar{b}, \qquad A(t) = \widehat{S}(t)D, \qquad A(t)x(t) = \widehat{S}(t)\overline{Q}_A\bar{b}, \tag{17}$$

where $\bar{b} = \{b^1\ b^2\ b^3\ \cdots\ b^n\}^{\mathrm{T}}$ is an $nm \times 1$ vector, $D = [d^{i1}\ d^{i2}\ d^{i3}\ \cdots\ d^{ij}]$, $i, j = 1, 2, 3, 4, \ldots, n$, is an $nm \times n$ matrix and $\overline{Q}_A$ is the $nm \times nm$ product operation matrix associated with $A(t)$ that is constructed from the elements of D. Finally, we can integrate equation (5) and express it in an approximate expanded polynomial form

$$x(t) = x_0 + \int_0^t A(\tau)x(\tau)\,\mathrm{d}\tau$$
$$\Longrightarrow \quad \widehat{S}(t)\bar{b} = \widehat{S}(t)\bar{b}_0 + \widehat{S}(t)\overline{G}^{\mathrm{T}}\overline{Q}_A\bar{b}. \tag{18}$$

Now, in order to find an approximation of the solution $x(t)$, all we need to do is solve the linear algebraic matrix equation

$$(\widehat{I} - \overline{G}^{\mathrm{T}}\overline{Q}_A)\bar{b} = \bar{b}_0 \tag{19}$$

for the unknown Chebyshev coefficients $\bar{b}$. Since the STM is constructed of n independent solutions (see equation (5)), if we solve the above equation n times with the appropriate initial conditions we finally have a polynomial approximation of $\Phi(t)$, from which the L–F transformation can be computed as described above. Details about the superconvergent nature of this expansion and the computational efficiency of this method can be found in Sinha and Wu (1991).

5. Nonlinear analysis

In the following, two important concepts, the time-periodic center manifold reduction and time-dependent normal form theory are briefly discussed. These are the basic tools for simplification of nonlinear dynamical systems with periodic coefficients. In order to apply these techniques, one must work with the transformed equation

$$\dot{y} = Ry + Q^{-1}(t)\big(f_2(y,t) + \cdots + f_k(y,t)\big). \tag{20}$$

This equation preserves the stability and bifurcation characteristics of the original problem given by equation (2).

5.1. Time-periodic center manifold reduction

In the case of a critical (nonhyperbolic) system, the nonlinear dynamic behavior can be studied in the manifold defined by those state variables that are associated with the critical eigenvalues. This submanifold is called the center manifold. As most frequently there are only a few critical states in a system, this method can lead to a very significant reduction of the dimension of the problem, most often, in the case of codimension one bifurcations, to just one or two dimensions.

Center manifold reduction is widely used for simplifying autonomous systems. Its generalization to systems with periodic coefficients is due to Malkin (1962) who suggested the basic idea, and Pandiyan and Sinha (1995) who made it applicable by developing a technique to actually compute the periodic center manifold relations.

The main idea is that a nonlinear transformation may be found which decouples the critical states from the stable ones in the nonlinear equation, thereby allowing us to study the critical dynamics on a system with much smaller dimension than the original. The theory is applicable to periodic equations with time-invariant linear parts of the form of equation (22). The linear part can be decoupled easily by applying the modal transformation

$$y = Mz, \tag{21}$$

which converts the matrix R into its Jordan canonical form as

$$\begin{aligned}\dot{z} &= Jz + M^{-1}Q^{-1}(t)\big(f_2(z,t) + \cdots + f_k(z,t)\big)\\ &= Jz + w_2(z,t) + \cdots + w_k(z,t),\end{aligned} \tag{22}$$

where the linear matrix J is the Jordan canonical form of R, and w_i, $i = 2, 3, \ldots, k$, are, in general, $2T$-periodic functions containing homogeneous monomials of z of order i. Let us assume that the matrix J in equation (22) has n_1 critical eigenvalues and n_2 eigenvalues that have negative real parts. Then the equation can be partitioned as

$$\begin{Bmatrix} \dot{z}_c \\ \dot{z}_s \end{Bmatrix} = \begin{bmatrix} J_c & 0 \\ 0 & J_s \end{bmatrix} \begin{Bmatrix} z_c \\ z_s \end{Bmatrix} + \begin{Bmatrix} w_{c2}(z_c, z_s, t) + \cdots + w_{ck}(z_c, z_s, t) \\ w_{s2}(z_c, z_s, t) + \cdots + w_{sk}(z_c, z_s, t) \end{Bmatrix}, \tag{23}$$

where the subscripts "c" and "s" represent the critical and stable variables, respectively. The nonlinear part of this equation is still coupled, the w_{ci} functions contain all z states, in general. The goal of center manifold reduction is to find a relationship between the critical states z_c and the stable states z_s, which, upon substitution into equation (23) decouples the critical part completely. According to the center manifold theorem, there exists a nonunique relation

$$z_s = H_c(z_c, t) = h_2(z_c, t) + \cdots + h_k(z_c, t) \tag{24}$$

between the stable and critical variables, such that the center manifold relations, $h_{ci}(z_c, t)$, are of the form

$$h_{ci}(z_c, t) = \sum_m h_{ci,(m_1,\ldots,m_{n_1})}(t) z_1^{m_1} \cdots z_{n_1}^{m_{n_1}},$$
$$m = \{m_1 \quad \cdots \quad m_{n_1}\}^T, \quad m_1 + \cdots + m_{n_1} = i, \quad i = 2, 3, \ldots, k, \tag{25}$$

where $h_{ci,(m_1,\ldots,m_{n_1})}(t)$ are time-periodic vector coefficient functions with period $2T$. If we substitute (24) into (23) we see that the center manifold relations can be obtained as a solution of the partial differential equation

$$\frac{\partial h_{ci}}{\partial t} + \frac{\partial h_{ci}}{\partial z_c}(J_c z_c + w_{c2} + \cdots + w_{ck}) = J_s z_s + w_{s2} + \cdots + w_{sk}, \tag{26}$$

where $\frac{\partial h_{ci}}{\partial z_c}$, $i = 2, 3, \ldots, k$, are Jacobian matrices of the partial derivatives. It is, in general, not possible to find exact solutions for equations of this complexity. However, the form of the center manifold relations in equation (25) and the periodicity of the w functions suggest an easy and efficient way to obtain approximate solutions with any desired accuracy. If we expand the known and unknown coefficient functions into a finite Fourier series, we can reduce the problem to a set of algebraic equations that can be solved easily. Let us assume the Fourier expansions in the forms

$$h_{ci}(z_c, t) = \sum_{\sum m_j = i} \sum_{j=1}^{n_1} \sum_{\nu=-q}^{q} h_{cij\nu(m_1,\ldots,m_{n_1})} e^{i\nu(\pi/T)t} z_1^{m_1} \cdots z_{n_1}^{m_{n_1}} e_j,$$

$$w_{ci}(z_c, t) = \sum_{\sum m_j = i} \sum_{j=1}^{n_1} \sum_{\nu=-q}^{q} a_{cij\nu(m_1,\ldots,m_{n_1})} e^{i\nu(\pi/T)t} z_1^{m_1} \cdots z_{n_1}^{m_{n_1}} e_j,$$

$$w_{si}(z_c, t) = \sum_{\sum m_j = i} \sum_{j=1}^{n_1} \sum_{\nu=-q}^{q} a_{sij\nu(m_1,\ldots,m_{n_1})} e^{i\nu(\pi/T)t} z_1^{m_1} \cdots z_{n_1}^{m_{n_1}} e_j, \tag{27}$$

where $h_{cij\nu(m_1,\ldots,m_{n_1})}$ are the unknown Fourier coefficients of the center manifold relations, $a_{cij\nu(m_1,\ldots,m_{n_1})}$ and $a_{sij\nu(m_1,\ldots,m_{n_1})}$ are the known Fourier coefficients of the nonlinear functions from equation (24) and the vector e_j is the jth member of the natural basis. Substituting these into the partial differential equation (26) leads to a set of algebraic equations that can be solved for the unknown $h_{cij\nu(m_1,\ldots,m_{n_1})}$

coefficients. Once the approximate center manifold relations are obtained, we can substitute them into the first half of equation (24) and decouple the critical states from the stable ones completely. In order to be consistent with the approximation, we neglect all the higher than kth-order terms as well as all the higher than qth-order harmonics in the Fourier series expansions. The resulting n_1-dimensional system has the form

$$\dot{z}_c = J_c z_c + w_{c2}^*(z_c, t) + \cdots + w_{ck}^*(z_c, t). \tag{28}$$

The center manifold theory states that this equation is dynamically equivalent to the original n-dimensional one, all stability and bifurcation properties are preserved.

5.2. *Time-dependent normal form theory*

The idea of normal form theory is to simplify a nonlinear equation by finding a series of invertible transformations that can eliminate as many nonlinear terms as possible, while preserving the original dynamic characteristics. The generalization of the classical time-invariant theory to systems with periodic coefficients was developed by Arnold (1988).

We start with our already greatly simplified equation on the center manifold, equation (28). According to the theory, there exists a sequence of near-identity transformations with T-periodic coefficients that we can use to try to cancel out each of the $w_{cj}^*(z_c, t)$, $j = 2, \dots, k$, terms. For the rth-order term, we assume the transformation as

$$z_c = v + h_{nr}(v, t), \tag{29}$$

where the unknown nonlinear part is assumed in the formal monomial form

$$h_{nr}(v, t) = \sum_{\sum m = r} h_{nr(m_1,\dots,m_{n_1})}(t) v_1^{m_1} \cdots v_{n_1}^{m_{n_1}}. \tag{30}$$

After substituting this transformation into equation (28) we obtain

$$\begin{aligned} \dot{v} &= J_c v + w_{c2}^*(v, t) + \cdots + w_{c,r-1}^*(v, t) \\ &\quad + \left(-\frac{\partial h_{nr}}{\partial v} J_c v + J_c h_{nr} - \frac{\partial h_{nr}}{\partial t} + w_{cr}^*(v, t) \right) + \mathrm{O}\big(|v|^{r+1}, t\big), \end{aligned} \tag{31}$$

where $\frac{\partial h_{nr}}{\partial v}$ is Jacobian derivative matrix. From this equation we can clearly see that this transformation does not affect the lower than rth-order terms. Also, all the rth-order terms are collected in the parentheses, and by examining them we can realize that the rth-order nonlinear terms can be eliminated if the partial differential equation,

$$\frac{\partial h_{nr}}{\partial v} J_c v - J_c h_{nr} + \frac{\partial h_{nr}}{\partial t} = w_{cr}^*(v, t), \tag{32}$$

is satisfied. This is the so-called *homological equation*. Similarly to the case of the center manifold relations, instead of trying to find a general solution, which in most cases may not be possible, we seek an approximate solution in a Fourier series expansion form,

$$h_{nr}(v,t) = \sum_{\sum m_j = r} \sum_{j=1}^{n_1} \sum_{\nu=-q}^{q} h_{nrj\nu(m_1,\ldots,m_{n_1})} \mathrm{e}^{\mathrm{i}\nu(\pi/T)t} v_1^{m_1} \cdots v_{n_1}^{m_{n_1}} e_j,$$

$$w_r^*(v,t) = \sum_{\sum m_j = r} \sum_{j=1}^{n_1} \sum_{\nu=-q}^{q} a^*_{rj\nu(m_1,\ldots,m_{n_1})} \mathrm{e}^{\mathrm{i}\nu(\pi/T)t} v_1^{m_1} \cdots v_{n_1}^{m_{n_1}} e_j. \tag{33}$$

In order to have a consistent approximation, we neglect terms higher than kth order and also Fourier harmonics higher than the qth order. Once these expressions are substituted into the homological equation, we obtain a set of linear algebraic equations for the unknown Fourier coefficients of the near-identity transformation. However, unlike the case of the center manifold relations, it may not always be possible to solve for all the unknowns. A term-by-term comparison of the Fourier coefficients provides the solution for the coefficients of the near-identity transformation as

$$h_{nrj\nu(m_1,\ldots,m_{n_1})} = \frac{a^*_{nrj\nu(m_1,\ldots,m_{n_1})}}{\mathrm{i}\nu(\pi/T) + \sum_{l=1}^{n_1} m_l\lambda_l - \lambda_j}, \tag{34}$$

where $\lambda_1, \ldots, \lambda_{n_1}$ are the eigenvalues of the Jordan matrix J_c. Obviously, when the denominator in this equation is zero, the particular $h_{nrj\nu(m_1,\ldots,m_{n_1})}$ coefficient cannot be solved for. Therefore, when the *solvability condition*,

$$\mathrm{i}\nu\frac{\pi}{T} + \sum_{l=1}^{n_1} m_l\lambda_l - \lambda_j \neq 0, \tag{35}$$

is satisfied, the corresponding nonlinear term can be eliminated, otherwise so-called resonant terms remain. If we apply a sequence of these transformations for $r = 2, 3, \ldots, k$, starting with the lowest order and always keeping terms that are higher order than the current r but lower than k, we can bring equation (28) into its simplest possible nonlinear form as

$$\dot{v} = J_\mathrm{c}v + \tilde{w}_2(v,t) + \cdots + \tilde{w}_k(v,t), \tag{36}$$

where the nonlinear functions, $\tilde{w}_{\mathrm{c}i}$, contain a significantly smaller number of terms than the original $w^*_{\mathrm{c}i}$ functions. In fact, it has been shown that if the original equation was hyperbolic (asymptotically stable or unstable), this normal form would become completely linear. However, we are now studying the nonhyperbolic (critical) cases. In the next subsections we will demonstrate that even in critical cases the normal form reduction yields very significant simplifications. For codimension one bifurcations, equation (36) becomes completely time-invariant and solvable in a closed form.

5.3. *Versal deformation of the normal form*

At this stage we should recall that the entire simplification has been done at the critical value of the bifurcation parameter only. When we substituted the critical value, α_c, into our original equation, we obtained a parameter-independent system, that was valid at the bifurcation point only. However, our main objective is to study the behavior of the system in the neighborhood of the critical point and obtain some information (quantitative as well as qualitative) about the post-bifurcation dynamics.

In order to achieve this, we need to reconstruct the parameter dependence in our normal form. Theoretically, this is done by constructing the versal deformation of the linear system matrix, J_c, of equation (36). A versal deformation is a special parameter-dependent family of matrices. The formal definition is given (Chow, Li and Wang, 1994) as follows.

Let $A_0 \in \mathbb{R}^{n\times n}$ be a given matrix. The parametric family of matrices $A(\lambda)$, $\lambda \in \Lambda \subseteq \mathbb{R}^k$, is a deformation of A_0 if $A(\lambda) : \Lambda \to \mathbb{R}^{n\times n}$ is a once continuously differentiable function and $A(0) = A_0$.

Let $A(\lambda)$ and $B(\mu)$ both be deformations of A_0, $\mu \in M \subseteq \mathbb{R}^l$. If there exists a deformation $C(\mu)$ of the identity matrix, I, with $\mu \in \widetilde{M} \subseteq M$ and there exists a once continuously differentiable mapping $\Phi : \widetilde{M} \to \Lambda$ with $\Phi(0) = 0$ such that $B(\mu) = C(\mu)A(\Phi(\mu))C^{-1}(\mu)$, $\mu \in \widetilde{M}$, then $B(\mu)$ is induced from $A(\lambda)$ by $C(\mu)$ and $\Phi(\mu)$.

A deformation $A(\lambda)$ of A_0 is a versal deformation if any deformation $B(\mu)$ of A_0 can be induced from it. It is a miniversal deformation if the dimension of its parameter space is the smallest among all versal deformations of A_0.

Let $J(\mu)$ be a miniversal deformation of J_c from the normal form equation (36), where μ is an m vector of small parameters. Then the versal deformation equation of the normal form on the center manifold can be formally written as

$$\dot{v} = J(\mu)v + \tilde{w}_2(v,t) + \cdots + \tilde{w}_k(v,t). \tag{37}$$

This equation describes the dynamics of our original system in a small neighborhood of the bifurcation point. However, since μ is an arbitrary parameter vector, with no connection to the original system parameters, α, this is only a qualitative description and we cannot say anything about the post-bifurcation behavior quantitatively.

In order to obtain a quantitatively meaningful result we need to find a relationship between the versal deformation parameter μ and the bifurcation parameter α. That, however, is not a trivial task. Essentially, it amounts to computing the

Floquét transition matrix and its eigenvalues in terms of the bifurcation parameter, since those eigenvalues directly relate to the eigenvalues of J through the Lyapunov–Floquét theorem. An exact relationship can only be found for a very special class of periodic system, called the commutative systems. For such systems the STM and the Lyapunov–Floquét transformation can be computed exactly in a closed form in terms of the parameters, and therefore, the relationship is known automatically. But for general periodic systems we need to seek an approximation. In the following subsections we suggest two ideas to obtain approximate α–μ relationships.

5.3.1. Finding an α–μ relationship by a sensitivity analysis of the Floquét transition matrix

Since equation (2) contains the bifurcation parameter α while the versal deformation equation (37) is given in terms of the parameter μ, we need to find the relationship $\mu = \mu(\alpha)$ satisfying the condition $\mu(\alpha_c) = 0$. One way to establish such a relationship is to compute the state transition matrix (STM) $\Phi(\alpha, t)$ associated with the linear part of equation (2) symbolically in terms of the parameter α. Since this requires obtaining exact solutions of equation (3), generally speaking, it is not possible. Sinha and Butcher (1997) developed a method to obtain approximations of the STM in terms of its parameters in a symbolic form. This technique is based on a Chebyshev polynomial expansion and Picard iterations. We outline the procedure only very briefly here. The STM is computed as a set of independent solutions of equation (3). The equivalent integral form of equation (3) can be written as

$$x(t) = x(0) + \int_0^t A(\alpha, \tau)x(\tau)\,d\tau. \tag{38}$$

After applying the method of Picard iteration we find the $(k + 1)$th approximation as

$$\begin{aligned} x^{(k+1)}(t) &= x(0) + \int_0^t A(\alpha, \tau_k)x^{(k)}(\tau_k)\,d\tau_k \\ &= \Bigg[I + \int_0^t A(\alpha, \tau_k)\,d\tau_k + \int_0^t A(\alpha, \tau_k)\int_0^{\tau_k} A(\alpha, \tau_{k-1})\,d\tau_{k-1}\,d\tau_k \\ &\quad + \cdots + \int_0^t A(\alpha, \tau_k)\cdots\int_0^{\tau_1} A(\alpha, \tau_0)\,d\tau_0 \cdots d\tau_k\Bigg]x(0), \end{aligned} \tag{39}$$

where $\tau_0, \ldots, \tau_k$ are dummy variables. The expression in the square brackets is an approximation to the fundamental solution matrix $\Phi(\alpha, t)$ since it is truncated after a finite number of terms (iterations). After the period is normalized by the transformation $t = T\tau$, 1-periodic system matrix $A(\alpha, \tau)$ is expanded into m shifted Chebyshev polynomials of the first kind. By substituting this expansion and utilizing the integral and product operational matrices associated with Chebyshev polynomials equation (39) yields a polynomial expression for the STM in terms of the system parameters.

Once the STM is computed symbolically, the Floquét transition matrix (FTM) $\Phi(\alpha, T)$ can also be found. Then, from the Lyapunov–Floquét theorem and the $2T$-periodic nature of the real L–F transformation it follows that

$$\Phi(\alpha, 2T) = \Phi^2(\alpha, T) = \mathrm{e}^{2R(\alpha)}. \tag{40}$$

Let $\lambda_i(\mu)$, $i = 1, \ldots, n_1$, denote those eigenvalues of $R(\mu) = R(\mu(\alpha))$, which are critical for $\alpha = \alpha_c$, and $\rho_i(\alpha)$ denote the eigenvalues of $\Phi(\alpha, 2T)$. Note that the formal dependence of the $\lambda_i(\mu)$ eigenvalues on the parameter μ is given by the miniversal deformation $J(\mu)$ of the matrix J_c. The relationship between the two sets of eigenvalues is given as

$$\rho_i = \mathrm{e}^{2\lambda_i T}. \tag{41}$$

At the bifurcation point, that is for $\mu = 0$, both the $\lambda_i(\mu)$, and the $\rho_i(\alpha)$ eigenvalues can be computed, and they satisfy equation (41). For $\mu \neq 0$ we assume that in a small neighborhood of the bifurcation point only the n_1 critical eigenvalues change when the system parameters are varied. This is a reasonable assumption, since the bifurcation phenomenon is determined by the equation on the center manifold only, and the stable states do not play a role. It can be verified by numerical simulation that, in fact, the critical eigenvalues move several orders faster as α is varied than the stable ones that are virtually stationary in the neighborhood of the bifurcation point. For $\mu \neq 0$ we still require $\lambda_i(\mu)$, $i = 1, \ldots, n_1$, to be eigenvalues of $R(\mu) = R(\mu(\alpha))$, and therefore it follows that these should also satisfy equation (41). This implies that

$$D = \det\left[\Phi(\alpha, 2T) - I\mathrm{e}^{2\lambda_i(\mu)T}\right] = 0 \tag{42}$$

has to be true for all $i = 1, \ldots, n_1$. This equation is nonlinear and unsolvable, in general. By expanding the determinant into Taylor series up to the first-order terms, however, we can obtain a set of linear algebraic equations given by

$$\mu_i \frac{\partial D}{\partial \mu_i}\bigg|_{\mu=0,\alpha=\alpha_c} + (\alpha_i - \alpha_{ci}) \frac{\partial D}{\partial \alpha_i}\bigg|_{\mu=0,\alpha=\alpha_c} = 0. \tag{43}$$

Solving equation (43) for the unknown μ_i, $i = 1, \ldots, n_1$, yields a linear relationship between the parameters μ and α. If the eigenvalues λ_i were complex, naturally the relationship will have complex coefficients (note that α is real, in general, coming from a physical problem). The relationship then can be substituted into the versal deformation equation (37) to obtain an approximate quantitative description of the dynamics of the bifurcation.

The idea can easily be extended to obtain higher-order relationships, also. We assume $\mu_i(\alpha) \approx a_{1i}(\alpha_i - \alpha_{ci}) + a_{2i}(\alpha_i - \alpha_{ci})^2 + \cdots$, where a_{ji}, $j = 1, 2, \ldots, p$, are unknown complex coefficients, in general. By expanding the determinant in equation (42) into Taylor series up to the pth-order terms, we obtain a set of n_1 equations as

$$\sum_{j=1}^{n_1}\left(\mu_j \frac{\partial D}{\partial \mu_j}\bigg|_{\mu=0,\alpha=\alpha_c} + (\alpha_j - \alpha_{cj})\frac{\partial D}{\partial \mu_j}\bigg|_{\mu=0,\alpha=\alpha_c}\right)$$
$$+ \sum_{j=1}^{n_1}\sum_{k=1}^{n_1}\left(\frac{1}{2}\mu_k\mu_j \frac{\partial^2 D}{\partial \mu_j\,\partial \mu_k}\bigg|_{\mu=0,\alpha=\alpha_c}\right.$$
$$\left. + \mu_j(\alpha_k - \alpha_{ck})\frac{\partial^2 D}{\partial \mu_j\,\partial \mu_k}\bigg|_{\mu=0,\alpha=\alpha_c}\right)$$
$$+ \sum_{j=1}^{n_1}\sum_{k=1}^{n_1}\left(\frac{1}{2}(\alpha_j - \alpha_{cj})(\alpha_k - \alpha_{ck})\frac{\partial^2 D}{\partial \alpha_j\,\partial \alpha_k}\bigg|_{\mu=0,\alpha=\alpha_c}\right)$$
$$+ \cdots = 0. \tag{44}$$

Substituting $\mu_i(\alpha) = a_{1i}(\alpha_i - \alpha_{ci}) + a_{2i}(\alpha_i - \alpha_{ci})^2 + \cdots$, this becomes a set of algebraic equations for the unknowns a_{ji}, $j = 1, 2, \ldots, p$, and an approximate relationship can be found.

Theoretically, the relationship based on the symbolic computation and the higher-order sensitivity analysis can be as accurate as desired. But, of course, there are drawbacks to this technique. The symbolic computation of the STM becomes a little tedious for large systems or for more than two parameters, and it can be very time consuming, requires fast computers, lot of memory and a good symbolic computational software (Mathematica for example).

5.3.2. Finding an approximate α–μ relationship by a curve fitting technique

In case the previously described symbolic method proves to be computationally too unmanageable, a simpler alternative method might be used. Here the idea is to compute the FTM numerically, using the Chebyshev polynomial expansion technique developed by Sinha and Wu (1991), for several different sets of the parameters α in the neighborhood of the critical α_c value. As pointed out before, the Chebyshev polynomial expansion technique (as opposed to Runge–Kutta-type numerical integration schemes) has been shown to be much more efficient due to the superconvergent nature of the Chebyshev polynomials. From the FTM one can compute the matrix $J(\alpha)$. Let $\alpha_k = \alpha_c + \eta_k$ denote the k different sets of parameters, where η_k are small, and let λ_k denote the vector of eigenvalues of $J(\alpha_k)$. Based on a similar argument as in the previous subsection, we assume that these eigenvalues have the form $\lambda_{ki} = a_{1i}\eta_{ki} + a_{2i}\eta_{ki}^2 + \cdots$, where a_{ji} are complex, in general. Usually k can be restricted to 3–5 for a reasonable approximation. Then, by fitting a curve in the least squares sense (of quadratic or higher order) to the μ_k–η_k pairs, we can find the coefficients a_{ji} of the desired relationship.

This procedure is much faster and simpler to compute for any dimensions and any number of parameters. However, it can only be as good an approximation as the least square fit. In an illustrative example later we will compare the two methods to see whether this or the sensitivity analysis yields better results.

5.4. Solution in the original (physical) variables

The versal deformation equation (37) may be simple enough to solve in a closed form. In fact, it will be shown in following sections that it becomes solvable for all the codimension one bifurcations. However, it is valid on a reduced manifold and transformed several times. In order to have a physically meaningful result (for design purposes, for example), we need to transform it back into its full dimension and original variables.

We assume that the general solution of equation (37) is obtained in the form

$$v(\mu, t, v_0, t_0), \tag{45}$$

where t_0, $v(t_0) = v_0$ are the initial conditions. This result can be expressed in terms of the original variable $x(t)$. The first step is to restore the nonlinear terms that have been eliminated by the normal form reduction by applying the sequence of near-identity transformations given by equation (29) as

$$\begin{aligned} z_c(\mu, t, z_{c0}, t_0) &= v(\mu, t, v_0, t_0) + h_{nr}\big(v(\mu, t, v_0, t_0), t\big), \\ z_{c0} &= v_0 + h_{nr}(v_0, t_0), \end{aligned} \tag{46}$$

for all $r = 2, 3, \ldots, k$. Next, the full n-dimensional system is reconstructed by obtaining the eliminated stable states from the center manifold relations (given in equation (24)) as

$$\begin{aligned} &z_s(\mu, t, z_{s0}, t_0) \\ &\quad = h_{c2}\big(z_c(\mu, t, z_{c0}, t_0), t\big) + \cdots + h_{ck}\big(z_c(\mu, t, z_{c0}, t_0), t\big), \\ &z_{s0} = h_{c2}(z_{c0}, t_0) + \cdots + h_{ck}(z_{c0}, t_0) \end{aligned} \tag{47}$$

and

$$z(\mu, t, z_0, t_0) = \begin{Bmatrix} z_c(\mu, t, z_{c0}, t_0) \\ z_s(\mu, t, z_{s0}, t_0) \end{Bmatrix}. \tag{48}$$

Finally, the modal and Lyapunov–Floquét transformations need to be applied and the solution in the original variables is found as

$$x(\mu, t, x_0, t_0) = Q(t)Mz(\mu, t, z_0, t_0), \quad x_0 = Q(t_0)Mz_0. \tag{49}$$

In the following sections the entire procedure described above will be applied to each of the codimension one bifurcations.

6. The codimension one bifurcations

In this section we apply the methodology described above for specific types of critical cases. We derive the normal forms for all of the one-parameter bifurcations. For each case we will also provide an example to illustrate how the method works in practice, give a comparison of the two different approximations for the versal deformation parameter and also compare the results with those obtained from numerical integration.

In this case there are basically three different scenarios in which a periodic system can loose its stability. A real Floquét multiplier can leave the unit circle at -1 or $+1$ or a complex pair of multipliers can become larger than 1 in magnitude.

6.1. Flip bifurcation

Let us assume that for some critical value, α_c, of the bifurcation parameters the Floquét transition matrix associated with the linear system matrix, $A(t)$, of equation (2) has one -1 eigenvalue (Floquét multiplier) and all its other eigenvalues have magnitudes less than one.

The first two steps of the analysis procedure are the application of the Lyapunov–Floquét transformation $Q(t)$ and the modal transformation M. Under our assumption, after these two transformations, equation (22) takes the form

$$\begin{Bmatrix} \dot{z}_1 \\ \dot{z}_s \end{Bmatrix} = \begin{bmatrix} 0 & 0 \\ 0 & J_s \end{bmatrix} \begin{Bmatrix} z_1 \\ z_s \end{Bmatrix} + \begin{Bmatrix} w_{12}(z_1, z_s, t) + w_{13}(z_1, z_s, t) \\ w_{s2}(z_1, z_s, t) + w_{s3}(z_1, z_s, t) \end{Bmatrix}. \tag{50}$$

The -1 Floquét multiplier is converted into a $\lambda_1 = 0$ eigenvalue of the constant matrix J. Note, that the nonlinear functions are truncated after the third order. The higher-order terms are neglected for simplicity of the following calculations. This can be done without losing any information, since it is a well demonstrated result in the literature of bifurcations that in the case of one codimension the cubic expansion provides correct qualitative description of the nonlinear phenomena and it is also a very good approximation quantitatively. So, from here on we will always neglect terms that are higher than cubic order. Also, we introduce a new notation for the nonlinear terms that will eliminate the use of at least one index and make it easier to keep track of them. We will denote the quadratic functions from equation (2) as $Q_x(x,t)$ and the cubic ones as $C_x(x,t)$. With the new notation equation (50) becomes

$$\begin{Bmatrix} \dot{z}_1 \\ \dot{z}_s \end{Bmatrix} = \begin{bmatrix} 0 & 0 \\ 0 & J_s \end{bmatrix} \begin{Bmatrix} z_1 \\ z_s \end{Bmatrix} + \begin{Bmatrix} \widetilde{Q}_1(z_1, z_s, t) + \widetilde{C}_1(z_1, z_s, t) \\ \widetilde{Q}_s(z_1, z_s, t) + \widetilde{C}_s(z_1, z_s, t) \end{Bmatrix}, \tag{51}$$

where $\widetilde{Q}(z,t) = M^{-1}Q^{-1}(t)Q_x(Q(t)Mz,t)$ and $\widetilde{C}(z,t) = M^{-1}Q^{-1}(t) \times C_x(Q(t)Mz,t)$.

In our assumed case of one -1 multiplier, because the Floquét multipliers all lie in the left half of the complex plane, the real L–F transformation is $2T$-periodic and has the symmetry property $Q(t+T) = -Q(t)$. This is also true for the inverse L–F transformation. Therefore, the nonlinear part of equation (51) is also $2T$-periodic and the symmetry of the transformation results in a special structure: the coefficients of the quadratic terms have zero time-averages over the period. (In fact, all nonlinear terms of even powers have zero averages over the period, such as the fourth-order terms, the sixth-order terms and so on.) The center manifold relations are assumed in a scalar form for each stable coordinate as

$$z_i = h_{c2,i}(z_1, t) + h_{c3,i}(z_1, t), \quad i = 2, 3, \ldots, n. \tag{52}$$

And after substituting these into equation (51), the reduced system on the one-dimensional center manifold becomes

$$\begin{aligned} \dot{z}_1 &= \widetilde{Q}_{1,(2,0,\ldots,0)}(t)z_1^2 \\ &\quad + \left(\sum_{i=2}^{n} \widetilde{Q}_{1,(1,0,\ldots,1_i,\ldots,0)}(t)h_{c2,i}(t) + \widetilde{C}_{1,(3,0,\ldots,0)}(t) \right) z_1^3 \\ &:= Q^*_{1,2}(t)z_1^2 + C^*_{1,3}(t)z_1^3, \end{aligned} \tag{53}$$

where $\widetilde{Q}_{1,(2,0,\ldots,0)}$ and $\widetilde{C}_{1,(3,0,\ldots,0)}$ are the $2T$-periodic coefficient functions of the quadratic term $z_1^2 z_2^0 \cdots z_n^0$ and the cubic term $z_1^3 z_2^0 \cdots z_n^0$, respectively. Similarly, $\widetilde{Q}_{1,(1,0,\ldots,1_i,\ldots,0)}$ is the coefficient function of the quadratic term $z_1^1 z_2^0 \cdots z_i^1 \cdots z_n^0$, for $i = 2, \ldots, n$. Equation (53) is already greatly simplified; it is now a single scalar equation. To further simplify and try to eliminate as many nonlinear terms as possible, we apply the normal form theory. We assume a sequence of near-identity transformations in the form

$$\begin{aligned} z_1 &= \tilde{v} + h_{n2}(t)\tilde{v}^2, \\ \tilde{v} &= v + h_{n3}(t)v^3. \end{aligned} \tag{54}$$

In order to solve for the transformations, we expand these and the nonlinear terms of the center manifold equation (53) into Fourier series in the form

$$h_{n2}(t) = \sum_{\nu=-q}^{q} h_{n2\nu} e^{i\nu(\pi/T)t}, \qquad h_{n3}(t) = \sum_{\nu=-q}^{q} h_{n3\nu} e^{i\nu(\pi/T)t},$$

$$Q_{1,2}^*(t) = \sum_{\nu=-q}^{q} b_{2\nu}^* e^{i\nu(\pi/T)t}, \qquad C_{1,3}^*(t) = \sum_{\nu=-q}^{q} a_{3\nu}^* e^{i\nu(\pi/T)t}. \tag{55}$$

The general resonance condition that determines weather a nonlinear term can be eliminated was given in equation (35) as

$$i\nu\frac{\pi}{T} + \sum_{l=1}^{n_1} m_l\lambda_l - \lambda_j \neq 0. \tag{56}$$

In this particular case, since we have just one critical eigenvalue, $\lambda_1 = 0$, the condition reduces to

$$i\nu\frac{\pi}{T} \neq 0. \tag{57}$$

The only way that this can be zero is when $\nu = 0$. Since $\nu = 0$ denotes the zeroth-order term of the Fourier series expansions, *which* corresponds to the time averages of the periodic coefficient functions $Q_{1,2}^*(t)$ and $C_{1,3}^*(t)$. Every other term can be eliminated by the normal form transformations. We mentioned already earlier that in the case of a -1 Floquét multiplier the time-average of the quadratic term is zero. Therefore, the normal form becomes

$$\dot{v} = a_{3,0}^* v^3. \tag{58}$$

Since $a_{3,0}^*$, being a Fourier coefficient, is a real constant number, this equation is completely time-invariant. As it is dynamically equivalent to the original problem (2), we can conclude from it that the equilibrium of the original system is asymptotically stable at the critical point if and only if the cubic coefficient $a_{3,0}^*$ is negative.

Our main purpose was to study the bifurcation phenomenon in the neighborhood of the critical point. To achieve this, we need to construct the versal deformation of equation (58) as

$$\dot{v} = \mu(\alpha)v + a_{3,0}^* v^3, \tag{59}$$

where the versal deformation parameter μ is a function of the original bifurcation parameter α and $\mu(\alpha_c) = 0$. This equation can be solved in a closed form and the solution is given by

$$v(t) = \left(\left(\frac{1}{v_0^2} + \frac{a_{3,0}^*}{\mu(\alpha)}\right)e^{-2\mu(\alpha)t} - \frac{a_{3,0}^*}{\mu(\alpha)}\right)^{-1/2}. \tag{60}$$

This equation describes the dynamics in the neighborhood of the critical point, for $\alpha \neq \alpha_c$. The trivial equilibrium $v = 0$, which corresponds to the original equilibrium or periodic solution $x = 0$, exists for all μ. There are two new nonzero equilibrium points born at $\mu = 0$, $v_{1,2} = \pm(-\mu(\alpha)/a^*_{3,0})^{1/2}$, that exist for $\mu \leqslant 0$ if $a^*_{3,0} > 0$ and for $\mu \geqslant 0$ if $a^*_{3,0} < 0$. When transformed back into the original variables, due to the periodic nature of the transformations, these two equilibrium solutions convert into one $2T$-periodic limit cycle in the original coordinates. In itself equation (60) describes a pitchfork bifurcation, but in the domain of the original variables it corresponds to a flip (or period doubling) bifurcation. For equation (2) a stable $2T$-periodic limit cycle exists for $\mu \geqslant 0$ if and only if the cubic coefficient is less than zero. An unstable limit cycle exits for $\mu \leqslant 0$ if and only if the cubic coefficient is greater than zero. See Figures 1 and 2 for an illustration.

There is one more point worth noting. During the normal form reduction, the only terms not eliminated were the time-averages of the nonlinear functions. This seems to yield the same result as the traditional averaging method. However, most important difference is that this "averaging" is performed after the application of the Lyapunov–Floquét transformation, therefore, the linear stability characteristics are preserved. Averaging methods generally average the linear part of the periodic equation as well as the nonlinear one, thereby introducing errors into the calculation of the stability boundaries. On the other hand, for systems with small parameters (small nonlinear terms), if averaging was performed after the application of the Lyapunov–Floquét transformation, it would yield the same result as the normal form reduction. The advantage of the normal form technique is that it is not restricted to systems with small parameters, but valid in the entire parameter range.

6.2. Transcritical and symmetry breaking bifurcations

Here we assume that for some critical value, α_c, of the system parameter the Floquét transition matrix associated with matrix $A(\alpha, t)$ in equation (2) has a $+1$ Floquét multiplier while all the rest of the eigenvalues have magnitudes less than one. There are three different bifurcations associated with a $+1$ multiplier. Two of these three, the transcritical and the symmetry breaking, can be bifurcations of equilibriums as well as bifurcations of periodic orbits. These two bifurcations can happen when the equilibrium, x_0, of the original equation (1) exists continuously in some neighborhood of the bifurcation parameter. Since we are discussing a codimension one bifurcation we can assume that we have a single scalar bifurcation parameter. Then the condition can be restated as follows. The transcritical and the symmetry breaking bifurcations can occur if the x_0 equilibrium solution exists for both $\alpha \leqslant \alpha_c$ and $\alpha > \alpha_c$. We assume that this condition is satisfied.

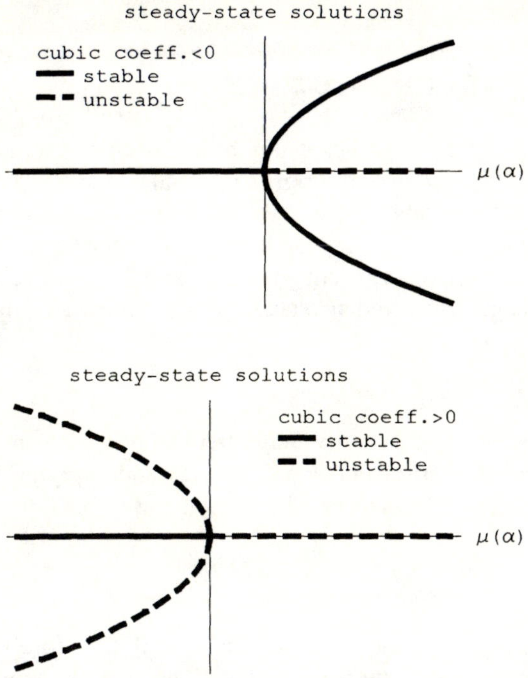

Figure 1. The flip bifurcation in the transformed coordinates: supercritical and subcritical case.

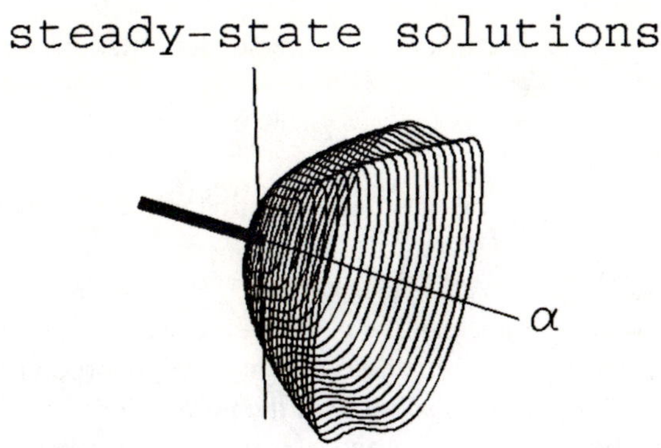

Figure 2. The flip bifurcation: supercritical limit cycles in the physical system.

Since both the $+1$ and the -1 Floquét multipliers correspond to a zero eigenvalue of the matrix J after the application of the Lyapunov–Floquét transformation, equation takes the same form as in the flip bifurcation case (equation (51)), namely

$$\left\{\begin{matrix} \dot{z}_1 \\ \dot{z}_s \end{matrix}\right\} = \begin{bmatrix} 0 & 0 \\ 0 & J_s \end{bmatrix} \left\{\begin{matrix} z_1 \\ z_s \end{matrix}\right\} + \left\{\begin{matrix} \widetilde{Q}_1(z_1, z_s, t) + \widetilde{C}_1(z_1, z_s, t) \\ \widetilde{Q}_s(z_1, z_s, t) + \widetilde{C}_s(z_1, z_s, t) \end{matrix}\right\}. \tag{61}$$

However, there is a very significant difference. In this case, since at least one of the Floquét multipliers is in the right half of the complex plane, the real L–F transformation is T-periodic and it does not have a special symmetry property. Therefore, the averages of the even powered nonlinear terms do not disappear. The center manifold relations are sought in the same form as (52), and the reduced equation on the center manifold also takes the same form as in the flip bifurcation case (equation (53)) as

$$\dot{z}_1 = Q^*_{1,2}(t) z_1^2 + C^*_{1,3}(t) z_1^3. \tag{62}$$

The difference shows itself clearly only after the normal form reduction. After expanding the nonlinearities into Fourier series and applying the near-identity transformations (same as equations (54) and (55)) we obtain the same solvability condition

$$\mathrm{i}\nu\frac{\pi}{T} \neq 0. \tag{63}$$

Just like before, this cannot be satisfied for $\nu = 0$, therefore the only resonant terms are time-averages of the periodic functions, and the reduced equation becomes

$$\dot{v} = b^*_{2,0} v^2 + a^*_{3,0} v^3, \tag{64}$$

where $a^*_{3,0}$ and $b^*_{2,0}$ are real constants. The versal deformation of this normal form can be constructed as

$$\dot{v} = \mu(\alpha) v + b^*_{2,0} v^2 + a^*_{3,0} v^3. \tag{65}$$

Theoretically, the most general versal deformation of the normal form (64) is

$$\dot{v} = \varsigma(\alpha) + \mu(\alpha) v + b^*_{2,0} v^2 + a^*_{3,0} v^3, \tag{66}$$

where ς is a constant parameter, like μ. However, we assumed that the equilibrium point, $x_0 = 0$, exists for all values of α in a neighborhood of α_c. From this it follows that $v = 0$ has to be a solution of the normal form (66) for all values of ς and μ in a neighborhood of $\varsigma = 0$ and $\mu = 0$. This can only be true if $\varsigma(\alpha) = 0$ for all α. This is why equation (65) contains only the $\mu(\alpha)v$ term.

The versal deformation equation (65) describes the behavior of our original system around the bifurcation point. There are two different types of bifurcations possible, depending on the coefficients $a^*_{3,0}$ and $b^*_{2,0}$. If the coefficient of the quadratic term, $b^*_{2,0}$, is not zero then this is the dominant term and the cubic term can be neglected, since v is small in the neighborhood of the critical point. In this case the equation reduces to

$$\dot{v} = \mu(\alpha)v + b_{2,0}^{*}v^{2}. \tag{67}$$

Beside the trivial solution $v = 0$, it has an other equilibrium point: $v_1 = -\mu(\alpha)/b_{2,0}^{*}$; and the general solution is given by

$$v(t) = \left(\left(\frac{b_{2,0}^{*}}{\mu(\alpha)} - \frac{1}{v_0}\right)e^{-\mu(\alpha)t} - \frac{b_{2,0}^{*}}{\mu(\alpha)}\right)^{-1}. \tag{68}$$

The nontrivial equilibrium exists for all $\mu(\alpha)$ in a neighborhood of the critical point. After transforming it back to the original coordinates, due to the T-periodic and nonsymmetric Lyapunov–Floquét transformation, the steady-state solution becomes a T-periodic asymmetric limit cycle. If $\mu(\alpha) \leqslant 0$, the trivial solution is unstable and the limit cycle is stable. If $\mu(\alpha) > 0$ it reverses: the limit cycle becomes unstable and the origin is stable. This is the so-called transcritical bifurcation.

If, on the other hand, the quadratic coefficient, $b_{2,0}^{*}$, is zero, the normal form is

$$\dot{v} = \mu(\alpha)v + a_{3,0}^{*}v^{3} \tag{69}$$

which is the same as in the case of the flip bifurcation with the nontrivial equilibrium solutions, $v_{1,2} = \pm(-\mu(\alpha)/a_{3,0}^{*})^{1/2}$ and a general solution that is the same as equation (60). Obviously, like in the flip bifurcation case, this nontrivial equilibrium exist only on one side of the critical point ether for $\mu(\alpha) < 0$ or for $\mu(\alpha) \geqslant 0$. But now the transformations that convert it back to the original variables are T-periodic and also nonsymmetric, therefore, the original system has an asymmetric limit cycle. If the cubic coefficient, $a_{3,0}^{*} \leqslant 0$, the limit cycle exists only if $\mu(\alpha) \geqslant 0$ and it is stable. If $a_{3,0}^{*} > 0$, the limit cycle exists only for $\mu(\alpha) < 0$ and it is unstable. This phenomenon is called the symmetry breaking bifurcation. See Figure 3 for an illustration of the supercritical case.

6.3. *Cyclic fold bifurcation*

The cyclic fold is a bifurcation of limit cycles. This is the bifurcation that causes the well-known jump phenomenon of resonance. Once again, we assume that for some critical value, α_c, of the system parameter the Floquét transition matrix associated with $A(\alpha, t)$ in equation (2) has a $+1$ multiplier while the other eigenvalues all have magnitudes less than one. To distinguish this case from the previous two types, which also occur when a multiplier becomes $+1$, we need an additional assumption. We assume that equation (1) has a known periodic solution, $x_0(t)$, which only exists on one side of the critical point, that is, it exists for either $\alpha \leqslant \alpha_c$ or $\alpha \geqslant \alpha_c$. Since the critical multiplier is the same as in the

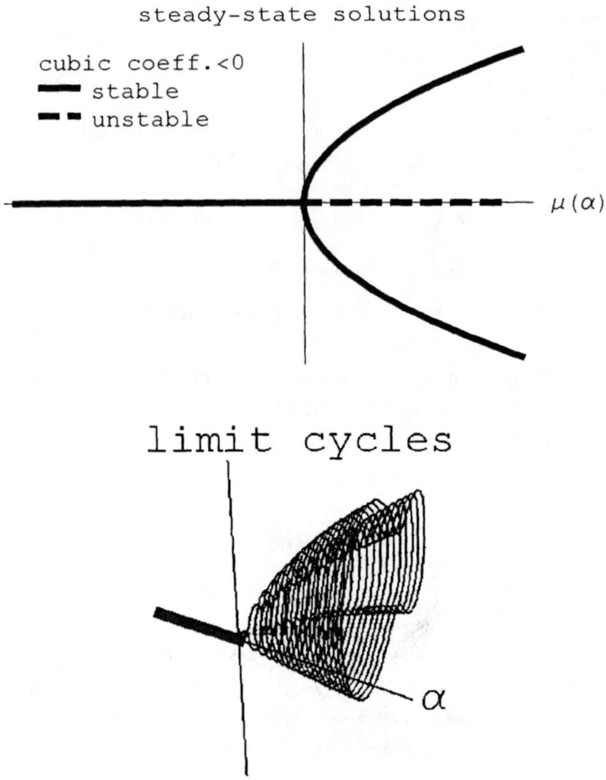

Figure 3. The symmetry-breaking bifurcation, supercritical case: steady-state solutions in the transformed coordinates and limit cycles in the physical system.

previous subsection, we obtain the same general versal deformation form

$$\dot{v} = \varsigma(\alpha) + \mu(\alpha)v + b_{2,0}^{*}v^{2} + a_{3,0}^{*}v^{3}. \tag{70}$$

However, due to our assumption that the trivial solution does not exists on both sides of the bifurcation point, the $\varsigma(\alpha)$ term is permitted. Since we are looking for a miniversal deformation with the smallest possible number of parameters, (namely one, in case of a codimension one bifurcation) we need to set $\mu(\alpha) = 0$. Due to the fact that in this case the original equation (1) is expanded about a periodic solution, the coefficient of the quadratic term is not zero, in general, so the cubic term can be neglected. Therefore, the versal deformation of the cyclic fold bifurcation is

$$\dot{v} = \varsigma(\alpha) + b_{2,0}^{*}v^{2}. \tag{71}$$

This equation has two nontrivial steady-state solutions, $v_{1,2} = \pm\sqrt{-\varsigma(\alpha)/b^*_{2,0}}$. These exist only if $\varsigma(\alpha)/b^*_{2,0} \leqslant 0$, so depending on the sign of the quadratic coefficient, they exist either for $\varsigma(\alpha) \leqslant 0$ or for $\varsigma(\alpha) \geqslant 0$. These equilibrium solutions transform into coexisting T-periodic limit cycles in the original domain. The general solution is given by

$$v(t) = \sqrt{\frac{\varsigma(\alpha)}{b^*_{2,0}}}\tan\left(\sqrt{\varsigma(\alpha)b^*_{2,0}}\,t + \arctan\left(\sqrt{\frac{\varsigma(\alpha)}{b^*_{2,0}}}\,v(0)\right)\right). \tag{72}$$

Now we can draw some conclusions about the dynamics of this bifurcation. The original solution, $x_0(t)$, has a coexisting pair and at the critical point the two cycles collide and disappear. One of these cycles is always stable and the other one always unstable, regardless of the value of the quadratic coefficient, $b^*_{2,0}$. This phenomenon is called the cyclic fold bifurcation. Another feature, which is interesting to note, is that the critical Floquét multiplier of both these limit cycles is invariant to the change in the bifurcation parameter. It remains one in a small neighborhood of α_c. In physical systems the cyclic fold bifurcation usually happens in during nonlinear resonance, and it generally occurs in pairs, hence the name "fold" (see Figure 4). It causes a sudden jump of the solution from one limit cycle to another.

6.4. Secondary Hopf bifurcation

For this situation we assume that for some critical value, α_c, of the system parameter the Floquét transition matrix associated with $A(\alpha, t)$ in equation (2) has one pair of complex eigenvalues with magnitudes equal to 1, while all the other eigenvalues have magnitudes less than 1. The pair of complex multipliers with magnitude one corresponds to a pair of purely imaginary eigenvalues of matrix J after the Lyapunov–Floquét transformation. Let us denote the critical eigenvalue pair of matrix J by $\pm\omega_c\mathrm{i}$. If ω_c is *not* an integer multiple of $2\pi/T$, where T is the period, then the normal form becomes time-invariant, otherwise, so-called resonant periodic terms corresponding to the "resonant" frequencies will remain. Let us consider the nonresonant case first. Under these assumptions and after the application of the L–F and modal transformations the system becomes

$$\begin{Bmatrix} \dot{z}_1 \\ \dot{z}_2 \\ \dot{z}_s \end{Bmatrix} = \begin{bmatrix} \mathrm{i}\omega_c & 0 & 0 \\ 0 & -\mathrm{i}\omega_c & 0 \\ 0 & 0 & J_s \end{bmatrix} \begin{Bmatrix} z_1 \\ z_2 \\ z_s \end{Bmatrix} + \begin{Bmatrix} \widetilde{Q}_1(z_1, z_2, z_s, t) + \widetilde{C}_1(z_1, z_2, z_s, t) \\ \widetilde{Q}_2(z_1, z_2, z_s, t) + \widetilde{C}_2(z_1, z_2, z_s, t) \\ \widetilde{Q}_s(z_1, z_2, z_s, t) + \widetilde{C}_s(z_1, z_2, z_s, t) \end{Bmatrix}. \tag{73}$$

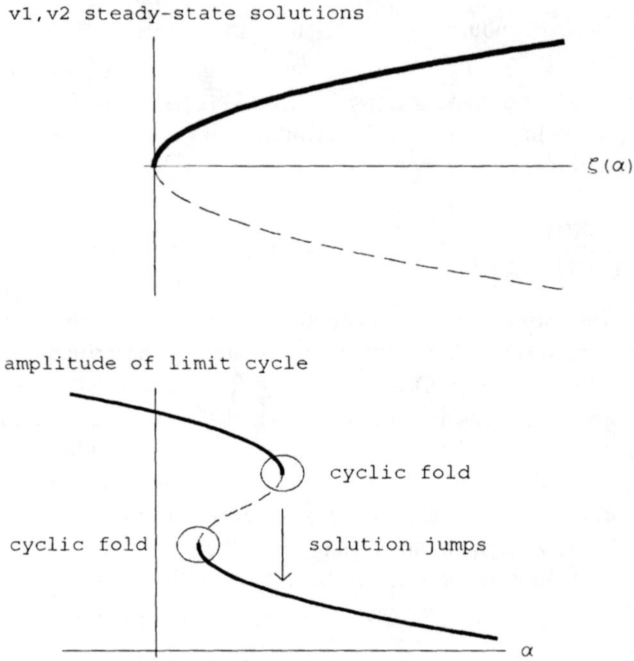

Figure 4. The cyclic fold bifurcation in the normal form and in the physical system.

We seek the center manifold relations in the form

$$z_i = h_{c2,i}(z_1, z_2, t) + h_{c3,i}(z_1, z_2, t), \quad i = 3, 4, \ldots, n, \tag{74}$$

where the quadratic and cubic terms can be defined as monomial functions with periodic coefficients of the form

$$\begin{aligned} h_{c2,i}(z_1, z_2, t) &= h_{c2,i,(2,0)}(t)z_1^2 + h_{c2,i,(1,1)}(t)z_1 z_2 + h_{c2,i,(0,2)}(t)z_2^2, \\ h_{c3,i}(z_1, z_2, t) &= h_{c3,i,(3,0)}(t)z_1^3 + h_{c3,i,(0,3)}(t)z_2^3 + h_{c3,i,(2,1)}(t)z_1^2 z_2 \\ &\quad + h_{c3,i,(2,1)}(t)z_1 z_2^2. \end{aligned} \tag{75}$$

If we substitute these expressions into equation (73) we obtain the reduced system on the two-dimensional center manifold as

$$\begin{Bmatrix} \dot{z}_1 \\ \dot{z}_2 \end{Bmatrix} = \begin{bmatrix} i\omega_c & 0 \\ 0 & -i\omega_c \end{bmatrix} \begin{Bmatrix} z_1 \\ z_2 \end{Bmatrix} + \begin{Bmatrix} Q_1^*(z_1, z_2, t) + C_1^*(z_1, z_2, t) \\ Q_2^*(z_1, z_2, t) + C_2^*(z_1, z_2, t) \end{Bmatrix}, \tag{76}$$

where the C^* functions contain terms with the h_c center manifold relations from equation (75) in them. Continuing the procedure, we expand the nonlinear functions into Fourier series and define a series of near-identity transformations given by

$$\begin{Bmatrix} z_1 \\ z_1 \end{Bmatrix} = \begin{Bmatrix} \tilde{v}_1 \\ \tilde{v}_2 \end{Bmatrix} + \begin{Bmatrix} h_{n2,1}(\tilde{v}_1, \tilde{v}_2, t) \\ h_{n2,2}(\tilde{v}_1, \tilde{v}_2, t) \end{Bmatrix},$$
$$\begin{Bmatrix} \tilde{v}_1 \\ \tilde{v}_2 \end{Bmatrix} = \begin{Bmatrix} v_1 \\ v_2 \end{Bmatrix} + \begin{Bmatrix} h_{n3,1}(v_1, v_2, t) \\ h_{n3,2}(v_1, v_2, t) \end{Bmatrix}, \tag{77}$$

where the quadratic and cubic functions are given in the same monomial form as the h_c functions in equation (75). With these transformations, the *solvability condition* (equation (35)) for the quadratic and the cubic terms is

$$\mathrm{i}\nu\frac{\pi}{T} + \sum_{l=1}^{2} m_l\lambda_l - \lambda_j \neq 0, \quad \lambda_1 = \mathrm{i}\omega_c, \lambda_2 = -\mathrm{i}\omega,$$
$$\omega_c \neq k\frac{\pi}{T}, \quad k = 1, 2, \frac{1}{2}, 3, \frac{1}{3}, \ldots, \tag{78}$$

where for the quadratic terms $\sum_{l=1}^{2} m_l = 2$ and for the cubic ones $\sum_{l=1}^{2} m_l = 3$. It is very easy to see that for the quadratic terms this condition can never be satisfied, so all of the terms can be removed. In case of the third-order terms, there is one resonant situation for each of the two coordinates, when the solvability condition can be zero. This happens when $\nu = 0$. It implies that the zeroth-order element of the Fourier series of the particular term (or in other words the time-average of the given periodic function) cannot be eliminated. Therefore, the normal form becomes

$$\begin{Bmatrix} \dot{v}_1 \\ \dot{v}_2 \end{Bmatrix} = \begin{bmatrix} \mathrm{i}\omega_c & 0 \\ 0 & -\mathrm{i}\omega_c \end{bmatrix} \begin{Bmatrix} v_1 \\ v_2 \end{Bmatrix}$$
$$+ \begin{Bmatrix} \left(\sum_{i=3}^{n}\left(\overline{\widetilde{Q}_{1,(1,0,\ldots,1_i,\ldots,0)}(t)h_{c2,i(1,1)}(t)} + \overline{\widetilde{Q}_{1,(0,1,\ldots,1_i,\ldots,0)}(t)h_{c2,i(2,0)}(t)}\right) \right. \\ \left. + \overline{\widetilde{C}_{1,(2,1,0,\ldots,0)}(t)}\right)v_1^2 v_2 \\ \left(\sum_{i=3}^{n}\left(\overline{\widetilde{Q}_{1,(1,0,\ldots,1_i,\ldots,0)}(t)h_{c2,i(0,2)}(t)} + \overline{\widetilde{Q}_{1,(0,1,\ldots,1_i,\ldots,0)}(t)h_{c2,i(1,1)}(t)}\right) \right. \\ \left. + \overline{\widetilde{C}_{1,(1,2,0,\ldots,0)}(t)}\right)v_1 v_2^2 \end{Bmatrix} \tag{79}$$

or with a new notation

$$\begin{Bmatrix} \dot{v}_1 \\ \dot{v}_2 \end{Bmatrix} = \begin{bmatrix} \mathrm{i}\omega_c & 0 \\ 0 & -\mathrm{i}\omega_c \end{bmatrix} \begin{Bmatrix} v_1 \\ v_2 \end{Bmatrix} + \begin{Bmatrix} (C_r + \mathrm{i}C_i)v_1^2 v_2 \\ (C_r - \mathrm{i}C_i)v_1 v_2^2 \end{Bmatrix}, \tag{80}$$

where C_r and C_i are real constants and the coefficients of the two equations are complex conjugates of each other.

The versal deformation of this equation can be constructed as

$$\begin{Bmatrix} \dot{v}_1 \\ \dot{v}_2 \end{Bmatrix} = \begin{bmatrix} \mu(\alpha) + \mathrm{i}\omega_c & 0 \\ 0 & \mu(\alpha)^* - \mathrm{i}\omega_c \end{bmatrix} \begin{Bmatrix} v_1 \\ v_2 \end{Bmatrix} + \begin{Bmatrix} (C_r + \mathrm{i}C_i)v_1^2 v_2 \\ (C_r - \mathrm{i}C_i)v_1 v_2^2 \end{Bmatrix}. \tag{81}$$

The versal deformation parameter is complex, in general, and "*" denotes its complex conjugate. However, in physical problems it can be observed that in the neighborhood of a secondary Hopf bifurcation the real part of the eigenvalues changes very rapidly while the imaginary part remains nearly constant. Therefore, in practical applications, in order to simplify the calculations, it may be reasonable to assume that $\mu(\alpha)$ is real. For now, we will continue with the most general case of a complex $\mu(\alpha)$. The versal deformation equation can be transformed into a real form in terms of polar coordinates as

$$\begin{aligned}\dot{R} &= \mathrm{Re}\big(\mu(\alpha)\big)R + C_\mathrm{r}R^3,\\ \dot{\Theta}R &= -\big(\mathrm{Im}\big(\mu(\alpha)\big) + \omega_\mathrm{c}\big)R - C_\mathrm{i}R^3.\end{aligned} \tag{82}$$

Since the radius equation is independently solvable, we can easily see that beside the trivial solution $R = 0$, we have one more steady-state solution: $R_{1,2} = \pm\sqrt{-\mu(\alpha)/C_\mathrm{r}}$. A steady-state solution for the radius implies that there exists a limit cycle, if $\mu(\alpha)/C_\mathrm{r} \leqslant 0$. The stability of this limit cycle depends on the sign of the cubic coefficient C_r. If $C_\mathrm{r} \geqslant 0$ there exists an unstable limit cycle for $\mu(\alpha) \leqslant 0$ and if $C_\mathrm{r} \leqslant 0$ there exists a stable limit cycle for $\mu(\alpha) \geqslant 0$. Now, in order to obtain the solution in the original coordinates, we need to apply all our transformations. As these are all T-periodic and Θ in general is not an integer multiple of the period T, after the transformations the limit cycle becomes a quasiperiodic limit set in the physical variables, either stable or unstable. This phenomenon, the birth of a quasiperiodic limit set at the critical point, is called the secondary Hopf bifurcation. See Figure 5 for an illustration of the supercritical case.

Now let us discuss the resonant case. If we assume that $(p/q)\omega_\mathrm{c} = \pi/T$ where p and q are relative prime integers, then during the normal form reduction there are terms with periodic coefficients that cannot be solved for, because the *solvability condition* (equation (35)) is not satisfied. It can be shown (see Chow, Li and Wang, 1994) that in this case the normal form becomes

$$\begin{aligned}\begin{Bmatrix}\dot{v}_1\\ \dot{v}_2\end{Bmatrix} &= \begin{bmatrix}\mu(\alpha) + \mathrm{i}\omega_\mathrm{c} & 0\\ 0 & \mu(\alpha)^* - \mathrm{i}\omega_\mathrm{c}\end{bmatrix}\begin{Bmatrix}v_1\\ v_2\end{Bmatrix}\\ &\quad + \begin{Bmatrix}(C_\mathrm{r} + \mathrm{i}C_\mathrm{i})v_1^2v_2 + d\mathrm{e}^{-\mathrm{i}q\omega_\mathrm{c}t}v_2^{q-1}\\ (C_\mathrm{r} - \mathrm{i}C_\mathrm{i})v_1v_2^2 + d^*\mathrm{e}^{\mathrm{i}q\omega_\mathrm{c}t}v_1^{q-1}\end{Bmatrix},\end{aligned} \tag{83}$$

where d is the Fourier coefficient of the given periodic term and "*" indicates the complex conjugate of the quantity. The order of the extra nonlinear term depends on q, which in turn depends on the ratio of π/T and ω_c. So we can obtain second- or third-order nonlinearities only if q is equal to 3 or 4. It is possible, through further transformations to obtain a time-invariant form of this periodic normal form, and obtain the local solution (Arrowsmith and Place, 1990). From the solution

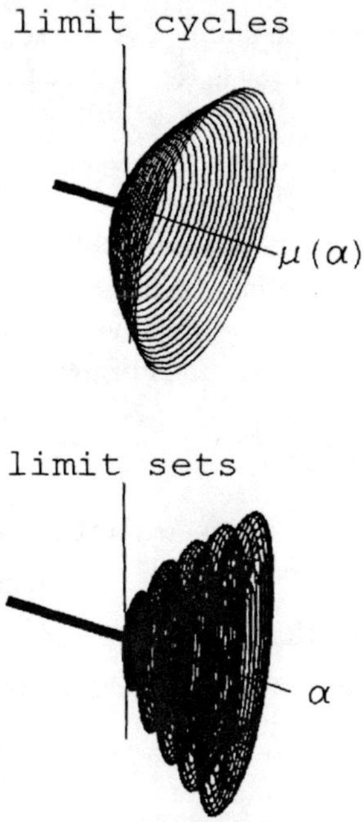

Figure 5. The supercritical secondary Hopf bifurcation: limit cycles of the normal form and the quasiperiodic limit sets in the physical coordinates.

it can be seen that even in this case the origin undergoes a Hopf bifurcation and the stability of the limit cycle depends on both C and d. The limit cycle corresponds to a quasiperiodic limit set in the physical coordinates. However, the local dynamics is much more complex in this situation. Depending on the order of the periodic nonlinear terms, there are other equilibrium points of the saddle type in the neighborhood of the origin, which transform into limit cycles in the original domain. Arrowsmith and Place (1990) also provide a detailed discussion of the dynamics for the cases of quadratic and cubic nonlinearities.

7. Applications

7.1. A system with an exact solution: an example of the flip bifurcation

This example is to illustrate the idea of versal deformation on a special case, when the relationship between the system parameter and the versal deformation parameter can be obtained exactly.

There is a special class of systems, called commutative systems, when the Lyapunov–Floquét transformation can be found in a closed form which makes the rest of the computations much simpler. A linear system matrix $A(t)$ is called commutative if there exists a matrix $B_c(t)$ such that

$$\frac{\mathrm{d}B_c(t)}{\mathrm{d}t} = A(t) \quad \text{and} \quad A(t)B_c(t) - B_c(t)A(t) = 0. \tag{84}$$

It has been shown (Lukes, 1982) that based on these properties the L–F transformation of a commutative system can be computed in a closed form. There are some special cases other than the commutative systems, when it may be possible to factor the STM such that the L–F transformation can be obtained in a closed form.

Consider the system given by

$$\begin{Bmatrix} \dot{x}_1 \\ \dot{x}_2 \end{Bmatrix} = \begin{bmatrix} -1+\alpha\cos^2 t & 1-\alpha\sin t\cos t \\ -1-\alpha\sin t\cos t & -1+\alpha\sin^2 t \end{bmatrix} \begin{Bmatrix} x_1 \\ x_2 \end{Bmatrix} + b \begin{Bmatrix} x_1^3 \\ x_2^3 \end{Bmatrix}, \tag{85}$$

where $\{x_1 \quad x_2\}^{\mathrm{T}}$ is the state-vector, α is the bifurcation parameter and b is an arbitrary real number. This system is periodic with period $T = \pi$. The state transition matrix is given by Mohler (1991) as

$$\Phi(t) = \begin{bmatrix} \mathrm{e}^{(\alpha-1)t}\cos t & \mathrm{e}^{-t}\sin t \\ -\mathrm{e}^{(\alpha-1)t}\sin t & \mathrm{e}^{-t}\cos t \end{bmatrix}. \tag{86}$$

This can be easily factored into

$$\Phi(t) = \begin{bmatrix} \cos t & \sin t \\ -\sin t & \cos t \end{bmatrix} \begin{bmatrix} \mathrm{e}^{(\alpha-1)t} & 0 \\ 0 & \mathrm{e}^{-t} \end{bmatrix} = Q(t)\mathrm{e}^{Rt}, \tag{87}$$

where $Q(t)$ is the Lyapunov–Floquét transformation, and it is real and $2T$-periodic. The Floquét transition matrix is given as

$$\Phi(T) = \begin{bmatrix} -\mathrm{e}^{(\alpha-1)\pi} & 0 \\ 0 & -\mathrm{e}^{-\pi} \end{bmatrix}. \tag{88}$$

We can immediately see from here that one of the Floquét multipliers is always stable while the other one can have a magnitude that equals to 1 if $\alpha = 1$. Therefore, $\alpha_c = 1$ is the critical value of the bifurcation parameter and the critical Floquét multiplier at this point is -1, indicating a flip bifurcation. The constant matrix R can also be easily computed as

$$R = \frac{1}{T}\ln\Phi(T) = \begin{bmatrix} \alpha-1 & 0 \\ 0 & -1 \end{bmatrix}. \tag{89}$$

Note, that since R is already in a diagonal form, there is no need for a modal transformation this time. After applying the Lyapunov–Floquét transformation

$x = Q(t)z$, equation (85) becomes

$$\begin{Bmatrix} \dot{z}_1 \\ \dot{z}_2 \end{Bmatrix} = \begin{bmatrix} \alpha - 1 & 0 \\ 0 & -1 \end{bmatrix} \begin{Bmatrix} z_1 \\ z_2 \end{Bmatrix} + \frac{1}{4}b \times \begin{Bmatrix} (3+\cos 4t)z_1^3 + 3\sin 4t z_1^2 z_2 + (3 - 3\cos 4t)z_1 z_2^2 - \sin 4t z_2^3 \\ \sin 4t z_1^3 + (3 - 3\cos 4t)z_1^2 z_2 - 3\sin 4t z_1 z_2^2 + (3+\cos 4t)z_2^3 \end{Bmatrix}. \tag{90}$$

If we examine the nonlinear part, we can readily see that in the center manifold relation, instead of the general formulation, we need to find only a few terms. So we seek the relation in a Fourier-series expansion form as

$$z_2 = \left(h_{c3,0} + h_{c3,-4}\mathrm{e}^{-4\mathrm{i}t} + h_{c3,4}\mathrm{e}^{4\mathrm{i}t}\right)z_1^3. \tag{91}$$

When we substitute this into equation (90) and solve for the Fourier coefficients of the center manifold relation, we obtain

$$h_{c3,0} = 0, \qquad h_{c3,-4} = b\left(-\frac{1}{34} + \frac{\mathrm{i}}{136}\right), \qquad h_{c3,4} = b\left(-\frac{1}{34} - \frac{\mathrm{i}}{136}\right). \tag{92}$$

Then, the center manifold relation in a trigonometric form becomes

$$z_2 = \frac{1}{68}b(\sin 4t + 4\cos 4t)z_1^3. \tag{93}$$

After substituting this into equation (90) and neglecting all the terms that are of order higher than cubic, the reduced equation on the center manifold takes the form

$$\dot{z}_1 = (\alpha - 1)z_1 + \frac{1}{4}b(3 + \cos 4t)z_1^3. \tag{94}$$

This is already a rather impressive simplification. In fact, this equation is already solvable in a closed form. However, for the sake of illustrating the entire procedure, let us simplify it further. The near-identity transformation is sought as

$$z_1 = v + \left(h_{n3,0} + h_{n3,-4}\mathrm{e}^{-4\mathrm{i}t} + h_{n3,4}\mathrm{e}^{4\mathrm{i}t}\right)v^3. \tag{95}$$

From the solvability condition, $\mathrm{i}\nu(\pi/T) \neq 0$, it follows that the $\nu = 0$ term, $h_{n3,0}$, is resonant and cannot be solved for. The other two coefficients are found to be $h_{n3,-4} = \frac{1}{32}b\mathrm{i}$ and $h_{n3,4} = -\frac{1}{32}b\mathrm{i}$. In the trigonometric form this yields

$$z_1 = v + \frac{1}{16}b\sin 4t\, v^3. \tag{96}$$

After applying this transformation we get the normal form as

$$\dot{v} = (\alpha - 1)v + \frac{3}{4}bv^3. \tag{97}$$

Observe that we have automatically obtained an exact versal deformation. It is because in the factorization of the state transition matrix, the α bifurcation parameter appears only in the e^{Rt} matrix which is diagonal and, from that, $R(\alpha)$ could be obtained analytically in a closed form. Therefore, the connection between the bifurcation parameter, α and the versal deformation parameter, $\mu(\alpha)$ is given right away as $\mu = \alpha - 1$.

The solution of this equation for a given initial condition $v(0) = v_0$ is

$$v(t) = \left(\left(\frac{1}{v_0^2} + \frac{3b}{4(\alpha-1)}\right)e^{-2(\alpha-1)t} - \frac{3b}{4(\alpha-1)}\right)^{-1/2}. \tag{98}$$

This equation has, beside the $v = 0$ solution, two nontrivial equilibrium points for $\alpha \leqslant 0$, $v_{1,2} = \pm\sqrt{-4(\alpha-1)/(3b)}$, which, after transforming back to the original states, becomes a $2T$-periodic limit cycle due to the fact that the real L–F transformation is $2T$-periodic in this case (both multipliers lie in the left half of the complex plane). The solution for this limit cycle in the original coordinates up to 5 Fourier terms is given as

$$\begin{aligned}
x_1(t) &= \frac{2\sqrt{b-\alpha b}}{\sqrt{3}}\cos t \\
&\quad + \frac{b^2\sqrt{b-\alpha b}(64 + b^4 + 3\alpha^2 b^4 - \alpha^3 b^4 - \alpha(64 + 3b^4))}{3264\sqrt{3}}\cos 3t \\
&\quad + \frac{b^2\sqrt{b-\alpha b}(-64 - b^4 - 3\alpha^2 b^4 + \alpha^3 b^4 - \alpha(64 + 3b^4))}{3264\sqrt{3}}\cos 5t \\
&\quad + \frac{(\alpha-1)^2 b^4\sqrt{b-\alpha b}(576 + (\alpha-1)^2 b^4)}{117504\sqrt{3}}\sin t \\
&\quad + \frac{b^2\sqrt{b-\alpha b}(12 - b^4 - 3\alpha^2 b^4 + \alpha^3 b^4 + \alpha(b^4 - 4))}{2448\sqrt{3}}\sin 3t \\
&\quad + \frac{b^2\sqrt{b-\alpha b}(396 + b^4 + 3\alpha^2 b^4 - \alpha^3 b^4 - \alpha(132 + b^4))}{2448\sqrt{3}}\sin 5t, \\
x_2(t) &= \frac{(\alpha-1)^2 b^4\sqrt{b-\alpha b}(576 + (\alpha-1)^2 b^4)}{117504\sqrt{3}}\cos t \\
&\quad + \frac{b^2\sqrt{b-\alpha b}(-12 + b^4 + 3\alpha^2 b^4 - \alpha^3 b^4 - \alpha(b^4 - 4))}{2448\sqrt{3}}\cos 3t \\
&\quad + \frac{b^2\sqrt{b-\alpha b}(396 + b^4 + 3\alpha^2 b^4 - \alpha^3 b^4 - \alpha(132 + b^4))}{2448\sqrt{3}}\cos 5t
\end{aligned}$$

$$
\begin{aligned}
&+ \frac{-2\sqrt{b-\alpha b}}{\sqrt{3}} \sin t \\
&+ \frac{b^2\sqrt{b-\alpha b}(64 + b^4 + 3\alpha^2 b^4 - \alpha^3 b^4 - \alpha(64 + 3b^4))}{3264\sqrt{3}} \sin 3t \\
&+ \frac{b^2\sqrt{b-\alpha b}(64 + b^4 + 3\alpha^2 b^4 - \alpha^3 b^4 - \alpha(64 + 3b^4))}{3264\sqrt{3}} \sin 5t.
\end{aligned} \tag{99}
$$

The fifth-order Fourier expansion gives a very good approximation, when compared with numerical results. This limit cycle is stable for $b < 0$, and unstable otherwise. In the neighborhood of the origin it exhibits a relatively slow quadratic growth, depending on the size of the parameter. The solution in terms of the original physical coordinates can be obtained by substituting all the transformations in the reverse order.

Because of the simplicity of this example the computations can be carried out even by hand. One can also use symbolic computation software, such as Mathematica, Matlab or Maple.

In general, however, it is not possible to solve a problem this easily. First of all, for general periodic systems the L–F transformation cannot be obtained in a closed form. Then, generally it is not possible to factor out the bifurcation parameter so nicely because even if we can compute the STM in terms of the bifurcation parameter we still cannot perform the factorization due to computational difficulties. Therefore, approximate techniques must be used to compute the L–F transformation and to find the versal deformation parameter. This will be illustrated in the following examples.

7.2. *A system with a small parameter: a comparison with averaging method*

Consider a system with two degrees of freedom given by the equation

$$
\begin{aligned}
&\ddot{x}_1 + \omega_1^2 x_1 + \varepsilon\big(e_{11}\dot{x}_1 + e_{12}\dot{x}_2 \\
&\qquad + \cos\Omega t(b_{11}x_1 + b_{12}x_2) + a_{11}x_1^3 + a_{13}x_1x_2^2\big) = 0, \\
&\ddot{x}_2 + \omega_2^2 x_2 + \varepsilon\big(e_{21}\dot{x}_1 + e_{22}\dot{x}_2 \\
&\qquad + \cos\Omega t(b_{21}x_1 + b_{22}x_2) + a_{21}x_2^3 + a_{23}x_2x_1^2\big) = 0,
\end{aligned} \tag{100}
$$

where ε is a small parameter. This system has been studied by Tso and Asmis (1974) using averaging method in the cases of parametric, internal and combined parametric and internal resonances. Here we study only the case of combined resonance via the proposed technique as well as via averaging method. Let

$$
\Omega_0 = 2\omega_1, \qquad \omega_2 = 3\omega_1 \quad \text{and} \quad \Omega = \Omega_0(1-\lambda), \tag{101}
$$

where λ is a small detuning parameter, and let us choose $\Omega/\Omega_0 = (1 - \lambda)$ to be the bifurcation parameter. For the parameter set $\varepsilon = 0.1$, $\omega_1 = 0.45$, $e_{11} = 0.1$, $e_{22} = 0.15$, $e_{12} = e_{21} = 0$, $b_{11} = 0.8$, $b_{12} = -1.2$, $b_{21} = 0.6$, $b_{22} = 0.1$, $a_{11} = 0.4$, $a_{13} = 0.3$, $a_{21} = 0.5$ and $a_{23} = 0.27$, the averaging method predicts that the loss of stability occurs at $\Omega/\Omega_0 = 0.91625$. From the state transition matrix computed either numerically through the Fortran IMSL numerical subroutines or in terms of Chebyshev polynomials, we get $\Omega/\Omega_0 = 0.91985$ for the bifurcation point, which differs from the averaging method result only in the third digit. At this critical point a flip bifurcation occurs and the normal form is the same as equation (59). It can be solved in a closed form to obtain the post-bifurcation limit cycles for both of the coordinates x_1 and x_2. The limit cycle amplitudes thus obtained are compared with those obtained from averaging method. Figure 6 shows a comparison for the amplitudes of the two coordinates as a function of the bifurcation parameter Ω/Ω_0. The growth of the amplitude exhibits quadratic characteristics according to both results, and the rate of the growth is about the same. However, because the location of the bifurcation point was predicted incorrectly by the averaging method, the two results are significantly different. Considering that at the correct bifurcation point averaging method predicts the amplitude to be about 0.22 instead of zero, we can see how a very small error in

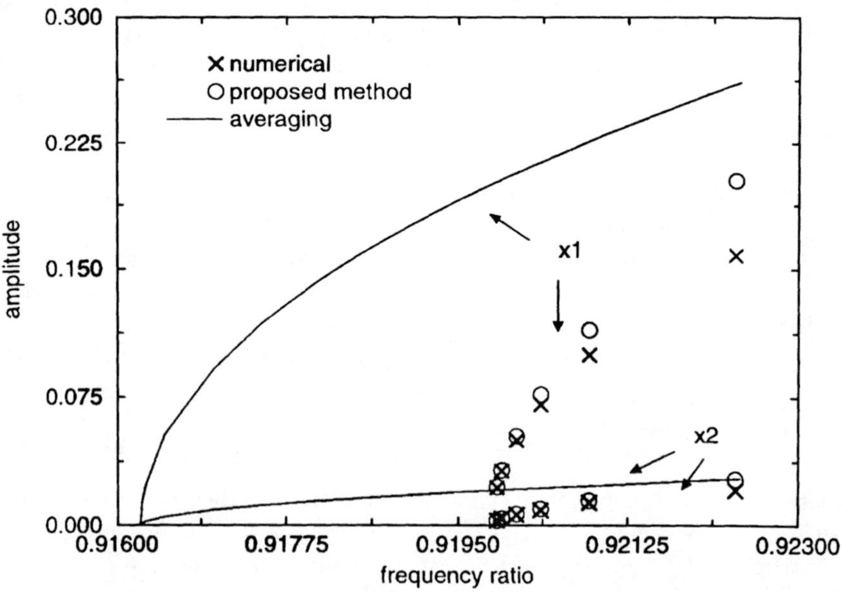

Figure 6. A system with a small parameter: comparison of bifurcation diagrams obtained by the proposed technique and averaging method. (From Dávid and Sinha, 2000.)

the location of the stability boundary leads to relatively serious inaccuracy of the post-bifurcation dynamics. The proposed method follows the numerical results closely near the bifurcation point and it provides much better approximations than averaging, even for small parameters.

7.3. A simple pendulum with periodic base excitation: an example of the symmetry breaking bifurcation

Let us consider a parametrically excited simple pendulum (Figure 7). The nonlinear equation of motion is given as

$$\ddot{\theta} + d\dot{\theta} + (a + b\sin\omega t)\sin\theta = 0, \tag{102}$$

where $a = g/L$, $b = -A\omega^2/L$ and $d = c/ML^2$. The behavior of the pendulum is generally studied in the a–b parameter space. The stability chart of this system is well known and it is shown in Figure 8; for both the undamped case ($d = 0$) and a damped ($d \neq 0$) case, and for $\omega = 2$. The areas shaded in grey refer to the regions where small parameter methods such as averaging and perturbation are applicable. We can see that even if we do have a small b parameter for the periodic part of the stiffness, as soon as there is any damping present in the system (even a relatively small one) these methods quickly loose their applicability. Also, normally, in averaging and perturbation methods approximate solutions are constructed in the neighborhood of resonance points, such as the 1:1 resonance when $\omega = \sqrt{a}$ and the 2:1 resonance $\omega = 2\sqrt{a}$. We can observe in the figure that as the values of b and d increase the stability boundaries curve away from

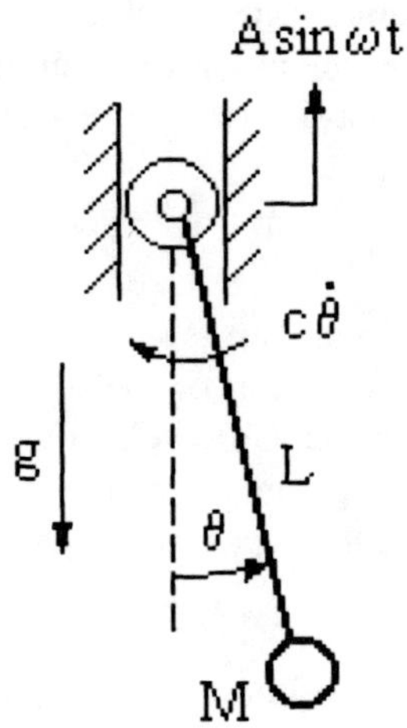

Figure 7. Parametrically excited simple pendulum.

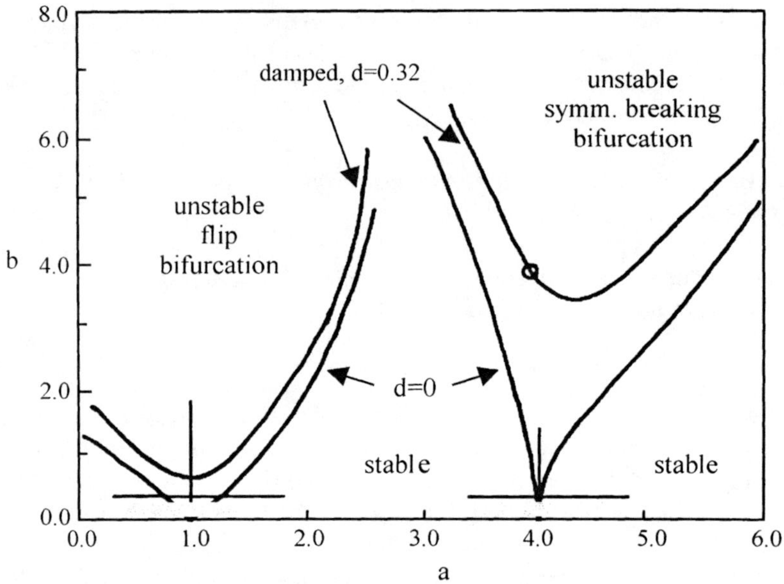

Figure 8. Stability chart of the undamped and damped simple pendulum comparing the applicability of the proposed technique and the averaging method.

the resonance points. For all these reasons, averaging and perturbation methods cannot be applied to study bifurcations in this system. For large parameters there is currently no other method available than the technique described above.

In equation (102) let us define state variables as $\{x_1 \quad x_2\}^{\mathrm{T}} = \{\theta \quad \dot{\theta}\}^{\mathrm{T}}$ and expand $\sin\theta$ into Taylor series about the zero equilibrium point up to the cubic order. Then we can rewrite equation (102) in the state-space form

$$\begin{Bmatrix} \dot{x}_1 \\ \dot{x}_2 \end{Bmatrix} = \begin{bmatrix} 0 & 1 \\ -(a + b\sin\omega t) & -d \end{bmatrix} \begin{Bmatrix} x_1 \\ x_2 \end{Bmatrix} + \begin{Bmatrix} 0 \\ \frac{1}{6}(a + b\sin\omega t)x_1^3 \end{Bmatrix}. \quad (103)$$

The stability chart in the a–b plane for the parameters $d = 0.31623$ and $\omega = 2$ is shown in Figure 9. The circled points on the stability boundary are the bifurcation points studied in this example. Observe that these points are located at large values of the periodic stiffness parameter b; the system is subjected to a strong parametric excitation. Further, these points are quite far away from the 1:1 resonance condition $\omega = \sqrt{a}$. First we investigate the dynamics as we fix the value of b and cross the boundary by varying the parameter a (indicated by the horizontal arrow in Figure 9). Next, we would like to compare post-bifurcation dynamics for different values of b, introducing the same change in a for all five points investigated (see the vertical arrow in Figure 9).

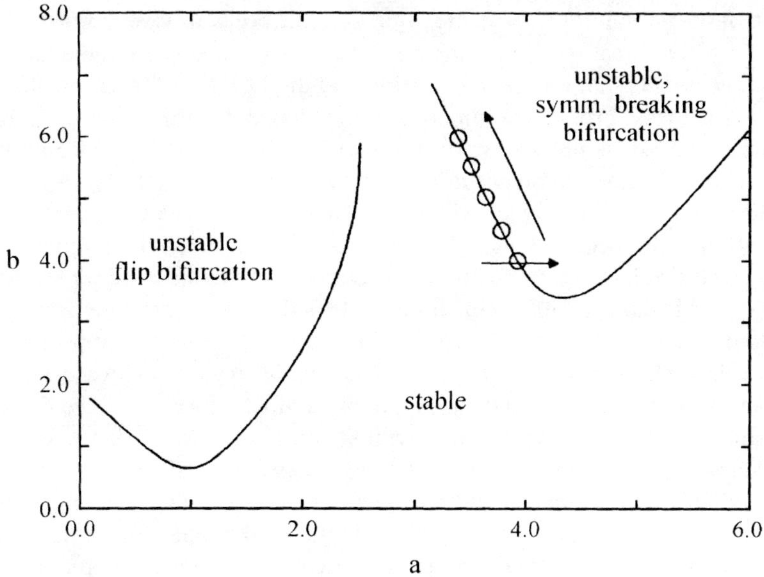

Figure 9. Stability chart of the simple pendulum for the parameters $d = 0.31623$ and $\omega = 2$. (From Dávid and Sinha, 2000.)

First, let parameter a be the bifurcation parameter and set the others as $b = 4$, $d = 0.31623$ and $\omega = 2$. For the critical value $a_c = 3.91778734$ the system undergoes a bifurcation: one of the Floquét multipliers of the linear part of equation (103) is $0.99999988 \approx 1$, the other one is 0.37029. Because the multipliers lie in the right half of the complex plane, the Lyapunov–Floquét transformation is real and T-periodic. After normalizing the time so that the period becomes 1, then applying the L–F and modal transformations at the critical point, we can apply the center manifold reduction to decouple the critical equation from the other. The center manifold equation can then be further simplified by normal form theory and the normal form becomes

$$\dot{v} = -0.1325v^3. \tag{104}$$

The negative coefficient in equation (104) indicates that the nonlinear system is asymptotically stable at the critical point and, also, that the bifurcation is supercritical. Thus, the versal deformation of this equation is

$$\dot{v} = \mu(a) - 0.1325v^3. \tag{105}$$

It is solvable in a closed from and the solution for the initial condition $v(0) = v_0$ is

$$v(t) = \left(\left(\frac{1}{v_0^2} + \frac{-0.1325}{\mu(a)}\right)e^{-2\mu(a)t} - \frac{-0.1325}{\mu(a)}\right)^{-1/2}. \tag{106}$$

Note that the normal form looks the same as in the previous example, where it described a flip bifurcation. Unlike the case of autonomous systems, here the normal form does not contain all the information and the L–F transformation also plays an important role. From equation (106) we can conclude the following about the dynamics in the neighborhood of the bifurcation. For $\mu < 0$ the $v = 0$ equilibrium is stable and there are no other limit sets in the neighborhood. For $\mu > 0$ the zero equilibrium becomes unstable and there is a stable nonzero equilibrium born. This equilibrium point becomes a stable T-periodic limit cycle in the original domain because the real L–F transformation is T-periodic in this case. So we say that the stability has been lost softly. This limit cycle is the boundary of the small attractive domain around the unstable equilibrium. By computing the amplitude of the limit cycle we have estimated the size of the "stable" region, and we can also estimate the rate of its growth as the bifurcation parameter increases. The limit cycle is obtained by transforming the nontrivial solution $v = \sqrt{\mu(a)/0.1325}$ back into the original $x(t)$ coordinates through the near-identity transformations, the center manifold relations, the modal and the Lyapunov–Floquét transformations. But, in order to have a physically meaningful result, we need to relate the arbitrary versal deformation parameter μ to the original bifurcation parameter a. First, following the technique described earlier, we establish a linear relationship between the two parameters by applying sensitivity analysis to the Floquét transition matrix. After the necessary calculations the linear sensitivity analysis yields

$$\mu = 0.868(a - a_c) \equiv 0.868\eta, \tag{107}$$

where $\eta = a - a_c$ is a small change in the bifurcation parameter from the critical value. We also compute a quadratic relationship using the same sensitivity analysis method, so that we can compare the accuracy of the two to results obtained by numerical integration. The quadratic relationship is found to be

$$\mu = 0.868\eta - 1.854\eta^2. \tag{108}$$

These two relationships were calculated using 32 Chebyshev polynomials in the expansion of linear part of equation (103) and 30 Picard iterations for the integration. These numbers were found to provide very good accuracy, as we can see in Table 1. Next we compute a quadratic relation between the parameters using our second proposed method: by fitting a curve in the least squares sense onto numerically obtained data points. This quadratic relationship is obtained as

$$\mu = 0.865\eta - 1.635\eta^2. \tag{109}$$

Table 1 shows a comparison of the versal deformation parameter computed by these three different rules and also obtained numerically, for several values of the bifurcation parameter. From this table we can draw three conclusions. First, in this case, the curve fitting technique provides the best result. Also, we can

Table 1 The bifurcation parameter η vs. the versal deformation parameter μ, obtained by four different techniques

Bifurcation parameter $a - a_c = \eta$	μ			
	Numerical value	Linear relation, sensitivity analysis	Quadratic relation, sensitivity analysis	Quadratic relation, curve fitting
0.0001	0.00008678	0.000086800	0.000086781	0.000086484
0.001	0.00086615	0.00086800	0.00086615	0.00085553
0.005	0.0042941	0.0043400	0.0042936	0.0042484
0.01	0.0084981	0.0086800	0.0084946	0.0084234
0.02	0.016645	0.017360	0.016618	0.016554
0.03	0.024460	0.026040	0.024371	0.024391
0.04	0.031958	0.034720	0.031754	0.031934
0.05	0.039155	0.043400	0.038765	0.039185
0.06	0.046065	0.052080	0.045405	0.046142
0.08	0.059069	0.069440	0.057574	0.059178
0.1	0.071057	0.086800	0.068260	0.071040
0.15	0.097048	0.130200	0.088484	not valid

see that by increasing η up to 0.1, the critical eigenvalue has grown from order $10^{-8} \approx 0$ to the order of 10^{-2}. The noncritical eigenvalue for this problem is of order 1. The versal deformation of the normal form has to remain on the center manifold, otherwise it does not provide correct approximation of the original dynamics. Therefore, the bifurcation parameter η can only be increased until the critical eigenvalue is still at least one order smaller than the stable eigenvalues. The actual limit value of η, of course, depends on the actual problem at hand. In this particular example, based on Table 1, it is about $\eta = 0.15$. Finally, we can observe that while the linear rule quickly becomes a poor approximation compared to the numerically obtained values, both quadratic relationships are accurate enough and there is no need to find higher-order approximations. This can be expected based on the fact that equation (103) itself is only a cubic approximation of the original physical problem. Figure 10 illustrates the limit cycles in the state space computed by transforming the solution (106) back to the physical coordinates. The limit cycles are obtained for two different values of the parameter η based on the quadratic relationship (109). Solutions from numerical integration (FORTRAN IMSL subroutine) are also plotted for a comparison. We can see that the solutions agree very well for small η's but for larger values it becomes a poorer approximation. This is expected, due to the local nature of the entire procedure.

Figure 11 compares the numerical and analytical amplitudes of the limit cycles for several values of the parameter η, for all three η–μ relationships. We can see that the results based on linear approximation diverge faster from the numerical

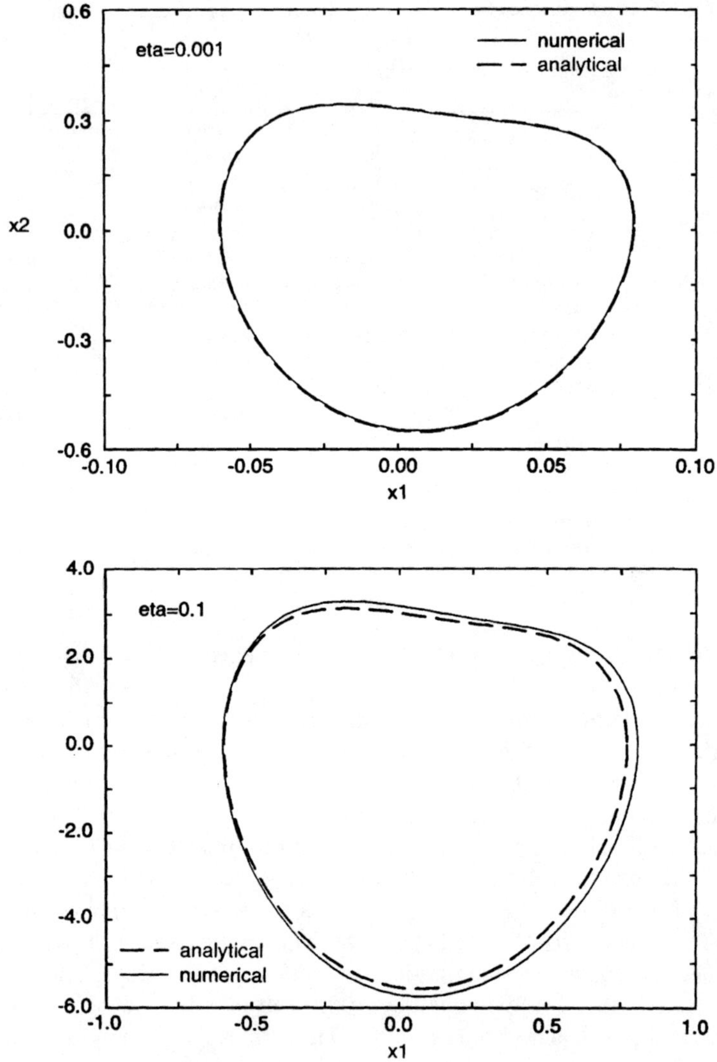

Figure 10. Comparison of the analytically and numerically obtained post-bifurcation limit cycles for the simple pendulum for $\eta = 0.001$ and for $\eta = 0.1$. (From Dávid and Sinha, 2000.)

solution, and for larger η's it does not even preserve the parabolic growth of the amplitude. We can also observe that even though the analytical solution with the quadratic approximation slowly diverges from the numerical one as η grows, it preserves the parabolic growth of the amplitude, which is typical in the case of

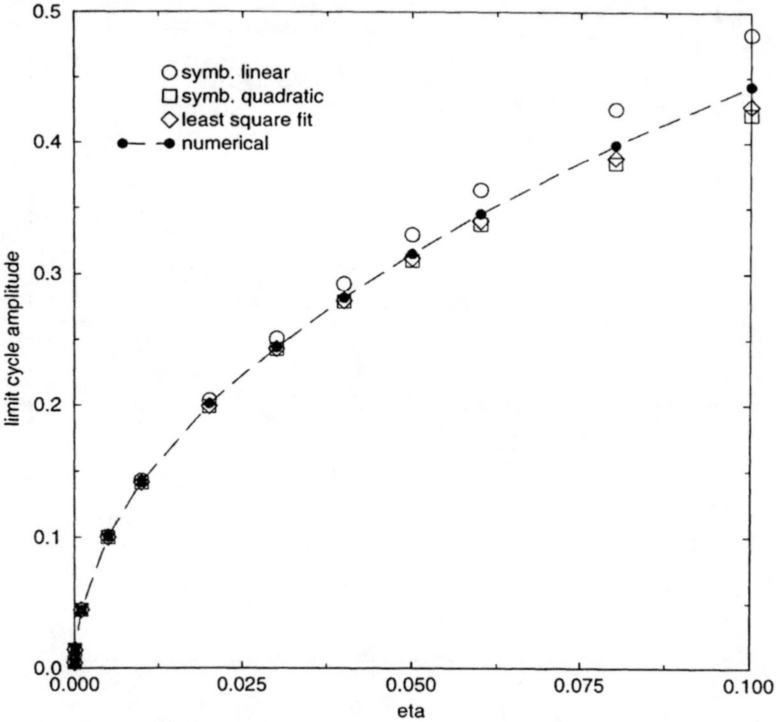

Figure 11. Comparison of the analytically and numerically obtained post-bifurcation limit cycle amplitudes of the simple pendulum, for several values of the bifurcation parameter, with the analytical results computed using the three different versal deformation relationships. (From Dávid and Sinha, 2000.)

symmetry breaking bifurcation in the cubic approximation. So from here in this example only the quadratic relationship based on the curve fitting method will be used for computational simplicity.

Next, our goal is to study how the dynamics changes as we follow the stability boundary toward larger values of the parameter b (i.e., as the parametric excitation becomes stronger). In order to do this, we compute the amplitudes of the post-bifurcation limit cycles for four different sets of parameters, always at the same distance from the boundary (at $\eta = 0.05$). The values for the parameter b are chosen to be 4.5, 5, 5.5 and 6 (as shown in Figure 9), and we keep the values of d and ω the same as before. First, the critical value of the bifurcation parameter, a_c is computed for each b, and then we follow the above described procedure to obtain the solutions for the limit cycles. The limit cycle amplitudes are shown in Figure 12 and the results are compared to those obtained by numerical integration. We can observe that the accuracy of the method remains about

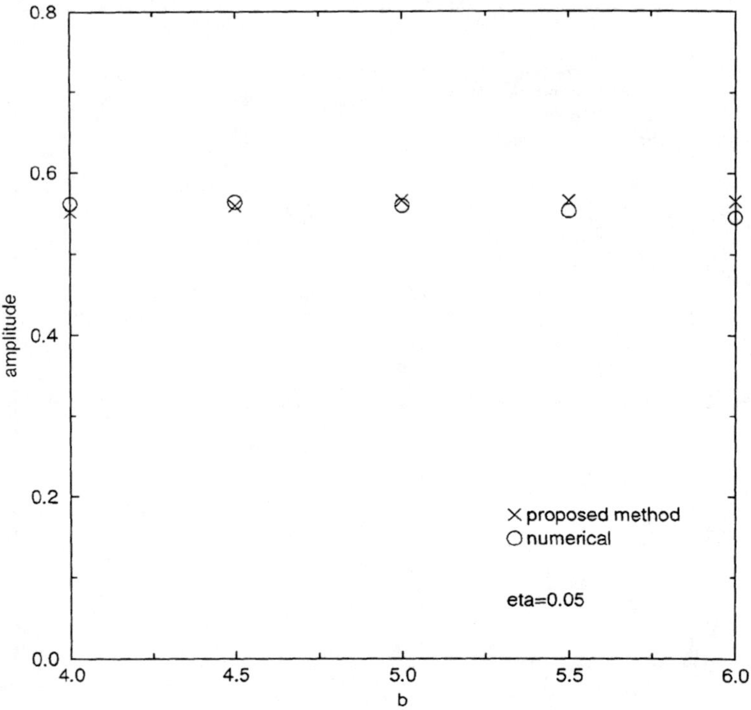

Figure 12. Comparison of the analytically and numerically obtained limit cycle amplitudes for different values of b along the stability boundary of the simple pendulum. (From Dávid and Sinha, 2000.)

the same as the parametric excitation becomes stronger. This was to be expected. Unlike in the averaging and perturbation methods, the accuracy of the approximation in this technique is independent of the system parameters. That is why it is a generally applicable method. However, the major limitation of this procedure lies in the fact that it is a local method; therefore, it gives good approximations only in a small neighborhood of the stability boundary. The opposite is true for averaging method; there the approximations become better the farther away from the critical point we go. So for systems with small parameters we now have a choice, depending on where we need to find the solutions, close to or away from the stability boundary, we can use one method or the other. However, for systems with large parameters, limited as it may be, our proposed method is still the only one.

One last observation: in Figure 12 we can also see that the amplitudes at the same distance from the stability boundary are the same in all five cases implying that the rate of growth of the attractive domain does not depend on how strong the parametric excitation is in this particular example.

7.4. *An example of the secondary Hopf bifurcation: a double inverted pendulum with a periodic follower load*

As an example of the secondary Hopf bifurcation (when a pair of the Floquét multipliers are complex and lie on the unit circle), we consider a double inverted pendulum subjected to a periodic load (Figure 13). The nonlinear equations of motion are given by Jin and Matsuzaki (1988) as

$$\begin{aligned}
&3\ddot{\phi}_1 + \cos(\phi_2 - \phi_1)\ddot{\phi}_2 - \sin(\phi_2 - \phi_1)\dot{\phi}_2^2 + (B_1 + B_2)\dot{\phi}_1 \\
&\quad - B_2\dot{\phi}_2 + 2\bar{k}\phi_1 - \bar{k}\phi_2 - \bar{p}(t)\sin(\phi_1 - \gamma\phi_2) = 0, \\
&\cos(\phi_2 - \phi_1)\ddot{\phi}_1 + \ddot{\phi}_2 + \sin(\phi_2 - \phi_1)\dot{\phi}_1^2 - B_2\dot{\phi}_1 + B_2\dot{\phi}_2 \\
&\quad - \bar{k}\phi_1 + \bar{k}\phi_2 - \bar{p}(t)\sin\big((1 - \gamma)\phi_2\big) = 0,
\end{aligned} \tag{110}$$

where $\bar{k} = k/ml^2$, $B_i = b_i/ml^2$ for $i = 1, 2$ and $\bar{p} = (P_1 + P_2\cos\omega t)/ml = p_1 + p_2\cos\omega t$. Let us first expand the nonlinear terms about the vertical equilibrium position into Taylor series up to the cubic order. Then we define the state variables $\{x_1 \quad x_2 \quad x_3 \quad x_4\}^{\mathrm{T}} = \{\phi_1 \quad \phi_2 \quad \dot{\phi}_1 \quad \dot{\phi}_2\}^{\mathrm{T}}$, and normalize the time so that the period of the coefficient functions becomes $T = 1$. With these changes

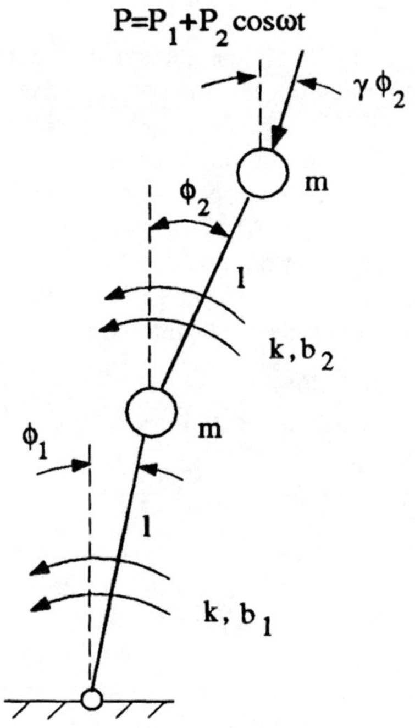

Figure 13. Double inverted pendulum subjected to a periodic follower load.

the equation can be rewritten in the state space as

$$\begin{Bmatrix} \dot{x}_1 \\ \dot{x}_2 \\ \dot{x}_3 \\ \dot{x}_4 \end{Bmatrix} = \begin{bmatrix} 0 & 0 & 1 & 0 \\ 0 & 0 & 0 & 1 \\ 0.5\bar{k}(\bar{p}-3) & 0.5\bar{k}(2-\bar{p}) & -0.5(b_1+2b_2) & b_2 \\ 0.5\bar{k}(5-\bar{p}) & \bar{k}(\bar{p}(1.5-\gamma)-2) & 0.5(b_1+4b_2) & -2b_2 \end{bmatrix} \times \begin{Bmatrix} x_1 \\ x_2 \\ x_3 \\ x_4 \end{Bmatrix} + \begin{Bmatrix} 0 \\ 0 \\ (-0.5(x_3^2+x_4^2)(x_1-x_2) - \bar{p}\bar{k}((x_1-\gamma x_2)^3 \\ \quad -(1-\gamma)^3 x_2^3)/12 - 0.25(x_1-x_2)^2(\bar{k}(\bar{p}-4)x_1 \\ \quad +\bar{k}(3+\bar{p}(\gamma-2))x_2 - (b_1+3b_2)x_3 + 3b_2x_4)) \\ (0.5(3x_3^2+x_4^2)(x_1-x_2) + \bar{p}\bar{k}((x_1-\gamma x_2)^3 \\ \quad -3(1-\gamma)^3 x_2^3)/12 + 0.25(x_1-x_2)^2(\bar{k}(2\bar{p}-7)x_1 \\ \quad +\bar{k}(5+\bar{p}(\gamma-3))x_2 - (2b_1+5b_2)x_3 + 5b_2x_4)) \end{Bmatrix}, \tag{111}$$

where $\bar{p} = p_1 + p_2 \cos 2\pi t$. We choose the constant part of the follower load P, p_1 to be the bifurcation parameter. For the parameter set $\bar{k} = 1$, $b_1 = b_2 = 0.01$, $p_2 = 1$, $p_{1c} = -3.2570215$, $\omega = 2$ and $\gamma = 0$ the linear system matrix has a pair of complex multipliers on the unit circle: $\mu_{1,2} = -0.9731 \pm 0.2305\text{i}$ and $|\mu_{1,2}| = 0.99999996 \approx 1$. This corresponds to a pair of purely imaginary eigenvalues in the transformed domain. After applying the Lyapunov–Floquét and the modal transformations, the center manifold reduction and the near-identity transformations, we find that the versal deformation equation on the two-dimensional center manifold is completely time-invariant and it is given by

$$\begin{Bmatrix} \dot{v}_1 \\ \dot{v}_2 \end{Bmatrix} = \begin{bmatrix} \mu(p_1) + 0.2326\text{i} & 0 \\ 0 & \mu(p_1)^* - 0.2326\text{i} \end{bmatrix} \begin{Bmatrix} v_1 \\ v_2 \end{Bmatrix} + \begin{Bmatrix} (-0.1367 - 0.2586\text{i})v_1^2 v_2 \\ (-0.1367 + 0.2586\text{i})v_1 v_2^2 \end{Bmatrix}, \tag{112}$$

where μ is a small change in the eigenvalues of the normal form due to a change in the bifurcation parameter $p_1 = p_{1c} + \eta = -3.2570215 + \eta$. In this example we compute the relationship between μ and η only one way, using the technique of fitting a quadratic curve onto numerically obtained data points. We have seen in the previous example that this method provided the best results and it is also the simplest to compute. The quadratic relationship is found to be

$$\mu = (0.2270 - 0.8047\text{i})\eta + (1.6000 - 4.7940\text{i})\eta^2. \tag{113}$$

Utilizing the complex change of variables

$$
\begin{aligned}
v_1 &= u_1 - \mathrm{i}u_2, \\
v_2 &= u_1 + \mathrm{i}u_2,
\end{aligned} \tag{114}
$$

we can transform equation (112) into a real form, and then, by introducing the polar coordinates

$$
\begin{aligned}
u_1 &= R\cos\theta, \\
u_2 &= R\sin\theta,
\end{aligned} \tag{115}
$$

we get

$$
\begin{aligned}
\dot{R} &= \mathrm{Re}(\mu)R - 0.1367R^3, \\
\dot{\theta}R &= -\big(\mathrm{Im}(\mu) + 0.2326\big)R + 0.2586R^3.
\end{aligned} \tag{116}
$$

Since the radius equation is independent, it can be solved very simply in a closed form, and then the phase equation can also be solved. The nontrivial steady-state solution $R = \sqrt{\mathrm{Re}(\mu)/0.1367}$ gives us right away the amplitude of the post-bifurcation limit cycle. This limit cycle exists for $\mu > 0$ and it is stable. If we transform it back to the complex variables we get the limit cycle solution of equation (112) as

$$
\begin{Bmatrix} v_1(t) \\ v_2(t) \end{Bmatrix} = \begin{Bmatrix} \sqrt{\frac{\mathrm{Re}(\mu)}{0.1367}} \exp\{(-\mathrm{Im}(\mu) - 0.2326 + \frac{0.2586}{0.1367}\mathrm{Re}(\mu))\mathrm{i}t\} \\ -\sqrt{\frac{\mathrm{Re}(\mu)}{0.1367}} \exp\{(\mathrm{Im}(\mu) + 0.2326 - \frac{0.2586}{0.1367}\mathrm{Re}(\mu))\mathrm{i}t\} \end{Bmatrix}. \tag{117}
$$

The solution for the limit set in the original x coordinates is computed by applying the near-identity transformations, center manifold reduction and the modal and Lyapunov–Floquét transformations. All these transformations are periodic with $T = 2$. As we can see from equation (117), the limit cycle it describes is a periodic solution with a period $T = 2\pi/(-\mathrm{Im}(\mu) - 0.2326 + 0.2586/0.1367\,\mathrm{Re}(\mu))$. As this is not a rational number in general, the solution will contain Fourier harmonics with frequencies that are not integer multiples of each other, therefore, it will not be a periodic limit cycle, but a quasiperiodic limit set. Since it is a stable limit set, we call it a quasiperiodic attractor. Figure 14(a) shows the Poincaré map of a transient solution converging to the attractor while Figure 14(b) shows the attractor itself in the state space and its Poincaré map for a given value of η (only one pair of states is shown out of the six possible pairs of all four states). It seems to be natural to characterize the motion with its simple Poincaré map instead of the fairly complicated state-space trajectory. Poincaré sections of the attractor for two different values of η (0.001 and 0.003) are compared to results obtained from numerical integration in Figure 15. From these figures we can conclude that the magnitude of the solutions is fairly accurate even for parameter values relatively

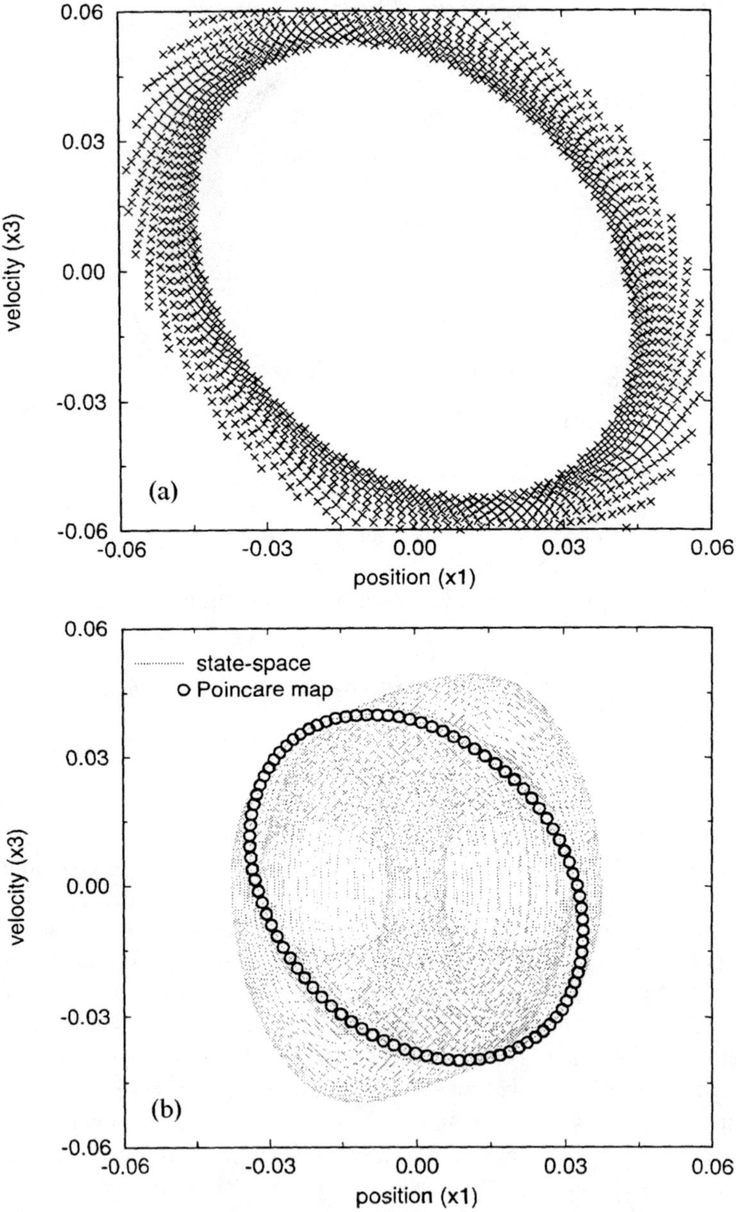

Figure 14. (a) Poincaré map of a transient solution of the double pendulum converging to a quasi-periodic attractor. (b) The attractor in the state space and its Poincaré map. (From Dávid and Sinha, 2000.)

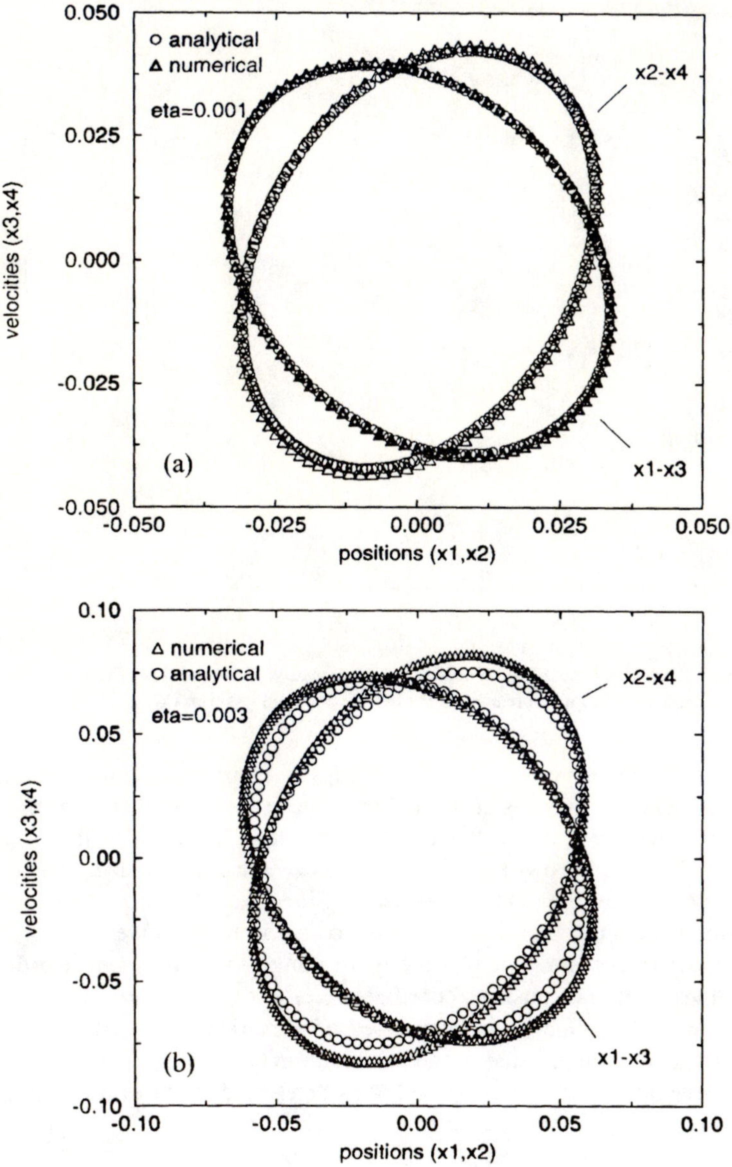

Figure 15. Comparison of the analytically and numerically obtained Poincaré maps of the quasiperiodic attractors of the double pendulum for (a) $\eta = 0.001$, (b) $\eta = 0.003$. (From Dávid and Sinha, 2000.)

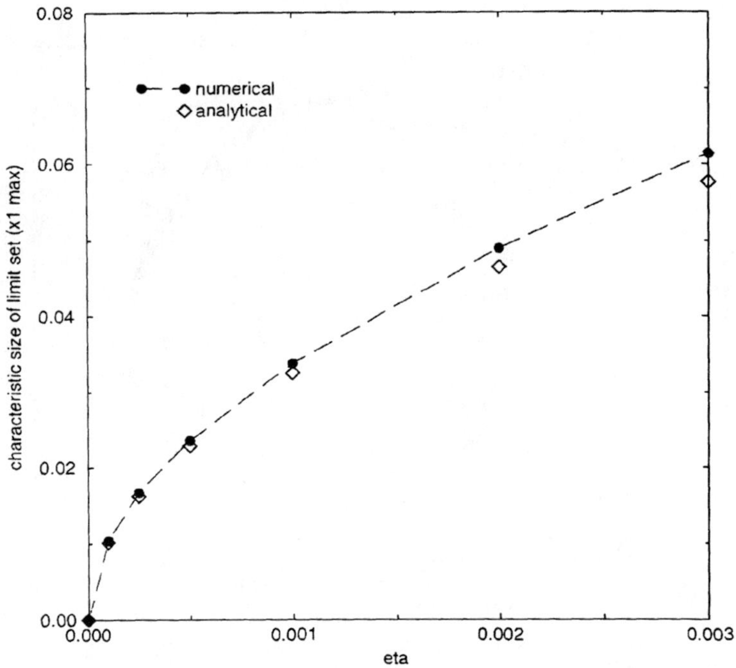

Figure 16. Comparison of the analytically and numerically obtained sizes of the attractive domain for the secondary Hopf bifurcation of the double pendulum. (From Dávid and Sinha, 2000.)

far from the critical point. However, in Figure 14(b) it is clearly seen that the numerical solution runs faster around the ellipse than the analytical one which implies that the error in the "phase" is larger than the error in the magnitude. The size and rate of growth of the attractive domain are significant because they indicate how softly the stability vanishes. Since we have a quasiperiodic four-dimensional attractor, it is not as obvious to determine what we mean by the size of the attractive domain. One physically meaningful way is to consider the maximum values of the two position coordinates, $x_1 = \Phi_1$ and $x_2 = \Phi_2$, as a measure of the size of the domain. A comparison of numerical and analytical results is given for the maximum value of the first position coordinate x_1 in Figure 16. The figure shows the quadratic growth of the attractive domain as a function of the change in the bifurcation parameter.

8. Summary and conclusions

In this chapter, a general method for a *quantitative* bifurcation analysis of nonlinear systems with time-periodic coefficients is presented. The procedure is based on the application of the Lyapunov–Floquét transformation, time-periodic center manifold reduction, time-dependent normal form theory and versal deformations of the normal forms. The great feature of the Lyapunov–Floquét transformation, that it converts the periodic system into a dynamically equivalent problem with a time-invariant linear part, allows us to apply the nonlinear simplification techniques mentioned above. The time-periodic center manifold theory is employed to separate the "critical" dynamics and reduce the dimension of the system to the dimension of its center manifold. Time dependent normal form theory is applied to further simplify the equations by reducing the number of nonlinear terms to a minimum. Finally, versal deformations of the normal forms are constructed by developing approximate relationships between the versal deformation parameter and the original bifurcation parameter. This allows us to perform a qualitative as well as quantitative study of the dynamics in the neighborhood of the critical point. For most of the codimension one bifurcations the normal form is completely time-invariant and solvable in a closed form. Even in the case of the resonant secondary Hopf bifurcation, there exists a transformation that converts the normal form into a time-invariant system; and then an analytical local steady-state solution can be obtained. After solving the versal deformation equation on the center manifold, the post-bifurcation steady-state solutions can be transformed back to the original physical state variables by substituting all the transformations in the reversed order. However, it is observed that while the transformation procedure introduces only a small error if we are close to the bifurcation point, the error becomes significant as we move farther from it. This error is partly due to the fact that the Lyapunov–Floquét transformation as well as the center manifold relations are computed at the critical point and are reasonably accurate only in a small neighborhood of this point; partly to the approximations we use in computing all our transformations, the Fourier-series and the Chebyshev polynomial expansions, and Taylor series expansions of inverse functions. Therefore, the method is useful for local analysis only. The comparisons with numerical integration in the illustrative examples clearly illustrate this nature of the technique. But they also demonstrate that as long as we are close enough to the critical point, we obtain excellent results regardless where the critical point lies in the parameter space and how strong the periodic excitation is.

The first example, the factorizable system, has been chosen for its simplicity. Because it was possible to calculate every step by hand, application of the entire procedure could be shown in detail. It was also our intent to illustrate the meaning of versal deformation, and show why it is that an exact quantitative versal deformation cannot be obtained for more general systems.

The second example, a system with a small parameter, was intended to compare the proposed method to one of the most popular existing techniques, the averaging method. Post-bifurcation steady-state solutions of periodic nonlinear systems have been traditionally obtained by averaging and perturbation methods. However, those techniques do not preserve the correct dynamics, because they treat the periodic part of linear term as a perturbation and average it out along with the nonlinearities. This leads to incorrect prediction of the bifurcation point in most cases and results in crude approximations. It is clearly illustrated by the example that, even in the case of a small parameter, close to stability boundaries the proposed method provides better approximations than the averaging method. On the other hand, if we need solutions for small parameter systems away from their stability boundaries, averaging and perturbation methods are still the best, especially if they are employed after applying the Lyapunov–Floquét transformation and thereby conserving the linear dynamic characteristics.

The simple pendulum problem demonstrated that the method is, indeed, applicable in case of strong parametric excitations (large parameters). It was shown that the accuracy of the predictions of the post-bifurcation dynamics depends only on the distance from the stability boundary, but it does not depend on the location of the parameters in the parameter space. The accuracy decreases as we move away from the stability boundary; but it is expected due to a series of approximations, as pointed out earlier.

The example of the double inverted pendulum with a periodic follower force demonstrated that even such complex post-bifurcation behavior as the existence of a quasiperiodic attractor can be accurately predicted by this method. Interestingly, in this case the size of the attractive domain is predicted more accurately than the frequencies that compose the quasiperiodic solution.

The method described in this chapter, due to its quantitative nature, can be used in actual engineering and physical applications to safely design systems that operate close to their stability boundaries.

Beyond analysis, this method may prove to be a useful tool for designing nonlinear feedback controllers to control bifurcations and even chaos in nonlinear systems with time-periodic coefficients.

From a theoretical point of view, nonlinear control becomes necessary in only one special situation, when the system has uncontrollable critical eigenvalues, and all its unstable eigenvalues are controllable. In practice, however, we may desire to take the effect of the nonlinearities into account even when dealing with a linearly controllable system. The motivations to do so include achieving better system performance and extending the local neighborhood in which the linear control is valid (including the design of robust or even global control systems).

Bifurcation control basically implies an attempt to modify nonlinear system characteristics, such as the type of bifurcation (super- or subcritical) and the size and rate of growth of post-bifurcation limit sets. This can be achieved by means of a purely nonlinear feedback controller. Sometimes performance can be significantly improved if a system is operated near a stability boundary, although it may not be safe to do so, especially if a subcritical bifurcation occurs when the system crosses the boundary. On the other hand, if small vibrations can be tolerated without causing system failure, then systems can be operated very close to the stability boundary when the bifurcation is of the supercritical type. Therefore, it may be desirable to design a controller that would change the bifurcation into a supercritical one, because in this case even if the system crosses over the boundary, the vibrations remain within a small attractive domain. The size and rate of growth of these attractive domains could be controlled as desired. Further, this type of bifurcation control may be more efficient than linear ones. It may require less control effort because it does not change the structure of the linearized system in the parameter space, while a linear controller delays the occurrence of the bifurcation by changing the location of the stability boundary.

This idea of bifurcation control has been well developed for autonomous systems (Abed and Fu, 1986, 1987; Abed, Wang and Chen, 1992). A recent work by Dávid and Sinha (2003) extends the method to time-periodic systems.

This type of purely nonlinear bifurcation control may be a good tool in controlling chaotic systems as well. The idea is based on controlling the bifurcations that lead to chaos. Since chaos in nonlinear systems always occurs through specific routes of bifurcations (such as the well known period-doubling cascade, for example) the onset of chaos may be delayed or even completely eliminated by changing certain characteristics of the bifurcations along the route. An example of bifurcation based chaos control for autonomous systems can be found in Abed, Wang and Chen (1992). Recently, Sinha and Dávid (2006) presented an extension of the technique for time-periodic systems.

An in-depth exploration of these ideas as well as application of the presented bifurcation analysis method to practical problems are, however, subjects for future research.

References

Abed, E.H., Fu, J.-H., 1986. Local feedback stabilization and bifurcation control, I. Hopf bifurcation. Systems Control Lett. 7, 11–17.

Abed, E.H., Fu, J.-H., 1987. Local feedback stabilization and bifurcation control, II. Stationary bifurcation. Systems Control Lett. 7, 467–473.

Abed, E.H., Wang, H.O., Chen, R.C., 1992. Stabilization of period doubling bifurcation and implications for control of chaos. In: Proc. 31st IEEE Conference on Decision and Control. Tucson, AZ, Dec. 1992, pp. 2119–2124.

Arnold, V.I., 1988. Geometrical Methods in the Theory of Ordinary Differential Equations. Springer-Verlag, New York.

Arrowsmith, D.K., Place, C.M., 1990. An Introduction to Dynamical Systems. Cambridge Univ. Press, Cambridge.

Bellman, R., 1970. Nonlinear Analysis, vol. 1. Academic Press, New York.

Bernussou, J., 1977. Point Mapping Stability. Pergamon, New York.

Birkhoff, G.D., 1966. Dynamical Systems. Amer. Math. Soc. Colloq. Publ., vol. 9. Am. Math. Soc., Providence.

Bogoliubov, N.N., Mitropolsky, Y.A., 1961. Asymptotic Methods in the Theory of Nonlinear Oscillations. Gordon & Breach, New York.
Bolotin, V.V., 1964. The Dynamic Stability of Elastic Systems. Holden-Day, San Francisco.
Butcher, E.A., Sinha, S.C., 1998. Symbolic computation of local stability and bifurcation surfaces for nonlinear time-periodic systems. Nonlinear Dynam. 17 (1), 1–21.
Chow, S.-N., Li, Ch., Wang, D., 1994. Normal Forms and Bifurcations of Planar Vector Fields. Cambridge Univ. Press, Cambridge.
Chow, S.-N., Wang, D., 1986. Normal Forms of Bifurcating Periodic Orbits. Contemp. Math., vol. 56. Am. Math. Soc., Providence, pp. 9–18.
Clairaut, A., 1754. Memoiré sur l'orbite apparent du soleil autour de la Terre, an ayant égard aux perturbations produites par les actions de la Lune et des Planetes principales. Mém. de l'Acad. des Sci. (Paris), 521–564.
Danby, J.M.A., 1964. Stability of the triangular points in the elliptical restricted problem of three bodies. Astron. J. 69 (2), 165–172.
Dávid, A., Sinha, S.C., 2000. Versal deformation and local bifurcation analysis of time-periodic nonlinear systems. Nonlinear Dynam. 21, 317–336.
Dávid, A., Sinha, S.C., 2003. Bifurcation control of nonlinear systems with time-periodic coefficients. J. Dynam. Syst. 125, 541–548.
Evan-Ivanowski, R.M., 1976. Resonance Oscillations in Mechanical Systems. Elsevier, Amsterdam.
Farooq, A., Homsy, G.M., 1996. Linear and nonlinear dynamics of a differentially heated slot under gravity modulation. J. Fluid Mech. 313, 1–38.
Flashner, H., Hsu, C.S., 1983. A study of nonlinear periodic systems via the point mapping method. Internat. J. Numer. Methods Engrg. 19, 185–215.
Floquét, G., 1883. Sur les équations differentials linéaries a coefficients périodiques. Ann. Sci. École Norm. Sup. 12, 47–88.
Friedmann, P.P., 1986. Numerical methods for determining the stability and response of periodic systems with applications to helicopter rotor dynamics and aeroelasticity. Comput. Math. Appl. 12A, 131–148.
Friedmann, P.P., 1987. Recent trends in rotary-wing aeroelasticity. Vertica 11, 139–170.
Glass, L., 1991. Cardiac arrhythmias and circle maps – a classical problem. Chaos 1, 13–19.
Grassia, P.G., Homsy, G.M., 1998. Thermocapillary flows with low frequency g-jitter. In: Proceedings of the 4th Microgravity Fluid Physics and Transport Phenomena Conference. Cleveland, Ohio, pp. 54–57.
Guevara, M.R., Shrier, A., Glass, L., 1990. Chaotic and complex cardiac rhythms. In: Zipes, D.P., Jalife, J. (Eds.), Cardiac Electrophysiology. Saunders, Philadelphia.
Guttalu, R.S., Flashner, H., 1989. Periodic solutions of nonlinear autonomous systems by approximate point mapping. J. Sound Vib. 129, 291–311.
Guttalu, R.S., Flashner, H., 1990. Analysis of dynamical systems by truncated point mapping and cell mapping. In: Schiehlen, W. (Ed.), Nonlinear Dynamics in Engineering Systems. Springer-Verlag, New York.
Hale, J., Koçak, H., 1991. Dynamics and Bifurcations. Springer-Verlag, New York.
Hawthorne, E.I., 1951. Sinusoidal variation of inductance in a linear series RCL circuit. Proc. Inst. Radio Eng. 39, 78–81.
Hill, G.W., 1886. On the part of the motion of the lunar perigee which is a function of the mean motions of the Sun and the Moon. Acta Math. 8, 1–36.
Hirschfelder, J.O., Wyatt, R.E., Coalson, R.D. (Eds.), 1989. Lasers, Molecules and Methods, Adv. Chem. Phys., vol. LXXIII. Wiley, New York.
Hsu, C.S., 1974. On approximating a general linear periodic system. J. Math. Anal. Appl. 45, 234–251.
Hsu, C.S., 1987. Cell-to-Cell Mapping. Springer-Verlag, New York.
Iooss, G., Joseph, D.D., 1990. Elementary Stability and Bifurcation Theory, 2nd ed. Springer-Verlag, New York.

Jin, J.-D., Matsuzaki, Y., 1988. Bifurcations in a two-degrees-of-freedom elastic system with follower forces. J. Sound Vib. 126 (2), 265–277.

Johnson, W., 1980. Helicopter Theory. Princeton Univ. Press, Princeton, NJ.

Lagrange, J.-L., 1788. Mécanique Analytique, vols. 1 and 2. Blanchard, Paris.

Lalanne, M., Ferraris, G., 1990. Rotordynamics Prediction in Engineering. Wiley, New York.

Lindh, K.G., Likins, P.W., 1970. Infinite determinant methods for stability analysis of periodic-coefficient differential equations. AIAA Journal 8, 680–686.

Lindstedt, A., 1882. Über die Integration einer für die Strorungstheorie wichtigen Differentialgleichung. Astron. Nachr. 103, 211–220.

Lindtner, E., Steindl, A., Troger, H., 1990. Generic one-parameter bifurcations in the motion of a simple robot. In: Mittelmann, H.D., Roose, D. (Eds.), Continuation Techniques and Bifurcation Problems. Birkhäuser, pp. 199–218.

Lukes, D.L., 1982. Differential Equations: Classical to Controlled. Academic Press, New York.

Lyapunov, A.M., 1896. Sur une serie relative a la theorie des equations differentielles lineaires a coefficients periodiques. Compt. Rend. 123 (26), 1248–1252.

Malkin, I.G., 1962. Some basic theorems of the theory of stability of motion in critical cases. In: Stability and Dynamic Systems, Amer. Math. Soc. Transl. Ser. 1, vol. 5. Am. Math. Soc., New York, pp. 242–290.

Mathieu, E., 1868. Memoire sur le mouvement vibratoire de une membrane de forme elliptique. J. Math. Pures Appl. 13, 137–203.

McKillip, R.M., 1985. Periodic control of individual-blade-control helicopter rotor. Vertica 9, 199–225.

Mingori, D.L., 1969. Effects of energy dissipation on the altitude stability of dual spin satellite. AIAA Journal 7, 20–27.

Mohler, R.H., 1991. Dynamics and Control. Nonlinear Systems, vol. 1. Prentice-Hall International, New Jersey.

Nayfeh, A.H., 1973. Perturbation Methods. Wiley, New York.

Nayfeh, A.H., 1985. Problems in Perturbation. Wiley, New York.

Nayfeh, A.H., 1993. Method of Normal Forms. Wiley, New York.

Nayfeh, A.H., Balachandran, B., 1995. Applied Nonlinear Dynamics. Wiley, New York.

Nayfeh, A.H., Mook, D.T., 1979. Nonlinear Oscillations. Wiley, New York.

Pandiyan, R., Sinha, S.C., 1995. Analysis of time-periodic nonlinear dynamical systems undergoing bifurcations. Nonlinear Dynam. 8, 21–43.

Poincaré, H., 1892. Les Méthodes Nouvelles de la Mécanique Céleste, vol. I. Gauthiers-Villars, Paris.

Powell, J.L., Crasemann, B., 1961. Quantum Mechanics. Addinson-Wesley, Boston, MA.

Richards, J.A., 1983. Analysis of Periodically Time-Varying Systems. Springer-Verlag, New York.

Rigney, D.R., Goldberger, A.L., 1989. Nonlinear mechanics of the heart's swinging during pericardial effusion. Amer. J. Phys. 257, H1292–H1305.

Roppo, M.N., Davis, S.H., Rosenblat, S., 1984. Bénard convection with time-varying heating. Phys. Fluids 27 (4), 796–803.

Roseau, M., 1987. Vibrations in Mechanical Systems: Analytical Methods and Applications. Springer-Verlag, New York.

Sanchez, N.E., Nayfeh, A.H., 1990. Nonlinear rolling motions of ships in longitudinal waves. Int. Shipbuild. Prog. 37 (411), 247–272.

Sanders, J.A., Verhulst, F., 1985. Averaging Methods in Nonlinear Dynamical Systems. Springer-Verlag, New York.

Sinha, S.C., Butcher, E.A., 1997. Symbolic computation of fundamental solution matrices for time-periodic dynamical systems. J. Sound Vib. 206 (1), 61–85.

Sinha, S.C., Dávid, A., 2006. Chaos Control Nonlinear Systems with Periodic Coefficients, special issue of Philos. Trans. R. Soc. Lond. Ser. A Math. Phys. Eng. Sci. on "Control of Chaos: Its

Suppression, Enhancement, Exploitation in Mechanical Systems, Invited Paper, submitted for publishing.

Sinha, S.C., Juneja, V., 1991. An approximate analytical solution for systems with periodic coefficients via symbolic computation. AIAA/ASME/ASCE/AHS/ASC 32nd Structures, Structural Dynamics and Materials Conference (A collection of papers Part I), pp. 790–797.

Sinha, S.C., Pandiyan, R., Bibb, J.S., 1996. Liapunov–Floquet transformation: Computation and applications to periodic systems. J. Vib. Acoust. 118, 209–219.

Sinha, S.C., Wu, D.-H., 1991. An efficient computational scheme for the analysis of periodic systems. J. Sound Vib. 15 (8), 345–375.

Stokes, G.G., 1847. On the theory of oscillatory waves. Cambridge Trans. 8, 441–473.

Streit, D.A., Krousgrill, C.M., Bajaj, A.K., 1989. Nonlinear response of flexible robotic manipulators performing repetitive tasks. ASME J. Dyn. Syst. 111 (3), 470–480.

Szasz, G., Flowers, G.T., 1999. Time-varying control of a bladed disk assembly using shaft based actuation. In: Proceedings of the 1999 Design Engineering Technical Conferences, 17th Biennial Conference on Mechanical Vibration and Noise, September 12–15, Las Vegas, Nevada.

Szebehely, V., 1967. Theory of Orbits: The Restricted Problem of Three Bodies. Academic Press, New York.

Thuraisingham, R.A., Meath, W.J., 1985. Phase and rotational averaged transitional probabilities for molecules in a sinusoidal field using Floquet formalism. Mol. Phys. 56, 193–207.

Tso, W.K., Asmis, K.G., 1974. Multiple parametric resonance in a nonlinear two degree of freedom system. Internat. J. Non-Linear Mech. 9, 269–277.

Verhulst, F., 1990. Nonlinear Differential Equations and Dynamical Systems. Springer-Verlag, Berlin.

Widmann, P.J., Gorman, M., Robbins, K.A., 1989. Nonlinear dynamics in a convection loop II: Chaos in laminar and turbulent flows. Physica D 36, 157–166.

Wu, M.-Y., 1978. Transformation of a linear time-varying system into a linear time-invariant system. Internat. J. Control 27 (4), 589–602.

Wu, D.-H., Sinha, S.C., 1994. A new approach in the analysis of linear systems with periodic coefficients for applications in rotorcraft dynamics. Aeronaut. J. R. Aeronaut. Soc. 98 (January), 9–16.

Yakubovich, V.A., Starzhinskii, V.M., 1975. Linear Differential Equations with Periodic Coefficients, Parts I and II. Wiley, New York.

Chapter 5

Modal Interactions in Asymmetric Vibrations of Circular Plates

Won Kyoung Lee

School of Mechanical Engineering, Yeungnam University, Gyongsan 712-749, Republic of Korea

E-mail: *wklee@yu.ac.kr*

Contents

1. Introduction 339
2. Governing equations 341
3. Solution 343
4. Steady-state responses and numerical examples 348
   4.1. The plate without elastic foundation ($K = 0$): the case of no internal resonance 348
   4.2. The plate on elastic foundation ($K > 0$): the case of internal resonance ($\omega_{NM} \approx 3\omega_{CD}$, where $N = 3C$) 352
   4.3. The plate on elastic foundation ($K > 0$): the case of internal resonance ($\omega_{NM} \approx 3\omega_{CD}$, where $N = C$) 361

Appendix A 370

Appendix B 371

   Case 1: $\omega_{32} \approx 3\omega_{11}$ and $\lambda \approx \omega_{11}$ 371

   Case 2: $\omega_{32} \approx 3\omega_{11}$ and $\lambda \approx \omega_{32}$ 373

Appendix C 376

References 376

1. Introduction

Circular plates experience mid-plane stretching when defected. The influence of this stretching on the dynamic response increases with the amplitude of the response. This situation can be described with nonlinear strain-displacement equations and a linear stress-strain law which give us the dynamic analogue of the

Edited Series on Advances in Nonlinear Science and Complexity
Volume 1 ISSN: 1574-6909 DOI: 10.1016/S1574-6909(06)01005-7

von Karman equations with geometric nonlinearity. Nonlinear dynamic responses of a circular plate subjected to harmonic excitations have been investigated by two approaches. One is to include symmetric vibrations and the other asymmetric vibrations. For symmetric responses, Sridhar, Mook and Nayfeh (1975) and Hadian and Nayfeh (1990) studied primary resonance of a circular plate with three-mode interaction. Lee and Kim (1995) studied combination resonances of the plate. In these studies the steady-state response can only have the superposition of standing wave components. For asymmetric responses, Sridhar, Mook and Nayfeh (1978) derived solvability conditions for modal interactions of a clamped circular plate without elastic foundation. These conditions are said to be general in the sense of two aspects. First, the conditions include asymmetric vibrations as well as symmetric vibrations. Second, the conditions include all of natural modes. They used these conditions to examine two cases. One is the case of the absence of internal resonance and the other is the case of the internal resonance involving four modes. They concluded that in the absence of internal resonance, the steady-state response can only have the form of a standing wave. When the frequency of excitation is near the highest frequency involved in the internal resonance, the steady-state response was said to be given by a superposition of the standing wave components of all the modes involved in the internal resonance, or a superposition of the standing wave components of all the lower modes and the traveling wave component of the highest mode involved in the internal resonance.

Yeo and Lee (2002) re-examined the analysis by Sridhar, Mook and Nayfeh (1978) to find that they had misderived the solvability conditions in applying the method of multiple scales. Yeo and Lee (2002) corrected the conditions and found that, in the absence of internal resonance, the steady-state response can have not only the form of a standing wave but also the form of a traveling wave, which is a remarkable contrast to the conclusion by Sridhar, Mook and Nayfeh (1978).

In order to investigate modal interactions of circular plates with internal resonance, Lee and Yeo (2003) considered a circular plate on an elastic foundation. In this study, the elastic foundation is considered to get a varied natural characteristic, which generates a desired commensurable relation between natural frequencies. Using the corrected solvability conditions, they investigated the case of internal resonance (a commensurable relationship between natural frequencies), $\omega_{NM} \approx 3\omega_{CD}$, in which the first subscript refers to the number of nodal diameters and the second subscript the number of nodal circles including boundary. In view of the corrected solvability condition, it has been found that, in order for two asymmetric modes to interact, the ratio of the numbers of nodal diameters of two modes must be either three to one ($N = 3C$) or one to one ($N = C$). They considered the case of $N = 3C$, which implies that the ratio of the numbers of nodal diameters of two modes is three to one. For a numerical example the case of internal resonance, $\omega_{32} \approx 3\omega_{11}$, was considered. When the frequency of excitation is near ω_{11}, there exist at most five stable steady-state responses. Four of them are superpositions of traveling wave components and one is a superposition of standing wave components. The result shows the interaction between modes corresponding to ω_{11} and ω_{32} by showing nonvanishing amplitudes of the mode not directly excited. When the frequency of excitation is near ω_{32}, similarly the interaction between modes is shown to exist. All of the responses with nonvanishing amplitudes of mode excited indirectly, however, turn out to be unstable, which is a peculiar phenomenon.

In order to investigate the effect of the number of nodal diameters on nonlinear interactions in asymmetric vibrations of the plate, Lee, Yeo and Samoilenko (2003) examined the one-to-one case ($N = C$), in which the modes have the same number of nodal diameters. The result shows very complicated interactions between two modes by telling existence of nonvanishing amplitudes of the mode not directly excited.

The three works by Yeo and Lee (2002), Lee and Yeo (2003), Lee, Yeo and Samoilenko (2003) are combined and reorganized in this chapter with permission from Elsevier.

2. Governing equations

The equations governing the free, undamped oscillations of nonuniform circular plates were derived by Efstathiades (1971). These equations are simplified to fit the special case of uniform plates, and damping and forcing terms are added. Then the nondimensionalized equations of motion of a circular plate on an elastic foundation shown in Figure 1 are given as follows (Leissa, 1969; Sridhar, Mook and Nayfeh, 1978; Nayfeh and Mook, 1979; Ghosh, 1997; Nayfeh, 2000):

$$\frac{\partial^2 w}{\partial t^2} + (\nabla^4 + K)w = \varepsilon\left[L(w, F) - 2c\frac{\partial w}{\partial t} + p^*(r, \theta, t)\right], \tag{1}$$

$$\nabla^4 F = \left(\frac{1}{r}\frac{\partial^2 w}{\partial r\,\partial\theta} - \frac{1}{r^2}\frac{\partial w}{\partial\theta}\right)^2 - \frac{\partial^2 w}{\partial r^2}\left(\frac{1}{r}\frac{\partial w}{\partial r} + \frac{1}{r^2}\frac{\partial^2 w}{\partial\theta^2}\right), \tag{2}$$

where

$$\begin{aligned} L(w, F) = {} & \frac{\partial^2 w}{\partial r^2}\left(\frac{1}{r}\frac{\partial F}{\partial r} + \frac{1}{r^2}\frac{\partial^2 F}{\partial\theta^2}\right) + \frac{\partial^2 F}{\partial r^2}\left(\frac{1}{r}\frac{\partial w}{\partial r} + \frac{1}{r^2}\frac{\partial^2 w}{\partial\theta^2}\right) \\ & - 2\left(\frac{1}{r}\frac{\partial^2 F}{\partial r\,\partial\theta} - \frac{1}{r^2}\frac{\partial F}{\partial\theta}\right)\left(\frac{1}{r}\frac{\partial^2 w}{\partial r\,\partial\theta} - \frac{1}{r^2}\frac{\partial w}{\partial\theta}\right), \end{aligned} \tag{3}$$

$\varepsilon = 12(1 - \nu^2)h^2/R^2$, c is the damping coefficient, p^* is the forcing function, K is the stiffness of the foundation, ν is Poisson's ratio, h is the thickness, R is the radius, w is the deflection of the middle surface, F is the force function which satisfies the in-plane equilibrium conditions (in-plane inertia is neglected) and

$$\nabla^4 \equiv \left(\frac{\partial^2}{\partial r^2} + \frac{1}{r}\frac{\partial}{\partial r} + \frac{1}{r^2}\frac{\partial^2}{\partial\theta^2}\right)^2. \tag{4}$$

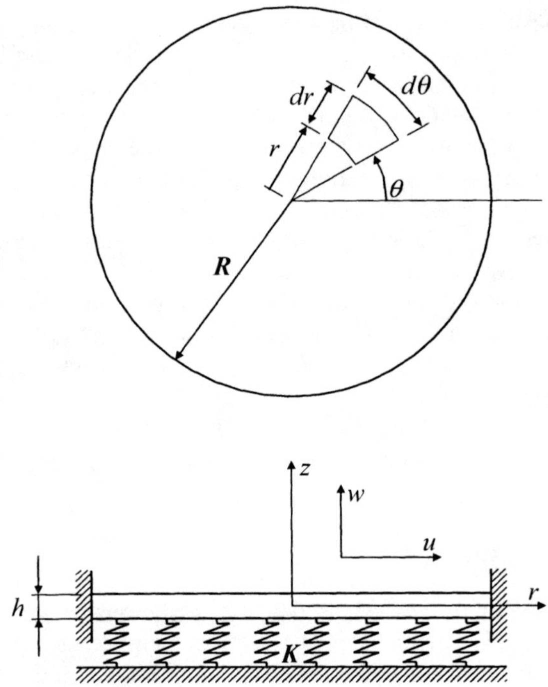

Figure 1. A schematic diagram of a clamped circular plate on an elastic foundation.

The boundary conditions are developed for the plates which are clamped along a circular edge. For all t and θ,

$$w = 0, \qquad \frac{\partial w}{\partial r} = 0 \quad \text{at } r = 1, \tag{5a, b}$$

$$\frac{\partial^2 F}{\partial r^2} - \nu\left(\frac{1}{r}\frac{\partial F}{\partial r} + \frac{1}{r^2}\frac{\partial^2 F}{\partial \theta^2}\right) = 0 \quad \text{at } r = 1, \tag{6a}$$

$$\frac{\partial^3 F}{\partial r^3} + \frac{1}{r}\frac{\partial^2 F}{\partial r^2} - \frac{1}{r^2}\frac{\partial F}{\partial r} + \frac{2+\nu}{r^2}\frac{\partial^3 F}{\partial r\,\partial \theta^2} - \frac{3+\nu}{r^3}\frac{\partial^2 F}{\partial \theta^2} = 0 \quad \text{at } r = 1. \tag{6b}$$

In addition, it is necessary to require the solution to be bounded at $r = 0$.

3. Solution

In order to re-examine the analysis by Sridhar, Mook and Nayfeh (1978) we expand w and F as follows:

$$w(r,\theta,t;\varepsilon)=\sum_{j=0}^{\infty}\varepsilon^j w_j(r,\theta,T_0,T_1,\ldots), \tag{7a}$$

$$F(r,\theta,t;\varepsilon)=\sum_{j=0}^{\infty}\varepsilon^j F_j(r,\theta,T_0,T_1,\ldots), \tag{7b}$$

where $T_n=\varepsilon^n t$.

Substituting equations (7) into equations (1) and (2), and equating coefficients of like powers of ε yield

$$D_0^2 w_0+(\nabla^4+K)w_0=0, \tag{8}$$

$$\nabla^4 F_0=\left(\frac{1}{r}\frac{\partial^2 w_0}{\partial r\,\partial\theta}-\frac{1}{r^2}\frac{\partial w_0}{\partial\theta}\right)^2-\frac{\partial^2 w_0}{\partial r^2}\left(\frac{1}{r}\frac{\partial w_0}{\partial r}+\frac{1}{r^2}\frac{\partial^2 w_0}{\partial\theta^2}\right), \tag{9}$$

$$\begin{aligned}&D_0^2 w_1+(\nabla^4+K)w_1\\&\quad=-2D_0D_1w_0-2cD_0w_0+p^*\\&\qquad+\frac{\partial^2 w_0}{\partial r^2}\left(\frac{1}{r}\frac{\partial F_0}{\partial r}+\frac{1}{r^2}\frac{\partial^2 F_0}{\partial\theta^2}\right)+\frac{\partial^2 F_0}{\partial r^2}\left(\frac{1}{r}\frac{\partial w_0}{\partial r}+\frac{1}{r^2}\frac{\partial^2 w_0}{\partial\theta^2}\right)\\&\qquad-2\left(\frac{1}{r}\frac{\partial^2 F_0}{\partial r\,\partial\theta}-\frac{1}{r^2}\frac{\partial F_0}{\partial\theta}\right)\left(\frac{1}{r}\frac{\partial^2 w_0}{\partial r\,\partial\theta}-\frac{1}{r^2}\frac{\partial w_0}{\partial\theta}\right),\end{aligned} \tag{10}$$

etc., where $D_n=\partial/\partial T_n$.

Substituting equations (7) into equations (5) and (6), and equating coefficients of like powers of ε, one obtains

$$w_j=0,\qquad \frac{\partial w_j}{\partial r}=0\quad \text{at } r=1, \tag{11a, b}$$

$$\frac{\partial^2 F_j}{\partial r^2}-\nu\left(\frac{\partial F_j}{\partial r}+\frac{\partial^2 F_j}{\partial\theta^2}\right)=0\quad\text{at } r=1, \tag{12a}$$

$$\begin{aligned}&\frac{\partial^3 F_j}{\partial r^3}+\frac{\partial^2 F_j}{\partial r^2}-\frac{\partial F_j}{\partial r}\\&\quad+(2+\nu)\frac{\partial^3 F_j}{\partial r\,\partial\theta^2}-(3+\nu)\frac{\partial^2 F_j}{\partial\theta^2}=0\quad\text{at } r=1,\end{aligned} \tag{12b}$$

for all j, θ and t. In addition, it is necessary to require w_j and F_j, for all j, to be bounded at $r=0$.

It follows from equations (8) and (11) that

$$w_0 = \sum_{m=1}^{\infty} \phi_{0m}(r) A_{0m} \mathrm{e}^{\mathrm{i}\omega_{0m} T_0} + \sum_{n,m=1}^{\infty} \phi_{nm}(r) \left\{ A_{nm} \mathrm{e}^{\mathrm{i}(\omega_{nm} T_0 + n\theta)} + B_{nm} \mathrm{e}^{\mathrm{i}(\omega_{nm} T_0 - n\theta)} \right\} + cc, \tag{13}$$

where the $\phi_{nm}(r)$ are the linear shape functions in the r direction given by

$$\phi_{nm} = \kappa_{nm} \left[J_n(\eta_{nm} r) - \frac{J_n(\eta_{nm})}{I_n(\eta_{nm})} I_n(\eta_{nm} r) \right]. \tag{14}$$

The κ_{nm} are chosen so that

$$\int_0^1 r\phi_{nm}^2 \, \mathrm{d}r = 1.$$

The functions J_n are Bessel functions of the first kind, of order n, and the functions I_n are modified Bessel functions of the first kind, of order n. The η_{nm} are the roots of $I_n(\eta)J_n(\eta) - I_n(\eta)J_n(\eta) = 0$, $\omega_{nm} = \eta_{nm}^2 + K$, the A_{nm} and the B_{nm} are complex functions of all T_n for $n \geqslant 1$ which are to be determined from the solvability conditions at the next level of approximation and cc represents the complex conjugate of the preceding terms. In ϕ_{nm} and ω_{nm}, the first subscript refers to the numbers of nodal diameters and the second subscript refers to the number of nodal circles including the boundary. The first summation of the right-hand side in equation (13) represents a superposition of symmetric standing waves. The second summation looks a superposition of asymmetric traveling waves, but it contains both traveling and standing waves depending on the relative values of the A_{nm} and B_{nm}. The solution can also be written in the following equivalent form:

$$w_0 = \sum_{n=-\infty}^{\infty} \sum_{m=1}^{\infty} \phi_{nm}(r) u_{nm}(T_0, T_1, \ldots) \mathrm{e}^{\mathrm{i}n\theta}, \tag{15}$$

where

$$u_{nm} = A_{nm} \mathrm{e}^{\mathrm{i}\omega_{nm} T_0} + \overline{B}_{nm} \mathrm{e}^{-\mathrm{i}\omega_{nm} T_0}, \tag{16}$$

$\phi_{-nm} = \phi_{nm}$ and $\omega_{-nm} = \omega_{nm}$. Because w_0 is real,

$$A_{-nm} = B_{nm}. \tag{17}$$

Substituting equation (15) into equation (9) leads to

$$\nabla^4 F_0 = \sum_{n,p=-\infty}^{\infty} \sum_{m,q=1}^{\infty} E(nm, pq) u_{nm} u_{pq} \mathrm{e}^{\mathrm{i}(n+p)\theta}, \tag{18}$$

where

$$E(nm, pq) = \frac{-np}{r^2}\left(\phi'_{nm} - \frac{\phi_{nm}}{r}\right)\left(\phi'_{pq} - \frac{\phi_{pq}}{r}\right) - \frac{1}{2r}(\phi'_{nm}\phi'_{pq})' + \frac{1}{2r^2}(p^2\phi''_{nm}\phi_{pq} + n^2\phi''_{pq}\phi_{nm})$$

and primes denote differentiation with respect to r.

An expansion for F_0 is assumed in the following form:

$$F_0 = \sum_{n=-\infty}^{\infty} U_n(r, T_0, T_1, \ldots)\mathrm{e}^{\mathrm{i}n\theta}. \tag{19}$$

Substituting equation (19) into equation (18), multiplying the result by $\mathrm{e}^{-\mathrm{i}a\theta}$, and integrating from $\theta = 0$ to 2π, we obtain

$$\nabla_a^4 U_a = \sum_{n=-\infty}^{\infty} \sum_{m,q=1}^{\infty} E(nm, pq)u_{nm}u_{pq}, \tag{20}$$

where

$$p = a - n \tag{21}$$

and

$$\nabla_a^4 = \left[\frac{\partial^2}{\partial r^2} + \frac{1}{r}\frac{\partial}{\partial r} - \frac{a^2}{r^2}\right]^2.$$

Then U_a is further expanded as

$$U_a = \sum_{n=1}^{\infty} v_{an}(T_0, T_1, \ldots)\psi_{an}(r), \tag{22}$$

where the ψ_{an} are the eigenfunctions of the following problem:

$$(\nabla_a^4 - \xi_{an}^4)\psi_{an} = 0 \quad \text{in } r = [0, 1],$$

where ψ_{an} is bounded at $r = 0$ and, from equations (12),

$$\psi''_{an} - \nu(\psi'_{an} - a^2\psi_{an}) = 0$$

and

$$\psi'''_{an} + \psi''_{an} - \psi'_{an} - a^2\big[(2 + \nu)\psi'_{an} - (3 + \nu)\psi_{an}\big] = 0$$

for all θ and t at $r = 1$. It follows that

$$\psi_{an} = \tilde{\kappa}_{an}\big[J_a(\xi_{an}r) - \tilde{c}_{an}I_a(\xi_{an}r)\big], \tag{23}$$

where the $\tilde{\kappa}_{an}$ are chosen so that

$$\int_0^1 r\psi_{an}^2 \,\mathrm{d}r = 1,$$

$$\tilde{c}_{an} = \frac{[a(a+1)(\nu+1) - \xi_{an}^2]J_a(\xi_{an}) - \xi_{an}(\nu+1)J_{a-1}(\xi_{an})}{[a(a+1)(\nu+1) + \xi_{an}^2]I_a(\xi_{an}) - \xi_{an}(\nu+1)I_{a-1}(\xi_{an})},$$

and ξ_{an} are the roots of

$$\begin{aligned} &a^2(a+1)(\nu+1)\big[J_a(\xi_{an}) - \tilde{c}_{an}I_a(\xi_{an})\big] \\ &\quad - a^2\xi_{an}(\nu+1)\big[J_{a-1}(\xi_{an}) - \tilde{c}_{an}I_{a-1}(\xi_{an})\big] \\ &\quad + a\xi_{an}^2\big[J_a(\xi_{an}) + \tilde{c}_{an}I_a(\xi_{an})\big] - \xi_{an}^3\big[J_{a-1}(\xi_{an}) + \tilde{c}_{an}I_{a-1}(\xi_{an})\big] = 0. \end{aligned}$$

Substituting equation (22) into equation (20), multiplying the result by $r\psi_{ab}$, and then integrating from $r = 0$ to 1, one obtains

$$v_{ab}(T_0, T_1, \ldots) = \sum_{n=-\infty}^{\infty} \sum_{m,q=1}^{\infty} G(nm, pq; ab)u_{nm}u_{pq}, \tag{24}$$

where

$$G(nm, pq; ab) = \xi_{ab}^{-4} \int_0^1 r\psi_{ab}E(nm, pq)\,\mathrm{d}r, \tag{25}$$

and p, a and n are related according to equation (21). It follows from equations (24), (22) and (19) that

$$F_0 = \sum_{a,n=-\infty}^{\infty} \sum_{b,m,q=1}^{\infty} \psi_{ab}G(nm, pq; ab)u_{nm}u_{pq}\mathrm{e}^{\mathrm{i}a\theta}, \tag{26}$$

where $p = a - n$.

Substituting equations (26) and (15) into equation (10) leads to

$$\begin{aligned} &D_0^2 w_1 + \big(\nabla^4 + K\big)w_1 \\ &\quad = \sum_{n=-\infty}^{\infty} \sum_{m=1}^{\infty} -2\mathrm{i}\omega_{nm}\phi_{nm}\big[(D_1A_{nm} + c_{nm}A_{nm})\mathrm{e}^{\mathrm{i}\omega_{nm}T_0} \\ &\qquad - \big(D_1\overline{B}_{nm} + c_{nm}\overline{B}_{nm}\big)\mathrm{e}^{-\mathrm{i}\omega_{nm}T_0}\big]\mathrm{e}^{\mathrm{i}n\theta} + p^*(r,\theta,t) \\ &\qquad + \sum_{a,n,c=-\infty}^{\infty} \sum_{m,d,q=1}^{\infty} G(nm, pq; ab)\widehat{E}(cd, ab)u_{cd}u_{pq}u_{nm}\mathrm{e}^{\mathrm{i}(a+c)\theta}, \end{aligned} \tag{27}$$

where modal damping has been assumed, p^* has been expanded as

$$p^*(r,\theta,t) = \sum_{n=-\infty}^{\infty}\sum_{m=1}^{\infty} P_{nm}\phi_{nm}\mathrm{e}^{\mathrm{i}(n\theta+\tau_{nm})}\cos\lambda T_0$$

and

$$\widehat{E}(cd,pq) = \frac{\phi''_{cd}}{r}\left(\psi'_{ab} - \frac{a^2}{r}\psi_{ab}\right) + \frac{\psi''_{ab}}{r}\left(\phi'_{cd} - \frac{c^2}{r}\phi_{cd}\right) + \frac{2ac}{r^2}\left(\psi'_{ab} - \frac{1}{r}\psi_{ab}\right)\left(\phi'_{cd} - \frac{1}{r}\phi_{cd}\right).$$

Because w_1 and w_0 satisfy the same boundary conditions, an expansion for w_1 is assumed in the form

$$w_1 = \sum_{n=-\infty}^{\infty}\sum_{m=1}^{\infty} H_{nm}(T_0, T_1, \ldots)\phi_{nm}\mathrm{e}^{\mathrm{i}n\theta}. \tag{28}$$

Substituting equation (28) into equation (27), multiplying the result by $r\phi_{kl}(r) \times \mathrm{e}^{-\mathrm{i}k\theta}$, and integrating the result from $r = 0$ to 1 and $\theta = 0$ to 2π, one obtains

$$\begin{aligned} &D_0^2 H_{kl} + \omega_{kl}^2 H_{kl} \\ &\quad = -2\mathrm{i}\omega_{kl}\left[(D_1 A_{kl} + c_{kl}A_{kl})\mathrm{e}^{\mathrm{i}\omega_{kl}T_0} - (D_1\overline{B}_{kl} + c_{kl}\overline{B}_{kl})\mathrm{e}^{-\mathrm{i}\omega_{kl}T_0}\right] \\ &\qquad + \frac{1}{2}P_{kl}\mathrm{e}^{\mathrm{i}\tau_{kl}}\left[\mathrm{e}^{\mathrm{i}\lambda T_0} + \mathrm{e}^{-\mathrm{i}\lambda T_0}\right] \\ &\qquad + \sum_{n,c=-\infty}^{\infty}\ \sum_{d,m,q=1}^{\infty} \Gamma(kl, cd, nm, pq) \\ &\qquad\qquad \times \sum_{j=1}^{\infty} S_j \mathrm{e}^{\mathrm{i}\Lambda_j T_0}, \quad k = 0, 1, \ldots,\ l = 1, 2, \ldots, \end{aligned} \tag{29}$$

where

$$\Gamma(kl, cd, nm, pq) = \sum_{b=1}^{\infty} G(nm, pq; ab)\int_0^1 r\phi_{kl}\widehat{E}(cd, ab)\,\mathrm{d}r, \tag{30a}$$

$$a = k - c, \qquad p = k - c - n, \tag{30b, c}$$

Λ_j are frequency combinations, and S_j are functions of A_{nm} and B_{nm}. Both Λ_j and S_j are listed in Appendix A. Up to now, the result may be said to be the same as one by Sridhar, Mook and Nayfeh (1978) if we ignore several misprints in the reference (Sridhar, Mook and Nayfeh, 1978).

Eliminating the secular terms (the coefficients of $e^{\pm i\omega_{kl}T_0}$) from the right-hand side of equation (29), we obtain the following solvability conditions:

$$\begin{aligned}
&-2i\omega_{kl}(D_1 A_{kl} + c_{kl}A_{kl}) \\
&\quad + A_{kl}\left\{\sum_{n=-\infty}^{\infty}\sum_{m=1}^{\infty}\gamma_{klnm}\left(A_{nm}\overline{A}_{nm} + B_{nm}\overline{B}_{nm}\right) - \gamma_{klkl}A_{kl}\overline{A}_{kl}\right\} \\
&\quad + 2(1-\delta_{k0})B_{kl}\left\{\sum_{m=1}^{\infty}\hat{\gamma}_{klkm}A_{km}\overline{B}_{km} - \hat{\gamma}_{klkl}A_{kl}\overline{B}_{kl}\right\} \\
&\quad + N_{kl}^{A} + R_{kl}^{A} = 0, \qquad (31a)\\
&2i\omega_{kl}\left(D_1\overline{B}_{kl} + c_{kl}\overline{B}_{kl}\right) \\
&\quad + \overline{B}_{kl}\left\{\sum_{n=-\infty}^{\infty}\sum_{m=1}^{\infty}\gamma_{klnm}\left(A_{nm}\overline{A}_{nm} + B_{nm}\overline{B}_{nm}\right) - \gamma_{klkl}B_{kl}\overline{B}_{kl}\right\} \\
&\quad + 2(1-\delta_{k0})\overline{A}_{kl}\left\{\sum_{m=1}^{\infty}\hat{\gamma}_{klkm}A_{km}\overline{B}_{km} - \hat{\gamma}_{klkl}A_{kl}\overline{B}_{kl}\right\} \\
&\quad + N_{kl}^{B} + R_{kl}^{B} = 0, \qquad (31b)
\end{aligned}$$

where δ_{k0} are the Kronecker delta, $R_{kl}^{A,B}$ are terms due to internal resonances, if any, $N_{kl}^{A,B}$ are terms due to the external excitation, if any, and γ_{klnm} and $\hat{\gamma}_{klkm}$ are constants given in Appendix A. It is noted that these solvability conditions are different from those by Sridhar, Mook and Nayfeh (1978). Terms including expressions $A_{kl}\overline{A}_{kl}$, $B_{kl}\overline{B}_{kl}$ and $2(1-\delta_{k0})$ in equations (31) are missed in their solvability conditions.

4. Steady-state responses and numerical examples

4.1. The plate without elastic foundation ($K = 0$): the case of no internal resonance

We consider a primary resonance in the absence of internal resonance, i.e., $R_{kl}^{A,B} = 0$. The frequency of excitation λ is near natural frequency ω_{GH}. We introduce a detuning parameter, σ, defined as follows:

$$\lambda = \omega_{GH} + \hat{\sigma}, \qquad \hat{\sigma} = \varepsilon\sigma, \qquad (32a, b)$$

and

$$N_{GH}^{A} = \frac{1}{2}P_{GH}e^{i(\sigma T_1 + \tau_{GH})}, \qquad N_{GH}^{B} = \frac{1}{2}P_{GH}e^{-i(\sigma T_1 - \tau_{GH})} \qquad (33a, b)$$

and

$$N_{kl}^{A,B} = 0 \quad \text{for } kl \neq GH. \tag{33c}$$

Next we let

$$A_{nm} = \frac{1}{2} a_{nm} e^{i\alpha_{nm}}, \qquad B_{nm} = \frac{1}{2} b_{nm} e^{i\beta_{nm}}, \tag{34a, b}$$

where a_{nm}, b_{nm}, α_{nm} and β_{nm} are real functions of T_1. Substituting equations (33) and (34) into (31) and separating the result into real imaginary parts yield

$$\omega_{kl}\left(a'_{kl} + c_{kl}a_{kl}\right) - \frac{1}{4}(1-\delta_{k0})b_{kl}\hat{s}^{s}_{kl} - \frac{1}{2}\delta_{kG}\delta_{lH}P_{GH}\sin\mu^{a}_{GH} = 0, \tag{35a}$$

$$\omega_{kl}\left(b'_{kl} + c_{kl}b_{kl}\right) + \frac{1}{4}(1-\delta_{k0})a_{kl}\hat{s}^{s}_{kl} - \frac{1}{2}\delta_{kG}\delta_{lH}P_{GH}\sin\mu^{b}_{GH} = 0, \tag{35b}$$

$$\omega_{kl}a_{kl}\alpha'_{kl} + \frac{1}{8}a_{kl}\left(s_{kl} - \gamma_{klkl}a^{2}_{kl}\right) + \frac{1}{4}(1-\delta_{k0})b_{kl}\hat{s}^{c}_{kl} + \frac{1}{2}\delta_{kG}\delta_{lH}P_{GH}\cos\mu^{a}_{GH} = 0, \tag{35c}$$

$$\omega_{kl}b_{kl}\beta'_{kl} + \frac{1}{8}b_{kl}\left(s_{kl} - \gamma_{klkl}b^{2}_{kl}\right) + \frac{1}{4}(1-\delta_{k0})a_{kl}\hat{s}^{c}_{kl} + \frac{1}{2}\delta_{kG}\delta_{lH}P_{GH}\cos\mu^{b}_{GH} = 0, \tag{35d}$$

where primes denote differentiation with respect to T_1,

$$s_{kl} = \sum_{n=-\infty}^{\infty}\sum_{m=1}^{\infty}\gamma_{klnm}\left(a^{2}_{nm} + b^{2}_{nm}\right), \tag{36a}$$

$$\hat{s}^{s}_{kl} = \sum_{m=1}^{\infty}\hat{\gamma}_{klkm}a_{km}b_{km}\sin(\alpha_{km} - \beta_{km} - \alpha_{kl} + \beta_{kl}), \tag{36b}$$

$$\hat{s}^{c}_{kl} = \sum_{m=1}^{\infty}(1-\delta_{ml})\hat{\gamma}_{klkm}a_{km}b_{km}\cos(\alpha_{km} - \beta_{km} - \alpha_{kl} + \beta_{kl}), \tag{36c}$$

$$\mu^{a}_{GH} = \sigma T_1 + \tau_{GH} - \alpha_{GH}, \qquad \mu^{b}_{GH} = \sigma T_1 - \tau_{GH} - \beta_{GH}. \tag{37a, b}$$

Terms including $\hat{s}$ in the system of equations (35), and terms of $\gamma_{klkl}a^{2}_{kl}$ and $\gamma_{klkl}b^{2}_{kl}$, respectively, in equations (35c) and (35d) make system (35) different from the corresponding system by Sridhar, Mook and Nayfeh (1978). Terms including $\hat{s}$ have something to do with internal resonance. Since we consider the case of no internal resonance, these terms do not affect the result at all. Terms of $\gamma_{klkl}a_{kl}^2$ and $\gamma_{klkl}b_{kl}^2$, therefore, are the effective difference between this and system by Sridhar, Mook and Nayfeh (1978).

Since in steady-state $a'_{kl} = b'_{kl} = 0$, from equations (35a) and (35b) we have the steady-state response with $a_{kl} = b_{kl} = 0$ when $kl \neq GH$. It can be easily proved that this is the only response for the case of $kl \neq GH$. Assuming that there exists any response with $a_{kl} \neq 0$ and $b_{kl} \neq 0$, dividing equations (35a) and (35b) by b_{kl} and a_{kl}, respectively, and adding the results, we have

$$\frac{c_{kl}(a_{kl}^2 + b_{kl}^2)}{a_{kl}b_{kl}} = 0. \tag{38}$$

For damping coefficients $c_{kl} \neq 0$, equation (38) tells that there does not exist any response with $a_{kl} \neq 0$ and $b_{kl} \neq 0$. Since this contradicts the hypothesis, we can conclude that when $kl \neq GH$, there only exist the response with $a_{kl} = b_{kl} = 0$.

For a numerical example we consider the case of a primary resonance $\lambda \approx \omega_{11}$, which is equivalent to the case of $G = H = 1$. The corresponding mode has one nodal diameter and no other nodal circle but the boundary. Substituting the relations of $G = H = 1$ into equations (35)–(37), we obtain a system of autonomous ordinary differential equations for the nondecaying amplitudes as follows:

$$a'_{11} = -c_{11}a_{11} + \frac{P_{11}}{2\omega_{11}}\sin\mu_{11}^a, \tag{39a}$$

$$b'_{11} = -c_{11}b_{11} + \frac{P_{11}}{2\omega_{11}}\sin\mu_{11}^b, \tag{39b}$$

$$a_{11}\mu_{11}^{a\prime} = \sigma a_{11} + \frac{\gamma_{1111}}{4\omega_{11}}a_{11}(a_{11}^2 + 2b_{11}^2) + \frac{P_{11}}{2\omega_{11}}\cos\mu_{11}^a, \tag{39c}$$

$$b_{11}\mu_{11}^{b\prime} = \sigma b_{11} + \frac{\gamma_{1111}}{4\omega_{11}}b_{11}(b_{11}^2 + 2a_{11}^2) + \frac{P_{11}}{2\omega_{11}}\cos\mu_{11}^b, \tag{39d}$$

where a_{11}, b_{11} and μ_{11}^a, μ_{11}^b denote amplitude and phase variables, respectively, corresponding to the mode with natural frequency ω_{11}. Parameters c_{11}, P_{11} and γ_{1111} denote damping coefficient, excitation amplitude and nonlinear coefficient, respectively, corresponding to the mode. It is worthwhile to note an invariance of system (39). More precisely speaking, interchange of a_{11} and b_{11}, and one of μ_{11}^a and μ_{11}^b give another solution of the system.

Since our main concern is steady-state responses of the plate, first of all, we consider equilibrium solutions ($a'_{11} = b'_{11} = \mu_{11}^{a\prime} = \mu_{11}^{b\prime} = 0$), each of which corresponds to a steady-state response. We can write the steady-state response to the first-order approximation as

$$\begin{aligned} w = \phi_{11}\{&a_{11}\cos(\lambda t - \mu_{11}^a + \theta + \tau_{11}) \\ &+ b_{11}\cos(\lambda t - \mu_{11}^b - \theta - \tau_{11})\} + \mathrm{O}(\varepsilon), \end{aligned} \tag{40}$$

where the response w is a superposition of the two traveling waves rotating clockwise and counterclockwise, respectively, and τ_{11} is a phase difference corresponding to θ. Form (40) can also be written as follows:

$$w = Z_1 \cos(\lambda t + \zeta_1)\phi_{11} \cos\theta + Z_2 \cos(\lambda t + \zeta_2)\phi_{11} \sin\theta + O(\varepsilon), \qquad (41)$$

which is the superposition of two standing waves. The constants Z_1, Z_2, ζ_1 and ζ_2 are given in Appendix A.

When $a_{11} = b_{11}$ and $\mu_{11}^a = \mu_{11}^b$, form (40) is reduced to a standing wave as follows:

$$w = 2\phi_{11} a_{11} \cos(\lambda t - \mu_{11}^a) \cos(\theta + \tau_{11}) + O(\varepsilon). \qquad (42)$$

In Figure 2 the amplitudes a_{11} and b_{11} are plotted as functions of detuning parameter $\hat{\sigma} = \varepsilon\sigma$ when $\omega_{11} = 21.2604$ (Leissa, 1969), $\nu = 1/3$, $\varepsilon = 0.001067$, $\varepsilon c = 0.01$, $\varepsilon P_{11} = 4$ and $\tau_{11} = 0$. Branches SS1, US1, US2 and SS2 represent the standing waves, while branches ST1, UT1 and UT2 represent traveling waves. Solid and dotted lines denote, respectively, stable and unstable responses. Except for the instability of branch US1, the response in the form of standing wave is the response of Duffing oscillator. The stable response in the form of traveling wave, $\{ST1_A, ST1_B\}$ represents $\{a_{11}, b_{11}\}$ or $\{b_{11}, a_{11}\}$. When $\hat{\sigma} < \hat{\sigma}_1$ and $\hat{\sigma}_1 < \hat{\sigma} < \hat{\sigma}_2$, standing and traveling waves, respectively, exist in reality. While standing and traveling waves coexist when $\hat{\sigma}_2 < \hat{\sigma} < \hat{\sigma}_3$, standing wave only exists when $\hat{\sigma} > \hat{\sigma}_3$. This result is remarkably different from one by Sridhar, Mook and Nayfeh (1978). They expected that the response is in the form of standing

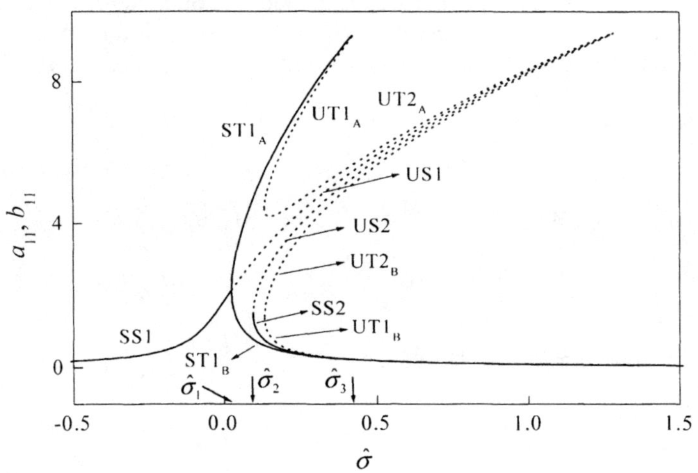

Figure 2. Variations of the amplitudes with detuning parameter $\hat{\sigma} = \varepsilon\sigma$ when $\varepsilon P_{11} = 4$ (——, stable; - - - - -, unstable).

wave, which is the response of Duffing oscillator. We believe that this difference comes from the correction of solvability conditions.

Considering the case of no internal resonance as the case of 1:1 internal resonance between two modes having shapes of $\phi_{11}\cos\theta$ and $\phi_{11}\sin\theta$ corresponding to one natural frequency ω_{11}, Nayfeh and Vakakis (1994) observed the coexistence of subharmonic standing and traveling waves in the case of subharmonic resonance. We believe that their result supports the validity of our observation.

In order to show the deflection of the plate we consider the case of $\hat{\sigma} = 0.1$, in which there exist three stable responses (one is a standing wave and two are traveling waves). The initial condition determines which deflection is to be realized. Figures 3–5 represent deflections corresponding to the stable responses of the plate for one period of excitation $T\ (= 2\pi/\lambda)$. A standing wave ($a_{11} = b_{11}$) is shown in Figure 3, in which we can see a nodal line at 5 minutes past 7 o'clock. Figures 4 and 5 represent traveling waves, which are rotating clockwise ($a_{11} > b_{11}$) and counterclockwise ($a_{11} < b_{11}$), respectively. It is noted that the dominant amplitude (a_{11} or b_{11}) determines the direction of the rotation. In these figures we can see that the period of deflection is the same as the one of excitation, which means the response of a primary resonance.

4.2. The plate on elastic foundation ($K > 0$): the case of internal resonance ($\omega_{NM} \approx 3\omega_{CD}$, where $N = 3C$)

In this case, the elastic foundation with the stiffness $K > 0$ is considered to get a varied natural characteristic, which generates a desired commensurable relation between natural frequencies. In order to consider the internal resonance condition $\omega_{NM} \approx 3\omega_{CD}$ ($N = 3C$) and the external resonance condition $\lambda \approx \omega_{GH}$ ($GH = CD$ or NM), the detuning parameters, σ_1 and σ_2, are introduced as follows:

$$\omega_{NM} = 3\omega_{CD} + \varepsilon\sigma_1, \qquad \lambda = \omega_{GH} + \varepsilon\sigma_2, \tag{43, 44}$$

where $\hat{\sigma}_i = \varepsilon\sigma_i$, $i = 1, 2$. In this case

$$R^A_{NM} = Q_{NM}A^3_{CD}e^{-i\sigma_1 T_1}, \qquad R^B_{NM} = Q_{NM}\overline{B}^3_{CD}e^{i\sigma_1 T_1}, \tag{45a, b}$$

$$R^A_{CD} = Q_{CD}\overline{A}^2_{CD}A_{NM}e^{i\sigma_1 T_1}, \qquad R^B_{CD} = Q_{CD}B^2_{CD}\overline{B}_{NM}e^{-i\sigma_1 T_1}, \tag{45c, d}$$

$$R^{A,B}_{kl} = 0 \quad \text{for } kl \neq CD, NM, \tag{45f}$$

$$N^A_{GH} = \frac{1}{2}P_{GH}e^{i(\sigma_2 T_1 + \tau_{GH})}, \qquad N^B_{GH} = \frac{1}{2}P_{GH}e^{-i(\sigma_2 T_1 - \tau_{GH})}, \tag{46a, b}$$

$$N^{A,B}_{kl} = 0 \quad \text{for } kl \neq GH, \tag{46c}$$

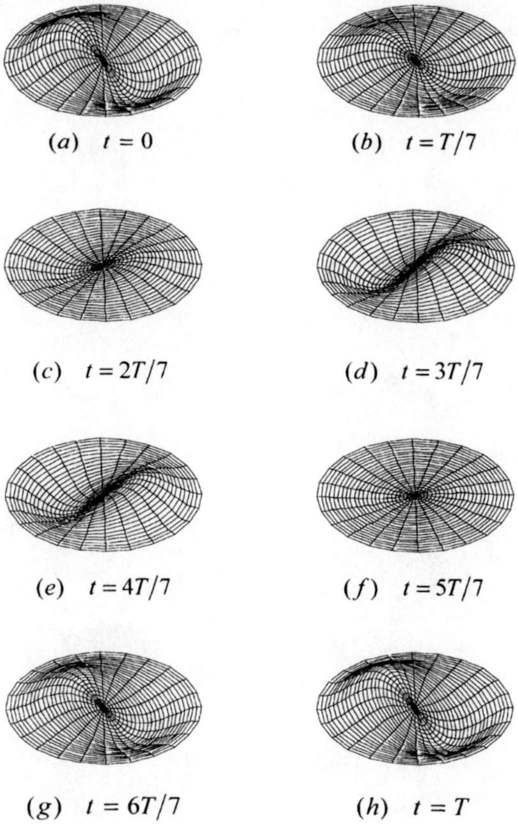

Figure 3. Deflections of the circular plate for one period of excitation T $(= 2\pi/\lambda)$ when $a_{11} = 1.1608$, $b_{11} = 1.1608$, $\mu_{11}^a = 3.0179$, $\mu_{11}^b = 3.0179$, $\omega_{11} = 21.2604$, $\hat{\sigma} = 0.1$ and $\tau_{11} = 0$. A standing wave ($a_{11} = b_{11}$).

where the Q_{NM} and Q_{CD} are constants given in Appendix B. Reminding

$$A_{nm} = \frac{1}{2} a_{nm} \mathrm{e}^{\mathrm{i}\alpha_{nm}}, \qquad B_{nm} = \frac{1}{2} b_{nm} \mathrm{e}^{\mathrm{i}\beta_{nm}} \tag{47a, b}$$

and substituting equations (45)–(47) into solvability conditions (31) and separating the result into real and imaginary parts, give

$$\omega_{kl}\left(a'_{kl} + c_{kl} a_{kl}\right) - \frac{1}{4}(1 - \delta_{k0}) b_{kl} \hat{s}^s_{kl} - \frac{1}{8} \delta_{kC} \delta_{lD} Q_{CD} a^2_{CD} a_{NM} \sin \tilde{\mu}_A$$
$$+ \frac{1}{8} \delta_{kN} \delta_{lM} Q_{NM} a^3_{CD} \sin \tilde{\mu}_A - \frac{1}{2} \delta_{kG} \delta_{lH} P_{GH} \sin \mu^a_{GH} = 0, \tag{48a}$$

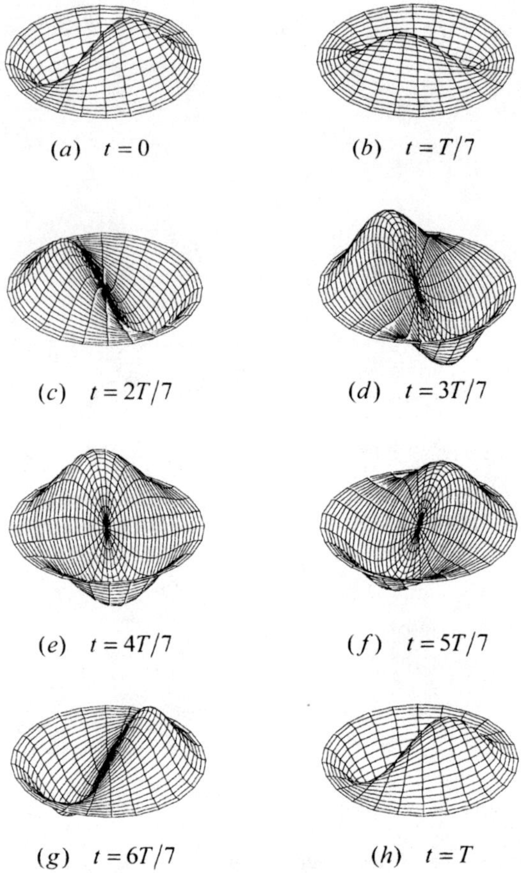

Figure 4. Deflections of the circular plate for one period of excitation T $(= 2\pi/\lambda)$ when $a_{11} = 4.7974$, $b_{11} = 0.7464$, $\mu_{11}^a = 0.5352$, $\mu_{11}^b = 0.07943$, $\omega_{11} = 21.2604$, $\hat{\sigma} = 0.1$ and $\tau_{11} = 0$. A traveling wave $(a_{11} > b_{11})$.

$$\omega_{kl}\left(b'_{kl} + c_{kl}b_{kl}\right) + \frac{1}{4}(1-\delta_{k0})a_{kl}\hat{s}^s_{kl} - \frac{1}{8}\delta_{kC}\delta_{lD}Q_{CD}b^2_{CD}b_{NM}\sin\tilde{\mu}_B$$
$$+ \frac{1}{8}\delta_{kN}\delta_{lM}Q_{NM}b^3_{CD}\sin\tilde{\mu}_B - \frac{1}{2}\delta_{kG}\delta_{lH}P_{GH}\sin\mu^b_{GH} = 0, \qquad (48b)$$

$$\omega_{kl}a_{kl}\alpha'_{kl} + \frac{1}{8}a_{kl}\left(s_{kl} - \gamma_{klkl}a^2_{kl}\right)$$
$$+ \frac{1}{4}(1-\delta_{k0})b_{kl}\hat{s}^c_{kl} + \frac{1}{8}\delta_{kC}\delta_{lD}Q_{CD}a^2_{CD}a_{NM}\cos\tilde{\mu}_A$$
$$+ \frac{1}{8}\delta_{kN}\delta_{lM}Q_{NM}a^3_{CD}\cos\tilde{\mu}_A + \frac{1}{2}\delta_{kG}\delta_{lH}P_{GH}\cos\mu^a_{GH} = 0 \qquad (48c)$$

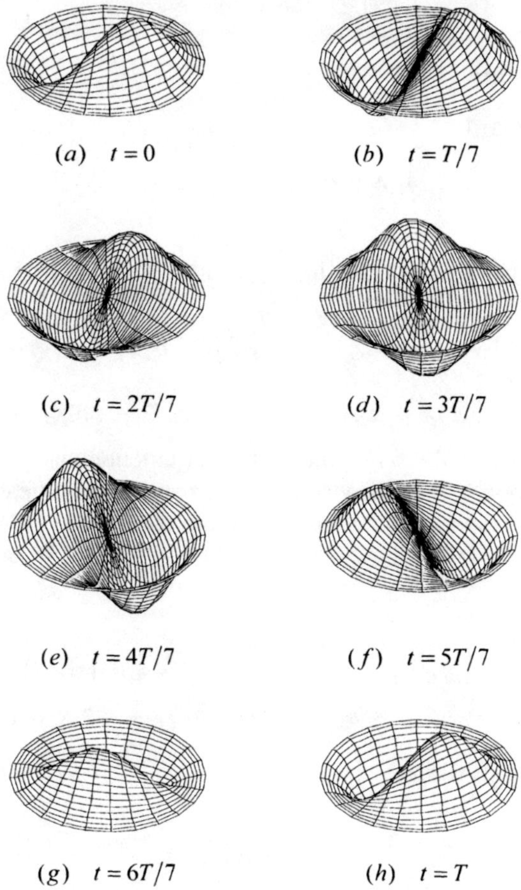

Figure 5. Deflections of the circular plate for one period of excitation T $(= 2\pi/\lambda)$ when $a_{11} = 0.7464$, $b_{11} = 4.7974$, $\mu_{11}^{a} = 0.07943$, $\mu_{11}^{b} = 0.5352$, $\omega_{11} = 21.2604$, $\hat{\sigma} = 0.1$ and $\tau_{11} = 0$. A traveling wave $(a_{11} < b_{11})$.

and

$$\omega_{kl} b_{kl} \beta'_{kl} + \frac{1}{8} b_{kl} \left(s_{kl} - \gamma_{klkl} b_{kl}^{2} \right) + \frac{1}{4}(1 - \delta_{k0}) a_{kl} \hat{s}_{kl}^{c} + \frac{1}{8} \delta_{kC} \delta_{lD} Q_{CD} b_{CD}^{2} b_{NM} \cos \tilde{\mu}_{B} + \frac{1}{8} \delta_{kN} \delta_{lM} Q_{NM} b_{CD}^{3} \cos \tilde{\mu}_{B} + \frac{1}{2} \delta_{kG} \delta_{lH} P_{GH} \cos \mu_{GH}^{b} = 0, \quad (48d)$$

where

$$s_{kl} = \sum_{n=-\infty}^{\infty}\sum_{m=1}^{\infty} \gamma_{klnm}\left(a_{nm}^2 + b_{nm}^2\right), \tag{49a}$$

$$\hat{s}_{kl}^{s} = \sum_{m=1}^{\infty} \hat{\gamma}_{klkm} a_{km} b_{km} \sin(\alpha_{km} - \beta_{km} - \alpha_{kl} + \beta_{kl}), \tag{49b}$$

$$\hat{s}_{kl}^{c} = \sum_{m=1}^{\infty} (1-\delta_{ml})\hat{\gamma}_{klkm} a_{km} b_{km} \cos(\alpha_{km} - \beta_{km} - \alpha_{kl} + \beta_{kl}), \tag{49c}$$

$$\mu_{GH}^{a} = \sigma_2 T_1 + \tau_{GH} - \alpha_{GH}, \qquad \mu_{GH}^{b} = \sigma_2 T_1 - \tau_{GH} - \beta_{GH}, \tag{50a, b}$$

$$\tilde{\mu}_A = \sigma_1 T_1 - 3\alpha_{CD} + \alpha_{NM}, \qquad \tilde{\mu}_B = \sigma_1 T_1 - 3\beta_{CD} + \beta_{NM}. \tag{50c, d}$$

Each equilibrium solution of the system of autonomous ordinary differential equations to be obtained from system (48) corresponds to a steady-state response. The steady-state response to the first-order approximation is given as follows:

$$w = w_{CD} + w_{NM} + O(\varepsilon), \tag{51}$$

where

$$\begin{aligned} w_{CD} = {} & \delta_{GC}\delta_{HD}\phi_{CD}\left\{a_{CD}\cos\left(\lambda t - \mu_{CD}^{a} + C\theta + \tau_{CD}\right)\right. \\ & \left. + b_{CD}\cos\left(\lambda t - \mu_{CD}^{b} - C\theta - \tau_{CD}\right)\right\} \\ & + \delta_{GN}\delta_{HM}\phi_{CD}\left\{a_{CD}\cos\left(\frac{\lambda}{3}t - \frac{\mu_{NM}^{a}}{3} - \frac{\tilde{\mu}_A}{3} + C\theta + \frac{\tau_{NM}}{3}\right)\right. \\ & \left. + b_{CD}\cos\left(\frac{\lambda}{3}t - \frac{\mu_{NM}^{b}}{3} - \frac{\tilde{\mu}_B}{3} - C\theta - \frac{\tau_{NM}}{3}\right)\right\}, \end{aligned} \tag{52}$$

$$\begin{aligned} w_{NM} = {} & \delta_{GC}\delta_{HD}\phi_{NM}\left\{a_{NM}\cos\left(3\lambda t - 3\mu_{CD}^{a} + \tilde{\mu}_A + N\theta + 3\tau_{CD}\right)\right. \\ & \left. + b_{NM}\cos\left(3\lambda t - 3\mu_{CD}^{b} + \tilde{\mu}_B - N\theta - 3\tau_{CD}\right)\right\} \\ & + \delta_{GN}\delta_{HM}\phi_{NM}\left\{a_{NM}\cos\left(\lambda t - \mu_{NM}^{a} + N\theta + \tau_{NM}\right)\right. \\ & \left. + b_{NM}\cos\left(\lambda t - \mu_{NM}^{b} - N\theta - \tau_{NM}\right)\right\}. \end{aligned} \tag{53}$$

Each of the w_{CD} and w_{NM} is the superposition of two traveling wave components. If $a_{CD} = b_{CD}$, one obtains $a_{NM} = b_{NM}$, $\mu_{CD}^{a} = \mu_{CD}^{b}$, $\mu_{NM}^{a} = \mu_{NM}^{b}$ and $\tilde{\mu}_A = \tilde{\mu}_B$. Then equations (52) and (53) can be reduced as follows:

$$\begin{aligned} w_{CD} = {} & 2\delta_{GC}\delta_{HD}\phi_{CD}a_{CD}\cos\left(\lambda t - \mu_{CD}^{a}\right)\cos(C\theta + \tau_{CD}) \\ & + 2\delta_{GN}\delta_{HM}\phi_{CD}a_{CD}\cos\left(\frac{\lambda}{3}t - \frac{\mu_{NM}^{a}}{3} - \frac{\tilde{\mu}_A}{3}\right)\cos\left(C\theta + \frac{\tau_{NM}}{3}\right) \end{aligned} \tag{54}$$

and

$$w_{NM} = 2\delta_{GC}\delta_{HD}\phi_{NM}a_{NM}\cos\left(3\lambda t - 3\mu^a_{CD} + \tilde{\mu}_A\right)\cos(N\theta + 3\tau_{CD}) + 2\delta_{GN}\delta_{HM}\phi_{NM}a_{NM}\cos\left(\lambda t - \mu^a_{NM}\right)\cos(N\theta + \tau_{NM}). \quad (55)$$

Now each of the w_{CD} and w_{NM} becomes a superposition of two standing wave components.

For a numerical example the case of $K = 1032$, which gives natural frequencies $\omega_{11} = 38.52$ and $\omega_{32} = 115.58$ is considered. Then there is an internal resonance condition $\omega_{32} \approx 3\omega_{11}$ and an internal detuning parameter $\varepsilon\sigma_1 = 0.007412$. Pursuing the internal resonance condition $\omega_{NM} \approx 3\omega_{CD}$ ($N = 3C$), gives the relation $C = 1$, $D = 1$, $N = 3$ and $M = 2$. Consider two primary resonance cases, $\lambda \approx \omega_{11}$ ($G = 1$ and $H = 1$) and $\lambda \approx \omega_{32}$ ($G = 3$ and $H = 2$). Corresponding autonomous systems obtained from system (48) are given in Appendix B. In Figures 6–11 the amplitudes a_{11}, b_{11}, a_{32} and b_{32} are plotted as functions of detuning parameter $\varepsilon\sigma_2 = \hat{\sigma}_2$ when $\{\nu, \varepsilon, \varepsilon c, \tau_{11}, \tau_{32}\} = \{1/3, 0.001067, 0.01, 0.0, 0.0\}$. Solid and dotted lines denote, respectively, stable and unstable responses. The abbreviations SS, US, ST and UT denote, respectively, stable standing, unstable standing, stable traveling and unstable traveling wave components. Numerical results were verified by using a software package AUTO (Doedel, 1986) which can perform bifurcation analysis and continuation of solutions for ordinary differential equations.

In the case of $\lambda \approx \omega_{11}$ ($G = 1$ and $H = 1$), Figure 6(a) and its partial enlargements, Figure 7, show that the response curve corresponding to standing waves is similar with the response curve of the Duffing oscillator, except that the upper branch changes its stability at pitchfork bifurcation points, $\hat{\sigma}_A$ (0.0219) and $\hat{\sigma}_G$ (0.2144). Figure 6(b) and its partial enlargements, Figure 8, show that the mode corresponding to ω_{32} is excited indirectly through the nonlinear interaction. If there were no nonlinear interaction, a_{32} and b_{32} would be zero. Figures 7 and 8 show that traveling wave components change their stability at Hopf bifurcation points, $\hat{\sigma}_D$ (0.0678) and $\hat{\sigma}_F$ (0.2099). These figures show that the response curves have four saddle-node bifurcation points, $\hat{\sigma}_B$ (0.0512), $\hat{\sigma}_C$ (0.0665), $\hat{\sigma}_E$ (0.0755) and $\hat{\sigma}_H$ (0.2162). When $\hat{\sigma}_C < \hat{\sigma}_2 < \hat{\sigma}_D$, there exist five stable steady-state responses. Those are from SS_2, ST_1, ST_2, ST_3 and ST_4. Since the overall deflection of the plate is a superposition of two wave components, respectively, due to modes excited directly (ω_{11}) and indirectly (ω_{23}), it will be one of five superpositions (one superposition of standing wave components and four superpositions of traveling wave components). The initial condition determines which deflection is to be realized.

In the case of $\lambda \approx \omega_{32}$ ($G = 3$ and $H = 2$), Figure 9 and its partial enlargements, Figure 10, show responses of directly excited mode, with response of $a_{11} = 0$ and $b_{11} = 0$, which means no interaction between two modes. The

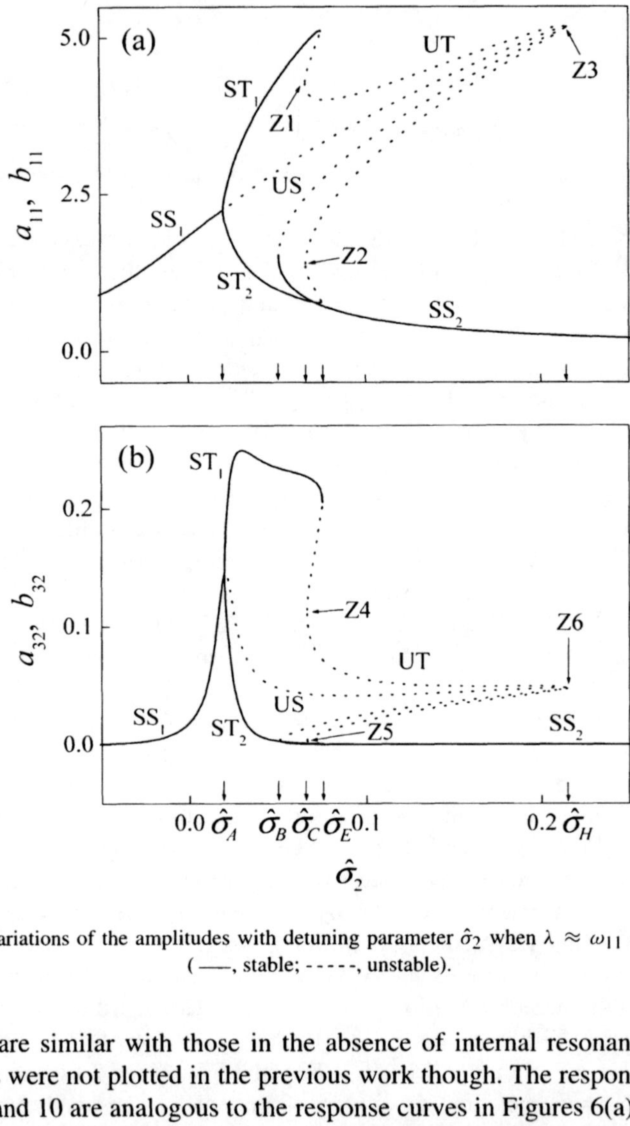

Figure 6. Variations of the amplitudes with detuning parameter $\hat{\sigma}_2$ when $\lambda \approx \omega_{11}$ and $\varepsilon P_{11} = 4$ (——, stable; - - - - -, unstable).

responses are similar with those in the absence of internal resonance. The enlargements were not plotted in the previous work though. The response curves in Figures 9 and 10 are analogous to the response curves in Figures 6(a) and 7. Figure 11 shows that there exist additional steady-state responses, all of which turn out to be unstable. In other words, no stable response with nonvanishing amplitudes of mode excited indirectly is found. It is believed that modal interaction via unstable responses is a peculiar phenomenon. Nonexistence of stable steady-state responses may imply the existence of quasiperiodic response or chaos. Exploring the entity of unstable responses in Figure 11, however, is beyond the scope of this work.

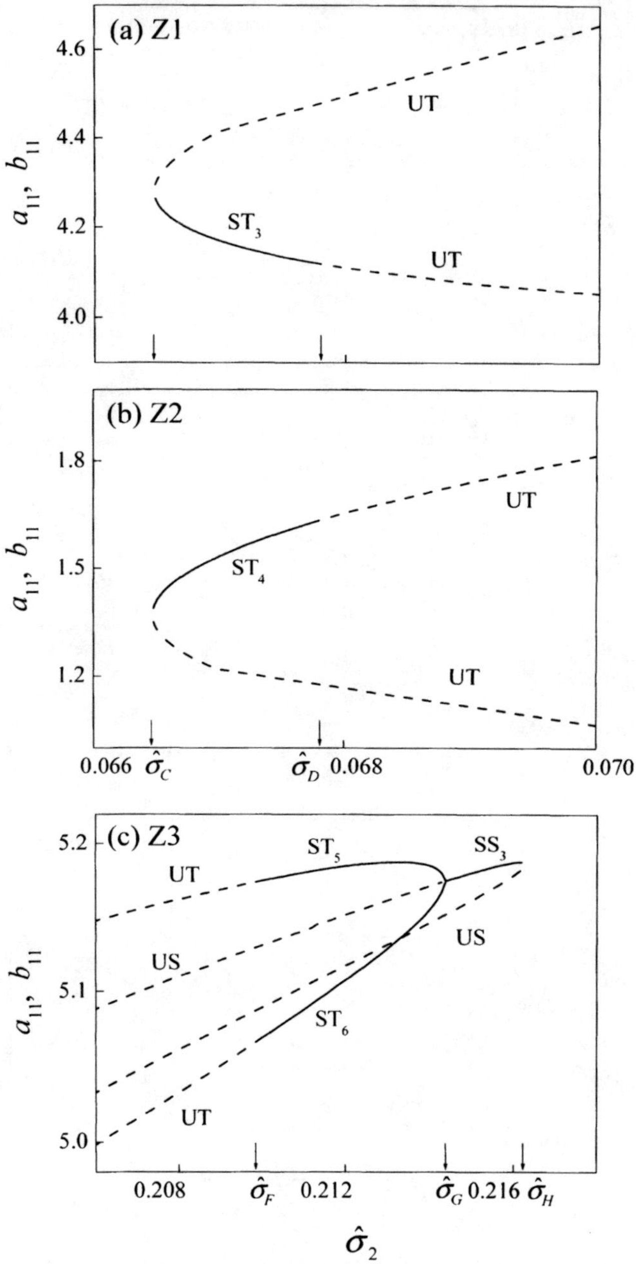

Figure 7. Variations of the amplitudes with detuning parameter $\hat{\sigma}_2$ when $\lambda \approx \omega_{11}$ and $\varepsilon P_{11} = 4$ (——, stable; - - - - -, unstable). Enlargements of the Z1, Z2 and Z3 in Figure 6(a).

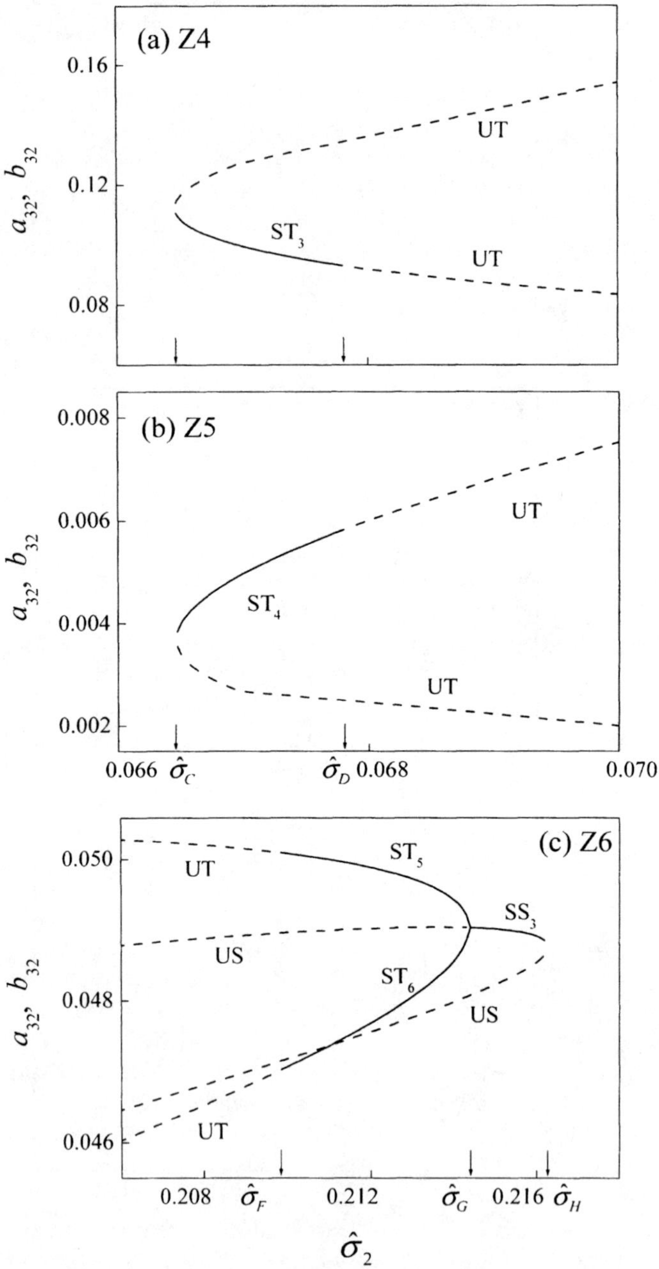

Figure 8. Variations of the amplitudes with detuning parameter $\hat{\sigma}_2$ when $\lambda \approx \omega_{11}$ and $\varepsilon P_{11} = 4$ (——, stable; - - - - -, unstable). Enlargements of the Z4, Z5 and Z6 in Figure 6(b).

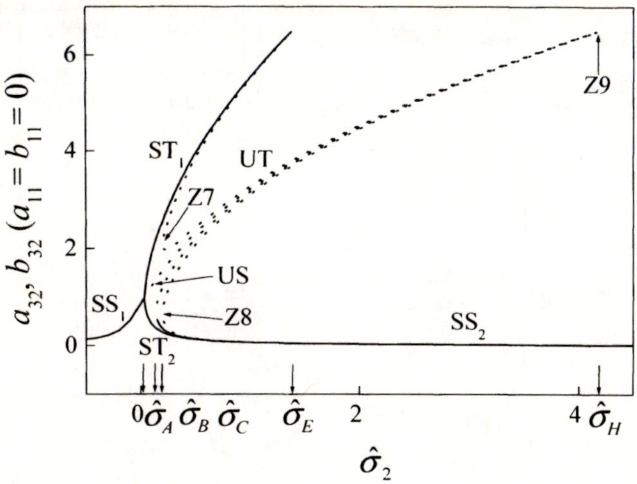

Figure 9. Variations of the amplitudes ($a_{11} = 0$ and $b_{11} = 0$) with detuning parameter $\hat{\sigma}_2$ when $\lambda \approx \omega_{32}$ and $\varepsilon P_{32} = 15$ (——, stable; - - - - -, unstable).

4.3. *The plate on elastic foundation ($K > 0$): the case of internal resonance ($\omega_{NM} \approx 3\omega_{CD}$, where $N = C$)*

It is noted that the corrected solvability conditions (31) are different from those by Sridhar, Mook and Nayfeh (1978) in several aspects. Terms including expressions γ_{klkl} and $2(1 - \delta_{k0})$ in equations (31) are added to their solvability conditions. The former (including γ_{klkl}) can contribute to the dynamics whether the modes involved are asymmetric or not. The latter (including $2(1 - \delta_{k0})$) can contribute to the dynamics only when the modes involved are asymmetric ($k \neq 0$). In this study we consider the internal resonance condition including two distinct natural frequencies (ω_{CD} and ω_{NM}) corresponding to asymmetric modes. Then the ratio of two frequencies must be three to one ($\omega_{NM} \approx 3\omega_{CD}$). Furthermore, the terms $R_{kl}^{A,B}$ (representing modal interactions) vanish unless the ratio of the numbers of nodal diameters of two modes is either three to one ($N = 3C$) or one to one ($N = C$). These facts have been clarified by observing equations (29) and Table A.1. In view of the three-to-one case investigated in Section 4.2, it has been found that terms including $2(1 - \delta_{k0})$ vanish even though all modes involved are asymmetric. We observed that in order for the terms not to vanish the ratio must be one to one. This is why we need to investigate the effect of the number of nodal diameters on nonlinear interactions in asymmetric vibrations of the plate by examining the one-to-one case ($N = C$).

In order to consider the internal resonance condition $\omega_{CM} \approx 3\omega_{CD}$ and the external resonance condition $\lambda \approx \omega_{CD}$, we introduce detuning parameters,

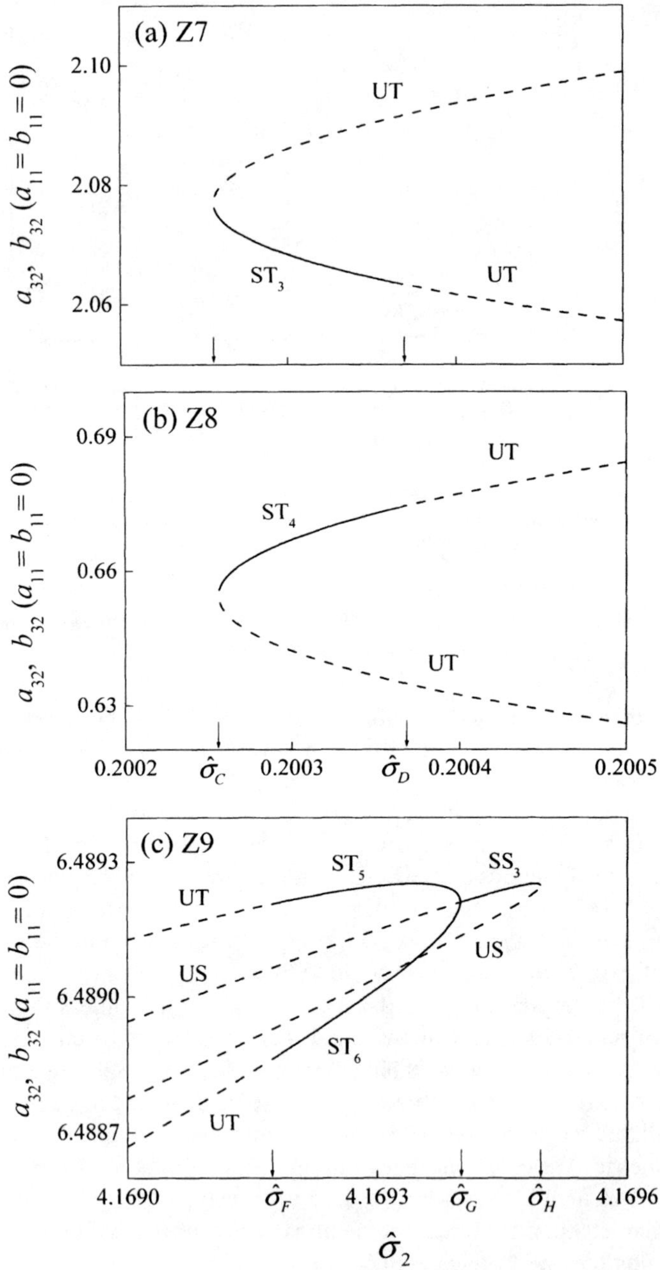

Figure 10. Variations of the amplitudes ($a_{11} = 0$ and $b_{11} = 0$) with detuning parameter $\hat{\sigma}_2$ when $\lambda \approx \omega_{32}$ and $\varepsilon P_{32} = 15$ (——, stable; - - - - -, unstable). Enlargements of the Z7, Z8 and Z9 in Figure 9.

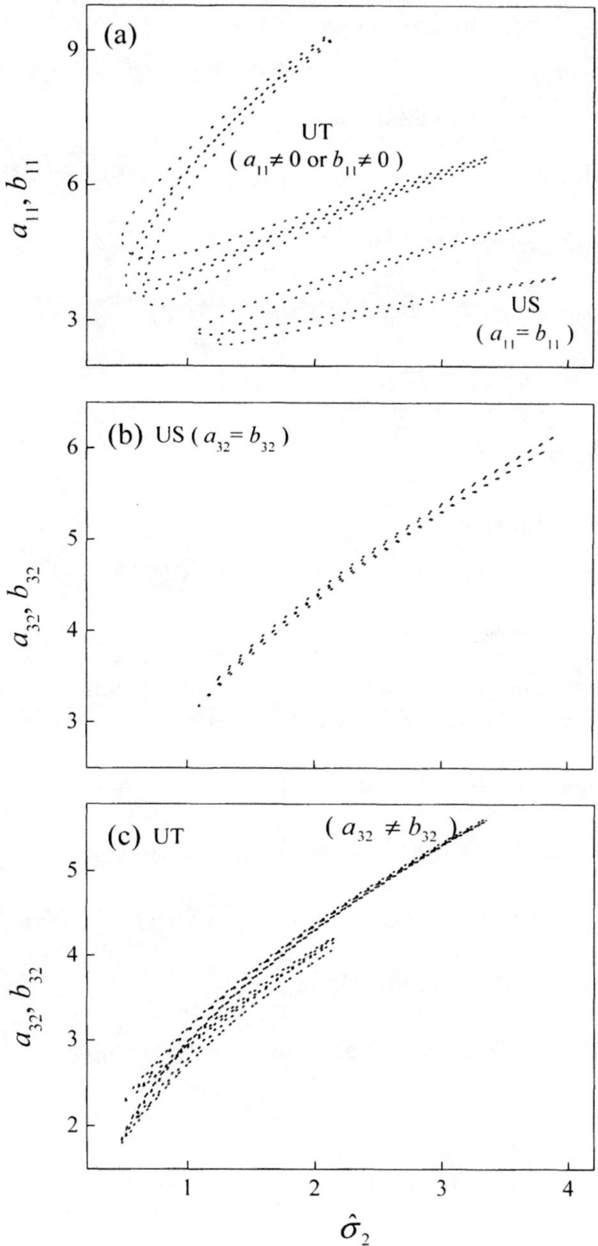

Figure 11. Variations of the amplitudes ($a_{11} \neq 0$ or $b_{11} \neq 0$) with detuning parameter $\hat{\sigma}_2$ when $\lambda \approx \omega_{32}$ and $\varepsilon P_{32} = 15$ (——, stable; - - - - -, unstable).

σ_1 and σ_2, as follows:

$$\omega_{CM} = 3\omega_{CD} + \hat{\sigma}_1, \qquad \lambda = \omega_{CD} + \hat{\sigma}_2, \tag{56, 57}$$

where $\hat{\sigma}_i = \varepsilon\sigma_i$, $i = 1, 2$. In this case

$$R_{CM}^A = Q_{CM} A_{CD}^2 B_{CD} \mathrm{e}^{-\mathrm{i}\sigma_1 T_1}, \qquad R_{CM}^B = Q_{CM} \overline{A}_{CD} \overline{B}_{CD}^2 \mathrm{e}^{\mathrm{i}\sigma_1 T_1}, \tag{58a, b}$$

$$R_{CD}^A = \left(Q_{CD} \overline{A}_{CD} \overline{B}_{CD} A_{CM} + Q_{CM} \overline{B}_{CD}^2 B_{CM}\right) \mathrm{e}^{\mathrm{i}\sigma_1 T_1}, \tag{58c}$$

$$R_{CD}^B = \left(Q_{CD} A_{CD} B_{CD} \overline{B}_{CM} + Q_{CM} A_{CD}^2 \overline{A}_{CM}\right) \mathrm{e}^{-\mathrm{i}\sigma_1 T_1}, \tag{58d}$$

$$R_{kl}^{A,B} = 0 \quad \text{for } kl \neq CD, CM, \tag{58f}$$

$$N_{CD}^A = \frac{1}{2} P_{CD} \mathrm{e}^{\mathrm{i}(\sigma_2 T_1 + \tau_{CD})}, \qquad N_{CD}^B = \frac{1}{2} P_{CD} \mathrm{e}^{-\mathrm{i}(\sigma_2 T_1 - \tau_{CD})}, \tag{59a, b}$$

$$N_{kl}^{A,B} = 0 \quad \text{for } kl \neq CD, \tag{59c}$$

where the Q_{CM} and Q_{CD} are constants given in Appendix C. Reminding

$$A_{nm} = \frac{1}{2} a_{nm} \mathrm{e}^{\mathrm{i}\alpha_{nm}}, \qquad B_{nm} = \frac{1}{2} b_{nm} \mathrm{e}^{\mathrm{i}\beta_{nm}}. \tag{60a, b}$$

and substituting equations (56)–(60) into solvability conditions (31) and separating the result into real and imaginary parts, we obtain

$$\begin{aligned}
&\omega_{kl}\left(a_{kl}' + c_{kl} a_{kl}\right) - \frac{1}{4}(1 - \delta_{k0}) b_{kl} \hat{s}_{kl}^s \\
&\quad - \frac{1}{8}\delta_{kC}\delta_{lD}\left(Q_{CD} a_{CD} b_{CD} a_{CM} \sin\tilde{\mu}_A + Q_{CM} b_{CD}^2 b_{CM} \sin\tilde{\mu}_B\right) \\
&\quad + \frac{1}{8}\delta_{kC}\delta_{lM} Q_{CM} a_{CD}^2 b_{CD} \sin\tilde{\mu}_A - \frac{1}{2}\delta_{kC}\delta_{lD} P_{CD} \sin\mu_{CD}^a = 0,
\end{aligned} \tag{61a}$$

$$\begin{aligned}
&\omega_{kl}\left(b_{kl}' + c_{kl} b_{kl}\right) + \frac{1}{4}(1 - \delta_{k0}) a_{kl} \hat{s}_{kl}^s \\
&\quad - \frac{1}{8}\delta_{kC}\delta_{lD}\left(Q_{CD} a_{CD} b_{CD} b_{CM} \sin\tilde{\mu}_B + Q_{CM} a_{CD}^2 a_{CM} \sin\tilde{\mu}_A\right) \\
&\quad + \frac{1}{8}\delta_{kC}\delta_{lM} Q_{CM} a_{CD} b_{CD}^2 \sin\tilde{\mu}_B - \frac{1}{2}\delta_{kC}\delta_{lD} P_{CD} \sin\mu_{CD}^b = 0,
\end{aligned} \tag{61b}$$

$$\begin{aligned}
&\omega_{kl} a_{kl} \alpha_{kl}' + \frac{1}{8} a_{kl}\left(s_{kl} - \gamma_{klkl} a_{kl}^2\right) + \frac{1}{4}(1 - \delta_{k0}) b_{kl} \hat{s}_{kl}^c \\
&\quad + \frac{1}{8}\delta_{kC}\delta_{lD}\left(Q_{CD} a_{CD} b_{CD} a_{CM} \cos\tilde{\mu}_A + Q_{CM} b_{CD}^2 b_{CM} \cos\tilde{\mu}_B\right) \\
&\quad + \frac{1}{8}\delta_{kC}\delta_{lM} Q_{CM} a_{CD}^2 b_{CD} \cos\tilde{\mu}_A + \frac{1}{2}\delta_{kC}\delta_{lD} P_{CD} \cos\mu_{CD}^a = 0,
\end{aligned} \tag{61c}$$

$$\omega_{kl} b_{kl} \beta'_{kl} + \frac{1}{8} b_{kl}\left(s_{kl} - \gamma_{klkl} b_{kl}^2\right) + \frac{1}{4}(1-\delta_{k0}) a_{kl} \hat{s}_{kl}^c$$
$$+ \frac{1}{8}\delta_{kC}\delta_{lD}\left(Q_{CD} a_{CD} b_{CD} b_{CM} \cos\tilde{\mu}_B + Q_{CM} a_{CD}^2 a_{CM} \cos\tilde{\mu}_A\right)$$
$$+ \frac{1}{8}\delta_{kC}\delta_{lM} Q_{CM} a_{CD} b_{CD}^2 \cos\tilde{\mu}_B + \frac{1}{2}\delta_{kC}\delta_{lD} P_{CD} \cos\mu_{CD}^b = 0, \quad (61d)$$

where

$$s_{kl} = \sum_{n=-\infty}^{\infty}\sum_{m=1}^{\infty} \gamma_{klnm}\left(a_{nm}^2 + b_{nm}^2\right), \quad (62a)$$

$$\hat{s}_{kl}^s = \sum_{m=1}^{\infty} \hat{\gamma}_{klkm} a_{km} b_{km} \sin(\alpha_{km} - \beta_{km} - \alpha_{kl} + \beta_{kl}), \quad (62b)$$

$$\hat{s}_{kl}^c = \sum_{m=1}^{\infty} (1-\delta_{ml}) \hat{\gamma}_{klkm} a_{km} b_{km} \cos(\alpha_{km} - \beta_{km} - \alpha_{kl} + \beta_{kl}), \quad (62c)$$

$$\mu_{CD}^a = \sigma_2 T_1 + \tau_{CD} - \alpha_{CD}, \qquad \mu_{CD}^b = \sigma_2 T_1 - \tau_{CD} - \beta_{CD}, \quad (63a, b)$$

$$\tilde{\mu}_A = \sigma_1 T_1 - 2\alpha_{CD} - \beta_{CD} + \alpha_{CM}, \quad (63c)$$

$$\tilde{\mu}_B = \sigma_1 T_1 - \alpha_{CD} - 2\beta_{CD} + \beta_{CM}. \quad (63d)$$

Each equilibrium solution of the system of autonomous ordinary differential equations to be obtained from system (61) is corresponding to a steady-state response. The steady-state response to the first-order approximation is given as follows:

$$w = w_{CD} + w_{CM} + \mathrm{O}(\varepsilon), \quad (64)$$

where

$$w_{CD} = \phi_{CD}\left\{a_{CD}\cos\left(\lambda t - \mu_{CD}^a + C\theta + \tau_{CD}\right) + b_{CD}\cos\left(\lambda t - \mu_{CD}^b - C\theta - \tau_{CD}\right)\right\}, \quad (65)$$

$$w_{CM} = \phi_{CM}\left\{a_{CM}\cos\left(3\lambda t - 2\mu_{CD}^a - \mu_{CD}^b + \tilde{\mu}_A + C\theta + \tau_{CD}\right) + b_{CM}\cos\left(3\lambda t - \mu_{CD}^a - 2\mu_{CD}^b + \tilde{\mu}_B - C\theta - \tau_{CD}\right)\right\}. \quad (66)$$

Each of the w_{CD} and w_{CM} is a superposition of two traveling wave components.

If $a_{CD} = b_{CD}$, $a_{CM} = b_{CM}$, $\mu_{CD}^a = \mu_{CD}^b$ and $\tilde{\mu}_A = \tilde{\mu}_B$, equations (65) and (66) can be reduced as follows:

$$w_{CD} = 2\phi_{CD} a_{CD}\cos\left(\lambda t - \mu_{CD}^a\right)\cos(C\theta + \tau_{CD}), \quad (67)$$

$$w_{CM} = 2\phi_{CM} a_{CM}\cos\left(3\lambda t - 3\mu_{CD}^a + \tilde{\mu}_A\right)\cos(C\theta + \tau_{CD}). \quad (68)$$

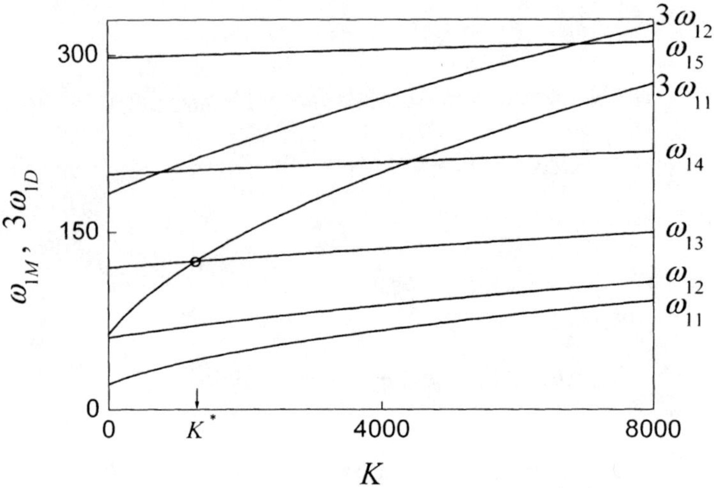

Figure 12. Variations of the natural frequencies ω_{1M} and $3\omega_{1D}$ with the stiffness of the foundation, K.

Now each of the w_{CD} and w_{CM} becomes a standing wave component.

Pursuing the internal resonance condition $\omega_{CM} \approx 3\omega_{CD}$, we consider the case of $C = 1$ for the convenience. In order to choose a proper value of the stiffness of the elastic foundation, K, we plot the variations of the natural frequencies ω_{1M} and $3\omega_{1D}$ with K in Figure 12. For a numerical example we choose the case of $K = 1239 \approx K^*$ (intersection of ω_{13} and $3\omega_{11}$ in Figure 12), which gives natural frequencies $\omega_{11} = 41.7733$ and $\omega_{13} = 125.348$. Then we have an internal resonance condition $\omega_{13} \approx 3\omega_{11}$ with the internal detuning parameter $\varepsilon\sigma_1 = 0.0278555$. Considering a primary resonance $\lambda \approx \omega_{11}$ (the lower mode is directly excited) and substituting the relations of $C = D = 1$ and $M = 3$ into equations (61)–(63), we obtain a simplified system of ordinary differential equations for the nondecaying amplitudes as follows:

$$\omega_{11}\left(a'_{11} + c_{11}a_{11}\right) + \frac{1}{4}\hat{\gamma}_{1113}b_{11}a_{13}b_{13}\sin(\alpha_{11} - \beta_{11} - \alpha_{13} + \beta_{13})$$
$$- \frac{1}{8}\left(Q_{11}a_{11}b_{11}a_{13}\sin\tilde{\mu}_A + Q_{13}b_{11}^2 b_{13}\sin\tilde{\mu}_B\right) - \frac{1}{2}P_{11}\sin\mu_{11}^a = 0, \tag{69a}$$

$$\omega_{11}\left(b'_{11} + c_{11}b_{11}\right) - \frac{1}{4}\hat{\gamma}_{1113}a_{11}a_{13}b_{13}\sin(\alpha_{11} - \beta_{11} - \alpha_{13} + \beta_{13})$$
$$- \frac{1}{8}\left(Q_{11}a_{11}b_{11}b_{13}\sin\tilde{\mu}_B + Q_{13}a_{11}^2 a_{13}\sin\tilde{\mu}_A\right) - \frac{1}{2}P_{11}\sin\mu_{11}^b = 0, \tag{69b}$$

$$\omega_{11}a_{11}\alpha'_{11} + \frac{1}{8}a_{11}\{\gamma_{1111}(a_{11}^2 + 2b_{11}^2) + 2\gamma_{1113}(a_{13}^2 + b_{13}^2)\}$$
$$+ \frac{1}{4}\hat{\gamma}_{1113}b_{11}a_{13}b_{13}\cos(\alpha_{11} - \beta_{11} - \alpha_{13} + \beta_{13})$$
$$+ \frac{1}{8}(Q_{11}a_{11}b_{11}a_{13}\cos\tilde{\mu}_A + Q_{13}b_{11}^2b_{13}\cos\tilde{\mu}_B) + \frac{1}{2}P_{11}\cos\mu_{11}^a = 0, \tag{69c}$$

$$\omega_{11}b_{11}\beta'_{11} + \frac{1}{8}b_{11}\{\gamma_{1111}(2a_{11}^2 + b_{11}^2) + 2\gamma_{1113}(a_{13}^2 + b_{13}^2)\}$$
$$+ \frac{1}{4}\hat{\gamma}_{1113}a_{11}a_{13}b_{13}\cos(\alpha_{11} - \beta_{11} - \alpha_{13} + \beta_{13})$$
$$+ \frac{1}{8}(Q_{11}a_{11}b_{11}b_{13}\cos\tilde{\mu}_B + Q_{13}a_{11}^2a_{13}\cos\tilde{\mu}_A) + \frac{1}{2}P_{11}\cos\mu_{11}^b = 0, \tag{69d}$$

$$\omega_{13}(a'_{13} + c_{13}a_{13}) - \frac{1}{4}\hat{\gamma}_{1311}a_{11}b_{11}b_{13}\sin(\alpha_{11} - \beta_{11} - \alpha_{13} + \beta_{13})$$
$$+ \frac{1}{8}Q_{13}a_{11}^2b_{11}\sin\tilde{\mu}_A = 0, \tag{70a}$$

$$\omega_{13}(b'_{13} + c_{13}b_{13}) + \frac{1}{4}\hat{\gamma}_{1311}a_{11}b_{11}a_{13}\sin(\alpha_{11} - \beta_{11} - \alpha_{13} + \beta_{13})$$
$$+ \frac{1}{8}Q_{13}a_{11}b_{11}^2\sin\tilde{\mu}_B = 0, \tag{70b}$$

$$\omega_{13}a_{13}\alpha'_{13} + \frac{1}{8}a_{13}\{2\gamma_{1311}(a_{11}^2 + b_{11}^2) + \gamma_{1313}(a_{13}^2 + 2b_{13}^2)\}$$
$$+ \frac{1}{4}\hat{\gamma}_{1311}a_{11}b_{11}b_{13}\cos(\alpha_{11} - \beta_{11} - \alpha_{13} + \beta_{13})$$
$$+ \frac{1}{8}Q_{13}a_{11}^2b_{11}\cos\tilde{\mu}_A = 0, \tag{70c}$$

$$\omega_{13}b_{13}\beta'_{13} + \frac{1}{8}b_{13}\{2\gamma_{1311}(a_{11}^2 + b_{11}^2) + \gamma_{1313}(2a_{13}^2 + b_{11}^2)\}$$
$$+ \frac{1}{4}\hat{\gamma}_{1311}a_{11}b_{11}a_{13}\cos(\alpha_{11} - \beta_{11} - \alpha_{13} + \beta_{13})$$
$$+ \frac{1}{8}Q_{13}a_{11}b_{11}^2\cos\tilde{\mu}_B = 0, \tag{70d}$$

$$\mu_{11}^a = \sigma_2T_1 + \tau_{11} - \alpha_{11}, \qquad \mu_{11}^b = \sigma_2T_1 - \tau_{11} - \beta_{11}, \tag{71a, b}$$

$$\tilde{\mu}_A = \sigma_1T_1 - 2\alpha_{11} - \beta_{11} + \alpha_{13}, \tag{71c}$$

$$\tilde{\mu}_B = \sigma_1T_1 - \alpha_{11} - 2\beta_{11} + \beta_{13}. \tag{71d}$$

Each equilibrium solution of a system of autonomous ordinary differential equations to be obtained from system (69) and (70) is corresponding to a steady-state response ($a'_{11} = b'_{11} = a'_{13} = b'_{13} = \mu^{a\prime}_{11} = \mu^{b\prime}_{11} = \tilde{\mu}'_A = \tilde{\mu}'_B = 0$).

In Figure 13 the amplitudes a_{11}, b_{11}, a_{13} and b_{13} are plotted as functions of the external detuning parameter $\varepsilon\sigma_2 = \hat{\sigma}_2$ when $\{\nu, \varepsilon, \varepsilon c, \varepsilon P_{11}, \tau_{11}\} =$

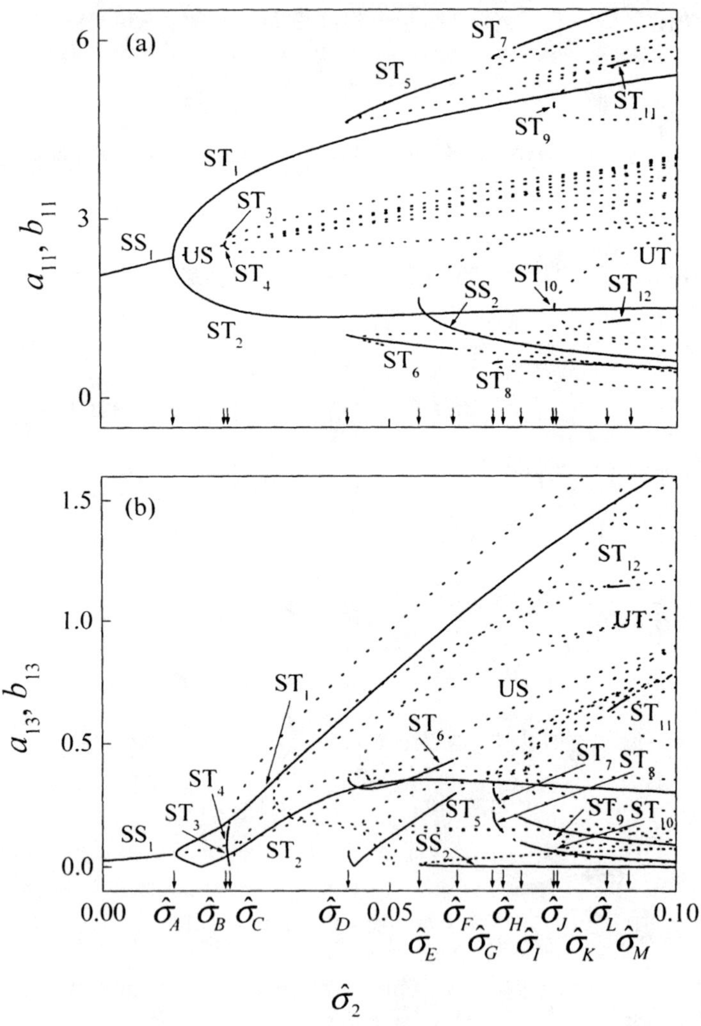

Figure 13. Variations of the amplitudes with detuning parameter $\hat{\sigma}_2$ when $\varepsilon P_{11} = 5$ (——, stable; - - - - -, unstable).

$\{1/3, 0.001, 0.001, 5.0, 0.0\}$. Solid and dotted lines denote, respectively, stable and unstable responses. The abbreviations SS, US, ST and UT denote, respectively, stable standing, unstable standing, stable traveling and unstable traveling wave components.

The response curves are shown to have two pitchfork, ten saddle-node and seven Hopf bifurcation points. At one of two pitchfork bifurcation points, $\hat{\sigma}_A$ (0.012797), the stable standing wave component SS_1 bifurcates into two stable traveling wave components ST_1 and ST_2, and one unstable standing wave component US. At the other pitchfork bifurcation point, $\hat{\sigma}_B$ (0.021699), the unstable standing wave component US appears to bifurcate into two stable traveling wave components ST_3 and ST_4, and one unstable standing wave component US. In fact, at $\hat{\sigma}_2 = 0.021675$ not identified in the figure, the unstable component US becomes a stable component, which bifurcates into ST_3 and ST_4. At four saddle-node bifurcation points out of 10, $\hat{\sigma}_D$ (0.042807), $\hat{\sigma}_E$ (0.055153), $\hat{\sigma}_G$ (0.067977) and $\hat{\sigma}_J$ (0.078646), one stable response and one unstable response generate. At the other saddle-node bifurcation points (not marked in the figure), two unstable responses generate.

It is observed that only traveling wave components experience Hopf bifurcations. The components change their stability at seven Hopf bifurcation points $\hat{\sigma}_C$ (0.022026), $\hat{\sigma}_F$ (0.060973), $\hat{\sigma}_H$ (0.067977), $\hat{\sigma}_I$ (0.072787), $\hat{\sigma}_K$ (0.078647), $\hat{\sigma}_L$ (0.088036) and $\hat{\sigma}_M$ (0.091731). For instance, when $\hat{\sigma}_J < \hat{\sigma}_2 < \hat{\sigma}_K$ and $\hat{\sigma}_L < \hat{\sigma}_2 < \hat{\sigma}_M$, there exist seven stable steady-state responses, which consist of six traveling wave components and one standing wave component. When $\hat{\sigma}_E < \hat{\sigma}_2 < \hat{\sigma}_F$, $\hat{\sigma}_G < \hat{\sigma}_2 < \hat{\sigma}_H$, $\hat{\sigma}_I < \hat{\sigma}_2 < \hat{\sigma}_J$, $\hat{\sigma}_K < \hat{\sigma}_2 < \hat{\sigma}_L$ and $\hat{\sigma}_2 > \hat{\sigma}_M$, there exist five stable steady-state responses, which consist of four traveling wave components and one standing wave component. When $\hat{\sigma}_B < \hat{\sigma}_2 < \hat{\sigma}_C$ and $\hat{\sigma}_D < \hat{\sigma}_2 < \hat{\sigma}_E$, there exist four stable steady-state responses, which are traveling wave components. When $\hat{\sigma}_F < \hat{\sigma}_2 < \hat{\sigma}_G$ and $\hat{\sigma}_H < \hat{\sigma}_2 < \hat{\sigma}_I$, there exist three stable steady-state responses, which consist of two traveling wave components and one standing wave component. When $\hat{\sigma}_F < \hat{\sigma}_2$, $\hat{\sigma}_H < \hat{\sigma}_2 < \hat{\sigma}_I$ and $\hat{\sigma}_K < \hat{\sigma}_2$, two pairs of traveling wave components, $\{ST_5, ST_6\}$, $\{ST_7, ST_8\}$ and $\{ST_9, ST_{10}\}$, respectively, become unstable. Another pair of stable components $\{ST_{11}, ST_{12}\}$ loses stability at $\hat{\sigma}_2 = \hat{\sigma}_L$ and $\hat{\sigma}_M$. These instabilities imply that there may exist quasiperiodic or chaotic responses generated from these pairs. Exploring this type of response, however, is beyond the scope of this work.

Conclusively, nonvanishing amplitudes of indirectly excited modes (a_{13} and b_{13}) tell us modal interactions between lower (ω_{11}) and higher (ω_{13}) modes. The characteristics of the responses are much more complicated than the three-to-one case ($N = 3C$) in Section 4.2.

Appendix A

Coefficients S_j and frequency combinations Λ_j in equations (29) are given in Table A.1.

$$\begin{aligned}\gamma_{klnm} &= \Gamma(kl, kl, nm, -nm) \\ &\quad + \Gamma(kl, -nm, kl, nm) + \Gamma(kl, nm, -nm, kl), \\ \hat{\gamma}_{klkm} &= \Gamma(kl, km, km, -kl) + \Gamma(kl, -kl, km, km) + \Gamma(kl, km, -kl, km).\end{aligned}$$

The steady-state response to the first-order approximation of equation (40) can be expressed as follows:

$$\begin{aligned}w &= \phi_{11}\{a_{11}\cos(\lambda t - \mu^a_{11} + \theta + \tau_{11}) \\ &\quad + b_{11}\cos(\lambda t - \mu^b_{11} - \theta - \tau_{11})\} + \mathrm{O}(\varepsilon) \\ &= \phi_{11}\{X_1\cos(\lambda t - \chi_1)\cos(\theta + \tau_{11}) \\ &\quad + X_2\cos(\lambda t - \chi_2)\sin(\theta + \tau_{11})\} + \mathrm{O}(\varepsilon) \qquad \text{(A.1)} \\ &= q_1(t)\phi_{11}\cos\theta + q_2(t)\phi_{11}\sin\theta + \mathrm{O}(\varepsilon), \qquad \text{(A.2)}\end{aligned}$$

where

$$\begin{aligned}q_1(t) &= Y_1\cos\lambda t + Y_2\sin\lambda t = Z_1\cos(\lambda t - \zeta_1), \\ q_2(t) &= Y_3\cos\lambda t + Y_4\sin\lambda t = Z_2\cos(\lambda t - \zeta_2), \\ X_1 &= \sqrt{a^2_{11} + b^2_{11} + 2a_{11}b_{11}\cos(\mu^a_{11} - \mu^b_{11})}, \\ \tan\chi_1 &= \frac{a\sin\mu^a_{11} + b\sin\mu^b_{11}}{a\cos\mu^a_{11} + b\cos\mu^b_{11}}, \\ X_2 &= \sqrt{a^2_{11} + b^2_{11} - 2a_{11}b_{11}\cos(\mu^a_{11} - \mu^b_{11})},\end{aligned}$$

Table A.1

j	S_j	Λ_j
1	$A_{cd}A_{nm}A_{pq}$	$\omega_{cd} + \omega_{nm} + \omega_{pq}$
2	$A_{cd}A_{nm}\overline{B}_{pq}$	$\omega_{cd} + \omega_{nm} - \omega_{pq}$
3	$A_{cd}\overline{B}_{nm}A_{pq}$	$\omega_{cd} - \omega_{nm} + \omega_{pq}$
4	$\overline{B}_{cd}A_{nm}A_{pq}$	$-\omega_{cd} + \omega_{nm} + \omega_{pq}$
5	$\overline{B}_{cd}\overline{B}_{nm}\overline{B}_{pq}$	$-\omega_{cd} - \omega_{nm} - \omega_{pq}$
6	$\overline{B}_{cd}\overline{B}_{nm}A_{pq}$	$-\omega_{cd} - \omega_{nm} + \omega_{pq}$
7	$\overline{B}_{cd}A_{nm}\overline{B}_{pq}$	$-\omega_{cd} + \omega_{nm} - \omega_{pq}$
8	$A_{cd}\overline{B}_{nm}\overline{B}_{pq}$	$\omega_{cd} - \omega_{nm} - \omega_{pq}$

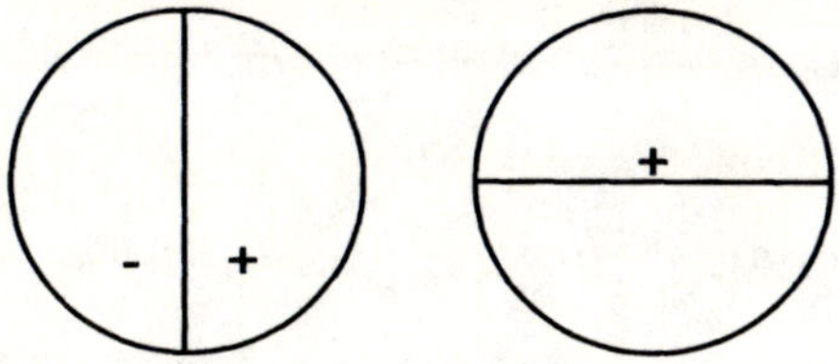

Figure A.1. Two mode shapes corresponding to ω_{11}. (a) $\phi_{11}\cos\theta$; (b) $\phi_{11}\sin\theta$.

$$\begin{aligned}
\tan\chi_2 &= \frac{a\cos\mu_{11}^a - b\cos\mu_{11}^b}{-a\sin\mu_{11}^a + b\sin\mu_{11}^b},\\
Y_1 &= a_{11}\cos(\mu_{11}^a - \tau_{11}) + b_{11}\cos(\mu_{11}^b + \tau_{11}),\\
Y_2 &= a_{11}\sin(\mu_{11}^a - \tau_{11}) + b_{11}\sin(\mu_{11}^b + \tau_{11}),\\
Y_3 &= a_{11}\sin(\mu_{11}^a - \tau_{11}) - b_{11}\sin(\mu_{11}^b + \tau_{11}),\\
Y_4 &= -a_{11}\cos(\mu_{11}^a - \tau_{11}) + b_{11}\cos(\mu_{11}^b + \tau_{11}),\\
Z_1 &= \sqrt{Y_1^2 + Y_2^2} = \sqrt{a_{11}^2 + b_{11}^2 + 2a_{11}b_{11}\cos(\mu_{11}^a - \mu_{11}^b - 2\tau_{11})},\\
Z_2 &= \sqrt{Y_3^2 + Y_4^2} = \sqrt{a_{11}^2 + b_{11}^2 - 2a_{11}b_{11}\cos(\mu_{11}^a - \mu_{11}^b - 2\tau_{11})},\\
\tan\zeta_1 &= \frac{Y_2}{Y_1}, \qquad \tan\zeta_2 = \frac{Y_4}{Y_3}.
\end{aligned}$$

Equation (A.2) is a superposition of two modes $\phi_{11}\cos\theta$ and $\phi_{11}\sin\theta$ shown in Figure A.1, corresponding to degenerate natural frequency ω_{11}.

Appendix B

In equations (45),

$$Q_{NM} = \Gamma(NM, CD, CD, CD), \tag{B.1}$$

$$\begin{aligned}
Q_{CD} &= \Gamma(CD, -CD, -CD, NM) + \Gamma(CD, -CD, NM, -CD)\\
&\quad + \Gamma(CD, NM, -CD, -CD).
\end{aligned} \tag{B.2}$$

Case 1: $\omega_{32} \approx 3\omega_{11}$ and $\lambda \approx \omega_{11}$

$$\omega_{11}a_{11}' = -\omega_{11}c_{11}a_{11} + \frac{1}{4}b_{11}\hat{s}_{11}^s + \frac{1}{8}Q_{11}a_{11}^2a_{32}\sin\tilde{\mu}_A + \frac{1}{2}P_{11}\sin\mu_{11}^a, \tag{B.3}$$

$$\omega_{11}b'_{11} = -\omega_{11}c_{11}b_{11} - \frac{1}{4}a_{11}\hat{s}^s_{11} + \frac{1}{8}Q_{11}b^2_{11}b_{32}\sin\tilde{\mu}_B + \frac{1}{2}P_{11}\sin\mu^b_{11}, \tag{B.4}$$

$$\omega_{32}a'_{32} = -\omega_{32}c_{32}a_{32} + \frac{1}{4}b_{32}\hat{s}^s_{32} - \frac{1}{8}Q_{32}a^3_{11}\sin\tilde{\mu}_A, \tag{B.5}$$

$$\omega_{32}b'_{32} = -\omega_{32}c_{32}b_{32} - \frac{1}{4}a_{32}\hat{s}^s_{32} - \frac{1}{8}Q_{32}b^3_{11}\sin\tilde{\mu}_B, \tag{B.6}$$

$$\begin{aligned}\omega_{11}a_{11}\mu^{a\prime}_{11} &= \omega_{11}\sigma_2 a_{11} + \frac{1}{8}a_{11}(s_{11} - \gamma_{1111}a^2_{11}) + \frac{1}{4}b_{11}\hat{s}^c_{11}\\ &\quad + \frac{1}{8}Q_{11}a^2_{11}a_{32}\cos\tilde{\mu}_A + \frac{1}{2}P_{11}\cos\mu^a_{11},\end{aligned} \tag{B.7}$$

$$\begin{aligned}\omega_{11}b_{11}\mu^{b\prime}_{11} &= \omega_{11}\sigma_2 b_{11} + \frac{1}{8}b_{11}(s_{11} - \gamma_{1111}b^2_{11}) + \frac{1}{4}a_{11}\hat{s}^c_{11}\\ &\quad + \frac{1}{8}Q_{11}b^2_{11}b_{32}\cos\tilde{\mu}_B + \frac{1}{2}P_{11}\cos\mu^b_{11},\end{aligned} \tag{B.8}$$

$$\begin{aligned}&\omega_{11}\omega_{32}a_{11}a_{32}\tilde{\mu}'_A\\ &\quad = \omega_{11}\omega_{32}\sigma_1 a_{11}a_{32} + \frac{3}{8}\omega_{32}a_{11}a_{32}(s_{11} - \gamma_{1111}a^2_{11})\\ &\qquad - \frac{1}{8}\omega_{11}a_{11}a_{32}(s_{32} - \gamma_{3232}a^2_{32}) + \frac{1}{4}(3\omega_{32}b_{11}a_{32}\hat{s}^c_{11} - \omega_{11}a_{11}b_{32}\hat{s}^c_{32})\\ &\qquad + \frac{1}{8}a^2_{11}(3\omega_{32}Q_{11}a^2_{32} - \omega_{11}Q_{32}a^2_{11})\cos\tilde{\mu}_A + \frac{3}{2}\omega_{32}P_{11}a_{32}\cos\mu^a_{11},\end{aligned} \tag{B.9}$$

$$\begin{aligned}&\omega_{11}\omega_{32}b_{11}b_{32}\tilde{\mu}'_B\\ &\quad = \omega_{11}\omega_{32}\sigma_1 b_{11}b_{32} + \frac{3}{8}\omega_{32}b_{11}b_{32}(s_{11} - \gamma_{1111}b^2_{11})\\ &\qquad - \frac{1}{8}\omega_{11}b_{11}b_{32}(s_{32} - \gamma_{3232}b^2_{32}) + \frac{1}{4}(3\omega_{32}a_{11}b_{32}\hat{s}^c_{11} - \omega_{11}b_{11}a_{32}\hat{s}^c_{32})\\ &\qquad + \frac{1}{8}b^2_{11}(3\omega_{32}Q_{11}b^2_{32} - \omega_{11}Q_{32}b^2_{11})\cos\tilde{\mu}_B + \frac{3}{2}\omega_{32}P_{11}b_{32}\cos\mu^b_{11}.\end{aligned} \tag{B.10}$$

When $k \neq 1, 3$ and $l \neq 1, 2$,

$$\omega_{kl}a'_{kl} = -\omega_{kl}c_{kl}a_{kl} + \frac{1}{4}(1 - \delta_{k0})b_{kl}\hat{s}^s_{kl}, \tag{B.11}$$

$$\omega_{kl}b'_{kl} = -\omega_{kl}c_{kl}b_{kl} - \frac{1}{4}(1 - \delta_{k0})a_{kl}\hat{s}^s_{kl}, \tag{B.12}$$

$$\omega_{kl}a_{kl}\alpha'_{kl} = -\frac{1}{8}a_{kl}(s_{kl} - \gamma_{klkl}a^2_{kl}) - \frac{1}{4}(1 - \delta_{k0})b_{kl}\hat{s}^c_{kl}, \tag{B.13}$$

$$\omega_{kl}b_{kl}\beta'_{kl} = -\frac{1}{8}b_{kl}(s_{kl} - \gamma_{klkl}b^2_{kl}) - \frac{1}{4}(1 - \delta_{k0})a_{kl}\hat{s}^c_{kl}. \tag{B.14}$$

In these equations,

$$s_{11} = \sum_{n=-\infty}^{\infty} \sum_{m=1}^{\infty} \gamma_{11nm} \left(a_{nm}^2 + b_{nm}^2 \right), \tag{B.15}$$

$$s_{32} = \sum_{n=-\infty}^{\infty} \sum_{m=1}^{\infty} \gamma_{32nm} \left(a_{nm}^2 + b_{nm}^2 \right), \tag{B.16}$$

$$\hat{s}_{11}^s = \sum_{m=2}^{\infty} \hat{\gamma}_{111m} a_{1m} b_{1m} \sin\left(\alpha_{1m} - \beta_{1m} + \mu_{11}^a - \mu_{11}^b - 2\tau_{11} \right), \tag{B.17}$$

$$\hat{s}_{11}^s = \sum_{m=2}^{\infty} \hat{\gamma}_{111m} a_{1m} b_{1m} \cos\left(\alpha_{1m} - \beta_{1m} + \mu_{11}^a - \mu_{11}^b - 2\tau_{11} \right), \tag{B.18}$$

$$\hat{s}_{32}^s = \sum_{\substack{m=1 \\ (m \neq 2)}}^{\infty} \hat{\gamma}_{323m} a_{3m} b_{3m} \times \sin\left(\alpha_{3m} - \beta_{3m} - \tilde{\mu}_A + \tilde{\mu}_B + 3\mu_{11}^a - 3\mu_{11}^b - 6\tau_{11} \right), \tag{B.19}$$

$$\hat{s}_{32}^c = \sum_{\substack{m=1 \\ (m \neq 2)}}^{\infty} \hat{\gamma}_{323m} a_{3m} b_{3m} \times \cos\left(\alpha_{3m} - \beta_{3m} - \tilde{\mu}_A + \tilde{\mu}_B + 3\mu_{11}^a - 3\mu_{11}^b - 6\tau_{11} \right), \tag{B.20}$$

$$s_{kl} = \sum_{n=-\infty}^{\infty} \sum_{m=1}^{\infty} \gamma_{klnm} \left(a_{nm}^2 + b_{nm}^2 \right). \tag{B.21}$$

When $k \neq 1, 3$ and $l \neq 1, 2$,

$$\hat{s}_{kl}^s = \sum_{m=1}^{\infty} \hat{\gamma}_{klkm} a_{km} b_{km} \sin(\alpha_{km} - \beta_{km} - \alpha_{kl} + \beta_{kl}), \tag{B.22}$$

$$\hat{s}_{kl}^c = \sum_{m=1}^{\infty} (1 - \delta_{ml}) \hat{\gamma}_{klkm} a_{km} b_{km} \cos(\alpha_{km} - \beta_{km} - \alpha_{kl} + \beta_{kl}). \tag{B.23}$$

Case 2: $\omega_{32} \approx 3\omega_{11}$ and $\lambda \approx \omega_{32}$

$$\omega_{11} a_{11}' = -\omega_{11} c_{11} a_{11} + \frac{1}{4} b_{11} \hat{s}_{11}^s + \frac{1}{8} Q_{11} a_{11}^2 a_{32} \sin \tilde{\mu}_A, \tag{B.24}$$

$$\omega_{11} b_{11}' = -\omega_{11} c_{11} b_{11} - \frac{1}{4} a_{11} \hat{s}_{11}^s + \frac{1}{8} Q_{11} b_{11}^2 b_{32} \sin \tilde{\mu}_B, \tag{B.25}$$

$$\omega_{32}a'_{32} = -\omega_{32}c_{32}a_{32} + \frac{1}{4}b_{32}\hat{s}^s_{32} - \frac{1}{8}Q_{32}a^3_{11}\sin\tilde{\mu}_A + \frac{1}{2}P_{32}\sin\mu^a_{32}, \tag{B.26}$$

$$\omega_{32}b'_{32} = -\omega_{32}c_{32}b_{32} - \frac{1}{4}a_{32}\hat{s}^s_{32} - \frac{1}{8}Q_{32}b^3_{11}\sin\tilde{\mu}_B + \frac{1}{2}P_{32}\sin\mu^b_{32}, \tag{B.27}$$

$$\omega_{32}a_{32}\mu^{a\prime}_{32} = \omega_{32}\sigma_2 a_{32} + \frac{1}{8}a_{32}(s_{32} - \gamma_{3232}a^2_{32}) + \frac{1}{4}b_{32}\hat{s}^c_{32} + \frac{1}{8}Q_{32}a^3_{11}\cos\tilde{\mu}_A + \frac{1}{2}P_{32}\cos\mu^a_{32}, \tag{B.28}$$

$$\omega_{32}b_{32}\mu^{b\prime}_{32} = \omega_{32}\sigma_2 b_{32} + \frac{1}{8}b_{32}(s_{32} - \gamma_{3232}b^2_{32}) + \frac{1}{4}a_{32}\hat{s}^c_{32} + \frac{1}{8}Q_{32}b^3_{11}\cos\tilde{\mu}_B + \frac{1}{2}P_{32}\cos\mu^b_{32}, \tag{B.29}$$

$$\begin{aligned}\omega_{11}\omega_{32}a_{11}a_{32}\tilde{\mu}'_A &= \omega_{11}\omega_{32}\sigma_1 a_{11}a_{32} + \frac{3}{8}\omega_{32}a_{11}a_{32}(s_{11} - \gamma_{1111}a^2_{11}) \\ &\quad - \frac{1}{8}\omega_{11}a_{11}a_{32}(s_{32} - \gamma_{3232}a^2_{32}) + \frac{1}{4}(3\omega_{32}b_{11}a_{32}\hat{s}^c_{11} - \omega_{11}a_{11}b_{32}\hat{s}^c_{32}) \\ &\quad + \frac{1}{8}a^2_{11}(3\omega_{32}Q_{11}a^2_{32} - \omega_{11}Q_{32}a^2_{11})\cos\tilde{\mu}_A - \frac{1}{2}\omega_{11}P_{32}a_{11}\cos\mu^a_{32},\end{aligned} \tag{B.30}$$

$$\begin{aligned}\omega_{11}\omega_{32}b_{11}b_{32}\tilde{\mu}'_B &= \omega_{11}\omega_{32}\sigma_1 b_{11}b_{32} + \frac{3}{8}\omega_{32}b_{11}b_{32}(s_{11} - \gamma_{1111}b^2_{11}) \\ &\quad - \frac{1}{8}\omega_{11}b_{11}b_{32}(s_{32} - \gamma_{3232}b^2_{32}) + \frac{1}{4}(3\omega_{32}a_{11}b_{32}\hat{s}^c_{11} - \omega_{11}b_{11}a_{32}\hat{s}^c_{32}) \\ &\quad + \frac{1}{8}b^2_{11}(3\omega_{32}Q_{11}b^2_{32} - \omega_{11}Q_{32}b^2_{11})\cos\tilde{\mu}_B - \frac{1}{2}\omega_{11}P_{32}b_{11}\cos\mu^b_{32}.\end{aligned} \tag{B.31}$$

When $k \neq 1, 3$ and $l \neq 1, 2$,

$$\omega_{kl}a'_{kl} = -\omega_{kl}c_{kl}a_{kl} + \frac{1}{4}(1-\delta_{k0})b_{kl}\hat{s}^s_{kl}, \tag{B.32}$$

$$\omega_{kl}b'_{kl} = -\omega_{kl}c_{kl}b_{kl} - \frac{1}{4}(1-\delta_{k0})a_{kl}\hat{s}^s_{kl}, \tag{B.33}$$

$$\omega_{kl}a_{kl}\alpha'_{kl} = -\frac{1}{8}a_{kl}(s_{kl} - \gamma_{klkl}a^2_{kl}) - \frac{1}{4}(1-\delta_{k0})b_{kl}\hat{s}^c_{kl}, \tag{B.34}$$

$$\omega_{kl} b_{kl} \beta'_{kl} = -\frac{1}{8} b_{kl} \left(s_{kl} - \gamma_{klkl} b_{kl}^2 \right) - \frac{1}{4} (1 - \delta_{k0}) a_{kl} \hat{s}_{kl}^c. \tag{B.35}$$

In these equations,

$$s_{11} = \sum_{n=-\infty}^{\infty} \sum_{m=1}^{\infty} \gamma_{11nm} \left(a_{nm}^2 + b_{nm}^2 \right), \tag{B.36}$$

$$s_{32} = \sum_{n=-\infty}^{\infty} \sum_{m=1}^{\infty} \gamma_{32nm} \left(a_{nm}^2 + b_{nm}^2 \right), \tag{B.37}$$

$$\hat{s}_{11}^s = \sum_{m=2}^{\infty} \hat{\gamma}_{111m} a_{1m} b_{1m} \times \sin\left(\alpha_{1m} - \beta_{1m} + \frac{1}{3}\tilde{\mu}_A - \frac{1}{3}\tilde{\mu}_B + \frac{1}{3}\mu_{32}^a - \frac{1}{3}\mu_{32}^b - \frac{2}{3}\tau_{32} \right), \tag{B.38}$$

$$\hat{s}_{11}^c = \sum_{m=2}^{\infty} \hat{\gamma}_{111m} a_{1m} b_{1m} \times \cos\left(\alpha_{1m} - \beta_{1m} + \frac{1}{3}\tilde{\mu}_A - \frac{1}{3}\tilde{\mu}_B + \frac{1}{3}\mu_{32}^a - \frac{1}{3}\mu_{32}^b - \frac{2}{3}\tau_{32} \right), \tag{B.39}$$

$$\hat{s}_{32}^s = \sum_{\substack{m=1 \\ (m \neq 2)}}^{\infty} \hat{\gamma}_{323m} a_{3m} b_{3m} \sin\left(\alpha_{3m} - \beta_{3m} + \mu_{32}^a - \mu_{32}^b - 2\tau_{32} \right), \tag{B.40}$$

$$\hat{s}_{32}^c = \sum_{\substack{m=1 \\ (m \neq 2)}}^{\infty} \hat{\gamma}_{323m} a_{3m} b_{3m} \cos\left(\alpha_{3m} - \beta_{3m} + \mu_{32}^a - \mu_{32}^b - 2\tau_{32} \right). \tag{B.41}$$

When $k \neq 1, 3$ and $l \neq 1, 2$,

$$s_{kl} = \sum_{n=-\infty}^{\infty} \sum_{m=1}^{\infty} \gamma_{klnm} \left(a_{nm}^2 + b_{nm}^2 \right), \tag{B.42}$$

$$\hat{s}_{kl}^s = \sum_{m=1}^{\infty} \hat{\gamma}_{klkm} a_{km} b_{km} \sin(\alpha_{km} - \beta_{km} - \alpha_{kl} + \beta_{kl}), \tag{B.43}$$

$$\hat{s}_{kl}^c = \sum_{m=1}^{\infty} (1 - \delta_{ml}) \hat{\gamma}_{klkm} a_{km} b_{km} \cos(\alpha_{km} - \beta_{km} - \alpha_{kl} + \beta_{kl}). \tag{B.44}$$

Appendix C

In equation (58),

$$Q_{CM} = 2\Gamma(CM, CD, CD, -CD) + \Gamma(CM, -CD, CD, CD), \tag{C.1}$$

$$Q_{CD} = 2\{\Gamma(CD, -CD, -CD, NM) + \Gamma(CD, -CD, NM, -CD) + \Gamma(CD, NM, -CD, -CD)\}. \tag{C.2}$$

References

Doedel, E., 1986. AUTO: Software for Continuation and Bifurcation Problems in Ordinary Differential Equations. California Institute of Technology, Pasadena.

Efstathiades, G.J., 1971. A new approach to the large-deflection vibrations of imperfect circular disks using Galerkin's procedure. J. Sound Vib. 16, 231–253.

Ghosh, A.K., 1997. Axisymmetric dynamic response of a circular plate on an elastic foundation. J. Sound Vib. 205, 112–120.

Hadian, J., Nayfeh, A.H., 1990. Modal interaction in circular plates. J. Sound Vib. 142, 279–292.

Lee, W.K., Kim, C.H., 1995. Combination resonances of a circular plate with three-mode interaction. Trans. ASME J. Appl. Mech. 62, 1015–1022.

Lee, W.K., Yeo, M.H., 2003. Non-linear interactions in asymmetric vibrations of a circular plate. J. Sound Vib. 263, 1017–1030.

Lee, W.K., Yeo, M.H., Samoilenko, S.B., 2003. The effect of the number of nodal diameters on non-linear interactions in two asymmetric vibration modes of a circular plate. J. Sound Vib. 268, 1013–1023.

Leissa, A.W., 1969. Vibration of Plates. U.S. Government Printing Office, Washington, DC. NASA-SP-160.

Nayfeh, A.H., 2000. Nonlinear Interactions. Wiley, New York.

Nayfeh, A.H., Mook, D.T., 1979. Nonlinear Oscillations. Wiley, New York.

Nayfeh, T.A., Vakakis, A.F., 1994. Subharmonic traveling waves in a geometrically non-linear circular plate. Internat. J. Non-Linear Mech. 29, 233–245.

Sridhar, S., Mook, D.T., Nayfeh, A.H., 1975. Non-linear resonances in the forced responses of plates, Part I: Symmetric responses of circular plates. J. Sound Vib. 41, 359–373.

Sridhar, S., Mook, D.T., Nayfeh, A.H., 1978. Non-linear resonances in the forced responses of plates, Part II: Asymmetric responses of circular plates. J. Sound Vib. 59, 159–170.

Yeo, M.H., Lee, W.K., 2002. Corrected solvability conditions for non-linear asymmetric vibrations of a circular plate. J. Sound Vib. 257, 653–665.

List of Contributors

A. Dávid, *szandra@comcast.net* (Ch. 4)

L. Hong, *Department of Mechanical Engineering, University of Delaware, Newark, DE 19716, USA* (Ch. 3)

W.K. Lee, *School of Mechanical Engineering, Yeungnam University, Gyongsan 712-749, Republic of Korea; wklee@yu.ac.kr* (Ch. 5)

A.C.J. Luo, *Department of Mechanical and Industrial Engineering, Southern Illinois University, Edwardsville, IL 62026, USA; aluo@siue.edu* (Ch. 2)

S.C. Sinha, *Nonlinear Systems Research Laboratory, Department of Mechanical Engineering, Auburn University, 202 Ross Hall, Auburn, AL 36849, USA; ssinha@eng.auburn.edu* (Ch. 4)

J.-Q. Sun, *Department of Mechanical Engineering, University of Delaware, Newark, DE 19716, USA; jqsun@udel.edu* (Ch. 3)

P. Yu, *Department of Applied Mathematics, The University of Western Ontario, London, Ontario, N6A 5B7, Canada; pyu@uwo.ca* (Ch. 1)

Author Index

Roman numbers refer to pages on which the author (or his/her work) is mentioned. Italic numbers refer to reference pages.

Abed, E.G., 79–81, *124*, 259, *275*
Abed, E.H., 24, *121*, 335, *335*
Abraham, R.H., 198, *274*
Aracil, J., 230, *275*
Arnold, L., 214, *274*
Arnold, V.I., 2, *121*, 284, 288, 294, *335*
Arrowsmith, D.K., 284, 285, 312, 313, *335*
Asmis, K.G., 317, *338*

Bajaj, A.K., 281, *338*
Balachandran, B., 259, *277*, 284, 288, *337*
Bapat, C.N., 128, *189*
Bautin, N.N., 38, 41, *121*
Baxendale, P., 214, *274*
Bellman, R., 283, *335*
Berns, D., 37, *121*
Bernussou, J., 284, *335*
Bi, Q., 3, 97, 99, *125*
Bibb, J.S., 285, 290, *338*
Birkhoff, G.D., 127, *189*, 284, *335*
Bleich, M.E., 24, *121*
Blows, T.R., 41, *123*
Blüel, R., 205, *277*
Bogoliubov, N.N., 283, *336*
Bolotin, V.V., 280, *336*
Brambilla, M., 212, *274*
Brandt, M.E., 24, *121*
Broucke, M., 128, *189*
Budd, C.J., 128, *189*
Bunin, A., 74, *123*
Bursal, F.H., 260, *274*
Butcher, E.A., 287, 297, *336*, *337*

Carroll, T.L., 73, *124*
Caughey, T.K., 127, 128, *190*, 214, *274*
Chajes, M., 259, *274*
Champney, A.R., 128, *189*
Chan, H.S.Y., 50, 53, *123*
Chen, G.R., 2, 4, 24–30, 32, 37, 73, 75–80, *121–125*, 259, *274*
Chen, H.B., 44, 46, 47, *122*
Chen, L.D., 129, 156, *190*
Chen, R.C., 335, *335*
Chen, Y.Y., 193, *274*
Chen, Z., 25, *122*
Cheng, A., 259, *274*
Chiang, W.L., 212, 214, *274*, *275*
Chin, C.M., 37, *124*
Chiu, H.M., 193, 215, 221, *274*, *276*
Choi, H.T., 193, *275*
Chow, S.-N., 285, 296, 312, *336*
Chua, L.O., 105, *122*
Chung, K.L., 197, *274*
Chung, K.W., 50, 53, *123*
Clairaut, A., 282, *336*
Coalson, R.D., 281, *336*
Comer, D.J., 193, *277*
Corless, R., 40, *125*
Crasemann, B., 281, *337*
Crauel, H., 214, *275*
Crespo, L.G., 193, 258, 260, 261, *275*
Crutchfield, J.P., 214, 215, 217, *275*
Cuesta, F., 230, *275*
Cuomo, K.M., 73, *122*

Danby, J.M.A., 281, *336*
Das, D., 230, *277*
Dávid, A., 286, 318, 321, 324–326, 330–332, 335, *336*, *337*
Davis, S.H., 281, *337*
Davison, M., 5, 92, *125*
Deb, S., 25, *122*
Den Hartog, J.P., 128, *189*

Deng, W., 128, *189*
di Bernardo, M., 128, *189*
Doedel, E., 357, *376*
Doi, S., 230, *275*
Dong, W.M., 212, 214, *274*, *275*
Dong, X., 2, 4, 73, *122*
Douay, A.C., 258, *275*

Edwards, D., 193, *275*
Efstathiades, G.J., 341, *376*
Endo, T., 258, *277*
Essex, C., 5, 37, 92, 113, *124*, *125*
Evan-Ivanowski, R.M., 280, *336*

Fang, T., 194, 226–229, *278*
Farmer, J.D., 214, 215, 217, *275*
Farooq, A., 280, *336*
Feigenbaum, M.J., 215, *275*
Ferraris, G., 280, *337*
Field, R.J., 104, *122*
Fillippov, A.F., 128, 132, *189*
Flandoli, F., 214, *275*
Flashner, H., 284, *336*
Floquét, G., 281, 287, *336*
Flowers, G.T., 280, *338*
Foale, S., 193, *275*
Friedman, Y., 232, *275*
Friedmann, P.P., 280, *336*
Fu, J.H., 24, *121*, 335, *335*

Gallas, J.A.C., 211, *275*
Gegg, B.C., 129, 158, 161, 163, 164, 186, *190*
Genesio, R., 259, *275*
Ghosh, A.K., 341, *376*
Glass, L., 281, *336*
Goldberger, A.L., 281, *337*
Gong, P.L., 241, *275*
Gorman, M., 281, *338*
Grassia, P.G., 281, *336*
Grebogi, C., 4, 73, *124*, 198, 202, 205, 211, 242, 259, *275*, *277*, *278*
Griffith, J.S., 262, *276*
Gua, S.M., 73, *122*
Guckenheimer, J., 2–4, 7, 82, *122*, 192, 245, *276*
Guder, R., 193, *276*
Guevara, M.R., 281, *336*
Gupta, P.B.D., 230, *277*
Guttalu, R.S., 284, *336*

Hackl, K., 259, *274*
Hadian, J., 340, *376*
Hagood, N.W., 258, *275*
Haken, H., 92, *122*
Hale, J., 12, *122*, 288, *336*
Hale, J.K., 192, *276*
Han, M., 38, 39, 42, 48, *122*, *125*
Han, R.P.S., 127, 128, *189*, *190*
Harb, A.M., 37, *124*
Hardy, Y., 103, *122*
Hassard, B.D., 2, 24, 30, *122*
Hawthorne, E.I., 281, *336*
He, Q., 194, 226–229, *278*
Hilbert, D., 38, 57, *122*
Hill, G.W., 281, *336*
Hirschfelder, J.O., 281, *336*
Holmes, P.J., 2–4, 7, 82, *122*, 128, *190*, 192, 245, *276*
Homsy, G.M., 280, 281, *336*
Hong, L., 194, 211, 241, *276*
Hopf, E., 4, *122*
Hou, L., 128, *190*
Hrinca, I., 258, *276*
Hsu, C.S., 193, 194, 196, 215, 217, 221, 226, 231–236, 260, *274*, *276*, *278*, 280, 284, *336*
Huang, Q., 50, 53, *123*
Huang, W.T., 44, 46, *123*
Huberman, B.A., 214, 215, 217, *275*
Hunt, B., 259, *277*

Ibrahim, R.A., 214, *276*
Inoue, J., 230, *275*
Iooss, G., 288, *336*

James, E.M., 47, *122*
Jiang, J., 193, *276*
Jin, J.-D., 327, *337*
Johari, R., 25, *122*
Johnson, W., 280, *337*
Joseph, D.D., 288, *336*
Juneja, V., 283, *338*

Kalenge, M.C., 41, *123*
Kan, I., 205, *277*

Kapitaniak, T., 214, *276*
Kazarinoff, N.D., 2, 24, 30, *122*
Kelly, F.P., 24, 25, *122*
Kennedy, J., 205, *276*
Kim, C.H., 340, *376*
Kim, M.C., 193, *276*
Kleczka, M., 128, *189*
Knuth, D.E., 215, *276*
Kobs, T., 258, *276*
Koçak, H., 192, *276*, 288, *336*
Koksch, N., 74, *123*
Komuro, M., 105, *122*
Kostelich, E., 259, *277*
Kotulski, Z., 214, *276*
Kowalczyk, K., 128, *189*
Krauskopf, B., 24, *124*
Kreuzer, E., 128, *189*, 193, *276*
Krousgrill, C.M., 281, *338*
Kukles, I.S., 38, *122*
Kumagai, S., 230, *275*
Kunze, M., 128, *189*

Lagrange, J.-L., 282, *337*
Lai, Y.C., 198, 205, *277*
Lalanne, M., 280, *337*
Lee, W.K., 340, 341, *376*
Leine, R.I., 128, *189*
Leissa, A.W., 341, 351, *376*
Leonov, G.A., 74–76, *123*
Lever, P.J.A., 260, *278*
Lewis, F.L., 260, *277*
Li, C.F., 50, 53, *123*
Li, C.G., 24–27, 29, 30, 32, *123*
Li, Ch., 285, 296, 312, *336*
Li, D., 75, 76, *123*
Li, J., 38, 46, 50, *123*, *124*
Li, J.B., 38, 50–53, 56, *123*
Liao, X., 74, *125*
Liao, X.F., 24–27, 29, 30, 32, *123*
Liao, X.X., 73, 86, *123*
Librescu, L., 6, 7, 9, 12, 18, 19, *123*, *125*
Liénard, A., 3, *123*
Likins, P.W., 281, *337*
Lin, C.F., 73, *122*
Lin, Y.K., 214, *277*
Lindh, K.G., 281, *337*
Lindstedt, A., 282, *337*
Lindtner, E., 281, *337*
Ling, S.J., 73, *124*
Liu, Y., 38, 46, *123*
Liu, Y.R., 44, 46, 47, *122*, *123*
Liu, Z., 38, 50, 53, *123*
Lloyd, N.G., 41, 47, *122*, *123*
Lorenz, E.N., 4, 73, 74, *123*
Lu, J., 75, 76, *123*, 259, *274*
Lü, J.H., 73, *122*, *123*
Lugiato, L.A., 212, *274*
Lukes, D.L., 283, 314, *337*
Lunel, S., 12, *122*
Luo, A.C.J., 127–129, 133, 153, 155, 156, 158, 161, 163, 164, 186, *189*, *190*
Lyapunov, A.M., 282, *337*
Lynch, S., 5, *123*

Malkin, I.G., 285, 292, *337*
Malkin, K.E., 38, *123*
Martizez, M., 73, *124*
Marzocca, P., 6, 7, 9, 12, 18, 19, *123*, *125*
Masri, S.F., 127, 128, *190*
Mathieu, E., 281, *337*
Matsumoto, T., 105, *122*, *124*
Matsuzaki, Y., 327, *337*
Maulloo, A., 24, *122*
May, R.M., 215, *277*
McDonald, S.W., 202, *277*
McKillip, R.M., 280, *337*
Mclachlan, K., 128, *189*
Meath, W.J., 281, *338*
Melnikov, V.K., 6, 53, *124*
Menon, S., 128, *190*
Meunier, C., 214, 230, 239, *277*
Michel, A., 128, *190*
Mikina, S.J., 128, *189*
Mingori, D.L., 281, *337*
Mitropolsky, Y.A., 283, *336*
Mohler, R.H., 314, *337*
Moiola, J.L., 4, 24, 37, 79, 80, *121*, *122*
Mook, D.T., 280, 284, *337*, 340, 341, 343, 347–351, 361, *376*

Namachchivaya, N.S., 245, *277*
Narducci, L.M., 212, *274*
Natsiavas, S., 128, *190*
Nayfeh, A.H., 3, 7, 37, *124*, 259, *277*, 280, 281, 284, 288, *337*, 340, 341, 343, 347–351, 361, *376*
Nayfeh, T.A., 352, *376*
Neufeld, R.W.J., 101, 102, *124*

Nicolis, G., 92, *124*
Nigam, N.C., 214, *277*
Nordmark, A.B., 128, *189*, *190*
Noyes, R.M., 104, *122*
Nusse, H.E., 205, *277*

Onoda, J., 258, *277*
Oppenheim, A.V., 73, *122*
Ott, E., 4, 73, *124*, 198, 202, 205, 242, 259, *275*, *277*

Pandiyan, R., 285, 290, 292, *337*, *338*
Pecora, L.M., 73, *124*
Peren, J.P., 73, *124*
Perko, L.M., 6, *124*
Pesheck, E., 94, 99, *124*
Pierre, C., 94, 99, *124*
Place, C.M., 284, 285, 312, 313, *335*
Poincaré, H., 3, *124*, 127, *190*, 281, *337*
Ponce, E., 230, *275*
Popp, K., 128, *190*
Popplewell, N., 128, *189*
Powell, J.L., 281, *337*
Prigogine, I., 92, *124*
Pugh, C., 128, *189*

Rand, D., 245, *276*
Reitmann, V., 74, *123*
Richards, J.A., 281, *337*
Rigney, D.R., 281, *337*
Risken, H., 233, *277*
Rizk, C.G., 193, *277*
Robbins, K.A., 281, *338*
Rong, H., 194, 226–229, *278*
Roppo, M.N., 281, *337*
Roseau, M., 280, *337*
Rosenblat, S., 281, *337*
Rössler, O.E., 73, 97, 103, *124*, 211, *277*

Samoilenko, S.B., 341, *376*
Sanchez, E.N., 73, *124*
Sanchez, N.E., 281, *337*
Sanders, J.A., 283, 284, *337*
Sandler, U., 232, *275*
Satpathy, P.K., 230, *277*
Schenk-Hoppe, K.R., 245, *277*
Schiehlen, W., 128, *189*
Senator, M., 128, *190*
Shaw, S.W., 94, 99, *124*, 128, *190*
Shieh, L.S., 73, *122*
Shih, H.T., 24, *121*
Shinozuka, M., 221, *277*
Shrier, A., 281, *336*
Sieber, J., 24, *124*
Silva, W.A., 6, 7, 18, *123*
Simic, S.N., 128, *189*
Sinha, S.C., 283, 285–287, 290–292, 297, 299, 318, 321, 324–326, 330–332, 335, *336–338*
Smale, S., 38, *124*
Smith, S.M., 193, *277*
Sobczyk, K., 214, *276*
Socolar, J.E.S., 24, *121*
Soliman, M.S., 205, *277*, *278*
Song, F., 193, *277*
Song, Y.X., 24, *124*
Soong, T.T., 214, *277*
Sprott, J.C., 73, *124*
Sridhar, S., 340, 341, 343, 347–351, 361, *376*
Srikant, R., 25, *122*
Starzhinskii, V.M., 288, 289, *338*
Steeb, W.H., 103, *122*
Steindl, A., 281, *337*
Stewart, H.B., 198, 211, *274*, *277*, *278*
Stewart, L., 73, *124*
Stokes, G.G., 282, *338*
Streit, D.A., 281, *338*
Strini, G., 212, *274*
Strogatz, S.H., 265, 268, *278*
Sun, J.Q., 193, 194, 215, 231, 235, 236, 258, 260, 261, *275*, *276*, *278*
Syrmos, V.L., 260, *277*
Szasz, G., 280, *338*
Szebehely, V., 281, *338*

Takatsuka, K., 230, *278*
Tamoaki, H., 258, *277*
Tan, D., 24, *122*
Tan, D.K.H., 25, *122*
Tél, T., 202, *277*
Tesi, A., 259, *275*
Thompson, J.M.T., 193, 198, 205, *275*, *277*, *278*
Thuraisingham, R.A., 281, *338*
Tian, Y.P., 24, *124*

Tomonaga, Y., 230, *278*
Tongue, B.H., 193, *278*
Troger, H., 281, *337*
Tsao, T.C., 193, *274*
Tso, W.K., 317, *338*
Tsuneda, A., 105, 106, 108, 109, *124*

Udwadia, F.E., 214, *278*
Ueda, Y., 198, 211, *278*
Ushio, T., 193, *278*
Utkin, V.I., 128, *190*

Vakakis, A.F., 352, *376*
Van Campen, D.H., 128, *189*
Van der Pol, B., 3, *124*
Verga, A.D., 214, 230, 239, *277*
Verhulst, F., 283, 284, *337*, *338*

Wan, Y.-H., 2, 24, 30, *122*
Wang, D., 285, 296, 312, *336*
Wang, F.Y., 260, *278*
Wang, H., 259, *275*
Wang, H.O., 4, 24, 79–81, *122*, *124*, 335, *335*
Wang, S., 50, 72, *124*
Wang, S.P., 50, *124*
Watanabe, N., 258, *277*
Wei, K., 259, *278*
Widmann, P.J., 281, *338*
Wiesenfeld, K., 211, *277*
Wiggins, S., 2, *124*
Wong, F.S., 212, 214, *274*, *275*
Wu, D.-H., 283, 290, 291, 299, *338*
Wu, J.H., 24, *124*
Wu, M.-Y., 283, *338*
Wu, X., 75, 76, *123*
Wyatt, R.E., 281, *336*

Xu, J., 7, 10, 15, *125*
Xu, J.X., 24, *124*, 193, 194, 211, 241, *275*, *276*
Xu, W., 194, 226–229, *278*

Yakubovich, V.A., 288, 289, *338*
Yang, C., 259, *274*
Yao, W., 5, 37, 92, 113, *124*, *125*
Yap, K.C., 259, *274*
Ye, H., 128, *190*
Yeo, M.H., 340, 341, *376*
Yorke, J.A., 4, 73, *124*, 198, 202, 205, 211, 242, 259, *275–278*
Yoshida, Y., 232, *278*
Yu, J., 50, *124*
Yu, J.B., 24–27, 29, 30, 32, *123*
Yu, P., 2–7, 9, 10, 12, 14, 15, 18, 19, 24, 25, 28, 37–40, 42–44, 47–50, 72–74, 77, 78, 80, 82, 92, 97, 99, 113, *122–125*
Yu, X.H., 24, *124*
Yuan, G., 259, *277*
Yuan, Y., 6, 7, 9, 10, 12, 15, 18, 19, *125*

Zhang, S., 73, *123*
Zhang, W., 3, 97, 99, *125*
Zhao, X.H., 50, 53, *123*
Zhou, T., 73, *123*
Zoladek, H., 47, *125*

Subject Index

A
accessible subdomain, 129, 187
active queue management schemes, 24
additive fuzzy noise, 245
asymptotic stability, 287
attractor invariant, 255

B
backward
– algorithm, 234
– search algorithm, 234
Bessel functions of the first kind, 344
bifurcation
– analysis, 14
– control, 73
– of fuzzy nonlinear oscillators, 240
– of limit cycles, 5, 53, 307
– of one-dimensional fuzzy systems, 235
– of nonlinear systems with small random disturbances, 212

C
cell mapping method, 193, 231
center manifold
– reduction, 7, 11, 12, 292
– theory, 38, 294
chaos
– control, 73
– synchronization, 73, 74
chaotic
– attractors, 202
– boundary crisis, 200
– crisis, 198
– saddle, 202
Chebyshev polynomial expansion, 290, 297
Chua system
– simple, 105
– smooth, 105
circular plate, 339, 341
closed set, 197
codimension
– one bifurcations, 301
– two bifurcations, 245
commutative systems, 297, 313
competitive mode, 92–94
complex center, 46
connectable domain, 129
corner points or vertex, 133
crises in deterministic systems, 198
cyclic
– fold bifurcation, 288, 307, 309
– group, 50

D
delay differential equations, 24
detuning parameters, 352, 361
discontinuous dynamic system, 131
disjoint, 132
domain accessibility, 129
domains of attraction, 195
double
– crises, 210, 211
– inverted pendulum, 327
Duffing oscillator, 357
– with small random excitations, 220
Duffing–Van der Pol oscillator, 245

E
equilibrium, 7
– cell, 194
extended center, 46
external
– mode, 93
– resonance, 352

F
flip, 288
– bifurcation, 301, 313
Floquét–Lyapunov theory, 74
Floquét multipliers, 288
– theory, 281, 282, 289
– transition matrix, 288, 298
formal monomial form, 294
friction-induced oscillator, 158

fundamental solution, 282
fuzzy
– attractor, 241, 255
– – chaotic, 242
– bifurcation, 230, 239, 242
– dynamical system, 231
– generalized cell mapping, 231
– – system, 232
– nonlinear dynamical systems, 233
– set, 231, 237

G
G-equivariant vector field, 50
G-invariant function, 50
Gaussian white noise, 221
general
– control database, 260
– resonance condition, 303
generalized cell mapping, 193, 195
global
– Lipschitz condition, 74
– ultimate boundedness of chaotic systems, 74
grazing flow, 141, 143
– of the first kind, 150
– of the second kind, 150
– of the third kind, 150

H
Hilbert's 16th problem, 6, 38
Hill's method, 281
homoclinic and heteroclinic
– bifurcation, 4, 61
– – values, 61
homological equation, 295
Hopf bifurcation, 2, 4, 6, 7, 10, 15, 26
– control, 28, 77
– controling, 79
– double, 7
– onset of, 30
– parameters, 59
– secondary, 288, 309, 327
– subcritical, 268
Hopf-type critical point, 39
hyperbolic flow, 153

I
inaccessible subdomain, 129, 187
indeterminate crisis, 202
instability flutter, 15
interior crisis, 200
internal
– mode, 93
– resonance, 340, 361
– – condition, 352

J
jump phenomenon of resonance, 307

K
*k*th-order singular point, 46

L
Liénard equation, 3
limit cycle, 3, 5–7
local singularity, 141, 187
logistic map, 215
Lorenz system, 79
Lyapunov–Floquét (L–F), 282, 289
– transformation, 289, 290
– – complex, 289, 290
– – real, 289, 290

M
marginal stability, 288
Markov chains, 233
Mathieu
– equation, 281
– oscillator, 240
membership grade, 232
modal
– interactions, 339
– system, 92
Monte Carlo simulation method, 214
multiple scales, 43
multiplicative fuzzy noise, 248

N
n-step membership distribution vector, 232
Neimark–Sacker bifurcation, 288
noisy crisis, 221
nonpassable boundary
– of the first kind, 138, 139
– of the second kind, 138, 140
nonresonant, 12
nonuniform circular plates, 341
normal form, 14, 296
– theory, 7, 38, 42

O
one-step transition membership, 232
– matrix, 232, 233
optimal
– control problem, 259
– semiactive control, 258, 259
Oregonator model, 104
oriented boundary, 132

P
pendulum, 319
period doubling, 288
periodic motions, 194
perturbation technique, 42
perturbed Hamiltonian system, 50
Picard iteration, 297
piecewise linear systems, 150
Poincaré map, 194
– fuzzy, 231
positive Lagrangian function, 259
primary resonance, 348
psychological model, 101

Q
quantitative bifurcation analysis, 332
quasiperiodic, 15

R
random
– coefficient, 215
– map, 217
reduction, 293
regular crisis, 198
resonant
– frequency, 309
– periodic term, 309
Routh–Hurwitz criterion, 287

S
saddle-node bifurcation, 2, 262
sampling point method, 232
self-cycling set, 197
semiactive control, 259
semipassable, 133
– boundary, 136
– – sets, 134
– from the domain Ω_i to Ω_j, 136
separable domain, 130
shifted Chebyshev polynomials, 290
Shur–Cohn criterion, 287
simple cell mapping, 193
singular, 133
– point method, 44
– sets, 132
sink boundary, 138
solvability condition, 295, 344, 348
source boundary, 138
stabilizing, 73
stable fuzzy equilibrium, 240
standard normal form, 234
state transition, 282
– matrix, 288
stochastic attractors, 214
subcritical, 14, 26
supercritical, 14, 26
– Pitchfork bifurcation, 265
symmetry breaking, 288
– bifurcation, 304, 319
synchronization, 89

T
tangency, 187
tangential to the boundary, 142
terminal cost, 259
thin plate, 97
time-dependent normal form theory, 291, 294
time-periodic center manifold, 291
– reduction, 292
topological matrix, 233
tracking, 73, 85
– and chaos synchronization, 84
transcritical bifurcation, 288, 304
transient
– cell, 195, 197
– sets, 197
transition probability matrix, 214, 215
transversality condition, 4
traveling wave, 341

U
universal domain, 129

V
Van der Pol's equation, 3
versal deformation
– equation, 296
– of the linear system matrix, 296
– of the normal form, 296
von Karman equation, 340

W
Wada boundary
– basin, 210
– fractal, 202
weak critical singular point, 46

Z
Z_{10}-equivariant planar vector field, 54
Z_q-equivariance, 52
Z_q-equivariant
– Hamiltonian vector field, 51, 52
– planar vector field, 50

LaVergne, TN USA
31 January 2011
214608LV00002B/25/P

9 780444 522290